Applications of Mathematics

G. I. Marchuk

Methods of
Numerical Mathematics

Second Edition

Translated by Arthur A. Brown

With 28 Illustrations

Springer-Verlag
New York Heidelberg Berlin

G. I. Marchuk
Presidium of the Academy of Sciences
    of the USSR
Leninprospekt 14
Moscow
USSR

Translator:
Arthur A. Brown
10709 Weymouth Street
Garrett Park, MD 20896
USA

Managing Editor:

A. V. Balakrishnan
Systems Science Department
University of California
Los Angeles, CA 90024
USA

AMS Subject Classification (1980): 65-XX

Library of Congress Cataloging in Publication Data
Marchuk, G. I. (Gurii Ivanovich), 1925–
    Methods of numerical mathematics.
    (Applications of mathematics; 2)
    Translation of Metody vychislitel'noĭ matematiki.
    Includes bibliographies and index.
    1. Numerical analysis.   I. Title.   II. Series.
    QA297.M3413   1981   519.4   81–16634
                                                    AACR2

The original Russian edition *Metody Vychislitel'noi Matematiki*
was published in 1973 by Nauka, Novosibirsk.

9 8 7 6 5 4 3 2 1

ISBN-13:978-1-4613-8152-5   e-ISBN-13:978-1-4613-8150-1
DOI: 10.1007/978-1-4613-8150-1

Preface to the First Edition

The present volume is an adaptation of a series of lectures on numerical mathematics which the author has been giving to students of mathematics at the Novosibirsk State University during the span of several years. In dealing with problems of applied and numerical mathematics the author sought to focus his attention on those complicated problems of mathematical physics which, in the course of their solution, can be reduced to simpler and theoretically better developed problems allowing effective algorithmic realization on modern computers.

It is usually these kinds of problems that a young practicing scientist runs into after finishing his university studies. Therefore this book is primarily intended for the benefit of those encountering truly complicated problems of mathematical physics for the first time, who may seek help regarding rational approaches to their solution.

In writing this book the author has also tried to take into account the needs of scientists and engineers who already have a solid background in practical problems but who lack a systematic knowledge in areas of numerical mathematics and its more general theoretical framework.

Consequently, the author has selected a form of exposition which, in his opinion, helps to attract the attention of a wide range of researchers to problems of numerical mathematics. This style has required certain concessions in the exposition, thus allowing concentration only on basic ideas and approaches. As for the details (sometimes important) and the possible generalizations (such as minimal smoothness requirements, constraints on the input data, etc.), they are obvious to the specialist and present useful exercises for a beginner.

Chapter 10 is an expanded version of the paper given by the author at the International Congress of Mathematicians in Nice (1970). This chapter gives some idea of the material considered in the previous chapters, and of various methods and problems of numerical mathematics that are of fundamental importance but have not found their way into this volume.

In the process of preparation for publication this book has undergone considerable changes in response to advice and comments obtained by the author from his colleagues and associates. Those whose help is gratefully acknowledged include M. M. Lavrentiev, V. I. Lebedev, I. Marek, M. K. Fage, and N. N. Yanenko. They have made a number of constructive comments regarding the exposition of individual chapters, especially the first and fifth. The changes in the second chapter, which are due to Yu. A. Kuznetsov, are so profound that the nature of his contribution in this part is essentially that of coauthorship. The author has also enjoyed valuable advice and comments from V. T. Vasil'ev, V. P. Il'in, A. N. Konovalov, V. P. Kochergin, V. V. Penenko, V. V. Smelov, U. M. Sultangazin, and others. G. S. Rivin did considerable work in editing the manuscript. To all these, as well as M. S. Yudin who took part in preparing the book for publication, the author expresses his deep gratitude.

Preface to the Second English Edition

The second edition is a re-written version of the first, remedying various ambiguities and typographical errors, and including new material that in the author's opinion extends the scope of the methods covered therein. This edition includes a new chapter dealing with optimization theory, which is today an indispensible part of the development of mathematical models and of methods for implementing them.

Part of the material in this edition was published in 1973, in a monograph having the same title. The present text is a manual differing essentially from the monograph in that it contains a number of new ideas and algorithms of methodological and practical interest. In particular it includes:

(1) new optimization algorithms based on variational methods;
(2) problems of automating the numerical processes by use of the so-called method of "fictive" domains;
(3) consideration of iterative algorithms for the splitting process for non-commutative operators;
(4) the method of incomplete factorization, and other topics.

That portion of the book dealing with the interpolation of functions by the use of splines has been extended, and in this edition forms a self-contained chapter. Also, a separate chapter has been devoted to the notions connected with Richardson's extrapolation methods for obtaining a higher order of approximation in the solution of problems. The portions of the book dealing with variational-difference methods contain a number of new ideas, e.g., the representation of continuous functions by piecewise-discontinuous bases, and the construction of bases that take into account the singularities of solutions, as well as other new notions. The chapter dealing with the solution of inverse problems has been expanded by the inclusion of new results in perturbation theory for the solution of nonlinear problems of mathematical physics and for the analysis of the sensitivity of mathematical models with respect to variations in the initial data. There are also other expansions of the text.

This new material should provide for a better understanding of the methods of numerical mathematics for the solution of complex problems in applied mathematics.

The author expresses his deep gratitude to Yu. A. Kuznetsov, whose contribution to the preparation of the book cannot be over-estimated. For the contributions of V. V. Smelov, V. P. Il'in, V. V. Shaidurov, V. I. Agoshkov, V. A. Vasilenko, A. M. Matsokin, V. A. Bulavskii, A. L. Buchheim, Yu. S. Zav'yalov, V. A. Kuzin, G. S. Rivin, V. A. Tsetsokho, and also V. I. Drobyshevich, V. P. Dymnikov, and V. V. Penenko, the author is also deeply grateful.

The manuscript was read by N. S. Bakhvalov, V. I. Lebedev, and M. K. Fage, who made many valuable comments which helped to improve the book. That portion of the text concerned with variational inequalities was written on the basis of material graciously made available to the author by the French mathematicians J. L. Lions and R. Glowinski. To all those mentioned, the author expresses his warmest thanks.

In studying the book, the reader is recommended to make use of the exercises contained in *Problems in Numerical Mathematics*, by Drobyshevich, Dymnikov, and Rivin, Moscow, Nauka, 1980, which corresponds to the expository material in this text.

The English translation has been reviewed by the author, and fully corresponds to the Russian original version. The author appreciates very much the cooperation of Springer-Verlag.

G. I. Marchuk

Contents

Introduction

Modern electronic computers have put into the hands of research workers an effective means for using mathematical models of complex problems in science and technology. In consequence, quantitative methods of research have spread into practically all fields of human endeavor, and mathematical models have become a tool of knowledge.

The role of mathematical models is far from being exhausted in studying natural laws. Their significance is constantly being increased by the natural tendency toward optimization of technical processes and technological systems for planning experiments. In the process of research, and in the desire to develop a detailed representation of the processes under study, we are driven to the construction of ever more complex mathematical models, which require refined and generally applicable mathematical methods. Mathematical models are implemented on an electronic computer by the methods of numerical mathematics, which are continually being perfected, in keeping with developments in computing technology.

Every reduction of the problems of mathematical physics or of technology usually comes down, in the end, to a set of algebraic equations having some definite structure. Therefore, the subject of numerical mathematics is, as a rule, connected with methods of reducing a problem to a system of algebraic equations and subsequently solving them.

The construction of a set of algebraic equations corresponding to a problem with continuously varying arguments relies, in general, on *a priori* information arising from the original problem. We may know, for instance, that the solution must belong to a given class of functions characterized by given smoothness properties, or by properties of the operator associated with the problem, or by properties of the boundary conditions, etc. Such information in many cases has a decisive influence on the choice of the numerical method

1

to be used for the solution of the corresponding algebraic equations. As a rule, the properties of the algebraic analog of the original problem must reflect our *a priori* information on the original constraints. This refers primarily to the operator of the problem and to the preservation of its properties during the reduction of the problem from one of continuous arguments to a discrete version.

Clearly, such a principle may be taken as a basic assumption in many problems. At the same time, we must note that the inheritance of the properties of operators during reduction opens up possibilities for the use of well-developed methods of functional analysis, which usually give us a simple and universal way of studying the effectiveness of the algorithms of numerical mathematics.

We now turn to a brief overview of the book in order to point out the weightings and the new ideas presented in it.

Chapter 1 is devoted to general questions in the theory of difference schemes. Along with the classical concepts such as approximation, countable stability, and convergence of solutions of difference equations, we present some important results connected with the general properties of basic and adjoint problems. These will be used in many later chapters. Section 1.1.2 contains contemporary algorithms for computing the bounds of the non-negative spectrum of matrices, and is of special interest. It is well known that the upper bound of the spectrum is found by a well-developed iterative process, and the implementation usually gives no trouble. The smallest eigenvalue—the lower bound of the spectrum—is usually difficult to compute.

The simplest method, theoretically, for finding the smallest eigenvalue is by estimating the maximal eigenvalue of the inverse operator and is of little use algorithmically. We present another approach, based on shifting the spectrum, which allows us to find the smallest eigenvalue rather easily. We have dwelt on this topic at some length because many numerical algorithms, especially those connected with the optimization of iterative processes, rely essentially on *a priori* information about the bounds of the spectrum.

In Chapter 2 we consider methods for constructing difference schemes, and we focus our attention on two approaches: the method of integral relationships and variational methods. Each of these approaches has advantages and weaknesses. We note only that they are not independent and under certain conditions lead to identical difference schemes approximating the original differential problems.

Nevertheless, it must be noted that the variational approach is to be preferred in many cases since it preserves the definiteness of the initial operators in the passage to the difference scheme. It is important to observe that this happens automatically in a wide class of problems.

We limit ourselves to the consideration of three methods of constructing difference schemes by the variational approach: namely, the methods of Ritz, Galërkin, and least-squares. These, of course, do not exhaust the great multiplicity of variational approaches, but they do provide an acquaintance with the general principles of the construction of difference schemes, which can be easily extended to other cases.

A few words on the method of finite elements. We may characterize it as a convenient way to construct difference schemes using the variational approach. At its methodological roots it is closely connected with Fourier analysis; instead of a basis of continuous functions (e.g., trigonometric functions, Legendre, or Hermite polynomials, etc.) we deal with polynomials which vanish outside a comparatively small region in the space of their arguments. These functions have been called finite elements.

The application of variational methods to the construction of difference schemes is not accidental. In fact, it follows from theory that a variational functional which adequately reflects definite laws of mechanics, mathematical physics, dynamics, etc., attains its extremal value on the solution of the problem that interests us. Therefore, if we are given a variational functional and a definite class of functions on which we are to find the minimum of the functional, the rest of the task consists of an algorithmic search for the function yielding an extremal of the functional.

If we restrict the class of admissible functions by imposing additional constraints, the minimizing function may be not a solution of the original problem but merely an approximation to the exact solution.

As the means for numerical technology becomes more powerful in the future, the role of variational methods for constructing solutions of problems of mathematical physics will continue to grow. Goal-directed methods of enumerating trial functions belonging to a wide class are beginning to appear, providing an effective means for finding extremal solutions. Thus, the use of variational methods for finding solutions to problems is ever more closely linked to the question of optimal organization of the algorithm for obtaining a solution to these problems with a given precision, i.e., to the theory of optimization.

Together with the classically formulated problems for the solution of tasks in science and technology, it is often necessary to deal with nonclassically formulated problems, for instance, those with constraints. Of course, the simplest of the constrained problems are classical, as in the case of boundary conditions for differential problems.

More complex problems with constraints demand a more complex mathematical apparatus for their solution. For instance, if we are required to find the deflection of a membrane under the action of various forces, while its position is constrained from above and below by given functions of the coordinates, the customary classical approach is powerless. However, if we set up a correspondence between this problem and some variational functional, and seek the minimum of the functional over the class of functions that each satisfy the given constraints, the minimizing function will be the solution to our problem.

A wide collection of studies in this direction has been carried out by French mathematicians. They have considered the so-called variational inequalities, which are specially adapted to the solution of problems with constraints. These questions are discussed in Section 2.8.

In Chapter 3 we deal with the interpolation of net functions. The interpolation problem arises whenever we must extend a function defined on the net to a continuous function over the whole region. Here we are concerned with the task of extending an approximate solution to the whole region, given its values on the vertices of the net, and with the task of reducing experimental data given on a discrete set of points.

The interpolation problem is a fundamental part of a system for automating construction project work, where the graphic presentation of information is at the very heart of the problem. The interpolation problem is not new, and classical methods are fully explained in the mathematical literature. A new direction in interpolation theory has been exploited in the last few decades; this is the use of the so-called spline interpolations to which Chapter 3 is essentially devoted.

Spline interpolations offer the best means of smooth completion of net functions on given classes of functions. The optimality of the spline is connected with its special extremal property. Since spline approximation is being used ever more widely in all areas of science and technology, it is necessary in our opinion that the reader should become acquainted with it.

Chapter 4 is essentially given over to iterative methods of solving linear algebraic equations. We discuss the general approaches to the solution of algebraic systems, and specific methods as well, in connection with peculiarities of the approximation of problems of mathematical physics by the use of difference and variational-difference methods. Although the literature on iterative methods is extensive and contains descriptions of many effective methods, we have considered not only the classical processes but also, and basically focused attention on, iterative methods optimizable by quadratic functionals. This constitutes our general approach to questions of optimization, for the development of numerical algorithms and for their implementation.

In the case of specific problems arising from the particular form of the matrices that arise in the numerical solution of problems in mathematical physics, we turn to methods of splitting matrices into the simplest within the general scheme of the iterative process. The splitting method is a natural development of the alternating direction method, playing an exceptional role in the numerical solution of problems in mathematical physics. It has numerous modifications and generalizations, some of which make use of variational principles.

Special attention should be paid to the direct methods of solving finite-difference equations, as discussed at the end of Chapter 4. These are primarily the fast Fourier transform and the method of cyclic reduction. Their application is relatively recent, and they are becoming increasingly popular.

Chapter 5 is devoted to methods of solving nonstationary problems. These essentially use the idea of splitting complex operators into simpler ones. We analyze not only methods well established in practice, such as the stabilization and predictor–corrector methods, but also, in some detail, the method of component-by-component splitting, which is more effective, in our opinion. The ideas are discussed in Sections 5.3.3 and 5.4.

The component-by-component method permits, at each time step, the reduction of a complex problem in mathematical physics to a sequence of very simple one-component problems. As a result, we arrive at an effective algorithm for implementation on an electronic computer which is absolutely stable and yields a second-order approximation in both time and space. It is applied to a wide class of nonstationary problems in mathematical physics.

In Chapter 6 we consider methods for increasing the precision of approximate solutions, developed by Richardson and Runge. A refinement of an approximate solution can be obtained in different ways, generally by using a higher-order approximation to the differential or integral equation in question. Richardson proposed to use a difference approximation of a comparatively low order of accuracy, but to apply it to a sequence of nets. Thus, if the initial difference equation corresponded to an approximation on a net with mesh length h, the next would correspond to a mesh length $h/2$, and so on. As a result, we arrive at difference equations defined on a sequence of nets. It turns out that, if a number of constraints are imposed on the operators and the initial values of the problem, a linear combination of the approximate solutions on the sequence of nets yields a solution with a higher order of precision than the initial solutions.

Richardson's extrapolation method, first proposed for the solution of ordinary differential equations, was successfully applied to boundary-value problems for equations of elliptic and parabolic type. Naturally, various singularities arise, and these are noted in the implementation schemes. It must be emphasized that Richardson's method can be applied to the solution of problems with a small parameter or for conditionally well-posed problems by using the method of regularization. In this case the Richardson method is based on the solution of problems with distinct parameters converging in the limit to some value. Thus the extrapolation method permits us to bring into numerical mathematics new ideas which successfully use various optimization algorithms for the solution of problems.

We must also point out the special place that has been set aside for this method for the solution of problems by variational-difference methods. In fact, we normally have two alternatives: either obtain a solution by difference equations with a very small step on the basis of rather coarse difference approximations, or use a scheme with a higher order of approximation and larger steps in the difference schemes.

The first method is simple but demands a larger volume of computation; the second is logically more difficult but demands fewer arithmetic operations. Therefore, neither is effective for problems of mathematical physics if highly precise results are required. The notion therefore arose of using the simplest variational-difference schemes, of first- or second-order approximation, but implementing them on a sequence of nets. A linear combination of the solutions of such problems, as we indicated above, will in many cases yield solutions of the desired order of approximation.

Chapter 7 is devoted to the formulation and numerical solution of inverse problems. The methods of mathematical modeling of complex problems in science and technology constantly place before the research worker problems of finding a solution by means of some functional of it, or of determining the form of the operator of the problem. This class of inverse problems is most difficult from the point of view of numerical mathematics, since it is usually concerned with the solution of problems that are ill-posed in the Hadamard sense. A whole line of research on ill-posed problems has developed in mathematics; the basic results have been achieved by the Soviet school. A. N. Tikhonov began the consideration of regularization processes for such problems, which soon found their basic and algorithmic formulations. The accent in Chapter 7 is on the formulation of inverse problems by specification of the structure of differential operators and initial conditions. Although the form of the differential operator is fixed, the coefficients are assumed to be unknown and to require determination. The theory of inverse problems is closely bound up with the use of basic and adjoint equations. A mathematical apparatus developed by the author turns out to be effective for estimating small perturbations of functionals of the solutions of problems in respect of variations in the initial parameters. We note that the theory can be applied to both linear and nonlinear problems of mathematical physics.

In Chapter 8 we discuss methods of optimization which have actively invaded the fields of mathematical modeling of technological processes, economics, and control. These are primarily the methods of linear and quadratic programming, which have made substantial advances from the algorithmic point of view. In this chapter we also describe approaches to nonlinear programming for convex functions and convex domains. The reader will also become acquainted with the general notions of dynamic programming and the Pontrjagin maximum principle. Finally, we consider the question of optimizing constrained problems of mathematical physics by the method of variational inequalities widely developed by the French mathematicians.

In Chapter 9 we consider applications of the methods developed in the preceding chapters to solve a number of concrete problems of mathematical physics.

Chapter 10 consists of a review of the methods of numerical mathematics. There is obviously not room in this book for the inclusion of a large number of recently developed algorithms. Many have been omitted because they are well described in textbooks or in the specialized literature.

Neither do we discuss such classical foundations of numerical analysis as cubic formulas, the methods of solution of ordinary differential equations, the simplest iterative methods, etc. Rather, we take up the new methods of modern numerical mathematics, such as the method of coarse particles, the method of integral relations, universal methods of linear algebra, etc. These are discussed in our review.

The review contained in Chapter 10 is accompanied by the statement of a number of problems of numerical mathematics and an analysis of their developmental trends. We believe that this will not only give the reader an orientation in problems of numerical mathematics, but also define the most actively developing fields.

Since this book is a teaching manual, we have attempted to free the main text of references to the literature, which might distract the reader from acquiring a systematic acquaintance with the material. This omission is partially remedied in Chapter 10, where references to the corresponding sources accompany the review of the methods.

A special section has been set aside for a conspectus of the literature. The citations have been systematized according to the various problems of numerical mathematics, so that the reader may more rapidly enter the field of problems that interest him.

The whole book has a single aim—to prepare the reader for the solution of complex problems in numerical and applied mathematics.

In conclusion, it should be said that the book makes wide use of new results obtained by Soviet and foreign authors, in particular the coworkers in the Computing Center of the Siberian Branch of the Academy of Sciences of the USSR, and primarily the work of Yu. A. Kuznetsov, V. I. Agoshkov, V. A. Vasilenko, A. M. Matsokin, V. V. Smelov, V. A. Tsetsokho, V. V. Shaidurov and others.

CHAPTER 1

Fundamentals of the Theory of Difference Schemes

This chapter surveys briefly those fundamentals of the theory of difference schemes that are used extensively in the following chapters. We restrict our theoretical considerations to the simplest and most easily interpreted cases, since our main purpose is to achieve familiarity with certain modern concepts in the construction of numerical algorithms in mathematical physics. For more refined and complex theoretical developments we refer the reader to the specialized bibliography given at the end of the book.

1.1 Basic Equations and Their Adjoints

Let us consider a region D in the n-dimensional Euclidean space E_n. We denote by $L_2(D)$ the Hilbert space of all real measurable square integrable functions

$$\int_D f^2(x)\,dx < \infty,$$

with the inner product

$$(f, g) = \int_D f(x)g(x)\,dx. \tag{1.1}$$

As usual, the norm of the function $f \in L_2(D)$ is defined by

$$\|f\| = (f, f)^{1/2}. \tag{1.2}$$

Let us now choose a subspace (a linear manifold) Φ of the Hilbert space $L_2(D)$ by imposing certain additional conditions (depending on the specific problem) which every element $\phi \in \Phi$ must satisfy. For example, we may require some specified smoothness conditions, conditions on the limit behavior at the boundary D, etc. These conditions, however, must be sufficient to guarantee that an operator A, if given, maps the subspace Φ into $L_2(D)$.

A linear operator A, defined on the linear manifold Φ, is called positive semidefinite if

$$(A\phi, \phi) \geq 0 \qquad (1.3)$$

for all $\phi \in \Phi$, with the equality sign possibly holding for a nonzero element ϕ. It is customary to write $A \geq 0$ in this case. If the equality sign above cannot hold for nonzero elements, that is,

$$(A\phi, \phi) > 0, \qquad \phi \neq 0, \qquad (1.4)$$

then we say that the operator A is positive and write $A > 0$. Finally in the case of the stronger inequality

$$(A\phi, \phi) \geq \gamma(\phi, \phi), \qquad \phi \in \Phi, \qquad (1.5)$$

where $\gamma > 0$ is a positive constant independent of ϕ, the operator A is called positive definite.

Note that if A is a square matrix of finite order, then, if it is positive, it is positive definite (Faddeev and Faddeeva [8]).†

The subspace Φ will be called the *domain* of the operator A and denoted by $\Phi(A)$.

Consider next the adjoint operator A^* defined by the Lagrange identity

$$(Ag, h) = (g, A^*h), \qquad (1.6)$$

where $g \in \Phi(A)$, $h \in \Phi(A^*)$.

The subspaces $\Phi(A)$ and $\Phi(A^*)$ of the Hilbert space $L_2(D)$ do not coincide in general, despite the fact that their elements are defined on the same region D in E_n.

The operator A is called *self-adjoint* if $Ah = A^*h$ for all $h \in \Phi(A)$ and $\Phi(A) = \Phi(A^*)$.

Let us note one important consequence regarding the properties of adjoint operators. Namely, if $\Phi(A) \equiv \Phi(A^*)$, then $A > 0$ implies $A^* > 0$.

A considerable role in analyzing algorithms is played by the Fourier expansions with respect to the eigenfunctions of operators and their adjoints.

Consider the following two spectral problems for $A \geq 0$:

$$Au = \lambda(A)u, \qquad A^*u^* = \lambda(A^*)u^*. \qquad (1.7)$$

Assume that each of the homogeneous equations (1.7) generates a complete set of eigenfunctions, $\{u_n\}$ and $\{u_n^*\}$, which are normalized as follows,

$$(u_n, u_m^*) = \begin{cases} 1, & n = m, \\ 0, & n \neq m, \end{cases} \qquad (1.8)$$

† Numbers in square brackets refer to the section of the bibliography in which the cited reference is listed.

and the corresponding eigenvalues $\lambda_n(A)$ are real. Then $\lambda_n(A) = \lambda_n(A^*)$. Suppose they belong to the interval $[\alpha, \beta]$

$$\alpha \le \lambda_n(A) \le \beta.$$

This complete set of eigenfunctions will be call a *biorthogonal basis*. Thus, under the assumption that the systems $\{u_n\}$ and $\{u_n^*\}$ are complete, arbitrary functions $f \in \Phi$ and $f^* \in \Phi^*$ can be represented in the form of a Fourier series

$$f = \sum_n f_n u_n, \qquad f^* = \sum_n f_n^* u_n^*, \tag{1.9}$$

where

$$f_n = (f, u_n^*), \qquad f_n^* = (f^*, u_n). \tag{1.10}$$

(In what follows, we use Φ, Φ^* instead of $\Phi(A), \Phi(A^*)$ for the sake of simplicity.)

Of great importance in the analysis of numerical algorithms are the norm estimates of operators. A *norm of an operator* A is defined as follows:

$$\|A\|^2 = \sup_{\substack{\phi \in \Phi \\ \phi \ne 0}} \frac{(A\phi, A\phi)}{(\phi, \phi)}. \tag{1.11}$$

(In order to simplify notation the qualification $\phi \ne 0$ will not be explicitly mentioned again.) Since

$$(A\phi, A\phi) = (\phi, A^*A\phi),$$

the square of the norm of A can also be expressed as

$$\|A\|^2 = \sup_{\phi \in \Phi} \frac{(\phi, A^*A\phi)}{(\phi, \phi)}. \tag{1.12}$$

The operator A^*A is symmetric and positive semidefinite. Consider the spectral problem

$$A^*A\Omega = \lambda(A^*A)\Omega. \tag{1.13}$$

This problem defines a family of eigenfunctions $\{\Omega_n\}$ and eigenvalues $\lambda_n(A^*A) \ge 0$. We will assume that $\{\Omega_n\}$ is a complete set. Then the function ϕ has the following Fourier expansion:

$$\phi = \sum_n \phi_n \Omega_n, \tag{1.14}$$

where

$$\phi_n = (\phi, \Omega_n). \tag{1.15}$$

Using the orthonormality of the functions Ω_n, the substitution of series (1.14) into (1.12) yields

$$\|A\|^2 = \sup_{\{\phi_n\} \in Q} \frac{\sum_n \lambda_n(A^*A)\phi_n^2}{\sum_n \phi_n^2}, \tag{1.16}$$

where Q is the space of Fourier coefficients. It is easy to see that

$$\frac{1}{\|A^{-1}\|^2} = \lambda_{\min}(A^*A) = \alpha(A^*A),$$

$$\|A\|^2 = \lambda_{\max}(A^*A) = \beta(A^*A),$$

(1.17)

where $\lambda_{\min}$ is the smallest and $\lambda_{\max}$ is the largest eigenvalue, respectively, in the set $\{\lambda_n(A^*A)\}$ for (spectral) problem (1.13). The quantity $\beta(A^*A) = \lambda_{\max}(A^*A)$ is usually called the *spectral radius* of the operator A^*A. In general, the spectral radius is defined as $\beta(A) = \sup\{|\lambda(A)|\}$. Note that for $\lambda(A) > 0$ the spectral radius $\beta(A) = \sup\{\lambda(A)\}$.

In the case of a self-adjoint operator A consider the spectral problem

$$Au = \lambda u.$$

(1.18)

We have

$$\|A\| = \beta(A).$$

(1.19)

It is not difficult to see that for a self-adjoint operator

$$\beta(A^2) = [\beta(A)]^2.$$

(1.20)

Consider some fixed operator C whose domain of definition Φ is everywhere dense in the Hilbert space $L_2(D)$. In other words, for any element $f \in L_2(D)$ there is an element $g \in \Phi$ such that $\|f - g\| \leq \varepsilon$, where ε is an arbitrarily small positive constant. If C is positive, then

$$(C\phi, \phi) > 0$$

(1.21)

for all nonzero $\phi \in \Phi$. Denote by $\Phi^* = \Phi(C^*)$ the domain of definition of the adjoint operator C^*. Assume Φ^* coincides with Φ. Then $C^*\phi$ exists for all $\phi \in \Phi$ and $(C^*\phi, \phi) = (\phi, C\phi) = (C\phi, \phi)$. Consequently,

$$(C\phi, \phi) = (\tfrac{1}{2}[C + C^*]\phi, \phi),$$

where $\tfrac{1}{2}[C + C^*]$ is now a symmetric positive operator. This allows one to introduce a new inner product in Φ, namely,

$$(f, g)_{\bar{C}} = (\bar{C}f, g),$$

and the norm

$$\|\phi\|_{\bar{C}}^2 = (C\phi, \phi) = (\bar{C}\phi, \phi),$$

where $\bar{C} = \tfrac{1}{2}[C + C^*]$. This norm will be called the *energy norm*. One can obtain the following significant estimate:

$$\|\phi\|_C^2 = \|\phi\|_{\bar{C}}^2 \leq \|\bar{C}\| \|\phi\|^2 = \beta(\bar{C})\|\phi\|^2,$$

(1.22)

where $\beta(\bar{C})$ is the largest eigenvalue of the operator $\bar{C}$.

In conclusion, let us note that in dealing with problems of mathematical physics and their adjoints it is often convenient to use functions from the

Sobolev space $W_2^l(D)$. This space is a Hilbert space of $L_2(D)$ functions whose generalized derivatives up to, and including, lth order are square integrable in D. The inner product in such a space is defined by the formula (see Sobolev [1], Vladimirov [2])

$$(u, v)_{W_2^l} = \sum_{k=0}^{l} \sum_{(k)} \int_D \frac{\partial^k u}{\partial x^k} \frac{\partial^k v}{\partial x^k} \, dD. \tag{1.23}$$

Here we have used the following notation for the partial derivatives:

$$\frac{\partial^k \phi}{\partial x^k} = \frac{\partial^{\alpha_1 + \cdots + \alpha_n}}{\partial x_1^{\alpha_1} \cdots \partial x_n^{\alpha_n}} \phi, \quad \alpha_1 + \cdots + \alpha_n = k.$$

The norm in the space $W_2^l(D)$ is defined by the relation

$$\|\phi\|_{W_2^l}^2 = (\phi, \phi)_{W_2^l}. \tag{1.24}$$

If the functions belong to the Sobolev space $W_2^l(D)$ and also satisfies the condition: $\phi = 0$ on the boundary ∂D of the domain D; the space is denoted by $\mathring{W}_2^l(D)$.

1.1.1 Norm Estimates of Certain Matrices

Let us consider a positive semidefinite matrix $A \geq 0$ on the Euclidean space. Then for any value of the parameter $\sigma \geq 0$ we have the following relation:

$$\|(E + \sigma A)^{-1}\| \leq 1. \tag{1.25}$$

For the proof of this important proposition we exploit the formula

$$\|(E + \sigma A)^{-1}\|^2 = \sup_\phi \frac{((E + \sigma A)^{-1}\phi, (E + \sigma A)^{-1}\phi)}{(\phi, \phi)}. \tag{1.26}$$

Let us introduce new elements

$$\psi = (E + \sigma A)^{-1}\phi.$$

Then

$$\|(E + \sigma A)^{-1}\|^2 = \sup_\psi \frac{(\psi, \psi)}{((E + \sigma A)\psi, (E + \sigma A)\psi)}$$

$$= \frac{1}{\inf_\psi \left[1 + 2\sigma \frac{(A\psi, \psi)}{(\psi, \psi)} + \sigma^2 \frac{(A\psi, A\psi)}{(\psi, \psi)} \right]}.$$

Since $A \geq 0$ on the elements ϕ, ψ, the last relation implies (1.25). If $A > 0$, then for $\sigma > 0$ we have immediately

$$\|(E + \sigma A)^{-1}\| < 1. \tag{1.27}$$

Kellogg's lemma [15]. *For any matrix $A \geq 0$ and for any $\sigma \geq 0$ one has*

$$\|(E - \sigma A)(E + \sigma A)^{-1}\| \leq 1. \tag{1.28}$$

PROOF. Let us define T by

$$T = (E - \sigma A)(E + \sigma A)^{-1},$$

and consider the expression for $\|T\|^2$

$$\|T\|^2 = \sup_{\phi} \frac{((E - \sigma A)(E + \sigma A)^{-1}\phi, (E - \sigma A)(E + \sigma A)^{-1}\phi)}{(\phi, \phi)}$$

$$= \sup_{\psi} \frac{((E - \sigma A)\psi, (E - \sigma A)\psi)}{((E + \sigma A)\psi, (E + \sigma A)\psi)}$$

$$= \sup_{\psi} \frac{(\psi, \psi) - 2\sigma(A\psi, \psi) + \sigma^2(A\psi, A\psi)}{(\psi, \psi) + 2\sigma(A\psi, \psi) + \sigma^2(A\psi, A\psi)} \leq 1.$$

Here the crucial role has been played by the positive semidefiniteness of the matrix A. The lemma is proved. $\qquad\square$

In the case when the matrix A is positive and $\sigma > 0$, the expression (1.28) is replaced by

$$\|(E - \sigma A)(E + \sigma A)^{-1}\| < 1. \tag{1.29}$$

1.1.2 Computing the Spectral Bounds of a Positive Matrix

Consider the problem of finding the largest and smallest eigenvalues of a matrix $A > 0$ with a positive spectrum. The approach below is due to Lyusternik [4].

Assume that the spectral problem

$$Au = \lambda u \tag{1.30}$$

defines a complete set of eigenfunctions $u_k \in \Phi$, and a set of eigenvalues $\lambda_k(A)$. (A fairly complete treatment of spectral problems can be found in the papers by Marek [8] and by Faddeev and Faddeeva [8].) Consider the iterative process

$$\phi^{(n+1)} = (1/c_n)A\phi^{(n)},$$

$$\phi^{(0)} = g,$$

where g is an arbitrary nonzero vector, and c_n is a normalizing factor which can be conveniently chosen in the form

$$c_n = \|\phi^{(n)}\|.$$

Thus

$$\phi^{(n+1)} = A\frac{\phi^{(n)}}{\|\phi^{(n)}\|}. \tag{1.31}$$

Let $0 < \alpha(A) = \lambda_1 \le \cdots \le \lambda_{m-1} < \lambda_m = \beta(A)$. Clearly, the following relation holds:

$$\beta(A) = \lim_{n \to \infty} \| \phi^{(n)} \|. \tag{1.32}$$

Indeed, because of the assumption of completeness of the system $\{u_n\}$ we have the representation

$$\phi^{(0)} = \sum_k g_k u_k,$$

where $g_k = (g, u_k^*)$, and $\{u_k^*\}$ are the eigenvectors of the matrix A^*. Using the recursive relation

$$\phi^{(n)} = A \frac{\phi^{(n-1)}}{\| \phi^{(n-1)} \|} = \cdots = \frac{A^n g}{\| A^{n-1} g \|}$$

we obtain

$$\lim_{n \to \infty} \| \phi^{(n)} \| = \lim_{n \to \infty} \frac{\| A^n g \|}{\| A^{n-1} g \|}.$$

Since

$$A^n g = \sum_k [\lambda_k(A)]^n g_k u_k,$$

then for large enough n

$$A^n g = \beta^n(A) g_m u_m \left\{ 1 + 0 \left[\left(\frac{\lambda_{m-1}}{\lambda_m} \right)^n \right] \right\},$$

where $\beta(A) = \lambda_m$ is the largest eigenvalue of the matrix A.

If we choose the vector of initial approximation g as an arbitrary linear combination of the eigenvectors which correspond to the eigenvalues distinct from $\beta(A)$, the actual computer-implemented process of sequential approximations still allows one to obtain $\beta(A)$, but at the expense of having all the basis components in the expansion of ϕ_n (because of the round-off errors).

From the last relation we have

$$\frac{\| A^n g \|}{\| A^{n-1} g \|} = \beta(A) + 0 \left[\left(\frac{\lambda_{m-1}}{\lambda_m} \right)^n \right],$$

and therefore

$$\beta(A) = \lim_{n \to \infty} \frac{\| A^n g \|}{\| A^{n-1} g \|}. \tag{1.33}$$

Next let us compute the smallest eigenvalue of the matrix A. Consider a new matrix†

$$B = \beta(A)E - A \tag{1.34}$$

† We could replace $\beta(A)$ by any larger quantity.

and the spectral problem

$$Bu = \lambda(B)u. \tag{1.35}$$

Clearly, $B \geq 0$. Considering relation (1.34), we see that A and B share the common basis $\{u_k\}$. Similarly as before, consider the iterative process

$$\psi^{(n+1)} = B \frac{\psi^{(n)}}{\|\psi^{(n)}\|}. \tag{1.36}$$

As a result we obtain

$$\beta(B) = \lim_{n \to \infty} \|\psi^{(n)}\|. \tag{1.37}$$

Note that from relation (1.34) and from the generality of the basis for the matrices A and B it follows that

$$\beta(B) = \beta(A) - \alpha(A).$$

Hence

$$\alpha(A) = \beta(A) - \beta(B). \tag{1.38}$$

Thus, for a matrix of the above form, the smallest eigenvalue is found as the difference between the largest eigenvalues of the respective matrices A and $B = \beta(A)E - A$. Since the largest eigenvalues $\beta(A)$ and $\beta(B)$ are found by the iterative process described above, the task of finding the smallest eigenvalue of A is, in principle, solved.

It is to be remarked, however, that in the case of ill-conditioned matrices A, the smallest eigenvalue $\alpha(A)$ is obtained as a difference of large numbers $\beta(A)$ and $\beta(B)$. For this reason the actual numerical algorithm can make errors, not only in the magnitude of $\alpha(A)$, but even in the sign. In order to avoid such errors, the computation of $\alpha(A)$ will be slightly changed. With this in mind, consider the iterative process

$$\psi^{(n+1)} = B \frac{\psi^{(n)}}{\|\psi^{(n)}\|}, \qquad \psi^{(0)} = h.$$

Note that the system of eigenvectors u_k of the matrix A, ordered through the natural ordering of the eigenvalues, is transformed into the ordered system of eigenvectors v_k of the matrix B in such a way that $v_k = u_{m-k+1}$ $(k = 1, 2, \ldots, m)$. Consider the expressions

$$\psi^{(n)} = B \frac{\psi^{(n-1)}}{\|\psi^{(n-1)}\|} = \cdots = \frac{B^n \psi^{(0)}}{\|B^{n-1}\psi^{(0)}\|} = \frac{\sum_{k=1}^{m} \lambda_k^n(B) h_k v_k}{\|\sum_{k=1}^{m} \lambda_k^{n-1}(B) h_k v_k\|},$$

$$A\psi^{(n)} = \frac{\sum_{k=1}^{m} \lambda_k^n(B) h_k A v_k}{\|\sum_{k=1}^{m} \lambda_k^{n-1}(B) h_k v_k\|},$$

where $h_k = (h, v_k^*)$. Passing to the limit in the preceding equations we have as $n \to \infty$

$$\lim_{n \to \infty} \psi^{(n)} = \beta(B) \frac{h_m v_m}{\|h_m v_m\|}, \qquad \lim_{n \to \infty} A\psi^{(n)} = \beta(B) \frac{h_m A v_m}{\|h_m v_m\|}.$$

Since $Av_m = Au_1 = \alpha(A)u_1 = \alpha(A)v_m$, we obtain the algorithm

$$\psi^{(n+1)} = B \frac{\psi^{(n)}}{\|\psi^{(n)}\|}, \tag{1.39}$$

$$\alpha(A) = \lim_{n \to \infty} \frac{\|A\psi^{(n)}\|}{\|\psi^{(n)}\|}. \tag{1.40}$$

The last formula no longer includes differences of large numbers and, as a rule, can be used effectively to find the spectral bounds with the aid of a computer.

It should be pointed out, however, that iterations (1.31), (1.36), and (1.39) converge slowly. In order to accelerate the convergence one can use various methods, the most feasible of which are the Chebyshev accelerations and the shift of origin method (Faddeev and Faddeeva [8], Gavurin [9], and Wilkinson [8]).

Let us note that in the case of symmetric matrices the computation of $\alpha(A)$ and $\beta(A)$ is handled more conveniently by using the energy norm.

Optimization of numerical processes and various theoretical estimates of algorithms often require knowledge of the norm of the operator A and of its inverse. The following relations hold for arbitrary operators:

$$\|A\|^2 = \sup_{\phi \in \Phi} \frac{(A\phi, A\phi)}{(\phi, \phi)} = \sup_{\phi \in \Phi} \frac{(A^*A\phi, \phi)}{(\phi, \phi)} = \beta(A^*A)$$

and

$$\|A^{-1}\|^2 = \sup_{\phi \in \Phi} \frac{(A^{-1}\phi, A^{-1}\phi)}{(\phi, \phi)} = \sup_{\phi \in \Phi} \frac{((AA^*)^{-1}\phi, \phi)}{(\phi, \phi)} = [\alpha(A^*A)]^{-1}.$$

Hence

$$\|A\| = \sqrt{\beta(A^*A)}, \tag{1.41}$$

$$\|A^{-1}\| = (\sqrt{\alpha(A^*A)})^{-1} \tag{1.42}$$

The numbers $\alpha(A^*A)$ and $\beta(A^*A)$ are obtained by means of the successive approximation method described above.

In conclusion, the following remark seems necessary: the algorithms for computing the spectral bounds of positive matrices, which are described above, facilitate the possibility of optimizing the iterative processes for solving problems of mathematical physics. They are based on well-developed methods which will be dealt with in Chapter 4. Such processes become constructive and allow for effective procedures in solving various problems of mathematical physics.

In addition to Lyusternik's method, Gershgorin's theorem† yields a useful estimate for spectral bounds: all the eigenvalues $\lambda(A)$ (generally complex) of an arbitrary matrix A, of order m, with elements a_{kl}, lie in the union of the discs

$$|z - a_{kk}| \le R_k, \qquad k = 1, 2, \dots, m,$$

where

$$R_k = \sum_{\substack{l=1 \\ l \ne k}}^{n} |a_{kl}|.$$

Note that one consequence of Gershgorin's theorem is the following estimate of the spectral radius of a matrix:

$$\beta(A) \le \max_k \sum_{l=1}^{n} |a_{kl}|.$$

In conclusion, let us observe that the increasing power of electronic computers for solving the problems of mathematical physics leads to more and more frequent use of methods for computing series of components of a solution corresponding to various leading eigenvalues of the spectral problems

$$Ag = \lambda g.$$

The method begins by using the algorithm described above to find the eigenvalue $\lambda_1 = \alpha$ and the corresponding eigenvector u_1. Suppose for simplicity that A is symmetric and has a simple spectrum $\{\lambda_n\}$. To compute λ_2 and u_2 we iterate as follows:

$$\psi^{(n+1)} = B \frac{\psi^{(n)}}{\|\psi^{(n)}\|}, \qquad B = \beta(A)E - A,$$

assuming that neither the null approximation $\phi^0 = h$, nor the succeeding approximations $\phi^{(n)}$ contain components corresponding to the eigenvector u_1. To validate this assumption we must orthogonalize in the sequence (1.39) with respect to the leading eigenvector in the following form

$$\bar{\psi}^{(n)} = \psi^{(n)} - a_1 u_1, \tag{1.43}$$

where a_1 is a constant derived from the condition that $\bar{\psi}^{(n)}$ is orthogonal to u_1:

$$(\bar{\psi}^{(n)}, u_1) = 0. \tag{1.44}$$

Taking the inner product of (1.43) and u_1, and using (1.44) we obtain

$$(\bar{\psi}^{(n)}, u_1) = (\psi^{(n)}, u_1) - a_1(u_1, u_1) = 0.$$

† A complete proof of Gershgorin's theorem can be found in Faddeev and Faddeeva, *Numerical Methods in Linear Algebra* [8].

Hence

$$a_1 = \frac{(\psi^{(n)}, u_1)}{(u_1, u_1)}. \tag{1.45}$$

As a result we arrive at the algorithm

$$\bar{\psi}^{(n)} = \psi^{(n)} - a_1 u_1,$$
$$\psi^{(n+1)} = B \frac{\bar{\psi}^{(n)}}{\|\bar{\psi}^{(n)}\|}. \tag{1.46}$$

As $n \to \infty$, the limit element in the process (1.46) is u_2 and the corresponding eigenvalue is given by the formula

$$\lambda_2 = \lim_{n \to \infty} \frac{\|A\bar{\psi}^{(n)}\|}{\|\bar{\psi}^{(n)}\|}.$$

The succeeding vectors and the corresponding eigenvalues may be defined in the same way.

In fact, suppose the first k eigenvectors $u_1, u_2, \ldots, u_k$ have been found, and we are required to find u_{k+1}. We use the following iterative process

$$\bar{\psi}^{(n)} = \psi^{(n)} - a_1 u_1 - a_2 u_2 - \cdots - a_k u_k,$$
$$\psi^{(n+1)} = B \frac{\bar{\psi}^{(n)}}{\|\bar{\psi}^{(n)}\|},$$

The constants $a_1, a_2, \ldots, a_k$ are computed from the condition that $\bar{\psi}^{(n)}$ is orthogonal to all the vectors $u_1, u_2, \ldots, u_k$. We arrive at the set of equations

$$a_1(u_1, u_1) = (\psi^{(n)}, u_1),$$
$$a_2(u_2, u_2) = (\psi^{(n)}, u_2),$$
$$\cdots\cdots\cdots\cdots\cdots\cdots$$
$$a_k(u_k, u_k) = (\psi^{(n)}, u_k).$$

It follows that

$$a_1 = \frac{(\psi^{(n)}, u_1)}{(u_1, u_1)}, \ a_2 = \frac{(\psi^{(n)}, u_2)}{(u_2, u_2)}, \ldots, a_k = \frac{(\psi^{(n)}, u_k)}{(u_k, u_k)}.$$

The limit element of this process as $n \to \infty$ will be

$$u_{k+1} = \lim_{n \to \infty} \bar{\psi}^{(n)}.$$

The corresponding eigenvalue λ_{k+1} is given by the formula

$$\lambda_{k+1} = \lim_{n \to \infty} \frac{\|A\bar{\psi}^{(n)}\|}{\|\psi^{(n)}\|}.$$

This process of sequential orthogonalization need not be applied at every step of the iteration. It is sufficient to do so at a large enough number of steps, to keep the round-off errors corresponding to the "suppressed" vectors from increasing too sharply. It is obviously desirable to orthogonalize the initial approximation $\psi^{(0)}$.

The above method illustrates the principal plan of the approach to finding the leading eigenvalues and the corresponding eigenvectors for a symmetric matrix with a simple spectrum.

Now let us suppose that the matrix A is nonsymmetric. Then we have two spectral problems at the same time

$$A\phi = \lambda\phi, \qquad A^*\phi^* = \lambda\phi^*. \tag{1.47}$$

We first assume that the eigenelements and the spectra of (1.47) are real and the eigenvalues simple. Since the bases $\{\phi_n\}$ and $\{\phi_n^*\}$ are biorthogonal, i.e.,

$$(\phi_l, \phi_k^*) = 0 \quad \text{if} \quad l \neq k,$$

the algorithm for constructing the leading eigenelements has the following form: we begin by finding the first eigenvectors

$$u_1 = \lim_{n \to \infty} B \frac{\psi^{(n)}}{\|\psi^{(n)}\|},$$

$$u_1^* = \lim_{n \to \infty} B^* \frac{\psi^{*(n)}}{\|\psi^{*(n)}\|}$$

$$(B = \beta(A)E - A, \; B^* = \beta(A^*)E - A^*)$$

and the minimal eigenvalue

$$\lambda_1 = \lim_{n \to \infty} \frac{\|A\psi^{(n)}\|}{\|\psi^{(n)}\|} = \lim_{n \to \infty} \frac{\|A^*\psi^{*(n)}\|}{\|\psi^{*(n)}\|}.$$

Then we find the succeeding eigenelements and eigenvalues, by the following iterative process:

$$\psi^{(n+1)} = B \frac{\bar{\psi}^{(n)}}{\|\bar{\psi}^{(n)}\|},$$

$$\psi^{*(n+1)} = B^* \frac{\bar{\psi}^{*(n)}}{\|\bar{\psi}^{*(n)}\|},$$

where

$$\bar{\psi}^{(n)} = \psi^{(n)} - a_1 u_1 - a_2 u_2 - \cdots - a_k u_k,$$

$$\bar{\psi}^{*(n)} = \psi^{*(n)} - b_1 u_1^* - b_2 u_2^* - \cdots - b_k u_k^*.$$

The constants $a_1, a_2, \ldots, a_k$ and $b_1, b_2, \ldots, b_k$ are computed from the conditions of orthogonality

$$(\bar{\psi}^{(n)}, u_i^*) = 0,$$
$$(\bar{\psi}^{*(n)}, u_i) = 0, \qquad i = 1, 2, \ldots, k.$$

Taking account of the biorthogonality conditions we arrive at the equations

$$(\psi^{(n)}, u_i^*) - a_i(u_i, u_i^*) = 0,$$
$$(\psi^{*(n)}, u_i) - b_i(u_i, u_i^*) = 0, \qquad i = 1, 2, \ldots, k.$$

Consequently,

$$a_i = \frac{(\psi^{(n)}, u_i^*)}{(u_i, u_i^*)}, \qquad b_i = \frac{(\psi^{*(n)}, u_i)}{(u_i, u_i^*)}, \qquad i = 1, 2, \ldots, k.$$

One must keep in mind the fact that today we have a different and much more powerful system of algorithms for solving the complete problem of eigenvalues on the basis of the QR-algorithm, which has given rise to a number of excellent programs and procedures for finding the eigenelements and the corresponding eigenvalues for the problems of linear algebra.[†] These compete successfully with the iterative methods for solving spectral problems, and seem likely, in the future, to form the basis for the solution of a broad class of problems, including those of higher-order matrices, where their effectiveness is dependent on the power of electronic computers, which is constantly increasing.

In judging the developmental trends of numerical mathematics, we suggest that, in the near future, the complete solution of the eigenvalue problem will resurrect the former power of the classical Fourier method in its application to the problems of mathematical physics that can be reduced to problems in linear algebra.

In fact, suppose given a problem in linear algebra

$$A\phi = f, \tag{1.48}$$

which has to be solved repeatedly with the same operator A for different right-hand sides f, and assume that we know the solution of the spectral problem

$$A\omega = \lambda\omega,$$
$$A^*\omega^* = \lambda\omega^*$$

in the form of the complete bases $\{\omega_n\}$ and $\{\omega_n^*\}$; it is then convenient to represent the solution of (1.48) and the right-hand side in the form of a Fourier sum

$$\phi = \sum_{n=1}^{m} \phi_n \omega_n,$$

$$f = \sum_{n=1}^{m} f_n \omega_n, \tag{1.49}$$

where

$$f_n = (f, \omega_n^*).$$

[†] These methods for solving spectral problems were developed by Francis [8], Kublanovskaya [8], Wilkinson [8], and others.

Using (1.49) in (1.48) and taking the scalar product with ω_n^*, we obtain the algebraic system

$$\lambda_n \psi_n = f_n, \qquad n = 1, 2, \ldots, m,$$

from which (for $\lambda \neq 0$) we find

$$\phi_n = \frac{f_n}{\lambda_n}.$$

The solution of problem (1.48) is thus obtained in the form

$$\phi = \sum_{n=1}^{m} \frac{f_n}{\lambda_n} \omega_n. \tag{1.50}$$

When we have available only a partial set (say $k < m$) of the eigenvectors, the Fourier method allows us to represent the solution of (1.48) in the form

$$\phi = \sum_{n=1}^{k} \frac{f_n}{\lambda_n} \omega_n + \xi. \tag{1.51}$$

Here ξ satisfies the equation

$$A\xi = \eta, \tag{1.52}$$

where

$$\eta = f - \sum_{n=1}^{k} f_n \omega_n.$$

It would appear at first glance that the problem (1.48) has been reduced to the analogous problem (1.52), and formally this is so. But if the order of the matrix A is so high that we can use only iterative methods (see Chapter 4), then it turns out that these methods, as applied to (1.52), are implemented more effectively on an electronic computer than they would be if applied to (1.48), The importance of the algorithm for finding the leading eigenelements of a spectral problem, as described in this section, consists precisely in the increase of effectiveness through reducing the order of the matrix on which the computer must work. This notion will be used many times in the remainder of the present text for the construction of numerical algorithms.

1.1.3 Eigenvalues and Eigenfunctions of the Laplace Operator

Let

$$A = -\Delta, \tag{1.53}$$

where $\Delta = \partial^2/\partial x^2 + \partial^2/\partial y^2$ is the Laplace operator. The operator A is defined on the set Φ of elements ϕ satisfying the following requirements.
First,

$$\phi = 0 \quad \text{on} \quad \partial D, \tag{1.54}$$

where ∂D stands for the boundary of the region D. For simplicity, we take $D = \{(x, y), 0 < x < 1, 0 < y < 1\}$.

Second, the functions $\phi(x) \in \Phi$ together with their first and second derivatives are continuous on the closed region $D + \partial D$.

Third, the set Φ of elements ϕ is a subspace of the Hilbert space $L^2(D)$ with the inner product

$$(a, b) = \int_D ab \, dD, \qquad (1.55)$$

where $a \in L_2(D)$, $b \in L_2(D)$, and with the norm

$$\|\phi\| = \sqrt{(\phi, \phi)}. \qquad (1.56)$$

We now show that under these conditions the operator A is symmetric. To this end, consider a function $\phi^* \in L_2(D)$ and the functional

$$(A\phi, \phi^*) = - \int_D \phi^* \Delta\phi \, dD. \qquad (1.57)$$

Note that these conditions guarantee boundedness of the functional $(A\phi, \phi^*)$ for any $\phi^* \in L_2(D)$. Assume next that the function ϕ^* is smooth enough so that one can use the second formula of Green. Then

$$(A\phi, \phi^*) = - \int_{\delta D} \left(\phi^* \frac{\partial\phi}{\partial n} - \phi \frac{\partial\phi^*}{\partial n} \right) ds - \int_D \phi \Delta\phi^* \, dD, \qquad (1.58)$$

where n is the external normal of D. If the function ϕ^* satisfies the boundary condition

$$\phi^* = 0 \quad \text{on} \quad \partial D, \qquad (1.59)$$

then, using conditions (1.54) and (1.59),

$$(A\phi, \phi^*) = - \int_D \phi \Delta\phi^* \, dD = (\phi, A\phi^*). \qquad (1.60)$$

This means $A = A^*$, i.e., A is symmetric. An analysis of the above shows that the function ϕ^* must be assumed to have continuous derivatives. Finally, we conclude that the operator A is self-adjoint on Φ. (The above development presupposes quite strong constraints on the problem to be solved. It can be shown that the Green formula, and hence all of our conclusions, hold true for any function ϕ from the Sobolev space $\overset{\circ}{W}_2^2$ which vanishes on the boundary ∂D.)

Next, let us investigate the positivity of A. For this purpose consider the functional

$$(A\phi, \phi) = - \int_D \phi \Delta\phi \, dD. \qquad (1.61)$$

With the help of the first formula of Green we get

$$(A\phi, \phi) = -\int_{\partial D} \phi \frac{\partial \phi}{\partial n} \, ds + \int_D \left[\left(\frac{\partial \phi}{\partial x} \right)^2 + \left(\frac{\partial \phi}{\partial y} \right)^2 \right] dD. \tag{1.62}$$

Since ϕ satisfies (1.54), it follows that for any $\phi \in \Phi$ not identically zero

$$(A\phi, \phi) = \int_D \left[\left(\frac{\partial \phi}{\partial x} \right)^2 + \left(\frac{\partial \phi}{\partial y} \right)^2 \right] dD > 0. \tag{1.63}$$

Let us use this example to illustrate the eigenvalue problem. It is well known (see Courant [2], Sobolev [1]) that an orthonormalized eigenfunction system of the problem

$$Au = \lambda u \quad \text{in} \quad D, \tag{1.64}$$

$$u = 0 \quad \text{on} \quad \partial D, \tag{1.65}$$

is complete and has the form

$$u_{mp} = 2 \sin m\pi x \sin p\pi x, \tag{1.66}$$

where $m = 1, 2, \ldots$ and $p = 1, 2, \ldots$. The eigenvalues of the operator A are of the form

$$\lambda_{mp}(A) = (m^2 + p^2)\pi^2 > 0. \tag{1.67}$$

Hence

$$2\pi^2 \leq \lambda_{mp}(A) \leq \infty.$$

Thus

$$\alpha(A) = 2\pi^2, \qquad \beta(A) = \infty \tag{1.68}$$

and $(A\phi, \phi) \geq 2\pi^2(\phi, \phi)$. Consequently, the operator A is positive definite and unbounded. Since the system of eigenfunctions is complete, any function from Φ can be represented in terms of Fourier series

$$\phi(x, y) = \sum_m \sum_p \phi_{mp} u_{mp}(x, y) = \sum_i \phi_i u_i(x, y), \tag{1.69}$$

and, in addition, since the system $\{u_{mp}\}$ is orthonormalized

$$\phi_i = (\phi, u_i), \tag{1.70}$$

where i is a new ordering index for the series.

Earlier we considered the Dirichlet problem for the Laplace operator in the square. For regions of more complicated form and operators with variable coefficients an explicit computation of the eigenvalues is impossible; the spectral bounds must be obtained by special methods.

1.1.4 Eigenvalues and Eigenvectors of the Finite-Difference Analog of the Laplace Operator

We shall consider the set of points (x_k, y_l), where $x_k = k/n$ and $y_l = l/n$; n is some given positive integer and k and l are arbitrary positive integers. A point set of this form will be called a *net*, and the points themselves *net points*. The quantity $h = 1/n$ is called the *mesh size*. We write

$$D_h = \{(x_k, y_l), 1 \le k < n, 1 \le l < n\},$$

and denote by ∂D_h the net consisting of vertices for which one of the co-ordinates is either zero or one. A function whose domain of definition is a net will be called a *net function*. The set of net functions ϕ^h defined on D_h will be denoted by Φ_h. Every function $\phi \in \Phi$ may be mapped into a net function $(\phi)_h$ by the rule: the value of $(\phi)_h$ at the vertex (x_k, y_l) is equal to $\phi(x_k, y_l)$. This correspondence is a linear operator from the subspace Φ into Φ_h, the set of net functions on D_h; this operator is called a *projection of the function ϕ on the net*.

Next, let A be a linear operator defined on the functions $\phi \in \Phi$. Then the function $\psi = A\phi$ can also be projected on the net by taking $(\psi)_h = (A\phi)_h$. The correspondence $(\phi)_h \rightarrow (A\phi)_h$ is again a linear operator, defined on the net functions $(\phi)_h$. We will call the obtained operator a *projection of A on the net* and denote it $(A)_h$. Projections of this kind lead to the finite difference analogs of the equations. Methods of their construction—as well as approximation problems, countable stability, and convergence of solutions of the approximating problems to the exact ones—will be considered in what follows.

Let ϕ^h be a net function with the components $\phi^h_{k,l}$ where $\phi^h_{k,l} = \phi^h(x_k, y_l)$ and let Δ^h be the finite difference analog of the Laplace operator on the uniform net $\Delta x = \Delta y = h$, defined as follows:

$$(\Delta^h \phi^h)_{k,l} = \frac{\phi^h_{k+1,l} + \phi^h_{k-1,l} + \phi^h_{k,l+1} + \phi^h_{k,l-1} - 4\phi^h_{k,l}}{h^2}. \quad (1.71)$$

Let us assume that the net function $\phi^h \in \Phi_h$ vanishes on the boundary of the net region, i.e.,

$$(\phi^h)_{k,l} = 0 \quad \text{on} \quad \partial D_h. \quad (1.72)$$

Next, let us introduce the difference operators indexed by k and l, as follows:

$$(\Delta_x \phi^h)_{k,l} = \frac{1}{h}(\phi^h_{k+1,l} - \phi^h_{k,l}),$$

$$(\nabla_x \phi^h)_{k,l} = \frac{1}{h}(\phi^h_{k,l} - \phi^h_{k-1,l}),$$

$$(\Delta_y \phi^h)_{k,l} = \frac{1}{y}(\phi^h_{k,l+1} - \phi^h_{k,l}),$$

$$(\nabla_y \phi^h)_{k,l} = \frac{1}{h}(\phi^h_{k,l} - \phi^h_{k,l-1}).$$

Consider the new difference operators A^h, A_x, and A_y which are defined by the following relations:

$$A^h = A_x + A_y,$$
$$A_x = -\Delta_x \nabla_x, \tag{1.73}$$
$$A_y = -\Delta_y \nabla_y.$$

Then we have

$$-\Delta^h = A_x + A_y = A^h.$$

The set of net points for which $k \in \{0, n\}$ or $l \in \{0, n\}$ is the boundary ∂D_h. Let us recall that at these net points the function ϕ^h vanishes [see conditions (1.72)]. We shall omit the index h for the net functions ϕ and ϕ^*.

Let us next consider the inner product

$$(a, b) = h^2 \sum_{k=1}^{n-1} \sum_{l=1}^{n-1} a_{k,l} b_{k,l}$$

and let

$$\|\phi\| = \sqrt{(\phi, \phi)}.$$

We construct the functional

$$(A^h\phi, \phi^*) = -h^2 \sum_{k=1}^{n-1} \sum_{l=1}^{n-1} [(\Delta_x \nabla_x \phi)_{k,l} + (\Delta_y \nabla_y \phi)_{k,l}] \phi^*_{k,l};$$

(here and below we omit the index h in the net functions ϕ and ϕ^* for simplicity). The following analogs of the first and second Green's formulas hold true (Ladyzhenskaya [2], Samarskii [3])

$$-\sum_{k=1}^{n-1} (\Delta_x \nabla_x \phi)_{k,l} \phi^*_{k,l} = \sum_{k=1}^{n} (\nabla_x \phi)_{k,l} (\nabla_x \phi^*)_{k,l},$$

$$-\sum_{k=1}^{n-1} (\Delta_x \nabla_x \phi)_{k,l} \phi^*_{k,l} = \sum_{k=1}^{n-1} (\Delta_x \nabla_x \phi^*)_{k,l} \phi_{k,l}. \tag{1.74}$$

The above identities are valid only for $\phi \in \Phi_h$, which satisfy the condition (1.72) and for $\phi^* \in \Phi_h^*$, which satisfy the relation

$$\phi^*_{k,l} = 0 \quad \text{on} \quad \partial D_h. \tag{1.75}$$

Similar identities hold also for sums over the index l. With the help of the second relation in equations (1.74) we obtain

$$(A^h\phi, \phi^*) = (\phi, A^h\phi^*).$$

From here we conclude that A^h is self-adjoint, i.e.,

$$A^h = (A^h)^* \quad \text{and} \quad \Phi(A^h) = \Phi(A^{h*}).$$

Next consider the functional

$$(A^h\phi, \phi) = -h^2 \sum_{k=1}^{n-1} \sum_{l=1}^{n-1} [(\Delta_x \nabla_x \phi)_{k,l} + (\Delta_y \nabla_y \phi)_{k,l}]\phi_{k,l}.$$

The first identity in equations (1.74) in k and l yields

$$(A^h\phi, \phi) = h^2 \sum_{k=1}^{n} \sum_{l=1}^{n} [((\nabla_x \phi)_{k,l})^2 + ((\nabla_y \phi)_{k,l})^2].$$

Hence

$$(A^h\phi, \phi) > 0,$$

provided ϕ is not the zero vector.

Finally, consider the spectral problem

$$\begin{aligned} A^h u &= \lambda u \quad \text{in} \quad D_h, \\ u &= 0 \quad \text{on} \quad \partial D_h. \end{aligned} \tag{1.76}$$

The components of the corresponding orthonormalized eigenvectors are of the form

$$u_{mp}^{kl} = 2 \sin m\pi kh \sin p\pi lh,$$
$$m = 1, 2, \ldots, n-1; p = 1, 2, \ldots, n-1. \tag{1.77}$$

Recall that

$$(u_{m_1 p_1}, u_{m_2 p_2}) = h^2 \sum_{k=1}^{n-1} \sum_{l=1}^{n-1} u_{m_1 p_1}^{kl} u_{m_2 p_2}^{kl}.$$

The indices k, l in equation (1.77) indicate the components, and m, p identify the eigenvectors, which can be ordered by writing

$$u_{m,p} = u_i, \qquad i = 1, 2, \ldots.$$

Since

$$-(\Delta_x \nabla_x u_{mp})_{k,l} = 2\left(\frac{4}{h^2} \sin^2 \frac{m\pi h}{2} \sin m\pi kh\right) \sin p\pi lh$$

and

$$-(\Delta_y \nabla_y u_{mp})_{k,l} = 2 \sin m\pi kh \left(\frac{4}{h^2} \sin^2 \frac{p\pi h}{2} \sin p\pi lh\right),$$

the eigenvalues become

$$\lambda_{mp}(A^h) = \frac{4}{h^2}\left(\sin^2 \frac{m\pi h}{2} + \sin^2 \frac{p\pi h}{2}\right). \tag{1.78}$$

Let us note that m and p are between unity and $n - 1$. Consequently $h = 1/n \le mh \le (n - 1)h = 1 - h$ and $h \le ph \le 1 - h$; therefore

$$\frac{8}{h^2} \sin^2 \frac{\pi h}{2} \le \lambda_i(A^h) \le \frac{8}{h^2} \cos^2 \frac{\pi h}{2}.$$

Here $\lambda_i(A^h)$ are the ordered eigenvalues $\lambda_{mp}(A^h)$. Since, as a rule, $(\pi h/2) \ll 1$, one can estimate

$$\sin^2 \frac{\pi h}{2} = \frac{\pi^2 h^2}{4} - O(h^4), \qquad \cos^2 \frac{\pi h}{2} = 1 - O(h^2).$$

Hence

$$\alpha(A^h) \le \lambda_i \le \beta(A^h), \tag{1.79}$$

where

$$\alpha(A^h) = \frac{1}{\|(A^h)^{-1}\|} \approx 2\pi^2, \qquad \beta(A^h) = \|A^h\| \approx \frac{8}{h^2}. \tag{1.80}$$

Using the basis of eigenvectors from equation (1.77), the vector ϕ can be expanded into the series

$$\phi = \sum_i \phi_i u_i, \tag{1.81}$$

where

$$\phi_i = (\phi, u_i). \tag{1.82}$$

1.2 Approximation

Consider a problem of mathematical physics in the operator form

$$\begin{aligned} A\phi &= f \quad \text{in} \quad D, \\ a\phi &= g \quad \text{on} \quad \partial D, \end{aligned} \tag{2.1}$$

where A is a linear operator, $\phi \in \Phi$ and $f \in F$. Here Φ and F are Hilbert spaces, the elements of which are defined on $D + \partial D$ and D, respectively, a is a linear operator which represents the boundary conditions, and $g \in G$, where G is a Hilbert space with elements defined on ∂D.

Along with equation (2.1) let us also consider the following equations in a finite-dimensional space of net functions

$$\begin{aligned} A^h \phi^h &= f^h \quad \text{in} \quad D_h, \\ a^h \phi^h &= g^h \quad \text{on} \quad \partial D_h, \end{aligned} \tag{2.2}$$

where A^h is a linear operator depending on the mesh size h; $\phi^h \in \Phi_h$, $f^h \in F_h$, and Φ_h, F_h are Euclidean spaces. D_h is the set of interior net points of the region D and ∂D_h is the set of net points which are used to approximate the boundary conditions; a^h is a linear operator, $g^h \in G_h$, G_h is a Euclidean space whose vectors are defined on ∂D_h.

In the net spaces F_h, G_h, and Φ_h we introduce the corresponding norms $\|\cdot\|_{F_h}$, $\|\cdot\|_{G_h}$, $\|\cdot\|_{\Phi_h}$. Let $(\cdot)_h$ be a linear operator which maps the element $\phi \in \Phi$ into the element $(\phi)_h \in \Phi_h$ in such a way that $\lim_{h\to 0} \|\phi_h\|_{\Phi_h} = \|\phi\|_\Phi$. We say that equation (2.2) is an *n-order approximation* of equation (2.1) on the solution ϕ if there exist positive constants $\bar{h}$, M_1, M_2 such that for all $h < \bar{h}$ we have

$$\|A^n(\phi)_h - f^h\|_{F_h} \leq M_1 h^{n_1},$$

$$\|a^h(\phi)_h - g^h\|_{G_h} \leq M_2 h^{n_2} \tag{2.3}$$

and $n = \min(n_1, n_2)$.

If the solution of equation (2.1) is sufficiently smooth, the approximation order can be conveniently found with the help of a natural norm on the space of continuous and differentiable functions. For this purpose one can usually use the Taylor expansions for the solution and other functions entering the problem statement. A more detailed account of these problems can be found in the texts by Godunov and Ryabenkii [3], Richtmyer [3], Kantorovich and Akilov [1], and Samarskii [3].

In what follows we will assume that problem (2.1) has already been reduced to (2.2), and, moreover, that the boundary condition from (2.2) has been used to eliminate the solution at the boundary points of the region $D_h + \partial D_h$. As a result we obtain the equivalent problem

$$\tilde{A}^h \tilde{\phi}^h = \tilde{f}^h. \tag{2.4}$$

where the domain of definition of the solution ϕ^h is now D_h. The behavior of the solution at the boundary is determined by equation (2.2) and by the solution of (2.4).

In some cases it is convenient to use form (2.4) of the approximation problem; otherwise we use (2.2).

Thus, as a result of the indicated reduction, and with the required approximation taken into account, the continuous problem (2.1) has been transformed into (2.4), a problem in linear algebra.

In the rest of the book we shall be primarily concerned with Hilbert spaces of net functions, and we shall suppose that the norm $\|\phi^h\|$ is defined, via the scalar product (ϕ^h, ψ^k), by the equation $\|\phi^h\| = (\phi^h, \phi^h)^{1/2}$. However, many of the concepts we introduce (approximations, etc.) are valid for Banach spaces, and in a number of our theorems and illustrative examples we shall introduce net function norms that are not defined by a scalar product, i.e., the investigation is actually proceeding in a Banach space.

EXAMPLE. Let us consider the following problem:

$$-\Delta\phi = f \quad \text{in} \quad D,$$

$$\phi = 0 \quad \text{on} \quad \partial D. \tag{2.5}$$

We assume the domain of definition D to be the square $\{0<x<1, 0<y<1\}$, and f to be a smooth function. Let $\bar{D}$ (the closure of D) be covered by a uniform net with the mesh size h. The net points of the region will be identified by the pair of indices (k, l), where the first index k ($0 \le k \le n$) corresponds to discretization of the x coordinate, and similarly, l ($0 \le l \le n$) corresponds to the discretization of the y coordinate. Consider the following approximations:

$$\phi_{xx} \to \Delta_x \nabla_x(\phi)^h, \qquad \phi_{yy} \to \Delta_y \nabla_y(\phi)^h,$$

where the difference operators Δ_x, Δ_x, $\pi\nabla_x$, and ∇_x have been defined in Section 1.1.4. The problem (2.5) can be approximated by the following one:

$$-[\Delta_x\nabla_x \phi^h + \Delta_x\nabla_y \phi^h] = f^h \quad \text{in} \quad D_h,$$

$$\phi^h = 0 \quad \text{on} \quad \partial D_h, \tag{2.6}$$

where ∂D_h is the set of net points which belong to the boundary. Now (2.6) can be written as follows:

$$-\Delta^h\phi^h = f^h \quad \text{in} \quad D_h,$$

$$\phi^h = 0 \quad \text{on} \quad \partial D_h, \tag{2.7}$$

where ϕ^h and f^h are vectors with the components $\phi^h_{k,l}$ and $f^h_{k,l}$, and

$$(\Delta^h\phi^h)_{k,l} = h^{-2}(\phi^h_{k+1,l} + \phi^h_{k-1,l} + \phi^h_{k,l+1} + \phi^h_{k,l-1} - 4\phi^h_{k,l}),$$

$$f^h_{k,l} = h^{-2} \int_{x_{k-1/2}}^{x_{k+1/2}} \int_{y_{l-1/2}}^{y_{l+1/2}} f \, dx \, dy,$$

$$x_{k\pm 1/2} = x_k \pm (h/2), \qquad y_{l\pm 1/2} = y_l \pm (h/2).$$

In the schemes presented here and later we adopt for $f^h_{k,l}$ some or other average of $f(x, y)$, computed by the methods given above. Generally speaking, this practice allows us to consider difference schemes for functions $f(x, y)$ that do not satisfy the smoothness conditions imposed for the occasion. In such cases we may also obtain corresponding estimates of the approximation errors. Let us introduce a solution space Φ_h. Let the elements from Φ_h be defined in the domain $D_h + \partial D_h = \{(x_k, y_l); 0 \le k \le n, 0 \le l \le n\}$. The vector f^h belongs to F_h with the domain of definition $D_h = \{(x_k, y_l); 1 \le k \le n - 1, 1 \le l \le n - 1\}$. By expanding the solution in a Taylor series in the vicinity of (x_k, y_l) and assuming that the derivatives with respect to (x, y), up to and including the fourth order, are bounded, we obtain

$$\phi(\bar{x}, \bar{y}) = \sum_{n=0}^{3} \frac{1}{n!} \left\{ \left[(\bar{x} - x_k) \frac{\partial}{\partial x} + (\bar{y} - y_l) \frac{\partial}{\partial y} \right]^n \phi \right\}_{k,l}$$
$$+ \frac{1}{4!} \left\{ \left[(\bar{x} - x_k) \frac{\partial}{\partial x} + (\bar{y} - y_l) \frac{\partial}{\partial y} \right]^4 \phi \right\}_{k+\Theta_1, l+\Theta_2},$$

where $(\bar{x}, \bar{y})$ is an arbitrary point of the domain

$$\{x_{k-1} \leq x \leq x_{k+1}, y_{l-1} \leq y \leq y_{l+1}\}, \qquad |\theta_1|, |\theta_2| < 1$$

and

$$x_{k+\theta_1} = x_k + \theta_1 h, \qquad y_{l+\theta_2} = y_l + \theta_2 h.$$

A similar expansion is obtained for the function $f(x, y)$,

As the norm in the space F_h we introduce the quantity

$$\|f^h\|_{F_h} = \max_{k,l} |f_{k,l}^h|.$$

and for G_h, a similar one. For $(\phi)_h$ we choose the vector whose components are the values of ϕ at the corresponding net points. Then, using the above expansions for ϕ and f, we obtain

$$\| -\Delta^h(\phi)_h - f^h \|_{F_h} \leq M_1 h^2, \tag{2.8}$$

where

$$M_1 = \tfrac{1}{6} \max_D (|\phi_x^{IV}|, |\phi_y^{IV}|).$$

In this case the approximation of the boundary conditions is exact.

It follows from this and from (2.8) that problem (2.7) is a second-order approximation of (2.5) on the solutions of (2.5) that possess bounded fourth-order derivatives.

So far we have considered only the approximation problem with respect to the space variables. However a similar procedure can be used to approximate the evolution equation

$$\frac{\partial \phi}{\partial t} + A\phi = f \quad \text{in} \quad D_t,$$

$$a\phi = g \quad \text{on} \quad \partial D_x D_t, \tag{2.9}$$

$$\phi = \phi^0 \quad \text{in} \quad D \quad \text{for} \quad t = 0.$$

(The term *evolution equation* always designates an equation of the above type, explicitly soluble for the first time derivative and with A containing no partials with respect to time.) The approximation procedure for problem (2.9) will be split into two stages. First, let us approximate the problem with respect to the space variables in the region $D_h + \partial D_h$. The result is a difference–differential equation: "difference" in the space variables and "differential" in the time variable.

In many instances of this difference–differential problem it is easy to exclude solutions on the boundary points of the domain $(D_h + \partial D_h) \times D_\tau$ by reference to difference conditions on the boundary. Supposing that this has been done we arrive at an evolution equation of the form

$$\frac{d\phi^h}{dt} + \Lambda\phi^h = f^h, \qquad (2.10)$$

where Λ, f^h, and ϕ^h are functions of time. From now on we will drop the index h in problem (2.10) as insignificant, assuming that we deal with a difference analog of the original problem of mathematical physics with respect to the space variables.

System (2.10) is clearly a system of ordinary differential equations for the components of the vector ϕ^h.

Thus, consider the following Cauchy problem:

$$\frac{d\phi}{dt} + \Lambda\phi = f,$$
$$\phi = g \quad \text{for} \quad t = 0. \qquad (2.11)$$

Assume that the operator Λ does not depend on time. Consider the simplest approximation methods for problem (2.11) with respect to time. The most convenient are the difference schemes with first- and second-order approximation in t.

Let us start with the simplest explicit first-order approximation scheme on the net D_τ:

$$\frac{\phi^{j+1} - \phi^j}{\tau} + \Lambda\phi^j = f^j, \qquad \phi^0 = g, \qquad (2.12)$$

where $\tau = t_{j+1} - t_j$, f^j is a projection of f. For simplicity, we will take $f^j = f(t_j)$.

The simplest implicit scheme is of the form

$$\frac{\phi^{j+1} - \phi^j}{\tau} + \Lambda\phi^{j+1} = f^j, \qquad \phi^0 = g, \qquad (2.13)$$

where we choose $f^j = f(t_{j+1})$. The approximations with respect to t in (2.12) and (2.13) are first-order approximations, as can be easily seen from the Taylor series expansion (assuming, of course, the existence of bounded first- and second-order derivatives of the solutions with respect to time).

Solving for ϕ^{j+1} in (2.12) and (2.13) yields the recursive relation

$$\phi^{j+1} = T\phi^j + \tau S f^j, \qquad (2.14)$$

where the *transition operator* T and the *source operator* S are defined as follows: for scheme (2.12), $T = E - \tau\Lambda$, $S = E$; for scheme (2.13) $T = (E + \tau\Lambda)^{-1}$, $S = T$.

Difference schemes of the above kind for evolution equations will be called *two-layer* schemes.

Of great interest in applications is the second-order approximation scheme of Crank and Nicholson

$$\frac{\phi^{j+1} - \phi^j}{\tau} + \Lambda \frac{\phi^{j+1} + \phi^j}{2} = f^j, \qquad \phi^0 = g, \qquad (2.15)$$

where $f^j = f(t_{j+1/2})$. Equation (2.15) can also be written in the form of (2.14) by taking

$$T = \left(E + \frac{\tau}{2} \Lambda \right)^{-1} \left(E - \frac{\tau}{2} \Lambda \right),$$

$$S = \left(E + \frac{\tau}{2} \Lambda \right)^{-1}.$$

The difference equations (2.12), (2.13), and (2.15) are in certain cases conveniently written as a system of two equations: one which approximates just the equation itself in $D_{h\tau}$, and the other which approximates the boundary conditions on $\partial D_{h\tau}$. In this case the difference analog of problem (2.9) becomes

$$\begin{aligned} L^{h\tau}\phi^{h\tau} &= f^{h\tau} \quad \text{in} \quad D_{h\tau}, \\ l^{h\tau}\phi^{h\tau} &= g^{h\tau} \quad \text{on} \quad \partial D_{h\tau}, \end{aligned} \qquad (2.16)$$

where $D_{h\tau} = D_h \times D_\tau$, $\partial D_{h\tau} = D_h \times \{0\} \cup \partial D_h \times D_\tau$. It is assumed that $L^{h\tau}$ approximates the operator

$$L = \frac{\partial}{\partial t} + \Lambda$$

and $l^{h\tau}$ approximates l on the interval $0 \le t \le T$. Similarly, $f^{h\tau}$ and $g^{h\tau}$ approximate f and g in the corresponding (different in general) norms, that is

$$\begin{aligned} \|(L\phi)_{h\tau} - L^{h\tau}(\phi)_{h\tau}\|_{F_{h\tau}} &\le M_1 h^n + N_1 \tau^p, \\ \|(l\phi)_{h\tau} - l^{h\tau}(\phi)_{h\tau}\|_{G_{h\tau}} &\le M_2 h^n + N_2 \tau^p, \\ \|(f)_{h\tau} - f^{h\tau}\|_{F_{h\tau}} &\le M_3 h^n + N_3 \tau^p, \\ \|(g)_{h\tau} - g^{h\tau}\|_{G_{h\tau}} &\le M_4 h^n + N_4 \tau^p. \end{aligned} \qquad (2.17)$$

The operator $(\cdot)_{h\tau}$ in these inequalities, as well as in equation (2.3), projects on the corresponding net space.

The canonical form (2.14) of the difference equations can also be written as

$$\tilde{L}^{h\tau}\tilde{\phi}^{h\tau} = \tilde{f}^{h\tau} \qquad (2.18)$$

by introducing vector functions and new operators with domains in $D_h \times D_\tau$, where D_τ is the set $\{t_j\}$.

In this manner, the evolution equation with its boundary conditions and initial data is reduced to a problem (2.18) in linear algebra. Note that the approximation schemes can be analyzed in terms of either the net D_h, or $D_h \times D_\tau$, depending on the choice. In particular, (2.18) may represent a boundary problem of elliptic type, an integral equation, etc., while the approximation condition can again be written in the form (2.17) with the approximation index h alone (h being the maximum from the set $\{\Delta x_i\}$ of steps in the space variables).

EXAMPLE. Consider the problem:

$$A\phi \equiv \frac{\partial \phi}{\partial t} - \Delta\phi = f \quad \text{in} \quad D \times D_t,$$

$$\phi = 0 \quad \text{on} \quad \partial D \times D_t, \tag{2.19}$$

$$\phi = g \quad \text{in} \quad D \quad \text{for} \quad t = 0.$$

Solutions are assumed to be defined on $(D + \delta D) \times D_t$, where D is a square as before, and $D_t = \{0 \le t \le T\}$. Consider D_h, ∂D_h, and D_τ along with D, ∂D, and D_t. Let D_τ be the set of points $\{t_j\}$, $t_{j+1} - t_j = \tau$. Then problem (2.19) can be approximated as follows:

$$A^{h\tau}\phi = f \quad \text{in} \quad D_h \times D_\tau,$$

$$\phi = 0 \quad \text{on} \quad \partial D_h \times D_\tau, \tag{2.20}$$

$$\phi^0 = g \quad \text{in} \quad D_h.$$

Consider the simplest explicit approximation

$$(A^{h\tau}\phi)^j_{k,l} \equiv \frac{\phi^{j+1}_{k,l} - \phi^j_{k,l}}{\tau} - (\Delta^h\phi^j)_{k,l} \tag{2.21}$$

$$f^j_{k,l} = \frac{1}{h^2} \int_{x_{k-1/2}}^{x_{k+1/2}} \int_{y_{l-1/2}}^{y_{l+1/2}} f(x, y, t_j) dx\, dy, \tag{2.22}$$

$$g_{k,l} = \frac{1}{h^2} \int_{x_{k-1/2}}^{x_{k+1/2}} \int_{y_{l-1/2}}^{y_{l+1/2}} g(x, y) dx\, dy. \tag{2.23}$$

where

$$\phi^j_{k,l} = \phi^{h\tau}(x_k, y_l, t_j).$$

Then

$$\phi^{j+1}_{k,l} = \phi^j_{k,l} + \tau(\Delta^h\phi^j)_{k,l} + \tau f^j_{k,l} \quad \text{in} \quad D_h \times D_\tau. \tag{2.24}$$

Also,

$$\phi^j_{k,l} = 0 \quad \text{on} \quad \partial D_h \times D_\tau,$$

$$\phi^0_{k,l} = g_{k,l} \quad \text{in} \quad D_h \times \{0\}. \tag{2.25}$$

The recursive relation (2.24) can be written as

$$\phi_{k,l}^{j+1} = T\phi_{k,l}^{j} + \tau f_{k,l}^{j}, \tag{2.26}$$

where $T = E + \tau\Delta^h = E - \tau(A_1 + A_2)$ is the transition operator, and the operators $A_i (A_1 = A_x, A_2 = A_y)$ are defined in (1.73). Suppose for simplicity that $F_h = \Phi_h$ and

$$\|\phi^h\|_{\Phi_h} = \sqrt{\sum_{k,l} |\phi_{k,l}|^2 h^2}.$$

Let us estimate the norm of T. For that let us find the largest eigenvalue of T:

$$Tu = \lambda(T)u \quad \text{in} \quad D_h,$$
$$u = 0 \quad \text{on} \quad \partial D_h. \tag{2.27}$$

The following relation holds true

$$\lambda_n(T) = 1 + \tau\lambda_n(\Delta^h).$$

Consequently, the norm of the operator T is expressed as

$$\|T\| = \max\left\{ \left|1 - \frac{8\tau}{h^2}\cos^2\frac{\pi h}{2}\right|, \left|1 - \frac{8\tau}{h^2}\sin^2\frac{\pi h}{2}\right| \right\}, \tag{2.28}$$

and if $(\tau/h^2) < \frac{1}{4}$, then $\|T\| < 1$.

Along with the explicit first-order approximation with respect to τ it is possible to consider the *implicit* first-order approximation with respect to τ and the second-order approximation with respect to h. Equation (2.21) is then replaced by

$$(A^{h\tau}\phi^{h\tau})_{k,l}^{j} \equiv \frac{\phi_{k,l}^{j+1} - \phi_{k,l}^{j}}{\tau} - (\Delta^h\phi^{j+1})_{k,l} \tag{2.29}$$

$f_{k,l}^{j}$ and $g_{k,l}$ are defined by (2.22) and (2.23), respectively. Now equations (2.20) can no longer be solved explicitly and instead we have to solve the operator equation

$$((E - \tau\Delta^h)\phi^{j+1})_{k,l} = \phi_{k,l}^{j} + \tau f_{k,l}^{j} \quad \text{in} \quad D_h \times D_\tau, \tag{2.30}$$

with

$$\phi_{k,l}^{j} = 0 \quad \text{on} \quad \partial D_h \times D_\tau,$$
$$\phi_{k,l}^{0} = g_{k,l} \quad \text{in} \quad D_h. \tag{2.31}$$

Let us write (2.30) in the form

$$\phi_{k,l}^{j+1} = (T(\phi^j + \tau f^j))_{k,l} \tag{2.32}$$

where

$$T = (E - \tau\Delta^h)^{-1}.$$

The norm of T in this case becomes

$$\|T\| = \max\left\{\frac{1}{1 + \dfrac{8\tau}{h^2}\cos^2\dfrac{\pi h}{2}}, \frac{1}{1 + \dfrac{8\tau}{h^2}\sin^2\dfrac{\pi h}{2}}\right\}, \tag{2.33}$$

and hence $\|T\| < 1$ for any τ and h.

Finally, let us consider the Crank–Nicholson approximation scheme. In this case the operators and functions in (2.20) will be defined as follows:

$$(A^{h\tau}\phi^{h\tau})_{k,l}^j \equiv \frac{\phi_{k,l}^{j+1} - \phi_{k,l}^j}{\tau} - \left(\Delta^h \frac{\phi^j + \phi^{j+1}}{2}\right)_{k,l} \tag{2.34}$$

and

$$f_{k,l}^j = \frac{1}{h^2}\int_{x_{k-1/2}}^{x_{k+1/2}}\int_{y_{l-1/2}}^{y_{l+1/2}} f(x, y, t_{j+1/2})dx\,dy,$$

$$g_{k,l} = \frac{1}{h^2}\int_{x_{k-1/2}}^{x_{k+1/2}}\int_{y_{l-1/2}}^{y_{l+1/2}} g(x, y)dx\,dy. \tag{2.35}$$

As a result we arrive at the following problem:

$$\left(\left(E - \frac{\tau}{2}\Delta^h\right)\phi^{j+1}\right)_{k,l} = \left(\left(E + \frac{\tau}{2}\Delta^h\right)\phi^j\right)_{k,l} + \tau f_{k,l}^j \quad \text{in} \quad D_h \times D_\tau, \tag{2.36}$$

$$\phi_{k,l}^j = 0 \quad \text{on} \quad \partial D_h \times D_\tau,$$

$$\phi_{k,l}^0 = g_{k,l} \quad \text{in} \quad D_h. \tag{2.37}$$

Equation (2.36) can be formally solved with respect to the unknowns $\phi_{k,l}^{j+1}$ in the form

$$\phi_{k,l}^{j+1} = (T\phi^j)_{k,l} + \tau(sf^j)_{k,l} \tag{2.38}$$

where

$$T = \left(E - \frac{\tau}{2}\Delta^h\right)^{-1}\left(E + \frac{\tau}{2}\Delta^h\right),$$

$$S = \left(E - \frac{\tau}{2}\Delta^h\right)^{-1}.$$

The norm of the transition operator is given by

$$\|T\| = \max\left\{\left|\frac{1 - \dfrac{4\tau}{h^2}\cos^2\dfrac{\pi h}{2}}{1 + \dfrac{4\tau}{h^2}\cos^2\dfrac{\pi h}{2}}\right|, \left|\frac{1 - \dfrac{4\tau}{h^2}\sin^2\dfrac{\pi h}{2}}{1 + \dfrac{4\tau}{h^2}\sin^2\dfrac{\pi h}{2}}\right|\right\}. \tag{2.39}$$

Since $\tau > 0$, $\|T\| < 1$.

1.3 Countable Stability

We will not strive for generality in defining the notion of countable stability. The reason is that our main objective is to study simple algorithmic methods and the properties of difference approximations of problems in mathe-material physics. Various aspects of stability theory and important generalized results can be found in a number of sources (Ryabenkii and Filippov [6], Lax [6], Richtmyer [3], Godunov and Ryabenkii [7], Yanenko [3], Isaacson and Keller [3], Richtmyer and Morton [3], Samarskii [3], and others).

Basic definitions and methods in the theory of stability will be clarified at first with the explicit difference scheme (2.12):

$$\phi^{j+1} = (E - \tau \Lambda)\phi^j + \tau f^j, \qquad \phi^0 = g, \tag{3.1}$$

the solution of which is sought for $0 \le \tau j \le T$.

Assume that the operator $\Lambda > 0$ induces a complete set of eigenfunctions $\{u_n\}$ along with the corresponding eigenvalues $\{\lambda_n > 0\}$, according to the spectral problem

$$\Lambda u = \lambda u.$$

We introduce the following Fourier series:

$$\phi^j = \sum_n \phi_n^j u_n, \qquad f^j = \sum_n f_n^j u_n, \qquad g = \sum_n g_n u_n, \tag{3.2}$$

where

$$\phi_n^j = (\phi^j, u_n^*), \qquad f_n^j = (f^j, u_n^*), \qquad g^n = (g, u_n^*),$$

and u_n^* are the eigenfunctions of the adjoint spectral problem. Using (3.2) in (3.1) and taking next the inner product of the result with the vectors u_n^*, we obtain the following expression for the Fourier coefficients:

$$\phi_n^{j+1} = (1 - \tau \lambda_n)\phi_n^j + \tau f_n^j. \tag{3.3}$$

Assuming that

$$\phi^0 = \sum_n g_n u_n,$$

we obtain the initial condition

$$\phi_n^0 = g_n. \tag{3.4}$$

Equations (3.3) and (3.4) can be solved by successive elimination of the unknowns. As a result, we have

$$\phi_n^j = r_n^j g_n + \tau \sum_{i=1}^{j} r_n^{j-i} f_n^{i-1}, \tag{3.5}$$

where

$$r_n = 1 - \tau \lambda_n. \tag{3.6}$$

From (3.5) it follows that for $\tau > 0$

$$|\phi_n^j| \le |r_n|^j |g_n| + \tau \sum_{i=1}^{j} |r_n|^{j-i} |f_n^{i-1}|.$$

Hence, taking $|f_n| = \max_j |f_n^j|$ rather than $|f_n^{i-1}|$ under the summation symbol, we have

$$|\phi_n^j| \le |r_n|^j |g_n| + \frac{1 - |r_n|^j}{1 - |r_n|} \tau |f_n|. \tag{3.7}$$

John von Neumann has introduced the so-called spectral criterion of stability, the essence of which is as follows (Richtmyer and Morton [3]): if for every Fourier coefficient ϕ_n^j from (3.2) one has

$$|\phi_n^j| \le C_{1n}|g_n| + C_{2n}|f_n|, n = 1, 2, \ldots, \tag{3.8}$$

where C_{1n}, C_{2n} are constants with a uniform bound for $0 \le j\tau \le T$, then the difference scheme (3.1) is countably stable. Let us see what hypothesis regarding the parameters in the difference scheme (2.12) is enough to guarantee the validity of relation (3.8). An analysis of (3.7) shows that the stability criterion (3.8) is satisfied if we require the following constraint on the parameter r_n:

$$|r_n| < 1, n = 1, 2, \ldots. \tag{3.9}$$

(Later we introduce a weaker assumption on the norm of the transition operator.)

Assume that the spectrum of the operator Λ is contained in the interval

$$0 < \alpha(\Lambda) \le \lambda_n(\Lambda) \le \beta(\Lambda).$$

According to (3.6), relation (3.9) will then hold true provided

$$\tau \le 2/\beta(\Lambda). \tag{3.10}$$

Thus (3.10) becomes a constructive condition for stability of the difference scheme (3.1). Let us note that condition (3.10) is only sufficient; the scheme remains stable, for instance, when

$$\tau = 2/\beta(\Lambda).$$

Relation (3.7) in this latter case becomes

$$|\phi_n^j| \le |g_n| + j\tau |f_n| \tag{3.11}$$

as can be easily seen. But $j\tau \le T$, where T is fixed. This means that for a small τ a large number of steps j is required: $j \to \infty$ as $\tau \to 0$, the upper end point of the time interval T being fixed. Again, we arrive at schemes stable in the sense of von Neumann.

Consider now some other difference schemes which are based on the implicit difference approximations. The implicit first-order approximation scheme (2.13) leads to an expression similar to (3.7):

$$|\phi_n^j| \leq |r_n|^j |g_n| + \frac{1 - |r_n|^j}{1 - |r_n|} \tau |r_n| |f_n|, \tag{3.12}$$

where

$$r_n = \frac{1}{1 + \tau \lambda_n(\Lambda)}.$$

If $\lambda_n(\Lambda) > 0$, this difference scheme is clearly stable for any $\tau > 0$, since

$$|r_n| < 1, n = 1, 2, \ldots.$$

Stability of this kind will be called *absolute*.

In the case of the Crank–Nicholson scheme (2.15) one obtains the following estimates for the Fourier coefficients of the solution:

$$|\phi_n^j| \leq |r_n|^j |g_n| + \frac{1 - |r_n|^j}{1 - |r_n|} \tau \mu_n |f_n|, \tag{3.13}$$

where

$$r_n = \frac{1 - \frac{\tau}{2} \lambda_n(\Lambda)}{1 + \frac{\tau}{2} \lambda_n(\Lambda)}, \qquad \mu_n = \frac{1}{1 + \frac{\tau}{2} \lambda_n(\Lambda)}.$$

Hence $|r_n| < 1$ for an arbitrary $\tau > 0$, provided that $\lambda_n(\Lambda) > 0$.

We make the following comments at this point. First, stability in the sense of von Neumann is based on the spectral analysis of the operator defined by the problem at hand. This means that in this approach the algorithm necessarily involves computation of the largest eigenvalue or the estimate of its upper bound. Second, the spectral stability criterion establishes stability of the solution with respect to each of the harmonics from the Fourier series, while nothing at all is said about the stability of solutions in terms of energy norms. At the same time the spectral norm of the solution ϕ^j often happens to be a unique characteristic. All of this has triggered an effort to give new definitions of stability which would be related to the norms of the operators. It is to be emphasized, however, that up to the present, stability analysis of the von Neumann type continues to play a prominent role in applications.

Let us now turn to a more general definition of the notion of countable stability. To this end, consider the following problem:

$$\frac{\partial \phi}{\partial t} + A\phi = f \quad \text{in} \quad D \times D_t$$

$$\phi = g \quad \text{for} \quad t = 0; \tag{3.14}$$

it can be approximated by the difference problem as follows:

$$\phi^{j+1} = T\phi^j + \tau S f^j \quad \text{on} \quad D_h \times D_\tau,$$

$$\phi^0 = g. \tag{3.15}$$

We will say that the difference scheme (3.15) is *stable*, if for any h (the parameter characterizing the difference approximation) and $j \leq T/\tau$, one has

$$\|\phi^j\|_{\Phi_h} \leq C_1 \|g\|_{G_h} + C_2 \|f^{h\tau}\|_{F_{h\tau}}, \tag{3.16}$$

where the constants C_1 and C_2 are uniformly bounded on $0 \leq t \leq T$ and are independent of τ, h, g, and f; G_h is the space to which g in (3.15) belongs.

The definition of countable stability is closely related to the notion of well-posed problems with a continuous argument (Godunov [2], Lavrentiev [2], Yanenko [3]). One may say that countable stability (for problems with a discrete argument) implies continuous dependence of the solutions on the input data.

Indeed, let $f^j = f^j_*, g = g_*$ be the input data for problem (3.15). Denote by ϕ_* the corresponding solution. Similarly, let ϕ_{**} correspond to the input data $f = f_* + \xi, g = g_* + \delta$. The difference $\varepsilon = \phi_{**} - \phi_*$ will satisfy

$$\varepsilon^{j+1} = T\varepsilon^j + \tau S\xi^j, \qquad \varepsilon^0 = \delta.$$

Along with this, the stability criterion assumes the form

$$\|\varepsilon^{j+1}\|_{\Phi_h} \leq C_1 \|\delta\|_{G_h} + C_2 \|\xi\|_{F_h}.$$

Hence, it follows that a small variation in the input data f, g results in a small variation of the solution ϕ.

It is easy to see that the definition of stability in the form of (3.16) already relates the solution itself with a *prior* knowledge concerning the input data for the problem. Although less specific, this definition is often more suitable for analyzing stability than the definition in the sense of von Neumann. From this point of view, let us consider the stability of scheme (2.12). First we rewrite the recursive relation (3.1) as follows:

$$\phi^{j+1} = T^{\phi^j} + \tau f^j, \qquad \phi^0 = g, \tag{3.17}$$

$$T = E - \tau\Lambda, \tag{3.18}$$

and Λ is an operator approximating A. The formal solution of (3.17) has the form

$$\phi^{j+1} = T^j g + \tau \sum_{i=1}^{j} T^{j-i} f^{i-1}. \tag{3.19}$$

Setting $G_h \equiv F_h$† and estimating (3.19) in the norm, we obtain the estimate

$$\|\phi^{j+1}\|_{\Phi_h} \leq \|T\|^j \|g\|_{G_h} + \tau \sum_{i=1}^{j} \|T\|^{j-i} \|f^{i-1}\|_{F_h}. \tag{3.20}$$

† This is done to simplify the exposition. Otherwise, instead of the single norm

$$\|T\| \equiv \|T\|_{F_h \to \Phi_h} = \operatorname*{Sup}_{\phi \in F_h} \frac{\|T\phi\|_{\Phi_h}}{\|\phi\|_{F_h}}$$

we should have to introduce two

$$\|T\|_{F_h \to \Phi_h} + \|T\|_{G_h \to \Phi_h} = \operatorname*{Sup}_{\phi \in G_h} \frac{\|T\phi\|_{\Phi_h}}{\|\phi\|_{G_h}}.$$

We replace $\| f^{i-1} \|$ under the summation sign by its maximal value over all j in a fixed time interval. Let

$$\| f^{h\tau} \| = \max_j \| f^j \|,$$

then

$$\| \phi^j \|_{\Phi_h} \leq \| T \|^j \| g \|_{G_h} + \frac{1 - \| T \|^j}{1 - \| T \|} \tau \| f^{h\tau} \|. \tag{3.21}$$

If we assume that

$$\| T \| < 1, \tag{3.22}$$

then scheme (2.12) will be stable in the sense of definition (3.16). Of course, (3.22) is a sufficient condition of stability. Sharper conditions could be obtained by exploiting the norms of powers of the transition operator, $\| T^i \|$ ($i = 1, 2, \ldots$). In this generality the problem was investigated by Lax and Richtmyer [7]. Weakening of the condition, however, brings additional difficulties in the constructive procedure of establishing the stability criteria. As a rule, it is the sufficient condition (3.22) that is used in practice.

Consider the case where the operator $\Lambda = \Lambda^* > 0$ in (3.18), and write

$$J[\phi] = \frac{(T\phi, T\phi)}{(\phi, \phi)}. \tag{3.23}$$

Then

$$J[\phi] = 1 - 2\tau \frac{(\Lambda\phi, \phi)}{(\phi, \phi)} + \tau^2 \frac{(\Lambda\phi, \Lambda\phi)}{(\phi, \phi)}.$$

Let

$$\phi = \sum_n \phi_n u_n,$$

where $\{u_n\}$ is the basis of the operator Λ. Then

$$J[\phi] = 1 - 2\tau\bar{\lambda} + \tau^2\overline{\lambda^2}, \tag{3.24}$$

where

$$\bar{\lambda} = \frac{\sum_n \lambda_n(\Lambda)\phi_n^2}{\sum_n \phi_n^2}, \qquad \overline{\lambda^2} = \frac{\sum_n [\lambda_n(\Lambda)]^2 \phi_n^2}{\sum_n \phi_n^2},$$

Let us find out which conditions must be satisfied by τ in order that $J[\phi] \leq 1$ that is

$$1 - 2\tau\bar{\lambda} + \tau^2\overline{\lambda^2} \leq 1.$$

Hence

$$\tau \leq 2 \frac{\bar{\lambda}}{\overline{\lambda^2}} = 2 \frac{\sum_n \lambda_n \phi_n^2}{\sum_n \lambda_n^2 \phi_n^2},$$

and thus, if $\beta(\Lambda) = \|\Lambda\| = \max_n \lambda_n(\Lambda) = \lambda_1(\Lambda)$, then

$$\tau \le \frac{2}{\lambda_1(\Lambda)} \frac{\phi_1^2 + \sum_{n \ne 1} \frac{\lambda_n(\Lambda)}{\lambda_1(\Lambda)} \phi_n^2}{\phi_1^2 + \sum_{n \ne 1} \frac{[\lambda_n(\Lambda)]^2}{[\lambda_1(\Lambda)]^2} \phi_n^2}. \tag{3.25}$$

Since

$$\frac{\phi_1^2 + \sum_{n \ne 1} \frac{\lambda_n(\Lambda)}{\lambda_1(\Lambda)} \phi_n^2}{\phi_1^2 + \sum_{n \ne 1} \frac{[\lambda_n(\Lambda)]^2}{[\lambda_1(\Lambda)]^2} \phi_n^2} \ge 1,$$

we obtain the following sufficient condition for $J[\phi] \le 1$:

$$\tau \le \frac{2}{\beta(\Lambda)}.$$

In this case we have (in agreement with the definition (1.11) of the norm of an operator)

$$\|T\|^2 = \sup_\phi (J[\phi]) \le 1,$$

and hence the computation is stable in the sense of definition (3.16). Let us note, that the two definitions of stability [i.e., the stability in the sense of von Neumann (3.10) and the one defined by (3.16)] coincide if the operator is self-adjoint. The relation between these two definitions is studied in the monographs by Godunov and Ryabenkii [3] and Richtmyer and Morton [3].

The stability of the implicit difference equations (2.13) and (2.15) can be handled in a similar fashion. In these cases we have

$$\|\phi^j\| \le \|T\|^j \|g\| + \frac{1 - \|T\|^j}{1 - \|T\|} \tau \|S\| \|f\|,$$

where $T = (E + \tau\Lambda)^{-1}$, $S = (E + \tau\Lambda)^{-1}$ for scheme (2.13), and $T = (E + \tau\Lambda/2)^{-1}(E - \tau\Lambda/2)$, $S = (E + \tau\Lambda/2)^{-1}$ for schemes (2.15).

It is not difficult to show that the difference schemes above are absolutely stable in the sense of definition (3.16), provided $\Lambda = \Lambda^* > 0$ and

$$\|\phi^i\|_{\Phi_h} = \sqrt{\sum_{k,l} |\phi_{k,l}^j|^2 h}.$$

Let us briefly discuss the limiting behavior. In dealing with difference analogs for evolution-type problems of mathematical physics we have to consider approximations with respect to time (the step size τ) and also with respect to the space variables (the grid size h). In other words, the transition operator $T = T(\tau, h)$ depends both on τ and h.

Construction of a stable algorithm for a given approximation method usually reduces to the problem of how to relate τ and h so as to achieve countable stability. The difference scheme becomes absolutely stable if it is stable for arbitrary choice of $\tau > 0, h > 0$. If, however, a certain dependence between τ and h is required in order to ensure the stability, the scheme will be termed *conditionally stable*.

Assume that τ and h are related according to the inequality

$$\tau \leq Ch^p, \tag{3.26}$$

where the constants C, p are given and are independent of τ and h. Let us note that such relations usually arise when considering the "shortest" perturbation. As a rule, they reflect the dependence between the minimal spatial and time scales of the events to be described by means of the difference scheme.

Of course, larger perturbations (say, of the order of several h's) will then be described more precisely.

Assume we need to increase the accuracy of the solution by formally refining the grid size h. Then we must simultaneously decrease the step size τ so that the above inequality is again satisfied. This means that we can even allow passing to the limit as $\tau \to 0, h \to 0$, provided (3.26) is not violated, that is

$$\frac{\tau}{h^p} = \text{const} \leq C.$$

Under such circumstances the norm of the transition operator T remains usually unchanged. Even if the scheme under consideration is absolutely stable, it is recommended that the limit as $\tau \to 0, h \to 0$ not be taken independently, but in such a way that the norm of the transition operator T stays constant. This ensures both that the process is stable and that the approximation is proper for the typical scales of the events considered.

The definitions of stability discussed above are not the only ones which are being used in the literature. For example, the scheme is stable if

$$\|T\| \leq 1 + O(\tau). \tag{3.27}$$

For small τ, such a definition allows for the exponential growth of round-off errors as time increases (Yanenko [3], Rozhdestvenskii and Yanenko [2]). There are yet other notions of stability (Strang [6], Godunov and Ryabenkii [3], Kreiss [6], Yanenko and Shokin [7], Samarskii [3], and others), which allow us to enlarge the class of difference schemes of interest in applications.

So far, our considerations regarding countable stability have used the assumption that the operator Λ does not depend on time. This is a natural assumption in many problems of mathematical physics. It also permits us to take into consideration a number of additional constructive approaches which are frequently used in numerical mathematics. Indeed, the stability problem is reduced to that of estimating the norm of the transition operator T. As shown in Section 1.1, the square of the norm of T coincides with the spectral radius of the positive self-adjoint operator T^*T, which can then be determined by the Kellogg iterative process, i.e.,

$$\|T\|^2 = \lim_{k \to \infty} \frac{(T^*T\phi^{(k)}, \phi^{(k)})}{(\phi^{(k)}, \phi^{(k)})}.$$

Here the elements $\phi^{(k)}$ are defined by

$$\phi^{(k+1)} = T\phi^{(k)}. \tag{3.28}$$

In this manner, the problem of finding the norm of T is reduced to the sequential procedure defined by the recursive relation (3.28). This is also the route along which most of the constructive computer-oriented work has been done. If T is self-adjoint, then

$$\|T\| = \beta(T).$$

Let us now make some particular comments. In studying the stability of difference schemes one sometimes uses a determination of the spectral radius for spatially periodic problems. For problems with periodic boundary conditions, the estimate of the spectral radius should clearly be made by means of Kellogg's method, the operators T being already constructed, to account for the actual boundary conditions, by Lyusternik's method.

If the operator Λ changes with time, the problem of stability becomes considerably more difficult. This is because the norm of the operator T also changes with time and so does the spectral radius. Therefore T has to be determined at each step in general. The best way to handle this situation is to try for absolutely stable analogs. Such schemes will be considered in Chapter 4.

In conclusion let us note that if the approximation of the evolution equation is investigated in terms of the space $D_h \times D_\tau$, then it is also useful to define stability in these terms. To be specific, let the original evolution problem be approximated by (2.16):

$$\begin{aligned} L^{h\tau}\phi^{h\tau} &= f^{h\tau} \quad \text{in} \quad D_h \times D_\tau, \\ l^{h\tau}\phi^{h\tau} &= g^{h\tau} \quad \text{on} \quad \partial D_h \times D_\tau. \end{aligned} \tag{3.29}$$

The stability criterion may then be taken in the following form:

$$\|\phi^{h\tau}\|_{\Phi_{h\tau}} \leq C_1 \|f^{h\tau}\|_{F_{h\tau}} + C_2 \|g^{h\tau}\|_{G_{h\tau}}, \tag{3.30}$$

where the constants C_1, C_2 are uniformly bounded for $0 \leq t \leq T$ and do not depend on h, τ, f, or g.

Such a quite general approach to stability analysis has been presented by Ryabenkii and Filippov [6].

Assume that the original problem is approximated by the difference equation with the boundary conditions already taken into account. Then a convenient form of the stability criterion is as follows:

$$\|\phi^{h\tau}\|_{\Phi_{h\tau}} \leq C\|f^{h\tau}\|_{F_{h\tau}}, \tag{3.31}$$

where C is bounded on the interval $0 \leq t \leq T$.

1.4 The Convergence Theorem

Filippov has defined stability for an arbitrary difference problem

$$L^{h\tau}\phi^{h\tau} = f^{h\tau}$$

as the uniform boundedness of the operators $(L^{h\tau})^{-1}$ and has proved that approximation and stability imply the convergence of the solution of the difference equation to the solution of the differential equation. For well-posed evolution problems Lax proposed a system of definitions of approximation and stability under which stability appears simultaneously with convergence when approximation exists. This is known as Lax's theorem (Richtmyer and Morton [3].)

The study of the convergence of the difference problem to the original problem, for both stationary and evolution problems of mathematical physics, is guided by a single set of principles. This means that we may follow the fundamental idea of the proof using, as an example, the stationary problem

$$A\phi = f \quad \text{in} \quad D, \tag{4.1}$$
$$a\phi = g \quad \text{on} \quad \partial D,$$

which is approximated by the difference equations

$$A^h\phi^h = f^h \quad \text{in} \quad D_h, \tag{4.2}$$
$$a^h\phi^h = g^h \quad \text{on} \quad \partial D_h.$$

The following *convergence theorem* holds:

Suppose that:

(1) *the difference scheme* (4.2) *approximates the initial problem* (4.1) *to order n on the solution* ϕ;
(2) A^h *and* a^h *are linear operators; and*
(3) *the difference scheme* (4.2) *is stable in the sense of* (3.30), *i.e., there exist positive constants* $\bar{h}$, C_1, *and* C_2 *such that for all* $h < \bar{h}$, $f^h \in F_h$, $g^h \in G_h$ *there exists a unique solution* ϕ^h *of the problem* (4.2) *satisfying the inequality*

$$\|\phi^h\|_{\Phi_h} \leq C_1\|f^h\|_{F_h} + C_2\|g^h\|_{G_h}. \tag{4.3}$$

Then the solution ϕ^h of the difference problem converges to the solution ϕ of the initial problem, i.e.,

$$\lim_{h \to 0} \|(\phi)_h - \phi^h\|_{\Phi_h} = 0,$$

and the following estimate of the rate of convergence is valid:

$$\|(\phi)_h - \phi^h\|_{\Phi_h} \leq (C_1 M_1 + C_2 M_2)h^n, \tag{4.4}$$

where M_1 and M_2 are the constants in (2.3).

PROOF. Let $\bar{h}$ be the smallest of the h appearing in the definitions of approximation and stability. Then stability implies that for arbitrary right-hand sides f^h and g^h, and for $h < \bar{h}$, there exists a unique solution ϕ^h, i.e., that for $h < \bar{h}$ we are entitled to consider the difference $(\phi)_h - \phi^h$. Since A^h is linear we may write

$$A^h[(\phi_h) - \phi^h] = A^h(\phi)_h - A^h \phi^h = A^h(\phi)_h - f^h.$$

Similarly,

$$a^h[(\phi)_h - \phi^h] = a^h(\phi)_h - g^h.$$

Since $h < \bar{h}$, stability and approximation imply that from

$$\begin{aligned} A^h[(\phi)_h - \phi^h] &= A^h(\phi)_h - f^h, \\ a^h[(\phi)_h - \phi^h] &= a^h(\phi)_h - g^h, \end{aligned} \tag{4.5}$$

we obtain

$$\begin{aligned} \|(\phi)_h - \phi^h\|_{\Phi_h} &\leq C_1 \|A^h(\phi)_h - f^h\|_{F_h} + C_2 \|a^h(\phi)_h - g^h\|_{G_h} \\ &\leq C_1 M_1 h^{n_1} + C_2 M_2 h^{n_2} \leq (C_1 M_1 + C_2 M_2)h^n. \end{aligned}$$

Without loss of generality, we may suppose, in deriving the last inequality, that $h < 1$. This completes the proof. $\qquad \square$

We note that the proof uses the linearity property only for the operators A^h and a^h. A generalization of the definition of stability to the nonlinear case and a statement of the corresponding theorem can be found in Godunov and Ryabenkii [3].

For the evolution problem we consider

$$\begin{aligned} \delta f^{h\tau} &= L^{h\tau}[(\phi)_{h\tau} - \phi^{h\tau}] \equiv L^{h\tau}(\phi)_{h\tau} - f^{h\tau}, \\ \delta g^{h\tau} &= l^{h\tau}[(\phi)_h - \phi^{h\tau}] \equiv l^{h\tau}(\phi)_{h\tau} - g^{h\tau}. \end{aligned} \tag{4.6}$$

From (4.6) and the stability condition (3.30) we have

$$\|(\phi)_{h\tau} - \phi^{h\tau}\|_{\Phi_h} \leq C_1 \|\delta f^{h\tau}\|_{F_{h\tau}} + C_2 \|\delta g^{h\tau}\|_{G_{h\tau}},$$

or, taking (2.17) into account

$$\|(\phi)_{h\tau} - \phi^{h\tau}\|_{\Phi_{h\tau}} \leq K_1 h^n + K_2 \tau^p, \tag{4.7}$$

where

$$K_1 = C_1 M_1 + C_2 M_2, \qquad K_2 = C_1 N_1 + C_2 N_2.$$

The estimate (4.7) proves the convergence of the difference solution to the exact solution, and gives a clear picture of the convergences with respect to the space mesh h and the time step τ.

The assumptions of the theorem include the rather strong requirement that C_1 and C_2 are independent of h and τ. Of particular inconvenience is the condition that C_1 and C_2 are independent of h, since in certain cases C_1 and C_2 may well go to infinity as $h \to 0$. Let

$$C_1^h = C_1/h^m, \qquad C_2^h = C_2/h^m,$$

where $m \geq 0$. The convergence of the approximate to the exact solution will then be estimated as follows:

$$\| \varphi^{h\tau} - (\varphi)_{h\tau} \|_{\Phi_{h\tau} = \Phi_{h\varepsilon}} \leq M h^{k-m} + N \tau^p h^{-m}.$$

If $k > m$ and $\tau^p h^{-m} \to 0$ as $\tau \to 0$, $h \to 0$, then convergence follows. Of course the convergence theorem can be formulated even when C_1 and C_2 depend on both h and τ (Strang [6, 7]).

CHAPTER 2

Methods of Constructing Difference Schemes for Differential Equations

The development of difference approximation schemes for various problems of mathematical physics has been approached by a number of well-known techniques. The most complete picture has been obtained for equations with sufficiently smooth coefficients and solutions, in which case it is possible to obtain high-accuracy approximation schemes. These particular schemes have been the subject of an ever-increasing interest, because the speed with which new complicated problems are emerging in science and technology has had a definite bearing on the evolutionary pace of computational means. In many problems it seems therefore reasonable to seek the approximate solutions (with a given accuracy) not at the expense of a formal increase in the dimensionality of the subspaces involved (for instance, by decreasing the mesh size), but rather by means of constructing more accurate approximations of the original problem using the *a priori* information about the smoothness of the solution (see also Chapter 6). This point of view has turned out to be quite fruitful in many cases and has led to satisfactory and quite universal methods based on the Ritz and Galërkin variational methods and the least-squares method. It is to be noted, however, that the class of problems which possess smooth solutions is somewhat small, and therefore, our main effort must be directed toward approximation methods suitable for problems with discontinuous coefficients. These problems come up, for instance, when studying diffusion, heat conduction, and hydrodynamics.

Therefore, we will sacrifice the opportunity of describing a number of original and fairly general results concerning difference approximations with high accuracy, and will rather pursue the idea of building a general framework for constructing the difference analogs of the equations which do not possess high smoothness properties. Naturally, all the approximations to be discussed later are automatically applicable to the problems with smooth solutions and parameters.

In order to become more familiar with the ways the scientific ideas in the area of difference approximation scheme evolve, we start with a detailed exposition of boundary problems for ordinary differential equations. After that, we will turn to more or less general approaches for solving two-dimensional and multi-dimensional problems of mathematical physics. We hope that the references in Chapter 10 to the original sources will help the reader to get a deeper and broader understanding of the theory and the algorithms.

2.1 Variational Methods in Mathematical Physics

In this section we shall consider variational methods for the approximate solution of the equations of mathematical physics. (A number of the results we shall quote are taken from the well-known monographs of Mikhlin [1] and Smirnov, vol. IV [5].) First, however, we shall cite as examples some simple problems, and note that their formulation is close to that of several problems of variational computation. This allows us later to study the essence of variational problems more closely.

2.1.1 Some Problems of Variational Calculation

Let us look at the very simple functional

$$J(u) = \int_{x_0}^{x_1} \pi(x, u, u') \, dx, \tag{1.1}$$

where $\pi = \pi(x, y, z)$ is a given function, continuous in x, y, and z, together with its derivatives, to and including the second order, in some region of three-dimensional Euclidean space.

We assume that $u(x)$ is continuous, has a continuous derivative $u'(x)$ in (x_0, x_1), and takes on the values

$$u(x_0) = u_0, \qquad u(x_1) = u_1. \tag{1.2}$$

on the ends of the interval $[x_0, x_1]$. We define an ε-neighborhood of $u = u(x)$ as the family of functions $u_1(x)$ satisfying the inequality

$$|u_1(x) - u(x)| \le \varepsilon \tag{1.3}$$

in the entire interval $[x_0, x_1]$. Then we formulate the following variational calculus problem: among the functions in an ε-neighborhood, having a continuous derivative and satisfying (1.2), find that one which yields an extremum of the functional $J(u)$. This is the problem with fixed endpoints of the curves $u = u(x)$.

We shall find the conditions that must be satisfied by $u(x)$, in order that it shall represent an extremal value in an ε-neighborhood. Consider a function $\eta(x)$ satisfying the condition

$$\eta(x_0) = \eta(x_1) = 0. \tag{1.4}$$

We construct a new function $u_\alpha(x) = u(x) + \alpha\eta(x)$, where α is a small parameter (hence we may suppose that $u_\alpha(x)$ also belongs to our ε-neighborhood). Substituting this function into the functional J, we find

$$J(u_\alpha) = \int_{x_0}^{x_1} \pi(x, u(x) + \alpha\eta(x), u'(x) + \alpha\eta'(x))\, dx,$$

and we shall look on $J(u_\alpha)$ as a function of the parameter α: $J(u_\alpha) = \Phi(\alpha)$. The first derivative of the function $\Phi(\alpha)$ at the point $\alpha = 0$ will be called the first variation of the functional J, and will be denoted by δJ:

$$\delta J(u) = \left.\frac{d\Phi}{d\alpha}\right|_{\alpha=0}.$$

The second variation $\delta^2 J$ of the functional J is defined as the second derivative of $\Phi(\alpha)$ at the point $\alpha = 0$:

$$\delta^2 J(u) = \left.\frac{d^2\Phi}{d\alpha^2}\right|_{\alpha=0}.$$

Taking account of the form of J, we find for δJ and $\delta^2 J$:

$$\delta J = \int_{x_0}^{x_1} (\pi_u \cdot \eta + \pi_{u'} \cdot \eta')\, dx, \tag{1.5}$$

$$\delta^2 J = \int_{x_0}^{x_1} (\pi_{u'u'}\eta'^2 + 2\pi_{uu'}\eta \cdot \eta' + \pi_{uu}\eta^2)\, dx \tag{1.6}$$

(Here we use the notation

$$\pi_u \equiv \frac{\partial \pi}{\partial u}, \qquad \pi_{uv} \equiv \frac{\partial^2 \pi}{\partial u\, \partial v}, \qquad u' = \frac{\partial u}{\partial x}\Big).$$

The necessary condition for $\Phi(\alpha)$ to have an extremal value at $\alpha = 0$ is $\Phi'(0) = 0$, i.e.,

$$\delta J(u) = \int_{x_0}^{x_1} (\eta \cdot \pi_u + \eta' \cdot \pi_{u'})\, dx = 0.$$

We integrate by parts, taking account of (1.4) and obtain

$$\delta J(u) = \int_{x_0}^{x_1} dx\eta(x)\left(\pi_u - \frac{d}{dx}\pi_{u'}\right). \tag{1.7}$$

Since $\eta(x)$ is arbitrary, we conclude that a curve $u(x)$ satisfying (1.2) and representing an extremum of the functional (1.1) must satisfy the differential equation

$$\pi_u - \frac{d}{dx}\pi_{u'} = 0, \tag{1.8}$$

which is customarily known as the Euler equation.

Note that if $u(x)$ yields a minimum (maximum) of J, we have $\Phi''(0) = \delta^2 J \geq 0 \, (\delta^2 J \leq 0)$.

As an illustration, we consider an example in which we set $u_0 = u_1 = 0$ in (1.2) and give $\pi(x, u, u')$ the form

$$\pi = \left(\frac{du}{dx}\right)^2 + ku^2 - 2fu, \tag{1.9}$$

where k and f are sufficiently smooth functions and $k > 0$. Then the Euler equation for our variational problem is

$$-\frac{d^2u}{dx^2} + ku = f(x), \tag{1.10}$$

(strictly speaking, in setting up the necessary condition for an extremum in the case (1.9) we should require that $u(x)$ have continuous first- and second-order derivatives).

Therefore, if a function in the domain of definition of the functional

$$J(u) = \int_{x_0}^{x_1} dx \left(\left(\frac{du}{dx}\right)^2 + ku^2 - 2fu\right), \tag{1.11}$$

and satisfying the conditions $u(x_0) = u(x_1) = 0$ represents an extremum of (1.11), it satisfies (1.10), i.e., it is the solution of a first boundary problem of the form

$$-\frac{d^2u}{dx^2} + ku = f(x), \tag{1.12}$$

$$u(0) = u(1) = 0. \tag{1.13}$$

The converse is also true, as we shall show later for functions of more general type: if $u(x)$ is a solution of (1.12) and (1.13) it represents an extremum of the functional (1.11) on the corresponding domain of definition.

Let us now consider another variational problem for the functional (1.1), as follows: among all curves $u = u(x)$ with end points lying on given verticals $x = x_0$, $x = x_1$, find that one which yields an extremum of (1.1). (This is the free end point problem.) Note that no other condition is imposed on the endpoints of the curve; they have merely to lie on the given verticals, parallel to the u-axis. Nevertheless, it turns out that if $u(x)$ corresponds to an extremum of $J(u)$ it must satisfy certain conditions in the limit for $x = x_0$ and $x = x_1$; these conditions are derived immediately from the conditions for an extremum of (1.1).

To prove this statement, let some curve $u(x)$ correspond to an extremum of $J(u)$ in comparison with all neighboring curves $u_\alpha(x) = u(x) + \alpha\eta(x)$ with free endpoints (i.e., as compared to the problem with fixed endpoints, $\eta(x)$ need not vanish at the points x_0 and x_1). The necessary conditions for an extremum lead, as before, to the relation

$$\delta J = \int_{x_0}^{x_1} (\pi_u \eta + \pi_{u'} \eta') \, dx = 0. \tag{1.14}$$

Integrating by parts, we find that

$$\int_{x_0}^{x_1} \eta(x) \left(\pi_u - \frac{d}{dx} \pi_{u'} \right) dx + \pi_{u'} \eta|_{x=x_1} - \pi_{u'} \eta|_{x=x_0} = 0. \tag{1.15}$$

Again, because $\eta(x)$ is arbitrary, we obtain the Euler equation

$$\pi_u - \frac{d}{dx} \pi_{u'} = 0, \tag{1.16}$$

and also the limiting conditions

$$\pi_{u'}|_{x=x_1} = 0, \qquad \pi_{u'}|_{x=x_0} = 0. \tag{1.17}$$

Conditions of the type of (1.17), necessary for an extremum, are often called natural boundary conditions (we shall study them in more detail when we take the Ritz method).

As another illustration, let $\pi(x, u, u')$ have the form (1.9) in the free endpoint problem. Then the equations (1.16), (1.17) take the form

$$-\frac{d^2 u}{dx^2} + ku = f(x), \tag{1.18}$$

$$\frac{du}{dx}(x_1) = \frac{du}{dx}(x_0) = 0, \tag{1.19}$$

Thus, the function $u(x)$ corresponding to an extremum of our functional is a solution of a second boundary problem of the form (1.18), (1.19). Hence we conclude that the boundary conditions (1.19) are natural, in contrast to the conditions (1.13). The converse is also true: if $u(x)$ is a solution of the problem (1.18), (1.19) it corresponds to an extremal value of a functional (1.11) in a free endpoint problem.

Up to now we have examined some very simple cases—one function u of one independent variable x. We may treat more general cases in like fashion: for instance, suppose we are given a functional

$$J = \iint_D \pi(x, y, u, u_x, u_y) \, dx \, dy,$$

where the function π and the boundary of the convex bounded region D exhibit the necessary smoothness. We state the problem as follows: to find a function $u(x, y)$, continuous together with its partial derivatives, to and including the second order, taking on given values on the boundary of D, and yielding an extremum of the functional J. Then, by analogy with the foregoing, we obtain Euler's equation in the following form:

$$\pi_u - \frac{\partial}{\partial x} \pi_{u_x} - \frac{\partial}{\partial y} \pi_{u_y} = 0.$$

The extension to the case of n variables is obvious.

Thus, we have arrived at the possibility of interpreting one and the same problem of mathematical physics, as either a problem in differential equations (Euler's equations), or as a problem in the variational calculation of the function yielding an extremum of some functional. In the latter case, the functions so computed will—if sufficiently smooth—be solutions of the corresponding Euler equations. Just as in the examples we have considered above, the problems studied in terms of Euler's equations (i.e., problems of the type of (1.12), (1.13) or (1.18), (1.19)) may be written in the form

$$Lu = f, \qquad u \in \Phi(L), \tag{1.20}$$

where $\Phi(L)$ is the domain of definition of the operator L. We have already noticed the equivalence of the problem (1.20) to the corresponding variational problem

$$J(u) = \min_{v \in \Phi(L)} J(v), \tag{1.21}$$

where

$$J(v) = (Lu, u) - 2(f, u) = \int_{x_0}^{x_1} \left(-\frac{d^2 u}{dx^2} + ku - 2f \right) u \, dx$$

$$= \int_{x_0}^{x_1} \left(\left(\frac{du}{dx} \right)^2 + ku^2 - 2fu \right) dx.$$

Let us now prove that the problems (1.20) and (1.21) are equivalent, using the abstract form of the operator L.

We consider the very general problem

$$Lu = f. \tag{1.22}$$

where L is a linear positive self-adjoint operator with its domain of definition $\Phi(L)$ everywhere dense in the Hilbert space H. Let the scalar product in H be denoted by (,) and let the range of values of L lie in H, let u lie in $\Phi(L)$, and let f be an element of H. Then the following statement holds true (cf. Mikhlin [1]): if the solution of (1.16) exists, it yields a minimum of the functional

$$J(u) = (Lu, u) - 2(u, f).$$

The proof is as follows. Let u_0 be a solution of (1.22), i.e.,

$$Lu_0 = f.$$

Let η be an arbitrary nonzero element of $\Phi(L)$ and let α be an arbitrary real number. We define an element v_α as

$$v_\alpha = u_0 + \alpha\eta.$$

Then

$$J(v_\alpha) = (L(u_0 + \alpha\eta), u_0 + \alpha\eta) - 2(u_0 + \alpha\eta, f).$$

Since L is self-adjoint

$$J(v_\alpha) = J(u_0) + 2\alpha(Lu_0 - f, \eta) + \alpha^2(L\eta, \eta).$$

We conclude that

$$J(v_\alpha) = J(u_0) + \alpha^2(L\eta, \eta).$$

Because L is positive, this condition implies that

$$J(v_\alpha) > J(u_0) \tag{1.23}$$

for all $\alpha \neq 0$. This means that the minimum of the functional $J(v)$ is attained on the solution $v_\alpha = u_0$.

The converse is also true. Namely, an element u_0 of the Hilbert space H which yields a minimum of the functional J and belongs to $\Phi(L)$ is a solution of the operator equation $Lu = f$.

Indeed, let $u_0 \in \Phi(L)$ be an element on which the minimum of the functional $J(u)$ is attained, and let η be an arbitrary element of $\Phi(L)$. It is well known that for all $u, v \in \Phi(L)$ the element $w = \alpha u + \beta v$ (α and β being constants) also belongs to $\Phi(L)$. Therefore, $v = u_0 + \alpha\eta \in \Phi(L)$. Since J attains its minimum at u_0, we have

$$J(u_0 + \alpha\eta) \geq J(u_0). \tag{1.24}$$

We assume α is real. If we combine (1.24) with the assumption that L is symmetric we reach the inequality

$$2\alpha(Lu_0 - f, \eta) + \alpha^2(L\eta, \eta) \geq 0.$$

Since α is arbitrary, this is possible only if

$$(Lu_0 - f, \eta) = 0. \tag{1.25}$$

Thus, the element $Lu_0 - f$ is orthogonal to all the elements of $\Phi(L)$; therefore,

$$Lu_0 - f = 0.$$

We shall introduce a number of auxiliary considerations with respect to the variational formulation of the problems of mathematical physics, and will describe some of the basic methods for solving the equations. We shall frequently illustrate these problems by the example of the elliptic differential equation

$$Lu \equiv -\sum_{i,j=1}^{2} \frac{\partial}{\partial x_i} A_{ij}(x) \frac{\partial u}{\partial x_j} + \sum_{i=1}^{2} B_i(x) \frac{\partial u}{\partial x_i} + q(x)u = f(x), \quad (1.26)$$

$$x = (x_1, x_2) \in D,$$

defined in a bounded region D with a boundary condition of the form

$$u = 0, \qquad x \in \partial D \tag{1.27}$$

(the first boundary problem), or

$$\frac{\partial u}{\partial N} \equiv \sum_{j,k=1}^{2} A_{jk}(x) \frac{\partial u}{\partial x_k} \cos(\nu, x_j) = 0, \tag{1.28}$$

where ν is the outward normal to ∂D (the second boundary problem).

Let the operator

$$L_0 = -\sum_{i,j=1}^{2} \frac{\partial}{\partial x_i} A_{ij}(x) \frac{\partial}{\partial x_j} \tag{1.29}$$

satisfy the following conditions:

(i)   its values lie in the Hilbert space $F = L_2(D)$;
(ii) its domain of definition $\Phi(L_0)$ consists of functions $u \in L_2(D)$ that satisfy either (1.27) or (1.28), with $L_0 u \in L_2(D)$;
(iii) it is self-adjoint in the sense of Lagrange and nonsingular—i.e., for all nonzero vectors $\xi = (\xi_1, \xi_2)$ the inequality

$$\inf_{x \in D} \sum_{i,j=1}^{2} A_{ij}(x)\xi_i\xi_j \geq \mu_0 \sum_{i=1}^{2} \xi_i^2 \tag{1.30}$$

holds true for some positive constant μ_0.

We shall suppose further (in the absence of special comment) that $q(x)$ is bounded and positive in D, and also that the solutions of (1.26), (1.27) or (1.26), (1.28) exist. In later discussion the reader may have questions as to the smoothness of the data, as to whether the solution is classical, generalized, etc.

First, we shall suppose that the solutions satisfy (1.26) almost everywhere and belong to the Sobolev space W_2^1, which consists of functions in $L_2(D)$ that have square-summable generalized first-order derivatives. The norm in W_2^1 is defined by the relation (cf. Section 1.1)

$$\|u\|_{W_2^1} = \left\{ \int_D u^2 \, dD + \int_D \left[\left(\frac{\partial u}{\partial x_1}\right)^2 + \left(\frac{\partial u}{\partial x_2}\right)^2 \right] dD \right\}^{1/2} < \infty.$$

We shall suppose further that the solutions of the first boundary problem (1.26), (1.27) belong to $\mathring{W}_2^1$, the subspace of W_2^1 consisting of those functions that vanish on the boundary of D.

We shall suppose that the initial data—for instance the smoothness of the coefficients and of the boundary of the region—are determined by the fact that the solutions belong to the indicated spaces.

In the second place, whenever we are considering specific problems, we shall assume the fulfillment of all necessary higher-order smoothness conditions on the solutions, the coefficients, and the right-hand sides of the equations.

These conventions allow us to focus our attention on the basic aim of our text—the principles for the construction of net analogs of partial differential equations.

2.1.2 The Ritz Method

The Ritz method is a well-known device for the solution of the equations of mathematical physics. We discuss it in its application to the operator equation

$$Lu = f \tag{1.31}$$

in the Hilbert space F, with scalar product (u, v), under the conditions that $f \in L$, that L has the domain of definition $\Phi(L)$ dense in F, and that L is self-adjoint and positive definite. The results obtained in Section 2.1.1. imply that the problem of finding a solution to (1.31) is equivalent to the problem of finding an element $u \in \Phi(L)$ for which the functional

$$J(u) = (Lu, u) - 2(f, u) \tag{1.32}$$

attains a minimum. We must note, however, that the assertion of equivalence says nothing about the existence of an element $u \in \Phi(L)$ that is a solution of (1.31) or yields a minimum of $J(u)$. Let us, therefore, make some changes in the formulation of the variational problem, so that the existence of a solution will be guaranteed.

We introduce a new scalar product in $\Phi(L)$, defining it by the relation

$$(\phi, \psi)_L = (L\phi, \psi), \qquad \phi, \psi \in \Phi(L) \tag{1.33}$$

and the corresponding norm

$$\|\phi\|_L = (\phi, \phi)_L^{1/2}.$$

If we complete $\Phi(L)$ in this norm, we obtain a complete Hilbert space F_L, called the energy space generated by the operator L. Every function in F_L belongs to F, but as a result of its completion F_L may contain elements not in $\Phi(L)$. (Therefore, the representation of the scalar product $(\phi, \psi)_L$ in the form (1.33) is not possible for arbitrary elements $\phi, \psi \in F_L$.)

Since we assumed in (1.32) that $u \in \Phi(L)$, we use (1.33) to represent $J(u)$ in the form

$$J(u) = (u, u)_L - 2(f, u). \tag{1.34}$$

In this form, we may consider $J(u)$ not only on the domain of definition of the operator L but also on the entire energy space F_L. We therefore extend the functional (1.34) to the whole space F_L, conserving our earlier notation, and seek its minimum there. It is easily seen that in this formulation the variational problem always has a unique solution. For, since L is positive definite by hypothesis, we have

$$(Lu, u) = (u, u)_L \geq \gamma^2 \|u\|^2 \qquad \forall u \in \Phi(L); \gamma = \text{const} > 0,$$

and because we completed $\Phi(L)$ to obtain F_L, the definiteness property $(u, u)_L \geq \gamma^2 \|u\|^2$ continues to hold for an arbitrary element $u \in F_L$. Now if we consider the functional (u, f) we observe that it is bounded in F_L:

$$|(u, f)| \leq \|u\| \cdot \|f\| \leq \frac{1}{\gamma} \|u\|_L \cdot \|f\| = C\|u\|_L.$$

Therefore, by Riesz's theorem there exists an element $u_0 \in F_L$ such that for all elements $u \in F_L$

$$(u, f) = (u, u_0)_L.$$

But then $J(u)$ may be written in the form

$$J(u) = (u, u)_L - 2(f, u)$$
$$= (u, u)_L - 2(u, u_0)_L \equiv \|u - u_0\|_L^2 - \|u_0\|_L^2 \qquad \forall u \in F_L. \qquad (1.35)$$

It follows from the last of these expressions that $J(u)$ attains its minimum value for $u = u_0$. We have already noted that u_0 is unique and belongs to F_L. It may, of course, turn out that $u_0 \in \Phi(L)$. Then it will be also a classical solution of the given problem, i.e., it will satisfy (1.31). If it belongs to F_L we call it a generalized solution of the equation $Lu = f$.

So, we have reduced the initial problem to one of minimizing the functional (1.34) over the energy space F_L. Now we shall consider the Ritz method for finding an approximate solution to this variational problem.

We introduce a sequence of finite-dimensional subspaces $F_h \subseteq F_L$, defined by an infinite sequence of parameters $h_1, h_2, \ldots, h_k, \ldots$ with $h_k \to 0$ as $k \to \infty$. We shall say that the sequence $\{F_h\}$ is complete in F_L if for all $u \in F_L$ and $\varepsilon > 0$ there exists an $\hat{h} = \hat{h}(u, \varepsilon) > 0$ such that

$$\inf_{w \in F_h} \|u - w\|_L < \varepsilon \qquad (1.36)$$

for all $h < \hat{h}$. In other words, the completeness of a sequence of subspaces $\{F_h\}$ means that every element of F_L can be arbitrarily closely approximated by elements of a space F_h.

Then the Ritz method may be formulated as follows: find an element $u^h \in F_h$ that minimizes $J(u)$ in F_h.

Then we may assert the following: under the above assumptions the sequence $\{u^h\}$ of Ritz approximations converges in F_L to the solution (generalized) u_0 of the problem. For, since each u^h corresponds on F_h to the minimum of $J(u)$, the relation (1.35) implies that for arbitrary $w \in F_h$

$$\|u_0 - u_h\|_L^2 = J(u_h) - J(u_0) \le J(w) - J(u_0) = \|u_0 - w\|_L^2.$$

Since $w \in F_h$ is arbitrary, we find by taking account of (1.36) that

$$\|u_0 - u^h\|_L \le \inf_{w \in F_h} \|u_0 - w\|_L \underset{h \to 0}{\to} 0 . \tag{1.37}$$

When the basis of F_h is known and consists of the functions $\{\phi_i^h\}_{i=1}^{N_h}$, the problem of finding $u^h \in F_h$ is equivalent to finding the coefficients $\{\alpha_i\}_{i=1}^{N_h}$ of the resolution

$$u_h = \sum_{i=1}^{N_h} \alpha_i \phi_i^h \tag{1.38}$$

from the condition that the functional J is minimized. As usual, we substitute (1.38) into J and set the derivatives $\partial J(u^h)/\partial \alpha_i$ $(i = 1, \dots, N_h)$, equal to zero. We arrive at a system of linear algebraic equations

$$A\alpha = g, \tag{1.39}$$

where α and g are N_h-dimensional vectors;

$$g_i = (f, \phi_i^h), \tag{1.40}$$

and $A = (a_{ij})$ is the Gramm matrix of the system of vectors $\{\phi_i\}$ with the scalar product belonging to F_L, i.e.,

$$a_{ek} = (\phi_e^h, \phi_k^h)_L = \int_D \left\{ \sum_{i,j=1}^{2} A_{ij}(x) \frac{\partial \phi_l^h}{\partial x_i} \frac{\partial \phi_k^h}{\partial x_j} + q(x)\phi_l^h \phi_k^h \right\} dD,$$

$$1 \le l; k \le N_h. \tag{1.41}$$

If, further, the basis functions ϕ_i^h belong to $\Phi(L)$, the a_{ij} may be represented also in the form

$$a_{ij} = (L\phi_i, \phi_j).$$

Since

$$a_{ij} = (\phi_i^h, \phi_j^h)_L = (\phi_j^h, \phi_i^h)_L = a_{ji}$$

the matrix A is symmetric, and because of the inequality for $\xi \neq 0$

$$(A\xi, \xi)_2 \equiv \sum_{i,j=1}^{N_h} a_{ij}\xi_i\xi_j = \left(\sum_{i=1}^{N_h} \xi_i \phi_i^h, \sum_{j=1}^{N_h} \xi_j \phi_j^h \right)_L \ge \gamma^2 \left\| \sum_{i=1}^{N_h} \xi_i \phi_i^h \right\|^2 > 0,$$

$$\tag{1.42}$$

it is positive definite.

Let us look at one of the most important practical problems in the application of the Ritz method—the separation of natural and principal boundary conditions.

In determining $\Phi(L)$, the domain of definition of the operator L, we often impose boundary conditions on $u \in \Phi(L)$. When we construct the energy space, we complete $\Phi(L)$ in the metric $\|\cdot\|_L$; as a result, F_L may turn out to contain elements that fail to satisfy some of the boundary conditions imposed on functions belonging to $\Phi(L)$. Those boundary conditions satisfied in $\Phi(L)$, and not necessarily satisfied in the energy space F_L, are called natural conditions; those necessarily satisfied in both spaces are called principal conditions.

The practical importance of the ability to distinguish these conditions lies in the fact that the basis functions $\{\phi_i^h\}$ in the Ritz method may be chosen from the energy space and not from $\Phi(L)$ alone; therefore, they need not satisfy the natural boundary conditions. This circumstance significantly facilitates the choice of the ϕ_i^h for the solution of many practically important problems, especially when D is multi-dimensional and has a complicated boundary. (Of course, we still have the problem of constructing the ϕ_i^h to satisfy the principal boundary conditions.)

We shall indicate an approach that allows us to decide, in every concrete problem, whether a boundary condition is natural or not. Suppose we wish to minimize a functional $J(u)$, and that there exists a function u_0 that minimizes $J(u)$ over a class of functions that generally fail to satisfy a given condition. Using a variational computation we find the conditions that u_0 must satisfy. If the given boundary condition belongs to this collection, it is natural. Thus, in Section 2.1.1 we displayed a process such that when we considered the problem (1.18), (1.19), the von Neumann conditions $du/dx(x_0)$ $= du/dx(x_1) = 0$ were natural. Therefore, in constructing a Ritz approximation to the solution of the free endpoint problem we may choose basis functions that fail to satisfy these conditions. At the same time, if we are solving (1.12), (1.13) the ϕ_i^h must satisfy (1.13)—i.e., the boundary conditions are principal.

Finally, we take note of a simple method that allows us to distinguish natural from principal boundary conditions, and is applicable to many boundary problems. Suppose that in (1.31) L is a differential operator of order $2m$, positive over $\Phi(L)$, and satisfies various homogeneous boundary conditions of the form $N_k u = 0$. Such a boundary condition will be natural, if $N_k u$ contains derivatives of u of order m and higher; it will be principal if $N_k u$ contains no derivative of u of order higher than $m - 1$.

Let us now consider the Ritz method for the solution of (1.26) with the conditions (1.27) or (1.28), under the supplementary assumption that the B_i ($i = 1, 2$) vanish identically in D. Then it is not hard to show that for the operator L in each of these problems, and for all ϕ and ψ in $\Phi(L)$, the following relation holds

$$(L\phi, \psi) \equiv \int_D \psi L\phi \, dD = \int_D \left\{ \sum_{i,j=1}^2 A_{ij}(x) \frac{\partial \phi}{\partial x_i} \frac{\partial \psi}{\partial x_j} + q(x)\phi\psi \right\} dD.$$

Now we set up a correspondence between the initial problems and the corresponding problems of finding an element of F_L on which the functional

$$J(u) = \int_D \left\{ \sum_{i,j=1}^{2} A_{ij}(x) \frac{\partial u}{\partial x_i} \frac{\partial u}{\partial x_j} + qu^2 - 2uf \right\} dD$$

attains its minimum. As we see, in both the problems (1.26), (1.27) and (1.26), (1.28), the minimand functional has the same form. However, the energy spaces differ. In dealing with the first boundary problem the scalar product and the norm in F_L are:

$$(\phi, \psi)_L = \int_D \left\{ \sum_{i,j=1}^{2} A_{ij}(x) \frac{\partial \phi}{\partial x_i} \frac{\partial \psi}{\partial x_j} + q(x)\phi\psi \right\} dD,$$

$$\|u\|_L = (\phi, \phi)_L^{1/2} < \infty,$$

but the elements of F_L satisfy a homogeneous Dirichlet boundary condition, which is therefore principal. Thus, taking account of (1.30) we see that F_L is, in this case, a subspace of $\mathring{W}_2^1$.

When we consider the second boundary problem, $(\phi, \psi)_L$ has the same form, but the von Neumann boundary condition turns out to be natural. Therefore, the energy space is a subspace of W_2^1, and if we apply the Ritz method the basis functions may be chosen in W_2^1 and need not satisfy the condition (1.28). This latter circumstance leads us to ask whether one cannot reduce a problem with principal boundary conditions to an approximation by a variational problem with natural conditions. It turns out that in many cases this can be done by the penalty method (cf. Lions [1, 2]; Babuška [5]). For instance, in the Dirichlet problem now under consideration we pose the problem of minimizing the following functional:

$$J_\varepsilon(u_\varepsilon) = \int_D \left\{ \sum_{i,j=1}^{2} A_{ij}(x) \frac{\partial u_\varepsilon}{\partial x_i} \frac{\partial u_\varepsilon}{\partial x_j} + q(x)u_\varepsilon^2 - 2fu_\varepsilon \right\} dD + \frac{1}{\varepsilon} \int_{\partial D} u_\varepsilon^2 \, ds,$$

where ε is a sufficiently small parameter. In the particular case (1.26), (1.27), that is, the Dirichlet problem for the Poisson equation

$$-\Delta u = f, \qquad u|_{\partial D} = 0,$$

we may set $\varepsilon = h^\sigma$, where $\sigma > 0$ and h is a small constant (cf. Babuška [5]). Then for $f \in W_2^k, k > 1 + \sigma/2$, we have the estimate

$$\|u - u_\varepsilon\|_{W_2^1} \leq c \cdot h^\mu \|f\|_{W_2^k},$$

where

$$\mu = \min\left(\frac{\sigma}{2}, k - 1 - \frac{\sigma}{2}\right).$$

Hence for a given k we may choose σ to maximize the order μ, which determines the rate of convergence.

2.1.3 The Galërkin Method

The main deficiency of the Ritz method is the fact that it is applicable only for equations with self-adjoint and positive definite operators. Another variational method, the so-called Galërkin (or Bubnov–Galërkin) method is free from this constraint. We will describe this method with an example of the equation

$$Lu = f \tag{1.43}$$

in Hilbert space, where we suppose $f \in F$ and $\Phi(L)$ is dense in F.

We shall suppose that $L = L_0 + K$, where L_0 is positive definite and symmetric. Suppose that $\Phi(L_0) \subseteq \Phi(K)$ and that L_0^{-1} is completely continuous in F. We introduce the energy space F_{L_0}, corresponding to the operator L_0, with scalar product $(u, v)_{L_0}$ and norm $\|u\|_{L_0} = (u, u)_{L_0}^{1/2}$. If we take the scalar product, in F, of (1.43) by an arbitrary function $v \in \Phi(L_0)$, we arrive at an identity that must be satisfied by the solution of (1.43):

$$(L_0 u, v) + (Ku, v) = (f, v). \tag{1.44}$$

Since $(L_0 u, v) = (u, v)_{L_0}$, (1.44) becomes

$$(u, v)_{L_0} + (Ku, v) = (f, v), \tag{1.45}$$

which permits a generalized statement of the problem for the equation (1.43). A generalized solution of (1.43) is a function $u_0 \in F_{L_0}$, satisfying (1.45) for all $v \in F_{L_0}$. Suppose such a generalized solution u_0 exists. If it turns out that $u_0 \in \Phi(L_0)$, the relation $(u, v)_{L_0} = (L_0 u, v)$ implies that

$$(L_0 u_0 + Ku_0 - f, v) = 0.$$

But since F_{L_0} is dense in F we conclude that u_0 satisfies the original equation (1.43).

As in the preceding subsection, we introduce a sequence of finite-dimensional subspaces $F_h \subset F_{L_0}$ ($h = h_1, h_2, \ldots$) with the bases $\{\phi_i^h\}_{i=1}^{N_h}$. Then the Galërkin approximation is to be found in the form

$$u^h = \sum_{i=1}^{N_h} \alpha_i \phi_i^h. \tag{1.46}$$

The coefficients α_i are to be chosen so that u^h satisfies (1.45) for all $v \in F_h$. Since $v \in F_h$ can be written in the form $\sum_{i=1}^{N_h} b_i \phi_i^h$ for some set of coefficients $\{b_i\}$, u^h is to be determined by a system of equations of the form

$$(u^h, \phi_s^h)_{L_0} + (Ku^h, \phi_s^h) = (f, \phi_s^h), \qquad s = 1, \ldots, N_h, \tag{1.47}$$

which may also be written as

$$A\alpha = g, \tag{1.48}$$

where

$$a_{ij} = (\phi_i^h, \phi_j^h)_{L_0} + (K\phi_j^h, \phi_i^h), \qquad g_i = (f, \phi_i^h), \qquad i, j = 1, \ldots, N_h.$$

After calculating the coefficients $\{\alpha_i\}$ we construct the solution via (1.46). As to the convergence of u^h, we may say the following: if (1.43) has no more than one generalized solution, if the sequence F_h is complete in F_{L_0} (per the definition in Section 2.1.2), and the operator $L_0^{-1}K$ is completely continuous in F_{L_0}, then the successive approximations u^h obtained by the Galërkin method converge to the exact generalized solution of (1.43). Note that $L_0^{-1}K$ will be completely continuous in F_{L_0} if K is bounded in F and L_0^{-1} is completely continuous there.

In this algorithm for the Galërkin method the basis functions ϕ_i^h may again be chosen to fail the natural boundary conditions.

Suppose the equation (1.26), $Lu = f$, is taken with the boundary condition (1.27). In the Ritz method the $\{B_i(x)\}_{i=1}^2$ were assumed to vanish identically; here they are not. Then, as we have already seen, we may take as the $\{\phi_i^h\}$ a set of linearly independent functions belonging to $\mathring{W}_2^1$, that is, satisfying a homogeneous Dirichlet condition and corresponding to the requirement that F_h be complete in F_{L_0}. Then the system (1.47) has the form

$$\int_D \left\{ \sum_{i,j=1}^2 A_{ij}(x) \frac{\partial u^h}{\partial x_i} \frac{\partial \phi_s^h}{\partial x_j} + \sum_{i=1}^2 B_i \frac{\partial u^h}{\partial x_i} \phi_s^h + q(x) u^h \phi_s^h - f \phi_s^h \right\} dD = 0$$

and the elements of the matrix A in (1.48) are

$$a_{ke} = \int_D \left\{ \sum_{i,j=1}^2 A_{ij}(x) \frac{\partial \phi_e^h}{\partial x_i} \frac{\partial \phi_k^h}{\partial x_j} + \sum_{i=1}^2 B_i \frac{\partial \phi_e^h}{\partial x_i} \phi_k^h + q(x) \phi_e^h \phi_k^h \right\} dD.$$

It can be shown that the Galërkin approximations converge if $F = L_2(D)$ (Mikhlin [1]).

Note that if $K = \Theta$ in (1.47) the Galërkin method leads to the same system (1.48) as in the Ritz method, i.e., these two methods coincide for positive definite self-adjoint operators L.

Let us now consider a modification of the Galërkin method in which L_0 is, in general, not symmetric and positive definite. Suppose there exists a bounded operator L_0^{-1}, defined over all of F. Then (1.43) is equivalent to

$$u + L_0^{-1}Ku = f', \qquad f' = L_0^{-1}f. \tag{1.49}$$

Let F_1 denote a Hilbert space with the scalar product $(u, v)_1 = (L_0 u, L_0 v)$ and norm $\|u\|_1 = \|L_0 u\|$. The Galërkin method for equation (1.49) may be formulated as follows: let F_h be a finite-dimensional subspace of F_1 with basis functions $\{\phi_i^h\}_{i=1}^{N_h}$. We look for an approximate solution in the form $u^h = \sum_{D=1}^{N_h} \alpha_i \phi_i^h$ where the unknown $\{\alpha_i\}_{i=1}^{N_h}$ are determined by the system of linear equations

$$(u^h, \phi_i^h)_1 + (L_0^{-1}Ku^h, \phi_i^h)_1 = (f', \phi_i^h)_1, \qquad i = 1, \ldots, N_h \tag{1.50}$$

We noted just now that if the equation $Lu = f$ has a unique solution, and F_h is complete in F_1 and $L_0^{-1}K$ is completely continuous in F_1, then the sequence u^h converges to the exact solution in both F and F_1.

The system (1.50) may be written in the equivalent form

$$(L_0 u^h, L_0 \phi_i^h) + (K u^h, L_0 \phi_i^h) = (f, L_0 \phi_i^h), \qquad i = 1, \ldots, N_h. \tag{1.51}$$

These equations also describe the well-known variational method of moments.

One of the complex problems in the method (1.50)—or (1.51)—is the choice of the basis functions. If we prescribe the functions $\{\phi_i^h\}_{i=1}^{N_h}$ in advance, with known approximation properties, it is often difficult to investigate the properties of the system $\{\psi_i^h\}_{i=1}^{N_h}$, where $\psi_i^h = L_0 \phi_i^h$. This difficulty, in turn, hampers the estimation of convergence rates, calculation of singularities, and other specific tasks.

Let us consider one of the algorithms for constructing basis functions in this method, which has been studied in connection with a number of boundary problems (Marchuk and Agoshkov [5]).

Let the range of values of the operator K and the function f lie in some subspace $F(K, f) \subset F$. We prescribe in $F(K, f)$ some initial system of coordinate functions $\{\psi_i^h\}_{i=1}^{N_h}$ with finite supports of order h, such that the sequence $\{\psi_i^h\}_{i=1}^{N_h}$ ($h = h_1, h_2, \ldots$) is complete in $F(K, f)$. We construct the sequence $\{\phi_i^h\}_{i=1}^{N_h}$, where $\phi_i^h = L_0^{-1} \psi_i^h$. These functions are linearly independent for all h. We adopt them as basis functions for the Galërkin solutions of (1.49).

Let us examine some properties of this algorithm. By construction, the functions $\{\phi_i^h\}_{i=1}^{N_h}$ have the singularities of the solution u that arise from the operator L_0, and, because of the special choice of the system $\{\psi_i^h\}_{i=1}^{N_h}$ we are able to take account of singularities of the function $\omega = f - Ku$, which are often known in advance. Then we may try, using a small number of initial basis functions, to obtain an effective approximation and hope for rapid convergence of the u^h to u.

If the solution of (1.49) depends on the variables x_i ($i = 1, \ldots, n$) and $F(K, f)$ consists of functions depending only on the x_i ($i = 1, \ldots, m < n$), then it suffices to introduce coordinate functions $\{\psi_i^h\}_{i=1}^{N_h}$ depending only on the x_i ($i = 1, \ldots, m < n$), and to approximate $\omega \in F(K, f)$ using these. The solution u itself will be approximated by the u^h with respect to all the variables. This fact leads in practice to a significant decrease in the number of coordinate functions and, consequently, to a decrease in the order of the system (1.50) that is to be solved; this is especially important in solving multidimensional problems of mathematical physics.

If we write (1.50) in the equivalent form:

$$\sum_{j=1}^{N_h} \alpha_j (\psi_j^h, \psi_i^h) = - \sum_{j=1}^{N_h} \alpha_j (K \phi_j^h, \psi_i^h) + (f, \psi_i^h), \qquad i = 1, \ldots, N_h \tag{1.52}$$

It is easily seen that because the ψ_j^h ($j = 1, \ldots, N_h$) are finite, the matrix on the left side of (1.52) is either band or sparse, which in many cases facilitates the solution of the system by iterative methods. The finiteness of the ψ_j^h ($j = 1, \ldots, N_h$) also simplifies the computation of the $\{\phi_i^h\}_{i=1}^{N_h}$, the matrix elements, and the values of (f, ψ_j^h).

In view of all we have said, we may assert that the Galërkin method, using special coordinate functions ϕ_i^h, may well turn out to be effective for the solution of certain boundary problems in which we are able to construct L_0^{-1} rapidly enough: for example, if L_0 is explicitly invertible, say the differential operator in the transport problem, the Laplace operator in the square or the circle, etc.

2.1.4 The Method of Least Squares

The least-squares method is widely used for the solution of boundary problems arising in mathematical physics. For the operator equation

$$Lu = f \tag{1.53}$$

in the Hilbert space F it uses the following scheme. Let F_h be a finite-dimensional subspace of F with the basis $\phi_1^h, \ldots, \phi_{N_h}^h$, where $F_h \subset \Phi(L)$. Note that, rather than L itself, we might consider an extension $\hat{L}$ such that $\Phi(\hat{L})$ is a complete space.

Then the approximate solutions (1.38) are constructed by the least-squares method, beginning with the equations

$$\frac{\partial}{\partial \alpha_i} \| Lu - f \| = 0, \qquad i = 1, \ldots, N_h. \tag{1.54}$$

These give rise to the system of linear equations (1.39) with the matrix $A = (a_{ij})$ and the vector $g = (g_i)$, where

$$a_{ij} = (L\phi_i^h, L\phi_j^h), \qquad g_i = (f, L\phi_i^h), \qquad 1 \le i, i \le N_h. \tag{1.55}$$

If L is nonsingular in $\Phi(L)$, the matrix is symmetric and positive definite.

Let us formulate the conditions sufficient for the convergence of the least-squares method.

The successive approximations u^h converge in F to the exact solution u of the equation (1.53) if: the solution is unique, the sequence LF_h is complete in $\Phi(L)$, and the operator L^{-1} exists and is bounded. (Note that the expression LF_h is meaningful, since by hypothesis $F_h \subset \Phi(L)$.)

To clarify the meaning of the second requirement in this theorem, we observe that the completeness of the spaces LF_h implies, as it did in Section 2.1.2, that for all $u \in \Phi(L)$ and $\varepsilon > 0$, we can find an $\hat{h} = \hat{h}(u, \varepsilon) > 0$ such that

$$\inf_{\omega \in F_h} \| Lu - L\omega \| < \varepsilon \tag{1.56}$$

for all F_h with $h < \hat{h}$. Further, it is clear that the uniqueness of the solution of the problem (1.43) and the requirements of the theorem will guarantee both

$$u^h \xrightarrow[h \to 0]{} u$$

and

$$Lu^h \xrightarrow[h \to 0]{} Lu.$$

When the least-squares method is applied to the problems of mathematical physics, the question of whether the solution satisfies the boundary conditions is more complicated than it is for the two preceding methods. We briefly sketch two possible approaches to the solution of this problem.

The first, and more obvious, approach is to require that the functions in F_h satisfy the boundary conditions exactly. In the mixed boundary condition problem for the elliptic equation (1.26) this turns out to be very complicated in its practical application.

The second possible approach is to use the *method of weights* as proposed by Bramble and Schatz [5] and others. The idea of this approach is as follows.

Corresponding to the $2m$ order partial differential equation

$$Lu = f \quad \text{in} \quad D, \tag{1.57}$$

with the boundary conditions

$$L_i u = f_i \quad \text{on} \quad \partial D, \quad i = 1, \ldots, m, \tag{1.58}$$

let us form the functional

$$J_h(u) = \|Lu - f\|^2 + \sum_{i=1}^{m} c_i(h)\|L_i u - f_i\|^2, \tag{1.59}$$

where $\{c_i(h)\}_{i=1}^{m}$ are positive functions of h which characterize the sequence of subspaces F_n. For the difference analog of the self-adjoint problem (1.57), (1.58) we have

$$c_i(h) = h^{-2(2m - m_i - 1/2)},$$

where m_i is the order of the highest derivative in the operator L_i. The approximations u^h are now thought of as the solutions of the variational problems

$$\inf_{u \in F_h} J_h(u) = J_h(u^h).$$

using the least-squares method.

The functions u^h converge to u as $h \to 0$, while both (1.57) and (1.58) are asymptotically satisfied. At the same time the functions in the subspaces F_h do not necessarily satisfy the boundary conditions.

2.2 The Method of Integral Identities

2.2.1 Method of Constructing Difference Equations for Problems with Discontinuous Coefficients on the Basis of an Integral Identity

As of now, there are several methods for solving boundary problems for ordinary differential equations. We shall look at some of those which have been discussed at length in the literature. We will illustrate these methods on very simple and conveniently chosen problems because the main purpose of the present text is to familiarize the reader with certain fundamentals of numerical mathematics. Let us note that the difference analogs of boundary problems for ordinary differential equations with smooth coefficients are understood with sufficient completeness by now. Our attention will be directed towards equations with discontinuous coefficients. Problems of this kind are encountered in many significant applications. The first schemes were obtained by Tikhonov and Samarskii [4].

In this section we will derive the finite difference equations for a diffusion, based on an integral identity due to the author [17].

Consider the diffusion equation for one-dimensional regions. It has the form

$$-\frac{d}{dx} p \frac{d\phi}{dx} + q\phi = f, \tag{2.1}$$

where $p = p(x) \geq p_0 > 0$ is the diffusion coefficient, $q = q(x) \geq 0$ is the absorption coefficient, and $f = f(x)$ is the source of the diffusing substance. Let us suppose that the functions are piecewise-continuous with discontinuities of the first kind.

We wish to find a continuous solution to (2.1) which has a differentiable "flow"

$$J = p \frac{d\phi}{dx}$$

and which satisfies the boundary conditions

$$\phi(0) = 0, \qquad \phi(1) = 0. \tag{2.2}$$

Let us choose two systems of net points over the range $[0, 1]$ of the variable x: the basic system $\{x_k\}_{k=0}^{n}$ and the auxiliary system $\{x_{k+1/2}\}_{k=0}^{n-1}$. The points from these two systems are mutually alternating in succession, i.e., $x_k < x_{k+1/2} < x_{k+1}$ (in this case $x_0 = 0$ and $x_n = 1$). In what follows we will assume that $x_{k+1/2} = (x_{k+1} + x_k)/2$.

Integrate (2.1) with respect to x from $x_{k-1/2}$ to $x_{k+1/2}$. As a result we obtain the equilibrium relation

$$-J_{k+1/2} + J_{k-1/2} + \int_{x_{k-1/2}}^{x_{k+1/2}} (q\phi - f)\, dx = 0, \tag{2.3}$$

where

$$J_{k\pm 1/2} = J(x_{k\pm 1/2}).$$

In order to find $J_{k\pm 1/2}$, proceed as follows: integrate (2.1) from $x_{k-1/2}$ to x:

$$p \frac{d\phi}{dx} = J_{k-1/2} + \int_{x_{k-1/2}}^{x} (q\phi - f)\, d\xi. \tag{2.4}$$

Divide expression (2.4) by p and then integrate in the limits (x_{k-1}, x_k). We obtain

$$\phi_k - \phi_{k-1} = J_{k-1/2} \int_{x_{k-1}}^{x_k} \frac{dx}{p} + \int_{x_{k-1}}^{x_k} \frac{dx}{p} \int_{x_{k-1/2}}^{x} (q\phi - f)\, d\xi. \quad (2.5)$$

Solving for $J_{k-1/2}$ in (2.5) yields

$$J_{k-1/2} = \frac{1}{\int_{x_{k-1}}^{x_k} \frac{dx}{p}} \left[\phi_k - \phi_{k-1} - \int_{x_{k-1}}^{x_k} \frac{dx}{p} \int_{x_{k-1/2}}^{x} (q\phi - f)\, d\xi \right]. \quad (2.6)$$

A similar expression is obtained for $J_{k+1/2}$ by taking $k + 1$ rather than k in (2.6). In this way we have managed to express the flows $J_{k\pm1/2}$ by means of known functions and the solution of the problem. The relation (2.6) is exact. A substitution of (2.6) and the corresponding $J_{k+1/2}$ into (2.3) results in

$$-\frac{\phi_{k+1} - \phi_k}{\int_{x_k}^{x_{k+1}} \frac{dx}{p}} + \frac{\phi_k - \phi_{k-1}}{\int_{x_{k-1}}^{x_k} \frac{dx}{p}} + \int_{x_{k-1/2}}^{x_{k+1/2}} (q\phi - f)\, dx$$

$$= -\frac{1}{\int_{x_k}^{x_{k+1}} \frac{dx}{p}} \int_{x_k}^{x_{k+1}} \frac{dx}{p} \int_{x_{k+1/2}}^{x} (q\phi - f)\, d\xi$$

$$+ \frac{1}{\int_{x_{k-1}}^{x_k} \frac{dx}{p}} \int_{x_{k-1}}^{x_k} \frac{dx}{p} \int_{x_{k-1/2}}^{x} (q\phi - f)\, d\xi. \quad (2.7)$$

Equation (2.7) is our basic identity to be used for obtaining the finite-difference equations.

Define the operator A on the domain Φ cf the solutions of (2.1) as follows:

$$(A\phi)_k = -\frac{1}{\Delta x_k} \left(\frac{\phi_{k+1} - \phi_k}{\int_{x_k}^{x_{k+1}} \frac{dx}{p}} - \frac{\phi_k - \phi_{k-1}}{\int_{x_{k-1}}^{x_k} \frac{dx}{p}} - \int_{x_{k-1/2}}^{x_{k+1/2}} q\phi\, dx \right.$$

$$- \frac{1}{\int_{x_k}^{x_{k+1}} \frac{dx}{p}} \int_{x_k}^{x_{k+1}} \frac{dx}{p} \int_{x_{k+1/2}}^{x} q\phi\, d\xi$$

$$\left. + \frac{1}{\int_{x_{k+1/2}}^{x_k} \frac{dx}{p}} \int_{x_{k-1}}^{x_k} \frac{dx}{p} \int_{x_{k-1/2}}^{x} q\phi\, d\xi \right). \quad (2.8)$$

Also, consider the vector f with the components

$$(f)_k = \frac{1}{\Delta x_k} \int_{x_{k+1/2}}^{x_{k+1/2}} f \, dx + \frac{1}{\Delta x_k} \left(\frac{1}{\int_{x_k}^{x_{k+1}} \frac{dx}{p}} \int_{x_k}^{x_{k+1}} \frac{dx}{p} \int_{x_{k+1/2}}^{x} f \, d\xi \right.$$

$$\left. - \frac{1}{\int_{x_{k-1}}^{x_k} \frac{dx}{p}} \int_{x_{k-1}}^{x_k} \frac{dx}{p} \int_{x_{k-1/2}}^{x} f \, d\xi \right), \qquad (2.8')$$

where

$$\Delta x_k = x_{k+1/2} - x_{k-1/2}, \qquad k = 1, \ldots, n-1.$$

(Do not confuse the above $(f)_k$ with $f(x_k)$.)

For the sake of simplicity we will assume from now on that the solutions of (2.1) are chosen from the class Φ, each function of which has certain smoothness properties and satisfies the boundary condition $\phi = 0$.

Using a more compact notation, equation (2.7) for $k = 1, \ldots, n-1$ can be written as

$$A\phi = f. \qquad (2.9)$$

Consider further various approximations of equation (2.9): to this end let us introduce the Euclidean norm

$$\|\phi^h\|_{F_h}^2 = \sum_{k=1}^{n-1} (\phi_k^h)^2 \Delta x_k, \qquad (2.10)$$

where F_h is the space of net functions of the form $\phi^h = (\phi_1^h, \ldots, \phi_{n-1}^h)'$, defined at points $x_1, x_2, \ldots, x_{n-1}$. Consider the following approximating problem:

$$A^h \phi^h = f^h, \qquad (2.11)$$

where

$$(A^h \phi^h)_k = -\frac{1}{\Delta x_k} \left(\frac{\phi_{k+1}^h - \phi_k^h}{\int_{x_k}^{x_{k+1}} \frac{dx}{p}} - \frac{\phi_k^h - \phi_{k-1}^h}{\int_{x_{k-1}}^{x_k} \frac{dx}{p}} - \phi_k^h \int_{x_{k-1/2}}^{x_{k+1/2}} q \, dx \right), \qquad (2.12)$$

$$(f^h)_k = \frac{1}{\Delta x_k} \int_{x_{k-1/2}}^{x_{k+1/2}} f \, dx,$$

for $k = 1, \ldots, n-1$, and $\phi_0^h = \phi_n^h = 0$. Using the triangle inequality we have

$$\|(A\phi)_h - A^h(\phi)_h\|_{F_h} \leq \|\xi^h\|_{F_h} + \|\eta^h\|_{F_h}, \quad \|(f)_h - f^h\|_{F_h} = \|\Theta^h\|_{F_h}, \qquad (2.13)$$

where

$$(\xi^h)_k = \frac{1}{\Delta x_k}\left(\int_{x_{k-1/2}}^{x_{k+1/2}} q\phi\, dx - \phi_k^h \int_{x_{k-1/2}}^{x_{k+1/2}} q\, dx\right),$$

$$(\eta^h)_k = -\frac{1}{\Delta x_k}\left(\frac{1}{\int_{x_k}^{x_{k+1}}\dfrac{dx}{p}}\int_{x_k}^{x_{k+1}}\frac{dx}{p}\int_{x_{k+1/2}}^{x} q\phi\, d\xi\right.$$

$$\left. -\frac{1}{\int_{x_{k-1}}^{x_k}\dfrac{dx}{p}}\int_{x_{k-1}}^{x_k}\frac{dx}{p}\int_{x_{k-1/2}}^{x} q\phi\, d\xi\right),$$

(here and below, for any continuous function u on $[0, 1]$ we adopt the symbol $(u)_h$ to denote the $(n-1)$-dimensional vector from F_h with the components $u(x_k)$);

$$(\Theta^h)_k = -\frac{1}{\Delta x_k}\left(\frac{1}{\int_{x_k}^{x_{k+1}}\dfrac{dx}{p}}\int_{x_k}^{x_{k+1}}\frac{dx}{p}\int_{x_{k+1/2}}^{x} f\, d\xi\right.$$

$$\left. -\frac{1}{\int_{x_{k-1}}^{x_k}\dfrac{dx}{p}}\int_{x_{k-1}}^{x_k}\frac{dx}{p}\int_{x_{k-1/2}}^{x} f\, d\xi\right).$$

Let us estimate the norms $\|\xi^h\|_{F_h}$, $\|\eta^h\|_{F_h}$, $\|\Theta^h\|_{F_h}$. To this end, assume that $q, f \in Q^{(2)}(0, 1)$ and $p \in Q^{(3)}(0, 1)$, where $Q^{(s)}(0, 1)$ is the space of piecewise-continuously differentiable functions (up to and including the order s), the possible discontinuities being those of the first kind at points $0 < y_1 < y_2 < \cdots < y_m < 1$. We will assume everywhere in what follows that the set $\{y_l\}_{l=1}^m$ belongs to the set of net points $\{x_k, k = 1, \ldots, n-1\}$. This assumption will be needed in analyzing the approximation error.

From the assumptions made it follows that the solution ϕ of problem (2.1) will be continuous, while on each of the segments $[y_l, y_{l+1}]$ $(l = 1, \ldots, m-1)$, the solution will have a fourth derivative, that is $\phi \in Q^{(4)}(0, 1)$. Let us now investigate the behavior of the components of ξ^h, η^h, and Θ^h under the assumption that $h \ll 1$, where $h = \max_{0 \le k \le n-1}|x_{k+1} - x_k|$. Expanding into the Taylor series in the vicinity of the net points, it is not difficult to show that the components of these vectors are majorized in modulus by the corresponding components of the vector ω^h, where

$$(\omega^h)_k = \begin{cases} Nh, & \text{if } x_k \text{ is one of the points } y_l\ (l = 1, 2, \ldots, m), \\ M(|\Delta x_{k+1/2} - \Delta x_{k-1/2}| + h^2) & \text{otherwise;} \end{cases}$$

M, N are positive constants. Here we have introduced the notation $\Delta x_{k+1/2} = x_{k+1} - x_k$. For proof let us assume that in the domain of definition of the solution there is a point of discontinuity of the coefficients, $x = x_l \, (1 \leq l \leq n)$ and that $\Delta x_{k+1/2} = \Delta x_{k-1/2}$ for $k \neq l$. Keeping in mind (2.10), we can write

$$\|\omega^h\|_{F_h}^2 = \sum_{\substack{k=1 \\ k \neq l}}^{n-1} (\omega^h)_k^2 \Delta x_k + (\omega^h)_l^2 \Delta x_l.$$

Suppose further that

$$\tilde{h} = \max\{\Delta x_k, 2(1 - x_{n-1/2}), 2x_{1/2}\}.$$

Taking account of the relation

$$1 - \tilde{h} \leq \sum_{k=1}^{n-1} \Delta x_k < 1$$

and using the above local estimates for $(\omega^h)_k$ in estimating the terms ω^h in the squared norm, we obtain the estimate

$$\|\omega^h\|_{F_h}^2 \leq M^2 h^4 + N^2 h^3,$$

whence

$$\|\omega^h\|_{F_h} \leq C h^{3/2}.$$

Hence we have the following estimate for the norms of approximation errors of ξ^h, η^h, and Θ^h:

$$\max(\|\xi^h\|_{F_h}, \|\eta^h\|_{F_h}, \|\Theta^h\|_{F_h}) \leq C h^{3/2} \tag{2.14}$$

(C being a positive constant independent of h), provided one of the two conditions below is satisfied: either the net is uniform on each of the intervals $[0, y_1], [y_1, y_2], \ldots, [y_m, 1]$; or the net is quasi-uniform, that is, the inequality $|\Delta x_{k+1/2} - \Delta x_{k-1/2}| \leq c h^2$ as $h \to 0$ is violated only finitely many times (c a positive constant). We could use a number of other conditions of this kind, but the two mentioned are most often met in practice.

Let us note that if the order of smoothness of any of the functions p, q, and f is decreased by one, the following estimate is obtained

$$\max(\|\xi^h\|_{F_h}, \|\eta^h\|_{F_h}, \|\Theta^h\|_{F_h}) \leq C_1 h.$$

The difference scheme (2.11), which we have considered, is rarely used in practice the way it stands, since the explicit integration of the functions p, q, and f can become very difficult. Therefore, as a rule, instead of (2.11) we use its simplified version:

$$(A^h \phi^h)_k = -\frac{1}{\Delta x_k} \left\{ p_{k+1/2} \frac{\phi_{k+1}^h - \phi_k^h}{\Delta x_{k+1/2}} - p_{k-1/2} \frac{\phi_k^h - \phi_{k-1}^h}{\Delta x_{k-1/2}} - (q\Delta x)_k \phi_k^h \right\},$$

$$(f^h)_k = \frac{1}{\Delta x_k} (f \Delta x)_k = f_k = \frac{f_{k+1/2}(x_k - x_{k-1/2}) + f_{k-1/2}(x_{k+1/2} - x_k)}{x_{k+1/2} - x_{k-1/2}}$$

$$k = 1, 2, \ldots, n-1.$$

It turns out that all the conclusions we have made, with regard to the size of the approximation error, still hold, provided all the corresponding assumptions on smoothness of the parameters also remain unchanged.

We will now turn to the convergence properties of (2.11) and (2.12), keeping the smoothness assumptions on $p, q,$ and f. We need only to prove the stability of (1.11) and then use the convergence theorem.

We first estimate the scalar product (ϕ^h, f^h) by the Cauchy–Bunyakovsky inequality:

$$(\phi^h, f^h) \leq \|\phi^h\|_{F_h} \|f^h\|_{F_h}, \tag{2.15}$$

where the scalar product is to be understood in the following sense:

$$(\chi, \psi) = \sum_{k=1}^{n-1} \Delta x_k \chi_k \psi_k, \qquad \chi, \psi \in F_h.$$

Let us investigate the left-hand side of (2.15) in more detail. Since $q(x) \geq 0$ and $p(x) \geq p_0 > 0$ by hypothesis, we have

$$(\phi^h, f^h) = (\phi^h, A^h \phi^h)$$

$$= \sum_{k=1}^{n} \frac{(\phi_k^h - \phi_{k-1}^h)^2}{\int_{x_{k-1}}^{x_k} \dfrac{dx}{p}} + \sum_{k=1}^{n-1} (\phi_k^h)^2 \int_{x_{k-1/2}}^{x_{k+1/2}} q\, dx \geq p_0 \sum_{k=1}^{n} \frac{(\phi_k^h - \phi_{k-1}^k)^2}{\Delta x_{k-1/2}} > 0. \tag{2.16}$$

This inequality follows from the fact that the vector ϕ^h is nonnull, since it is the solution of the inhomogeneous problem (2.11) with a nonsingular matrix A^h.

Noting that $\phi_0^h = 0$ we may write

$$\phi_k^h = \sum_{j=1}^{k} (\phi_j^h - \phi_{j-1}^h) = \sum_{j=1}^{k} \frac{\phi_j^h - \phi_{j-1}^h}{\sqrt{\Delta x_{j-1/2}}} \sqrt{\Delta x_{j-1/2}},$$

whence the Cauchy–Bunyakovsky inequality for the sum yields

$$(\phi_k^h)^2 = \left(\sum_{j=1}^{k} \frac{\phi_j^h - \phi_{j-1}^h}{\sqrt{\Delta x_{j-1/2}}} \sqrt{\Delta x_{j-1/2}} \right)^2$$

$$\leq \left(\sum_{j=1}^{k} \frac{(\phi_j^h - \phi_{j-1}^h)^2}{\Delta x_{j-1/2}} \right) \left(\sum_{j=1}^{k} \Delta x_{j-1/2} \right) \leq \sum_{j=1}^{n} \frac{(\phi_j^h - \phi_{j-1}^h)^2}{\Delta x_{j-1/2}}.$$

Finally,

$$\sum_{k=1}^{n-1} (\phi_k^h)^2 \Delta x_k \leq \sum_{j=1}^{n} \frac{(\phi_j^h - \phi_{j-1}^h)^2}{\Delta x_{j-1/2}} \sum_{k=1}^{n-1} \Delta x_k < \sum_{j=1}^{n} \frac{(\phi_j^h - \phi_{j-1}^h)^2}{\Delta x_{j-1/2}}. \tag{2.17}$$

From (2.15)–(2.17), we find that

$$\|\phi^h\|_{F_h} \|f^h\|_{F_h} \geq (\phi^h, f^h) \geq p_0 \sum_{j=1}^{n} \frac{(\phi_j^h - \phi_{j-1}^h)^2}{\Delta x_{j-1/2}} \geq p_0 \|\phi^h\|_{F_h}^2,$$

so that

$$\|\phi^h\|_{F_h} \le \frac{1}{p_0} \|f^h\|_{F_h}.$$

This inequality proves the stability of the difference algorithm, by definition. Therefore, using the convergence theorem with the norm (2.10) we obtain the estimate

$$\|\varepsilon^h\|_{F_h} \le Kh^{3/2}, \qquad \varepsilon^h = (\phi)_h - \phi^h,$$

where $K \ge 3C/p_0$ is a positive constant.

By drawing certain network analogs of the imbedding theorems we can clarify the estimate $\|\varepsilon^h\|_{F_h} \le Kh^{3/2}$. First we note that, as we proved earlier for $\phi_0^h = \phi_n^h = 0$,

$$(\phi_k^h)^2 \le \sum_{j=1}^n \frac{(\phi_j^h - \phi_{j-1}^h)^2}{\Delta x_{j-1/2}} = \sum_{j=1}^n \left(\frac{\phi_j^h - \phi_{j-1}^h}{\Delta x_{j-1/2}}\right)^2 \cdot \Delta x_{j-1/2}$$

Accordingly, if

$$C_1 \le \frac{\Delta x_{j-1/2}}{\Delta x_j} \le C_2, \qquad 0 < C_1, C_2$$

are constants independent of j, we have

$$(\phi_k^h)^2 \le \sum_{j=1}^n \left(\frac{\phi_j^h - \phi_{j-1}^h}{\Delta x_{j-1/2}}\right)^2 \cdot \frac{\Delta x_{j-1/2}}{\Delta x_j} \Delta x_j$$

$$\le C_2 \sum_{j=1}^n \left(\frac{\phi_j^h - \phi_{j-1}^h}{\Delta x_{j-1/2}}\right)^2 \cdot \Delta x_j \equiv C_2 \|\phi^h\|_{\mathring{W}_2^1, h}^2.$$

From this we obtain the following relation for the net functions (the net analog of the imbedding of $\mathring{W}_2^1(0, 1)$ in $C(0, 1)$ in the one-dimensional case)

$$\|\phi^h\|_{C_h} = \max_{k=1,\ldots,n-1} |\phi_k^h| \le C\|\phi^h\|_{\mathring{W}_2^1, h}, \qquad C = \text{const} < \infty.$$

We also apply the latter inequality to obtain a more precise estimate of the error $\varepsilon^h = (\phi)_h - \phi^h$. To this end we write out the identity

$$A^h \varepsilon^h = \xi^h + \eta^h + \Theta^h.$$

We take the scalar product with ε^h

$$(A^h \varepsilon^h, \varepsilon^h) = (\xi^h + \eta^h + \Theta^h, \varepsilon^h).$$

Then, since (cf. (2.16))

$$(A^h \varepsilon^h, \varepsilon^h) \ge P_0 \sum_{k=1}^n \frac{(\phi_k^h - \phi_{k-1}^h)^2}{\Delta x_{k-1/2}} \ge P_0 C_1 \|\varepsilon^h\|_{\mathring{W}_2^1, h}$$

that

$$|(\xi^h + \eta^h + \Theta^h, \varepsilon^h)| = \left| \sum_{k=1}^{n-1} \Delta x_k (\xi_k^h + \eta_k^h + \Theta_k^h) \varepsilon_k^h \right|$$

$$\leq \|\varepsilon^h\|_{C_h} \sum_{k=1}^{n-1} \Delta x_k |\xi_k^h + \eta_k^h + \Phi_k^h|$$

$$\equiv \|\varepsilon^h\|_{C_h} \|\xi^h + \eta^h + \Theta^h\|_{L_{1,h}},$$

we have

$$\|\varepsilon^h\|_{\overset{\circ}{W}_2^1, h}^2 \leq C \|\xi^h + \eta^h + \Theta^h\|_{L_{1,h}}.$$

Drawing on the above imbedding theorem, we obtain the inequality

$$\|\varepsilon^h\|_{\overset{\circ}{W}_2^1, h} \leq C \|\xi^h + \eta^h + \Theta\|_{L_{1,h}}.$$

But because we have assumed the necessary smoothness of the solution and the initial data, and the quasi-uniformity of the net, we have

$$\|\xi^h + \eta^h + \Theta^h\|_{L_{1,h}} \leq 3Nmh^2 + CMh^2 \sum_{k=1}^{n-1} \Delta x_k,$$

where $C = \text{const} < \infty$. Therefore, for sufficiently small h and finite $m < \infty$, we arrive at the desired estimate

$$\|\varepsilon^h\|_{C_h} \leq C \|\varepsilon^h\|_{\overset{\circ}{W}_2^1, h} \leq O(h^2).$$

2.2.2 The Variational Form of an Integral Identity

In Section 2.2.1 we considered a method for constructing difference equations on the basis of an integral identity. Agoshkov [5] has shown recently that the integral identity method can be looked on as a variational method. Consider the differential equation

$$-\frac{d}{dx} p(x) \frac{du}{dx} + q(x)u = f(x), \qquad x \in (a, b), \tag{2.18}$$

with the boundary conditions

$$u(a) = u(b) = 0. \tag{2.19}$$

Assume that

$$p(x) > 0, \qquad q(x) \geq 0, \qquad p(x), q(x) \in L_\infty(a, b), \qquad f(x) \in L_2(a, b).$$

We apply the integral identity method to the problem (2.18), (2.19) and we find

$$
\frac{u(x_k) - u(x_{k+1})}{\displaystyle\int_{x_k}^{x_{k+1}} \frac{dx}{p(x)}} + \frac{u(x_k) - u(x_{k-1})}{\displaystyle\int_{x_{k-1}}^{x_k} \frac{dx}{p(x)}} + \int_{x_{k-1/2}}^{x_{k+1/2}} (qu - f)\, dx
$$

$$
= -\frac{1}{\displaystyle\int_{x_k}^{x_{k+1}} \frac{dx}{p(x)}} \int_{x_k}^{x_{k+1}} \frac{dx}{p(x)} \int_{x_{k+1/2}}^{x} (qu - f)\, d\xi
$$

$$
+ \frac{1}{\displaystyle\int_{x_{k-1}}^{x_k} \frac{dx}{p(x)}} \int_{x_{k-1}}^{x_k} \frac{dx}{p(x)} \int_{x_{k-1/2}}^{x} (qu - f)\, d\xi, \qquad (2.20)
$$

where

$$
a = x_0 < x_{1/2} < x_1 < x_{3/2} < \cdots < x_{N-3/2} < x_{N-1} < x_{N-1/2} < x_N = b
$$

is some point set. If we approximate the integrals appearing in (2.20) we obtain a corresponding set of difference equations.

Our immediate aim is to show that (2.20) may be written in a form (which we shall call the "variational form of an integral identity") close to the variational equation of Galërkin's method but permitting us to use discontinuous basis functions in constructing approximate solutions, and yielding a proof that the approximate solution converges under fairly general assumptions about the input data.

We perform some simple transformations on (2.20). Let

$$
\psi = qu - f, \qquad \rho_k(x) = \int_{x_k}^{x} \frac{dx'}{p(x')} \bigg/ \int_{x_k}^{x_{k+1}} \frac{dx'}{p(x')}.
$$

Then

$$
-\frac{1}{\displaystyle\int_{x_k}^{x_{k+1}} \frac{dx}{p(x)}} \int_{x_k}^{x_{k+1}} \frac{dx}{p(x)} \int_{x_{k+1/2}}^{x_k} \psi(\xi)\, d\xi
$$

$$
= -\frac{1}{\displaystyle\int_{x_k}^{x_{k+1}} \frac{dx}{p(x)}} \int_{x_k}^{x_{k+1}} d\left(\int_{x_k}^{x} \frac{dx'}{p(x')} \right) \int_{x_{k+1/2}}^{x} \psi(\xi)\, d\xi
$$

$$
= -\int_{x_{k+1/2}}^{x_{k+1}} \psi(\xi)\, d\xi + \frac{1}{\displaystyle\int_{x_k}^{x_{k+1}} \frac{dx}{p(x)}} \int_{x_k}^{x_{k+1}} \psi(x) \int_{x_k}^{x} \frac{d\xi}{p(\xi)}\, dx
$$

$$
= -\int_{x_{k+1/2}}^{x_{k+1}} \psi(\xi)\, d\xi + \int_{x_k}^{x_{k+1}} \rho_k(x)\psi(x)\, dx.
$$

Similarly, we find that

$$\frac{1}{\int_{x_{k-1}}^{x_k} \frac{dx}{p(x)}} \int_{x_{k-1}}^{x_k} \frac{dx}{p(x)} \int_{x_{k-1/2}}^{x} \psi(\xi)\, d\xi = -\int_{x_{k-1}}^{x_{k-1/2}} \psi(x)\, dx + \int_{x_{k-1}}^{x_k} \tilde{\rho}_k(x)\psi(x)\, dx$$

where

$$\tilde{\rho}_k(x) = \int_x^{x_k} \frac{dx'}{p(x')} \bigg/ \int_{x_{k-1}}^{x_k} \frac{dx}{p(x)}.$$

Given these transformations, (2.20) becomes

$$\frac{u(x_k) - u(x_{k+1})}{\int_{x_k}^{x_{k+1}} \frac{dx}{p(x)}} + \frac{u(x_k) - u(x_{k-1})}{\int_{x_{k-1}}^{x_k} \frac{dx}{p(x)}}$$

$$+ \int_{x_{k-1}}^{x_k} (1 - \tilde{\rho}_k(x))\psi(x)\, dx + \int_{x_k}^{x_{k+1}} (1 - \rho_k(x))\psi(x)\, dx = 0.$$

Let

$$Q_k(x) = \begin{cases} 1 - \int_x^{x_k} \frac{d\xi}{p(\xi)} \bigg/ \int_{x_{k-1}}^{x_k} \frac{d\xi}{p(\xi)}, & x \in (x_{k-1}, x_k), \\[2mm] 1 - \int_{x_k}^{x} \frac{d\xi}{p(\xi)} \bigg/ \int_{x_k}^{x_{k+1}} \frac{d\xi}{p(\xi)}, & x \in (x_k, x_{k+1}), \\[2mm] 0, & x \notin (x_{k-1}, x_{k+1}); \end{cases} \qquad (2.21)$$

and then the identity (2.20) can be written in the form

$$\frac{u(x_k) - u(x_{k+1})}{\int_{x_k}^{x_{k+1}} \frac{dx}{p(x)}} + \frac{u(x_k) - u(x_{k-1})}{\int_{x_{k-1}}^{x_k} \frac{dx}{p(x)}} + (qu, Q_k) = (f, Q_k),$$

$$k = 1, \dots, N-1, \qquad (2.22)$$

where

$$(\phi, \psi) = \int_b^a \phi\psi\, dx, \qquad \|\phi\| = (\phi, \phi)^{1/2}.$$

Noting that

$$\left(p\frac{du}{dx}, \frac{dQ_k}{dx} \right) = \frac{u(x_k) - u(x_{k+1})}{\int_{x_k}^{x_{k+1}} \frac{dx}{p(x)}} + \frac{u(x_k) - u(x_{k-1})}{\int_{x_{k-1}}^{x_k} \frac{dx}{p(x)}},$$

we rewrite (2.22) as follows:

$$\left(p\frac{du}{dx}, \frac{dQ_k}{dx} \right) + (qu, Q_k) = (f, Q_k), \qquad k = 1, \dots, N-1. \qquad (2.23)$$

Next, we introduce some interpolant $u_I(x)$ of the function $u(x)$, such that $u_I(x_i) = u(x_i)$ $(i = 0, \ldots, N)$ and the derivative du_I/dx is meaningful. Then, from the equation

$$\left(p \frac{du_I}{dx}, \frac{dQ_k}{dx}\right) = \left(p \frac{du_I}{dx}, \frac{dQ_k}{dx}\right)$$

we find that the equation (2.23) is equivalent to the following:

$$\left(p \frac{du_I}{dx}, \frac{dQ_k}{dx}\right) + (qu, Q_k) = (f, Q_k), \qquad k = 1, \ldots, N-1. \qquad (2.24)$$

Thus, we have shown that the identity (2.20) is equivalent to (2.22), (2.23), and (2.24). We may, therefore, regard the integral identity method as a variational method, and use (2.22)–(2.24) together with (2.20) for constructing approximate solutions of the problem, employing a rather wide choice of basis functions. The relation (2.23) is precisely the well-known variational equation used in the Galërkin method for constructing an approximate solution with the $Q_k(x)$ as basis functions. It is clear from (2.22) that one may use discontinuous basis functions (so long as the discontinuities avoid the points x_i $(i = 0, \ldots, N)$). Application of the identity (2.24) yields a number of estimates in the uniform metric.

Let us use the identities (2.22)–(2.24) to construct approximate solutions to the problem (2.18), (2.19) according to various choices of the basis functions (cf. Agoshkov [5]). First, we note that upon substitution of variables

$$y = \int_a^x \frac{d\xi}{p(\xi)}$$

the problem (2.18), (2.19) becomes

$$-\frac{d^2u}{dy^2} + p(y)q(y)u = \tilde{f}(y), \qquad 0 < y < T = \int_a^b \frac{d\xi}{p(\xi)}, \quad u(0) = u(T) = 0.$$

Therefore, for $f(x) \in L_2(a, b)$ we have $u(y) \in W_2^2(0, T)$.† With this observation in mind, we construct an approximate solution in the form

$$u^h(x) = \sum_{i=1}^{N-1} a_i Q_i(x),$$

where the unknowns a_i are defined uniquely as the solutions of the system

$$\left(p \frac{du^h}{dx}, \frac{dQ_k}{dx}\right) + (qu^h, Q_k) = (f, Q_k), \qquad k = 1, \ldots, N-1,$$

that is, the approximate solution is obtained by the standard Galërkin method with the $\{Q_i(x)\}$ as basis functions.

† $W_2^2(0, T)$ is the Hilbert space of functions with the finite norm

$$\|u\|_{W_2^2(0, T)} = \left(\int_0^T dx \left\{\left|\frac{d^2u}{dx^2}\right|^2 + \left|\frac{du}{dx}\right|^2 + |u|^2\right\}\right)^{1/2} < \infty.$$

Let us estimate the rate of convergence of the $u^h(x)$ to the exact solution as $h = \max_i |x_i - x_{i-1}| \to 0$. We begin with some problems of approximation, for which we use the $Q_i(x)$ and the "roof-functions'

$$\phi_i(y) = \begin{cases} \dfrac{y - y_{i-1}}{(y_i - y_{i-1})}, & y \in (y_{i-1}, y_i), \\[2ex] \dfrac{y_{i+1} - y}{(y_{i+1} - y_i)}, & y \in (y_i, y_{i+1}), \\[2ex] 0, & y \notin (y_{i-1}, y_{i+1}), \end{cases}$$

into which the $Q_i(x)$ are transformed by the change of variable

$$y = \frac{\displaystyle\int_a^x d\xi}{p(\xi)}.$$

Suppose that on the interval $[0, T]$ we have introduced the net $y_0 = 0 < y_1 < \cdots < y_{N-1} < y_N = T$; $H_i = y_i - y_{i-1}$, $H = \max_i H_i$; suppose also that we are given the system of roof-functions $\{\phi_i(y)\}_{i=1}^{N-1}$. Now assume that for some $u(y) \in W_2^2(0, T) \cap \mathring{W}_2^1(0, T)$ we have chosen an approximating function in the form

$$u_I(y) = \sum_{i=1}^{N-1} u(y_i)\phi_i(y).$$

To estimate the difference $u(y) - u_I(y)$ we use the following identity for $y \in (y_{i-1}, y_i)$:

$$\begin{aligned} u(y) - u_I(y) &= \int_{y_{i-1}}^y dy' \frac{d}{dy}(u - u_I)(y') = \int_{y_{i-1}}^y dy' \frac{du}{dy}(y') \\ &\quad - \frac{1}{H_i} \int_{y_{i-1}}^{y_i} dy'' \frac{du(y'')}{dy} \\ &= \frac{1}{H_i} \int_{y_{i-1}}^{y_i} dy'' \int_{y_{i-1}}^y dy' \left(\frac{du}{dy}(y') - \frac{du}{dy}(y'') \right) \\ &= \frac{1}{H_i} \int_{y_{i-1}}^{y_i} dy'' \int_{y_{i-1}}^y dy' \int_{y''}^{y'} d\xi \frac{d^2 u}{d\xi^2}(\xi), \end{aligned}$$

from which we easily derive the following approximation estimates:

(A) $\qquad |u(y) - u_I(y)| \le H_i \displaystyle\int_{y_{i-1}}^{y_i} d\xi \left| \frac{d^2 u}{d\xi^2}(\xi) \right|, \qquad y \in (y_{i-1}, y_i)$

$$\max_{y \in (y_{i-1}, y_i)} |u(y) - u_I(y)| \le H_i^2 \left\| \frac{d^2 u}{dy^2} \right\|_{L_\infty(y_{i-1}, y_i)},$$

$$\max_{y \in [0, T]} |u(y) - u_I(y)| \le H^2 \left\| \frac{d^2 u}{dy^2} \right\|_{L_\infty(0, T)};$$

(B) $\displaystyle\int_{y_{i-1}}^{y_i} dy |u(y) - u_I(y)|^2 \leq H_i^3 \left(\int_{y_{i-1}}^{y_i} d\xi \frac{d^2 u}{d\xi^2}(\xi) \right)^2 \leq H_i^4 \int_{y_{i-1}}^{y_i} d\xi \left(\frac{d^2 u}{d\xi^2} \right)^2,$

$$\int_0^T dy |u(y) - u_I(y)|^2 \leq H^4 \int_0^T dy \left(\frac{d^2 u}{dy^2} \right)^2;$$

(C) $\displaystyle\frac{d(u - u_I)}{dy}(y) = \frac{1}{H_i} \int_{y_{i-1}}^{y_i} dy'' \int_{y''}^y d\xi \frac{d^2 u}{d\xi^2}(\xi);$

$$\left| \frac{d(u - u_I)}{dy} \right| \leq \int_{y_{i-1}}^{y_i} d\xi \left| \frac{d^2 u}{d\xi^2} \right|; \quad \int_{y_{i-1}}^{y_i} dy \left| \frac{d(u - u_I)}{dy} \right|^2 \leq H_i^2 \int_{y_{i-1}}^{y_i} d\xi \left| \frac{d^2 u}{d\xi^2} \right|^2,$$

$$\int_0^T dy \left| \frac{d(u - u_I)}{dy} \right|^2 \leq H^2 \int_0^T d\xi \left| \frac{d^2 u}{d\xi^2} \right|^2.$$

Thus, we have shown that if $u(y) \in W_2^2(0, T) \cap \overset{\circ}{W}_2^1(0, T)$, then

$$\| u - u_I \|_{W_2^k(0, T)} \leq C \cdot H^{2-k} \left\| \frac{d^2 u}{dy^2} \right\|_{L_2(0, T)} \leq C \cdot H^{2-k} \| u \|_{W_2^2(0, T)}.$$

$k = 0, 1$

If, moreover, $d^2 u/dy^2 \in L_\infty(0, T)$ we have also that

$$\| u - u_I \|_{C(0, T)} \leq C \cdot H^2 \left\| \frac{d^2 u}{dy^2} \right\|_{L_\infty(0, T)}.$$

We shall now approximate the solution of (2.18), (2.19) by using the functions $Q_i(x)$. Let $u_I(x) = \sum_{i=1}^{N-1} u(x_i) Q_i(x)$. We estimate the difference $u(x) - u_I(x)$, making the change of variable $y = \int_a^x d\xi/p(\xi)$ and writing $y_i = \int_a^{x_i} d\xi/p(\xi)$, $H_i = y_i - y_{i-1}$, $T = \int_a^b d\xi/p(\xi)$:

$$\int_a^b dx \left| u(x) - \sum_{i=1}^{N-1} u(x_i) Q_i(x) \right|^2$$

$$\leq \operatorname*{supvrai}_x p(x) \int_a^b \frac{dx}{p(x)} \left| u(x) - \sum_{i=1}^{N-1} u(x_i) Q_i(x) \right|^2$$

$$\leq C \int_0^T dy \left| u(y) - \sum_{i=1}^{N-1} u(y_i) \phi_i(y) \right|^2 \leq C H_i^4 \int_0^T dy \left| \frac{d^2 u}{dy^2} \right|^2$$

$$\leq C \frac{\max_i (x_i - x_{i-1})^4}{\operatorname{infvrai}_x p(x)} \int_0^T dy \left| \frac{d^2 u}{dy^2} \right|^2 \leq C h^4 \int_0^T dy \left| \frac{d^2 u}{dy^2} \right|^2.$$

Since for the problem in question

$$\int_0^T dy \left| \frac{d^2 u}{dy^2} \right|^2 \leq C \cdot \int_a^b dx \left| \frac{d}{dx} p(x) \frac{du}{dx} \right|^2 \leq C \cdot \int_a^b |f(x)|^2 dx,$$

$$\left\| \frac{d^2 u}{dy^2} \right\|_{L_\infty(0, T)} \leq C \left\| \frac{d}{dx} p(x) \frac{du}{dx} \right\|_{L_\infty(a, b)} \leq C \| f \|_{L_\infty(a, b)},$$

we find the approximation estimate in $L_2(a, b)$

$$\|u - u_I\|_{L_2(a,b)} \leq Ch^2 \|f\|_{L_2(a,b)}.$$

Similarly we find the estimate for $d(u - u_I)/dx$:

$$\int_a^b dxp \cdot \left| \frac{d(u - u_I)}{dx} \right|^2 = \int_a^b \frac{dx}{p(x)} \left| \frac{d(u - u_I)}{dx/p(x)} \right|^2$$

$$= \int_0^T dy \left| \frac{d(u - u_I)(y)}{dy} \right|^2 \leq C \cdot h^2 \int_0^T dy \left| \frac{d^2u}{dy^2} \right|^2 \leq Ch^2 \|f\|_{L_2(a,b)}.$$

If we assume that in (2.18) $f \in L_\infty(a, b)$ we will have $f(y) \in L_\infty(0, T)$ and $d^2u/dy^2 \in L_\infty(0, T)$. Therefore, in view of the results we have just obtained, we also have an estimate of the form

$$\|u - u_I\|_{C(a,b)} \leq C \max_i H_i^2 \left\| \frac{d^2u}{dy^2} \right\|_{L_\infty(0,T)} \leq Ch^2 \|f\|_{L_\infty(a,b)}^2.$$

Now, knowing the order of the approximation to the exact solution, we use the $\{Q_i(x)\}_{i=1}^{N-1}$ to estimate the rate of convergence of the approximate solution $u^h(x) = \sum_{i=1}^{N-1} a_i Q_i(x)$ to the exact solution as $h \to 0$. An estimate in the energy norm or in the norm of $W_2^1(a, b)$ (in the present case they are equivalent) is easily obtained. We write the identity

$$\|u - u^h\|_A^2 \equiv \left(p \frac{d(u - u^h)}{dx}, \frac{d(u - u^h)}{dx} \right) + (q(u - u^h), u - u^h)$$

$$= \left(p \frac{d(u - u^h)}{dx}, \frac{d(u - u_I)}{dx} \right) + (q(u - u^h), u - u_I),$$

where $u_I(x) = \sum_{i=1}^{N-1} u(x_i)Q_i(x)$ and develop a simple estimate, applying our results about the approximation

$$\|u - u^h\|_A^2 \leq \|u - u^h\|_A \cdot \|u - u_I\|_A;$$

$$\|u - u^h\|_A^2 \leq \|u - u_I\|_A^2 \leq Ch^2 \int_0^T dy \left| \frac{d^2u}{dy^2} \right|^2 \leq Ch^2 \|f\|_{L_2(a,b)}^2.$$

Thus, we have shown that

$$\|u - u^h\|_A \leq C \cdot h \|f\|_{L_2(a,b)},$$

$$\|u - u^h\|_{W_2^1(a,b)} \leq C \cdot h \|f\|_{L_2(a,b)},$$

(since $\|u - u^h\|_{W_2^1(a,b)} \leq C\|u - u^h\|_A$).

Let us now estimate the rate at which $u^h(x)$ converges to $u(x)$ in the uniform metric. We shall use the fact that (2.24) holds for the exact solution, where $u_I = \sum_{i=1}^{N-1} u(x_i)Q_i(x)$. Then we have

$$\left(p \frac{d(u_I - u^h)}{dx}, \frac{d(u_I - u^h)}{dx}\right) + (q(u - u^h), u_I - u^h) = 0,$$

$$\left(p \frac{d(u_I - u^h)}{dx}, \frac{d(u_I - u^h)}{dx}\right) + (q(u - u^h), u - u^h)$$

$$= (q(u - u^h), u - u_I) \leq (q(u - u^h), u - u^h)^{1/2}(q(u - u_I), u - u_I)^{1/2}.$$

This last relation, together with the fact that $\|u - u_I\| \leq Ch^2$, imply the estimate

$$\left(p \frac{d(u_1 - u^h)}{dx}, \frac{d(u_1 - u^h)}{dx}\right) + (q(u - u^h), u - u^h) \leq Ch^4.$$

And since

$$\|u_1 - u^h\|_{C(a, b)} \leq C\left(p \frac{d(u_1 - u^h)}{dx}, \frac{d(u_I - u^h)}{dx}\right)^{1/2},$$

we have also

$$\|u_I - u^h\|_{C(a, b)} \leq C \cdot h^2.$$

But we have already proved that $f(x) \in L_\infty(a, b)$ implies $\|u - u_I\|_{C(a, b)} \leq Ch^2\|f\|_{L_\infty(a, b)}$. Therefore, applying the triangle inequality we obtain the estimate

$$\|u - u^h\|_{C(a, b)} \leq Ch^2.$$

in the uniform metric.

Let us now examine two different choices of the basis functions $\{\phi_i^h(x)\}$. Let $h_i = x_{i+1/2} - x_{i-1/2}$, $h = \max_i h_i$. Let $\phi_i^h(x)$ be the characteristic function of the interval $(x_{i-1/2}, x_{i+1/2})$, $\phi_0^h(x)$ that of the interval $(x_0, x_{1/2})$, and $\phi_N^h(x)$ that of the interval $(x_{N-1/2}, x_N)$. We treat the $\{\phi_i^h\}_{i=0}^N$ as basis functions and seek our approximate solution in the form

$$u^h(x) = \sum_{i=0}^{N} a_i \phi_i^h(x),$$

with $a_0 = a_N = 0$. Then

$$u^h(x) = \sum_{i=1}^{N-1} a_i \phi_i^h(x),$$

where the $\{a_i\}_{i=1}^{N-1}$ are given by the relations (cf. (2.22)):

$$\frac{u^h(x_k) - u^h(x_{k+1})}{\int_{x_k}^{x_{k+1}} \frac{dx}{p(x)}} + \frac{u^h(x_k) - u^h(x_{k-1})}{\int_{x_{k-1}}^{x_k} \frac{dx}{p(x)}} + (qu^h, Q_k) = (f, Q_k),$$

$$k = 1, \ldots, N - 1. \qquad (2.25)$$

Let us derive an *a priori* estimate for the solution of (2.25). We define an interpolant for the approximate solution as follows:

$$u_I^h(x) = \sum_{i=1}^{N-1} a_i Q_i^h(x).$$

Then (2.25) may be written in the equivalent form:

$$\left(p\frac{du_I^h}{dx}, \frac{dQ_k}{dx}\right) + (qu^h, Q_k) = (f, Q_k), \qquad k = 1, \ldots, N-1.$$

Multiplying this equation by a_k and adding for $k = 1, \ldots, N-1$ we find

$$\left(p\frac{du_I^h}{dx}, \frac{du_I^h}{dx}\right) + (qu^h, u_I^h) = (f, u_I^h)$$

or, equivalently,

$$\left(p\frac{du_I^h}{dx}, \frac{du_I^h}{dx}\right) + (qu^h, u^h) = (f, u_I^h) - (qu^h, u_I^h - u^h).$$

From this we obtain

$$\left(p\frac{du_I^h}{dx}, \frac{du_I^h}{dx}\right) + (qu^h, u^h) \le \|f\| \cdot \|u_I^h\| + (qu^h, u^h)^{1/2} \cdot (q(u_I^h - u^h), u_I^h - u^h)^{1/2}.$$

Since

$$\|u_I^h\| \le \mathrm{const}\left(p\frac{du_I^h}{dx}, \frac{du_I^h}{dx}\right)^{1/2}$$

the inequality

$$|a \cdot b| \le \varepsilon a^2 + \frac{1}{4\varepsilon}b^2, \qquad \varepsilon > 0$$

implies that

$$\left(p\frac{du_I^h}{dx}, \frac{du_I^h}{dx}\right) + (qu^h, u^h) \le \mathrm{const}\{\|f\|^2 + \|u_I^h - u^h\|^2\}.$$

Noting that

$$\int_{x_{i-1/2}}^{x_{i+1/2}} dx \cdot (u_I^h - u^h)^2 = \int_{x_{i-1/2}}^{x_{i+1/2}} dx\left(\int_{x_i}^x \frac{d(u_I^h - u^h)}{dx}\right)^2$$

$$\equiv \int_{x_{i-1/2}}^{x_{i+1/2}} dx \cdot \left(\int_{x_i}^x \frac{du_I^h}{dx}\right)^2 \le O(h^2) \cdot \int_{x_{i-1/2}}^{x_{i+1/2}} p(x)\left|\frac{du_I^h}{dx}\right|^2 dx,$$

$$\|u_I^h - u^h\|^2 \le O(h^2)\left(p\frac{du_I^h}{dx}, \frac{du_I^h}{dx}\right),$$

we obtain the *a priori* estimate

$$\left(p\frac{du_I^h}{dx}, \frac{du_I^h}{dx}\right) + (qu^h, u^h) \le \frac{\text{const}\|f\|^2}{1 - O(h^2)}. \tag{2.26}$$

It follows from (2.26) that for sufficiently small h the system (2.25) has the unique solution $\{a_i\}_{i=1}^{N-1}$.

Now let us estimate the approximation error. We note that for $u^h(x)$ and $u_I^h(x)$ the equation

$$\left(p\frac{du_I^h}{dx}, \frac{dQ_k}{dx}\right) + (qu^h, Q_k) = (f, Q_k)$$

holds, and we consider the function

$$u_I(x) = \sum_{i=1}^{N-1} u(x_i)Q_i(x),$$

which we substitute in (2.24) for $u_I(x)$. Then we have

$$\left(p\frac{d(u_I - u_I^h)}{dx}, \frac{dQ_k}{dx}\right) + (q(u - u^h), Q_k) = 0,$$

$$\left(p\frac{d(u_I - u_I^h)}{dx}, \frac{d(u_I - u_I^h)}{dx}\right) + (q(u - u^h), u - u^h)$$

$$= (q(u - u^h), -(u_I - u_I^h) + (u - u^h))$$

$$\le c \cdot (q(u - u^h), u - u^h)^{1/2} \cdot \|(u_I - u_I^h) - (u - u^h)\|,$$

$$\left(p\frac{d(u_I - u_I^h)}{dx}, \frac{d(u_I - u_I^h)}{dx}\right) + (q(u - u^h), u - u^h)$$

$$\le c \cdot \{\|u_I^h - u^h\|^2 + \|u - u_I\|^2\}, \qquad c = \text{const} > 0.$$

As we already know,

$$\|u^h - u_I^h\|^2 \le O(h^2)\left(p\frac{du_I^h}{dx}, \frac{du_I^h}{dx}\right) \le \frac{O(h^2)}{1 - O(h^2)}.$$

Moreover we have already proved that

$$\|u - u_I\|^2 \le O(h^2)\left(p\frac{du}{dx}, \frac{du}{dx}\right) \le O(h^2).$$

Then we find that the following estimates hold:

$$\left[\left(p\frac{d(u_I - u_I^h)}{dx}, \frac{d(u_I - u_I^h)}{dx}\right) + (q(u - u^h), u - u^h)\right]^{1/2} \le \frac{O(h)}{1 - O(h)} \le Ch$$

$$\max_i |u(x_i) - u^h(x_i)| + (q(u - u^h), u - u^h)^{1/2} \le \frac{O(h)}{1 - O(h)} \le Ch, \tag{2.27}$$

for sufficiently small h. We note that in the above arguments we needed only the assumption that $p(x)$, $q(x)$, $f(x) \in L_\infty(a, b)$. Generally speaking, this requirement can be weakened in many cases. For instance, we may require of $f(x)$ only that it belong to $L_1(a, b)$, i.e., that $\|f\|_{L_1} = \int_a^b dx |f(x)| < \infty$. Nevertheless, we require here that $f \in L_\infty(a, b)$, which we need in order to obtain estimates in the uniform norm (cf. the end of this section).

We may also construct an approximate solution with our second choice of basis functions—the coordinate roof-functions $\{\phi_i^h(x)\}_{i=1}^{N-1}$, which are piecewise-linear on the interval $[a, b]$; $\phi_i^h(x)$ vanishes outside the interval (x_{i-1}, x_{i+1}), and $\phi_i^h(x_i) = 1$. Then we seek a solution in the form

$$u^h(x) = \sum_{i=1}^{N-1} a_i \phi_i^h(x),$$

where the unknowns a_i are determined by the system of linear algebraic equations

$$\left(p\frac{du^h}{dx}, \frac{dQ_i}{dx}\right) + (qu^h, Q_i) = (f, Q_i), \qquad i = 1, \ldots, N-1, \qquad (2.28)$$

or, which amounts to the same thing, by the system

$$\left(p\frac{du_I^h}{dx}, \frac{dQ_i}{dx}\right) + (qu^h, Q_i) = (f, Q_i), \qquad i = 1, \ldots, N-1,$$

where

$$u_I^h = \sum_{i=1}^{N-1} a_i Q_i(x).$$

We shall assume that $p(x)$ is piecewise-linear with possible discontinuities of the first kind at the points $\tilde{x}_j$ ($j = 1, \ldots, J \le N$), which coincide with some of the net points x_i.

Since

$$\left(p\frac{du_I^h}{dx}, \frac{du_I^h}{dx}\right) + (qu^h, u^h) = (f, u_I) + (qu^h, u^h - u_I^h),$$

$$\left(p\frac{du_I^h}{dx}, \frac{du_I^h}{dx}\right) + (qu^h, u^h) \le c \cdot \{\|f\|^2 + \|u^h - u_I^h\|^2\},$$

we need an estimate of $\|u^h - u_I^h\|$ in order to complete our a priori estimate for the approximate solution. We consider $u^h - u_I^h$ on (x_{i-1}, x_i):

$$(u^h - u_I^h)(x) = (u^h(x_i) - u^h(x_{i-1})) \left[\frac{x - x_{i-1}}{h_i} - \frac{\int_{x_{i-1}}^x \frac{d\xi}{p(\xi)}}{\int_{x_{i-1}}^{x_i} \frac{dx}{p(x)}}\right]$$

$$\equiv (u^h(x_i) - u^h(x_{i-1})) \cdot \Phi_i(x).$$

Since $\Phi_i(x_{i-1}) = 0$, we have for $p(x) \in C^{(1)}(x_{i-1}, x_i)$

$$\Phi_i(x) = \int_{x_{i-1}}^{x} \frac{d\Phi_i}{dx'} \, dx' = \frac{1}{h_i} \int_{x_{i-1}}^{x} dx' \left[\int_{x_{i-1}}^{x_i} \frac{d\xi}{p(\xi)} \bigg/ h_i - \frac{1}{p(x')} \right] \bigg/ \left(\int_{x_{i-1}}^{x_i} \frac{d\xi}{p(\xi)} \bigg/ h_i \right)$$

$$|\Phi_i(x)| \leq \frac{\|p\|_{L_\infty}}{\text{infvrai}\,|p(x)|^2} \cdot \max_{\xi, x \in (x_{i-1}, x_i)} |p(x) - p(\xi)| \leq O(h).$$

Accordingly,

$$|u^h - u_I^h| \leq O(h) \cdot |u^h(x_i) - u^h(x_{i-1})| = O(h) \cdot \left| \int_{x_{i-1}}^{x_i} \frac{du_I^h}{dx} \, dx \right|$$

$$\leq O(h^{3/2}) \cdot \left\| \frac{du_I^h}{dx} \right\|_{L_2(x_{i-1}, x_i)} \leq O(h^{3/2}) \cdot \left(p \frac{du_I^h}{dx}, \frac{du_I^h}{dx} \right)_{L_2(x_{i-1}, x_i)},$$

$$\|u^h - u_I^h\| \leq O(h^4) \cdot \left(p \frac{du_I^h}{dx}, \frac{du_I^h}{dx} \right).$$

From these inequalities we obtain our *a priori* estimate:

$$\left(p \frac{du_I^h}{dx}, \frac{du_I^h}{dx} \right) + (qu^h, u^h) \leq c \cdot \|f\|^2 + O(h^4) \left(p \frac{du_I^h}{dx}, \frac{du_I^h}{dx} \right),$$

$$\left(p \frac{du_I^h}{dx}, \frac{du_I^h}{dx} \right) + (qu^h, u^h) \leq \frac{c \cdot \|f\|^2}{1 - O(h^4)}. \tag{2.29}$$

It follows from (2.29) that (2.28) has a unique solution for sufficiently small h. We can now estimate the approximation error. Let

$$u_I(x) = \sum_{i=1}^{N-1} u(x_i) Q_i(x).$$

Then we have the identities

$$\left(p \frac{d(u_I - u_I^h)}{dx}, \frac{d(u_I - u_I^h)}{dx} \right) + (q(u - u^h), u_I - u_I^h) = 0,$$

$$\left(p \frac{d(u_I - u_I^h)}{dx}, \frac{d(u_I - u_I^h)}{dx} \right) + (q(u - u^h), u - u^h)$$

$$= (q(u - u^h), (u - u_I) + (u_I^h - u^h)),$$

which imply the inequality;

$$\left(p \frac{d(u_I - u_I^h)}{dx}, \frac{d(u_I - u_I^h)}{dx} \right) + (q(u - u^h), u - u^h)$$

$$\leq c \cdot \{\|u - u_I\|^2 + \|u^h - u_I^h\|^2\} \leq \left[O(h^4) + \frac{O(h^4)}{1 - O(h^4)} \right] \|f\|^2.$$

Thus we have proved that if $q(x) \in L_\infty(a, b)$, $f(x) \in L_2(a, b)$, and $p(x)$ is a piecewise-smooth function of class $C^{(1)}$, with possible discontinuities of the first kind, then for sufficiently small h the system (2.28) has a unique solution u, and u^h-converges to u as $h \to 0$; moreover, the following estimates hold:

$$\left[\left(p \frac{d(u_I - u_I^h)}{dx}, \frac{d(u_I - u_I^h)}{dx} \right) + (q(u - u^h), u - u^h) \right]^{1/2} \leq \frac{O(h^2) \| f \|}{1 - O(h^2)} \leq Ch^2,$$

$$\max_i |u(x_i) - u^h(x_i)| + (q(u - u^h), u - u^h)^{1/2} \leq \frac{O(h^2) \| f \|}{1 - O(h^2)} \leq Ch^2.$$

As in the preceding cases, we could show that if $f(x) \in L_\infty(a, b)$, we have an estimate of the form

$$\| u - u^h \|_{C(a, b)} \leq \frac{O(h^2)}{1 - O(h^2)} \| f \|_{L_\infty(a, b)} \leq Ch^2$$

as $h \to 0$.

This approach to the construction of approximate solutions, based on the variational form of the integral identity, is applicable to many other problems of mathematical physics (cf. Agoshkov [5], [17]).

2.3 Difference Schemes for Equations with Discontinuous Coefficients Based on Variational Principles

Starting with the works of Lions [1, 2], Oganesyan and Rukhovets [5], Aubin [5], Birkhoff, Schultz and Varga [5], Babuška [5], and others, there has emerged in the recent years a sweeping scientific flow of new methods of constructing difference equations on the basis of variational principles. This direction has been enriched by a number of interesting ideas, of which the most important has been the idea of introducing the test functions with finite support, i.e., the functions which vanish everywhere except on relatively small sets (say, of the order of the mesh size). As it turns out, it is often convenient to seek the solution of a given problem as a linear combination of functions with finite support and to choose subsequently the coefficients so as to minimize a suitable functional related to the variational principle. This methodology has been used for various classes of problems and has led to a very effective algorithm of constructing difference schemes. We will illustrate this algorithm on a one-dimensional diffusion problem.

2.3.1 Simple Difference Equations for a Diffusion Based on the Ritz Method

As pointed out earlier the Ritz method can only be used for problems with self-adjoint operators. In order to find an approximate solution of the self-adjoint boundary-value problem [equations (2.1) and (2.2) from Section 2.2], let us introduce the following variational functional:

$$J = \int_0^1 \left[p\left(\frac{d\phi}{dx}\right)^2 + q\phi^2 - 2f\phi \right] dx. \tag{3.1}$$

As we have seen in Section 2.2, the minimum of the functional (3.1) is achieved on the solution of equations (2.1) and (2.2). Hence we will construct an approximate solution on the net D_h so as to choose its unspecified parameters in correspondence to the minimizations of (3.1). With this in mind, let us seek the approximate solution in the form of a continuous, piecewise-linear function belonging to $\mathring{W}_2^1$:

$$\phi^h(x) = \frac{x_{k+1} - x}{\Delta x_{k+1/2}} \phi_k + \frac{x - x_k}{\Delta x_{k+1/2}} \phi_{k+1}, \qquad x_k \leq x \leq x_{k+1}. \tag{3.2}$$

where $\Delta x_{k+1/2} = x_{k+1} - x_k$, and ϕ_k designates the values of the approximate solution of the problem at the net points. These values are not *a priori* known; however, they are to be found by minimizing the functional $J(\phi)$. We first introduce L, on the interval $x_k \leq x \leq x_{k+1}$, two linear functions $\omega_1(x)$ and $\omega_2(x)$ as follows:

$$\omega_1(x) = \frac{x - x_k}{\Delta x_{k+1/2}}, \qquad \omega_2(x) = \frac{x_{k+1} - x}{\Delta x_{k+1/2}}. \tag{3.3}$$

The interpolation formula (3.2) can then be rewritten as

$$\phi^h(x) = \omega_1(x)\phi_{k+1} + \omega_2(x)\phi_k. \tag{3.4}$$

The functions $\omega_1(x)$ and $\omega_2(x)$ depend on their interval of definition in general. Therefore, we perhaps should use the notation $\omega_{1,k+1/2}$ and $\omega_{2,k+1/2}$. But, in what follows, this correspondence will be ultimately apparent at any transformation stage; therefore the additional indices of the functions ω_1 and ω_2 will be omitted for the sake of simplicity. Note that, in agreement with the *a priori* information about the solution of equations (2.1) and (2.2), we construct the approximate solution $\phi^h(x)$ so as to be continuous on the interval $0 \leq x \leq 1$. The coefficients $p(x) \geq p_0 > 0$, $q(x) \geq 0$, and $f(x)$ are assumed to be piecewise-continuous functions with possible discontinuities of the first kind. We will assume that the points of discontinuity coincide with the points x_k. For convenience, the functional $J(\phi^h)$ to be minimized is taken in the form

$$J(\phi^h) = \sum_{k=0}^{n-1} \int_{x_k}^{x_{k+1}} \left[p\left(\frac{d\phi}{dx}\right)^2 + q\phi^2 - 2f\phi \right] dx. \tag{3.5}$$

Substituting (3.4) into (3.5), and noting the relations

$$\frac{d\omega_1}{dx} = \frac{1}{\Delta x_{k+1/2}}, \qquad \frac{d\omega_2}{dx} = -\frac{1}{\Delta x_{k+1/2}}, \qquad x_k < x < x_{k+1},$$

we obtain

$$J(\phi^h) = \sum_{k=0}^{n-1} \int_{x_k}^{x_{k-1}} \left[p\,\frac{\phi_{k+1}^2 - 2\phi_{k+1}\phi_k + \phi_k^2}{\Delta x_{k+1/2}^2} \right.$$

$$\left. + q(\omega_1^2\phi_{k+1}^2 + 2\omega_1\omega_2\phi_{k+1}\phi_k + \omega_2^2\phi_k^2) - 2f(\omega_1\phi_{k+1} + \omega_2\phi_k) \right] dx. \tag{3.6}$$

The values of ϕ_k $(k = 1, 2, \ldots, n - 1)$, will be chosen so as to minimize the functional $J(\phi^h)$. The conditions for minimum are as follows:

$$\frac{\partial J}{\partial \phi_k} = 0, \qquad \frac{\partial^2 J}{\partial \phi_k^2} > 0, \qquad k = 1, \ldots, n - 1. \tag{3.7}$$

Differentiating (3.6) with respect to ϕ_k and putting the result equal to zero, we get

$$\frac{p_{k-1/2}}{\Delta x_{k-1/2}}(\phi_k - \phi_{k-1}) - \frac{p_{k+1/2}}{\Delta x_{k+1/2}}(\phi_{k+1} - \phi_k) + q_{k+1/2}^{1,2}\phi_{k+1}$$

$$+ (q_{k+1/2}^{2,2} + q_{k-1/2}^{1,1})\phi_k + q_{k-1/2}^{1,2}\phi_{k-1} = F_k, \tag{3.8}$$

where

$$p_{k+1/2} = \frac{1}{\Delta x_{k+1/2}} \int_{x_k}^{x_{k+1}} p\,dx; \qquad q_{k+1/2}^{i,j} = \int_{x_k}^{x_{k+1}} \omega_i\omega_j q\,dx;$$

$$F_k = \int_{x_{k-1}}^{x_k} f\omega_1\,dx + \int_{x_k}^{x_{k+1}} f\omega_2\,dx.$$

To specify the boundary problem completely, we complement the difference equations of (3.8) by the boundary conditions

$$\phi_0 = 0, \qquad \phi_n = 0.$$

Equation (3.8) along with the boundary conditions can be rewritten as a *three-point difference problem*

$$a_k\phi_{k-1} - b_k\phi_k + c_k\phi_{k+1} = -F_k, \tag{3.9}$$

$$\phi_0 = 0, \qquad \phi_n = 0,$$

where

$$a_k = \frac{p_{k-1/2}}{\Delta x_{k-1/2}} - q_{k-1/2}^{1,2}; \qquad c_k = \frac{p_{k+1/2}}{\Delta x_{k+1/2}} - q_{k+1/2}^{1,2};$$

$$b_k = \frac{p_{k+1/2}}{\Delta x_{k+1/2}} + \frac{p_{k-1/2}}{\Delta x_{k-1/2}} + (q_{k+1/2}^{1,1} + q_{k-1/2}^{2,2}). \tag{3.10}$$

In particular, consider the case when the functions p and q are piecewise-constant on each of the intervals $x_k \le x \le x_{k+1}$. Then it is not difficult to obtain

$$a_k = \frac{p_{k-1/2}}{\Delta x_{k-1/2}} - \frac{\Delta x_{k-1/2}}{6} q_{k-1/2}; \qquad c_k = \frac{p_{k+1/2}}{\Delta x_{k+1/2}} - \frac{\Delta x_{k+1/2}}{6} q_{k+1/2};$$

$$b_k = \frac{p_{k+1/2}}{\Delta x_{k+1/2}} + \frac{p_{k-1/2}}{\Delta x_{k-1/2}} + \tfrac{1}{3}(\Delta x_{k+1/2} q_{k+1/2} + \Delta x_{k-1/2} q_{k-1/2}). \tag{3.11}$$

Here

$$p_{k+1/2} = p(x_{k+1/2}); \qquad q_{k+1/2} = q(x_{k+1/2}).$$

Assume that the solution of problem (3.9) has been found. Then the continuous solution $\phi^h(x)$ is restored with the help of relations (3.4). The significant property of the variational approach is the fact that we—a priori—construct the approximate solution by means of an interpolation polynomial, and the solution is found in the form of a continuous, piecewise-linear function, which is, by construction, the best approximation of the exact solution in the class of continuous, piecewise-linear functions.

Of course, the variational construction method can be used to obtain approximate solutions of any order of accuracy. For this, it is necessary to choose more accurate interpolation formulas than those from (3.4). Here on can formally exploit the interpolation formulas of Lagrange.

Next, let us turn to the analysis of approximating the difference equations obtained. For this we rewrite (3.8) as

$$(A^h \phi)_k = -\frac{1}{\Delta x_k} \left\{ p_{k+1/2} \frac{\phi_{k+1} - \phi_k}{\Delta x_{k+1/2}} - p_{k-1/2} \frac{\phi_k - \phi_{k-1}}{\Delta x_{k-1/2}} \right.$$

$$\left. - [q_{k+1/2}^{1,2} \phi_{k+1} + (q_{k+1/2}^{2,2} + q_{k-1/2}^{1,1})\phi_k + q_{k-1/2}^{1,2} \phi_{k-1}] \right\} = f_k^h \tag{3.12}$$

where

$$f_k^h = \frac{1}{\Delta x_k} F_k. \tag{3.13}$$

Here $\Delta x_k = (\Delta x_{k+1/2} + \Delta x_{k-1})/2$. Since the solution $\phi(x)$, and all the coefficients p, q, and f are smooth on the intervals $x_k \leq x \leq x_{k+1}$, we can show, using the Taylor series expansion, that

$$\|(A\phi)_h - A^h(\phi)_h\| \leq Mh^\alpha,$$

$$\|(f)_h - f^h\| \leq Nh^\alpha,$$

where the quantity α is determined by the properties of the net (see Section 2.2). Further analysis is identical to that of Section 2.2 and is therefore omitted.

2.3.2 Constructions of Simple Difference Schemes Based on the Galërkin (Finite Elements) Method

Consider problem (2.1), (2.2) on the interval $0 \leq x \leq 1$. Cover this region by the system of intervals $x_{k-1} \leq x \leq x_k$, and for each k introduce the following function from $\mathring{W}_2^1$:

$$\omega_k(x) = \begin{cases} 0, & 0 \leq x \leq x_{k-1}, \\ \omega_1(x), & x_{k-1} \leq x \leq x_k, \\ \omega_2(x), & x_k \leq x \leq x_{k+1}, \\ 0, & x_{k+1} \leq x \leq x_n = 1. \end{cases} \tag{3.14}$$

We can see (Figure 2.1) that the functions $\omega_k(x)$ $(k = 1, \ldots, n - 1)$ are defined on the whole interval $0 \leq x \leq 1$.

They are continuous and zero everywhere except for the interval $x_{k-1} \leq x \leq x_{k+1}$, on which they consist of two linear segments with the maximum attained at the point $x = x_k$. Such functions possess a sort of completeness property; namely, the system is complete in the sense that any continuous, piecewise-linear function $\phi(x)$ with possible corners at the net points x_k can be represented as a linear combination of these functions:

$$\phi(x) = \sum_k \phi_k \omega_k(x), \tag{3.15}$$

where the "Fourier coefficients" ϕ_k are actually the values of ϕ at the points x_k.

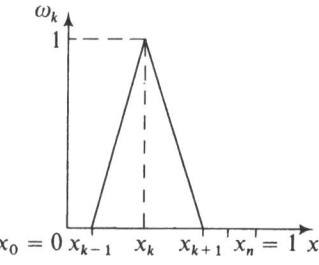

Figure 2.1

Let us also note that the functions $\omega_k(x)$ have certain orthogonality properties, though not in the usual sense. Indeed, consider the scalar product

$$(g, h) = \int_0^1 gh \, dx,$$

then for each $\omega_k(x)$

$$\int_0^1 \omega_k(x)\omega_n(x) \, dx = \begin{cases} 0, & n \le k - 2, \\ \frac{1}{6}\Delta x_{k-1/2}, & n = k - 1, \\ \frac{1}{3}(\Delta x_{k-1/2} + \Delta x_{k+1/2}), & n = k, \\ \frac{1}{6}\Delta x_{k+1/2}, & n = k + 1, \\ 0, & n \ge k + 2. \end{cases} \tag{3.16}$$

From (3.16) we have that $\omega_k(x)$ are orthogonal to all $\omega_n(x)$ except ω_{k-1}, ω_k, and ω_{k+1}. This is a specific feature of the basis chosen. Having defined the functions $\omega_k(x)$, we are now going to exploit them to obtain the equations in finite differences. To this end, consider again the diffusion equation

$$-\frac{d}{dx} p \frac{d\phi}{dx} + q\phi = f \quad \text{in} \quad D,$$

$$\phi(0) = 0, \qquad \phi(1) = 0. \tag{3.17}$$

With reference to the Galërkin method, let us take the scalar product of ω_k and equation (3.17). We obtain

$$\int_0^1 \left(-\frac{d}{dx} p \frac{d\phi}{dx} + q\phi - f \right)\omega_k(x) \, dx = 0. \tag{3.18}$$

Rewrite the functional in this equation to get

$$\int_0^1 \left[p \frac{d\phi}{dx} \cdot \frac{d\omega_k}{dx} + (q\phi - f)\omega_k \right] dx = 0. \tag{3.19}$$

(We have used integration by parts and the fact that $\omega_k(0) = \omega_k(1) = 0$.) Equation (3.19) can be further rewritten as

$$\sum_k \int_{x_{k-1}}^{x_{k+1}} \left[p \frac{d\phi}{dx} \cdot \frac{d\omega_k}{dx} + (q\phi - f)\omega_k \right] dx = 0. \tag{3.20}$$

Taking into account the form of the function ω_k, it is not difficult to obtain

$$\int_{x_{k-1}}^{x_k} p \frac{d\phi}{dx} \cdot \frac{d\omega_k}{dx} \, dx = \frac{p_{k-1/2}}{\Delta x_{k-1/2}} (\phi_k - \phi_{k-1})$$

$$\int_{x_k}^{x_{k+1}} p \frac{d\phi}{dx} \cdot \frac{d\omega_k}{dx} \, dx = -\frac{p_{k+1/2}}{\Delta x_{k+1/2}} (\phi_{k+1} - \phi_k), \tag{3.21}$$

and

$$\int_{x_{k-1}}^{x_k} q\phi\omega_k \, dx = q_{k-1/2}^{1,2} \phi_{k-1} + q_{k-1/2}^{1,1} \phi_k,$$

$$\int_{x_k}^{x_{k+1}} q\phi\omega_k \, dx = q_{k+1/2}^{2,2} \phi_k + q_{k+1/2}^{1,2} \phi_{k+1}.$$
(3.22)

The notation here corresponds to that of (3.8).

Substitute next (3.21) and (3.22) into (3.19). There results

$$\frac{p_{k-1/2}}{\Delta x_{k-1/2}} (\phi_k - \phi_{k-1}) - \frac{p_{k+1/2}}{\Delta x_{k+1/2}} (\phi_{k+1} - \phi_k) + q_{k-1/2}^{1,2} \phi_{k-1}$$

$$+ (q_{k-1/2}^{1,1} + q_{k+1/2}^{2,2})\phi_k + q_{k+1/2}^{1,2} \phi_{k+1} = F_k,$$
(3.23)

where

$$F_k = \int_{x_{k-1}}^{x_k} f\omega_1 \, dx + \int_{x_k}^{x_{k+1}} f\omega_2 \, dx = \int_{x_{k-1}}^{x_{k+1}} f\omega_k \, dx.$$
(3.24)

In order to have a complete picture, one only has to add the boundary conditions

$$\phi_0 = 0, \qquad \phi_n = 0.$$
(3.25)

The formulation is complete.

An examination of the difference equations (3.23) and (3.8) shows that they are identical. In other words, the Ritz and the Galërkin methods on the basis of finite functions give the same difference analogs for the self-adjoint problems. We point out, however, that the Galërkin method can be applied regardless of whether the problem is self-adjoint or not; thus, it has a broader range of applicability.

If the equations have smooth coefficients, the variational principle yields difference approximations with higher-order accuracy.

2.4 Principles for the Construction of Subspaces for the Solution of One-Dimensional Problems by Variational Methods

This section is primarily intended for the benefit of those who wish to get a deeper insight into some additional possibilities in the variational approach to solving problems of mathematical physics. To help the exposition we bring in some additional tools from functional analysis.

We begin with a more general approach to the construction of net equations by variational methods, which will allow us to obtain difference approximations of a high order of accuracy. Then we develop a method for constructing a basis made up of trigonometric functions, for the solution of problems with discontinuous piecewise-smooth parameters.

2.4.1 A General Approach to the Construction of Subspaces of Piecewise-Polynomial Functions

Subspaces consisting of piecewise-polynomial functions of high order are very effectively used in variational methods for solving the one-dimensional diffusion equation

$$Lu \equiv -\frac{d}{dx} p(x) \frac{du}{dx} + r(x) \frac{du}{dx} + q(x)u = f, \quad a < x < b, \qquad (4.1)$$

with the boundary conditions

$$u(0) = u(1) = 0. \qquad (4.2)$$

Such subspaces have been considered by Aubin [5], Babuška [5], Strang and Fix [5], Bramble and Schatz [5]. We shall consider one process for constructing a subspace of this kind (cf. Varga [1], Strang and Fix [5]).

We begin by introducing $m = (p + 1)/2$ basis functions $\phi_i(x)$, where p is an odd positive integer. We require that each of the functions vanishes outside the interval $-1 \leq x \leq 1$ and is a polynomial of degree p in each of the intervals $[-1, 0]$ and $[0, 1]$. These polynomials are defined by prescribing m constraints at the endpoints of the given intervals: at the points $x = -1$ and $x = 1$, the function $\phi_i(x)$ and all its derivatives of order less than m shall vanish, and at $x = 0$ the only nonzero derivative shall be

$$\left. \frac{d^{i-1}\phi_i(x)}{dx^{i-1}} \right|_{x=0} = 1, \quad 1 \leq i \leq m. \qquad (4.3)$$

We assume that the functions $\{\phi_i(x)\}_{i=1}^m$ have been determined from these conditions. We now introduce on the interval $[a, b]$ a uniform net with mesh size h and net points at $a = x_0 < x_1 = a + h < \cdots < a + Nh = b$, where $h = (b - a)/N$ and N is some positive integer. We form the set of functions

$$u^h(x) = \sum_{i=1}^m \sum_{j=0}^N u_{ij}^h \phi_{ij}^h(x), \qquad (4.4)$$

where

$$\phi_{ij}^h(x) = \phi_i\left(\frac{x - a}{h} - j\right), \quad 0 \leq j \leq N; 1 \leq i \leq m. \qquad (4.5)$$

We refer to this set of functions as $F_h \equiv H_h^{(m)}$; its basis is the system $\{\phi_{ij}^h\}_{i=1,m}^{j=0,N}$.

Note that (4.3) yields a useful property of interpolations using functions from $H_h^{(m)}$: we find from (4.4) that at the net points the values of $u^h(x)$ and its first $m - 1$ derivatives are defined by the u_{ij}^h:

$$\frac{d^{i-1}u^h(a + jh)}{dx^{i-1}} = u_{ij}^h. \tag{4.6}$$

Let us consider various examples of these subspaces.

1. $p = 1 \ (m = 1)$.

Here we have one basis function $\phi_1(x)$, a first-degree polynomial on $[-1, 0]$ and $[0, 1]$:

$$\phi_1(x) = \begin{cases} a + bx, & -1 \le x \le 0, \\ c + dx, & 0 \le x \le 1, \end{cases} \tag{4.7}$$

with the defining constraints

$$a + bx \Big|_{x=-1} = 0, \qquad c + dx \Big|_{x=1} = 0,$$

$$a + bx \Big|_{x=0} = 1, \qquad c + dx \Big|_{x=0} = 1. \tag{4.8}$$

These imply that $a = b = 1, c = -d = 1$, so that $\phi_1(x)$ is the standard roof-function (See Figure 2.1), and $H_h^{(1)}$ is the space of piecewise-linear functions that we have already studied in Section 2.3.2.

2. $p = 3 \ (m = 2)$.

In this case we have two basis functions $\phi_1(x)$ and $\phi_2(x)$. First, we construct $\phi_1(x)$. It is represented on $[0, 1]$ as

$$\phi_1(x) = a_0 + a_1x + a_2x^2 + a_3x^3. \tag{4.9}$$

The unknowns $a_0, \ldots, a_3$ are defined by the constraints

$$\begin{aligned} \phi_1(0) &= 1, & \phi_1(1) &= 0, \\ \phi_1'(0) &= 0, & \phi_1'(1) &= 0, \end{aligned} \tag{4.10}$$

which give us

$$a_0 = 1, \qquad a_1 = 0, \qquad a_2 = -3, \qquad a_3 = 2. \tag{4.11}$$

Therefore,

$$\phi_1(x) = \begin{cases} 0, & \text{for } |x| \ge 1 \\ 1 - 3x^2 + 2x^3 = (1 - x)^2(1 + 2x), & 0 \le x \le 1, \\ \phi_1(-x), & \text{for } -1 \le x \le 0. \end{cases} \tag{4.12}$$

(See Figure 2.2.)

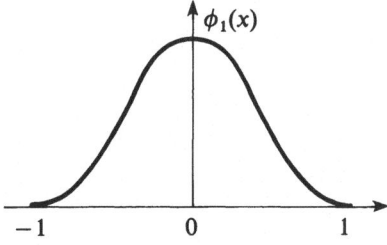

Figure 2.2

We construct $\phi_2(x)$ in the same way. On $[0, 1]$ it is written as

$$\phi_2(x) = b_0 + b_1 x + b_2 x^2 + b_3 x^3.$$

The coefficients b_i ($0 \le i \le 3$) are determined by the constraints

$$\begin{cases} \phi_2(0) = 0, & \phi_2(1) = 0, \\ \phi_2'(0) = 1, & \phi_2'(1) = 0. \end{cases} \tag{4.13}$$

These yield

$$\phi_2(x) = \begin{cases} 0, & \text{for} \quad |x| > 1, \\ (1 - x)^2 x, & 0 \le x \le 1. \\ -\phi_2(-x), & -1 \le x \le 0 \end{cases} \tag{4.14}$$

(See Figure 2.3.)
The functions in $H_h^{(2)}$ have the form

$$u^h(x) = \sum_{j=0}^{N} (a_j \phi_{1, j}^h(x) + b_j \phi_{2, j}^h(x)). \tag{4.15}$$

Using these, we can easily define a quasi-interpolant u_I^h, for instance for the solution of (4.1), (4.2), by writing

$$u_I^h(x) = \sum_{j=0}^{N} (u(a + jh)\phi_{1, j}^h(x) + u'(a + jh)\phi_{2, j}^h(x)). \tag{4.16}$$

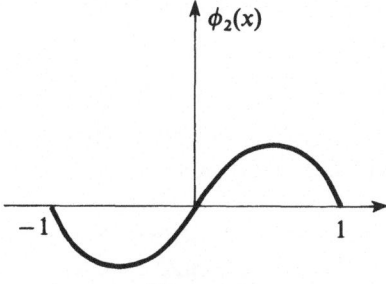

Figure 2.3

This again gives us an explicit representation for the physical meaning of the coefficient values in the functions $\phi_{1,j}^h(x)$ and $\phi_{2,j}^h(x)$ entering the approximation $u^h(x)$ constructed by the Galërkin method with the basis $\{\phi_{i,j}^h\}_{i=1,2}^{j=0,N}$.

We will now describe yet another method of constructing the space F_h. As before, let us take the interval $[a, b]$ on the real line and cover it by the net such that its net points $a = x_0 < x_1 < \cdots < x_N < x_{N+1} = b$ satisfy $x_k = a + kh\,(k = 0, 1, \ldots, N + 1), h = (b - a)/(N + 1)$. Denote by $M_N^m(a, b)$ a set of functions $g(x)$ satisfying the following two properties. First $g(x)$ is an mth-order polynomial on each of the intervals $[x_k, x_{k+1}]$; second, for any $0 \le k \le N$ and $0 \le j \le m$ we have

$$g(x_{k,j}) = d_{k,j},$$

where $x_{k,j} = x_k + (h/m)_j$ and $d_{k,j}$ are given numbers; in addition $g(a) = g(0) = 0$, that is, $d_{0,0} = d_{N,m} = 0$. Hence it follows that the function $g \in M_N^m(a, b)$ is a piecewise-polynomial function belonging to $\mathring{W}_2^1$; in other words, the function $g(x)$ is continuous, with possible discontinuities in the first derivative at the points $\{x_k\}_{k=1}^N$. Let us now take a look at the explicit construction of $g(x)$, given $\{d_{k,j}\}$.

Let us choose an arbitrary $k\,(0 \le k \le N)$, and construct the function $g(x)$ on $[x_k, x_{k+1}]$ (denote it by $g_k(x)$). By a well-known result from the theory of approximations, there exists a unique mth order polynomial, passing through the $(m + 1)$ points $d_{k,0}, \ldots, d_{k,m}$. It coincides with the Lagrange polynomial

$$g_k(x) = \sum_{i=0}^m d_{k,i} \prod_{\substack{l=0 \\ l \ne i}}^m \frac{(x_{k,l} - x)}{(x_{k,l} - x_{k,i})}. \tag{4.17}$$

For the case $m = 1$ we obtain a usual linear function

$$g_k(x) = d_k \frac{(x_{k+1} - x)}{(x_{k+1} - x_k)} + d_{k+1} \frac{(x_k - x)}{(x_k - x_{k+1})}. \tag{4.18}$$

It follows from this, in particular, that the space $M_N^1(a, b)$ is the same as $H_N^{(1)}(a, b)$, and therefore the basis obtained for $H_N^{(1)}$ can be used for $M_N^1(a, b)$. We will not consider the problem of a basis for $M_N^m(a, b)$ for large m.

Thus, we have constructed two kinds of sets of subspaces F_h (defined by the parameter h) for the space $F = \mathring{W}_2^1(a, b)$, each sequence being complete in $\mathring{W}_2^1(a, b)$.

2.4.2 Constructing a Basis Using Trigonometric Functions and Applying It in Variational Methods

We shall explain the fundamental principles by a concrete example, namely the problem (2.1), (2.2) of Section 2.2, which reduces to the minimization of the functional (3.1).

We shall allow the parameters p, q, and f to have discontinuities of the first kind at the points $\{y_i\}_{i=1}^m$ and we shall suppose that on the intervals $y_i < x < y_{i+1}$ they are sufficiently smooth. Let the smoothness of the parameters determine the smoothness of the solution ϕ to the vth order on the above intervals or—what amounts to the same thing—$\phi \in Q^{(v)}[0, 1]$. (See Section 2.2.)

If v is large enough, it is natural for us to construct the basis $\{\omega_k\}_{k=1}^\infty$ by using the good approximation properties of the trigonometric polynomials having periods larger than the corresponding intervals of smoothness. In more detail, this assertion is as follows (cf. Smelov [5]).

If the function $f(x) \in C^{(v)}[0, 1]$ is defined on the interval $0 \le x \le 1$, it may always be extended (not uniquely) to a function on the entire real axis, in such a way that the extended function $\tilde{f}(x)$ will have the following important properties (Fichtenholtz [2]):

$$\tilde{f}(x + T) = \tilde{f}(x), \qquad -\infty < x < \infty,$$

$$\tilde{f} \in C^{(v)}[x_0, x_0 + T] \quad \text{for every} \quad x_0 \in (-\infty, \infty),$$

where $T > 1$. The rate of convergence of the Fourier series for the function $\tilde{f}(x)$ (in $C[0, 1]$ and in $L_2[0, 1]$) is subject to the estimate

$$\|f - \tau_N\| \le \text{const} \frac{\ln N}{N^v}, \tag{4.19}$$

where

$$\tau_N = \sum_{k=0}^N \left(a_k \cos \frac{2\pi kx}{T} + b_k \sin \frac{2\pi kx}{T} \right).$$

Without worsening this estimate, we may adjust the coefficients a_k and b_k so that we have in addition $\tau_N(0) = f(0)$, $\tau_N(1) = f(1)$.

Returning to the problem (2.1), (2.2) and taking account of our assertion just now formulated, we approximate the solution, on each of the intervals of smoothness $y_i \le x \le y_{i+1}$, by a trigonometric polynomial

$$\phi_{N_i}(x) = \sum_{k=0}^{2N_i} C_k^i T_k^i(x), \qquad i = 0, 1, \ldots, m,$$

where, for instance,

$$T_{2k-1}^i(x) = \sin \frac{k\pi(x - y_i)}{t_i}, \qquad T_{2k}^i(x) = \cos \frac{k\pi(x - y_i)}{t_i}, \qquad t_i = y_{i+1} - y_i.$$

$$\tag{4.20}$$

Consider the system of functions

$$\omega_k^i(x) = \begin{cases} T_k^i(x), & \text{if} \quad x \in [y_i, y_{i+1}], \\ 0, & \text{outside this interval.} \end{cases} \tag{4.21}$$

We may regard the system of functions (4.21) ($i = 0, \ldots, m; k = 0, 1, \ldots$) as a sequence of basis elements (discontinuous for even values of k) and seek to approximate the solution of the problem (3.1) in the form

$$\phi(x) = \sum_{i=1}^{m} \sum_{k=0}^{2N_i} C_k^i \omega_k^i(x), \qquad 0 \le x \le 1. \tag{4.22}$$

Since the basis functions (4.21) are discontinuous the linear combination of them (4.22) does not automatically guarantee the continuity of ϕ and $p\, d\phi/dx$ at the points $\{y_i\}_{i=1}^{m}$ nor the fulfillment of the boundary conditions.†

The variational method based on the minimization of the functional (3.1) requires satisfaction of the forced conditions

$$\phi(0) = \phi(1) = 0, \qquad \phi(x_i - 0) = \phi(x_i + 0), \qquad i = 1, 2, \ldots, m$$

but not the natural ones

$$p\,\frac{d\phi}{dx}\bigg|_{x_i - 0} = p\,\frac{d\phi}{dx}\bigg|_{x_i + 0}, \qquad i = 1, 2, \ldots, m;$$

since the latter are automatically satisfied on the element $\phi^0 \in \Phi$ that solves the variational problem. Therefore, we must require the functions (4.22) to satisfy the forced conditions

$$\phi_{N_0}(0) = \phi_{N_m}(1) = 0, \qquad \phi_{N_{i-1}}(x_i) = \phi_{N_i}(x_i), \qquad i = 1, \ldots, m. \tag{4.23}$$

It is not hard to show that the distance between the approximate solution $\phi(x)$ and the exact solution $\phi^0(x)$ of the variational problem is subject to the following estimate:

$$\|\phi - \phi^0\|_{L_2[0, 1]} \le \text{const}\,\frac{\ln N}{N^{\nu - 1}}, \qquad N = \min_i N_i. \tag{4.24}$$

We shall consider the exact solution $\phi^0(x)$ only on the interval of smoothness $y_i \le x \le y_{i+1}$. Outside the interval $[y_i, y_{i+1}]$, we extend the function $\phi^0(x)$ to a function $\psi^i(x)$ defined by the following conditions (cf. our prior extension of $f(x)$ to $\tilde{f}(x)$):

$$\psi^i(x) \equiv \phi^0(x) \quad \text{for} \quad y_i \le x \le y_{i+1},$$

$$\psi^i \in C^{(\nu)}[x_0, x_0 + T_i] \quad \text{for all} \quad x_0 \in (-\infty, \infty),$$

$$\psi^i(x + T_i) = \psi^i(x), \qquad -\infty < x < \infty,$$

where $T_i = 2(y_{i+1} - y_i)$.

---

† The boundary conditions are easily satisfied if on the intervals $[0, x_1]$ and $[x_m, 1]$ we use the definining equations

$$\omega_k^0(x) = \sin\frac{k\pi x}{2x_1}, \qquad \omega_k^m(x) = \sin\frac{k\pi(1 - x)}{2t_m}.$$

instead of (4.20).

We approximate each of the functions $\psi^i(x)$ $(i = 0, 1, \ldots, m)$ by a finite segment of its Fourier expansion

$$\sum_{k=0}^{2N_i} a_k^i T_k^i(x).$$

Noting that for $v \geq 2$ the Fourier expansion of $\psi^i(x)$ admits of term-by-term differentiation, we may write the following estimates:

$$\|R_i\| \leq \text{const} \frac{\ln N_i}{N_i^v}, \qquad \left\|\frac{dR_i}{dx}\right\| \leq \text{const} \frac{\ln N_i}{N_i^{v-1}},$$

where

$$R_i(x) = \psi^i(x) - \sum_{k=0}^{2N_i} a_k^i T_k^i(x),$$

and the norm $\|\cdot\|$ may be taken as either $\|\cdot\|_C$ or $\|\cdot\|_{L_2}$. Without worsening the estimates, we may adjust the coefficients a_k^i so that $R_i(y_i) = R_i(y_{i+1}) = 0$. Assuming that such a correction has been made, we construct the following continuous function, which satisfies the boundary conditions (2.2):

$$\psi(x) = \sum_{i=0}^{m} \sum_{k=0}^{2N_i} a_k^i \omega_k^i(x), \qquad 0 \leq x \leq 1.$$

Because the operator A in the problem (2.1), (2.2) is positive definite we have for all functions $u(x) \in \Phi(A)$

$$(Au, u) \geq \gamma \|u\|_{L_2}, \qquad \gamma > 0$$

If we denote by V the finite-dimensional space of continuous functions of the form (4.22), satisfying the zero boundary conditions, we shall have

$$\|\phi^0 - \phi\|_{L_2}^2 \leq \frac{1}{\gamma}(A(\phi^0 - \phi), \phi^0 - \phi) = \frac{1}{\gamma} \min_{u \in v}(A(\phi^0 - u), \phi^0 - u)$$

$$\leq \frac{1}{\gamma}(A(\phi^0 - \psi), \phi^0 - \psi) = \frac{1}{\gamma} \int_0^1 \left[p(x)\left(\frac{dR}{dx}\right)^2 + q(x)R^2(x)\right] dx,$$

where $R(x) = \phi^0(x) - \psi(x)$.

The estimates we derived earlier for R_i and dR_i/dx allows us to estimate $R(x)$ and its derivative:

$$\|R\| \leq \text{const} \frac{\ln N}{N^v}, \qquad \left\|\frac{dR}{dx}\right\| \leq \text{const} \frac{\ln N}{N^{v-1}}, \qquad N = \min_i N_i.$$

Finally, we obtain the estimate (4.24).

Let us briefly describe the algorithm for numerical implementation.

We apply the method of Lagrange multipliers to the minimization of the functional (3.1) over the functions (4.22) under the supplementary conditions (4.23) and find the following system of equations:

$$\sum_{k=0}^{2N_i} \alpha_k^i (A\omega_k^i, \omega_j^i) + \sum_{s=i}^{i+1} \lambda_s \beta_{js}^i = (f, \omega_j^i), \qquad i = 0,1,\ldots,m; j = 0, 1,\ldots, 2N_i,$$

$$(4.25)$$

$$\sum_{i=s-1}^{s} \sum_{k=0}^{2N_i} \alpha_k^i \beta_{ks}^i = 0, \qquad s = 0, 1,\ldots, m + 1,$$

where

$$\beta_{ks}^i = \begin{cases} \omega_k^{s-1}(x_s), & i = s - 1, \\ -\omega_k^s(x_s), & i = s; \omega_k^{-1}(x) \equiv \omega_k^{m+1}(x) \equiv 0, \\ 0, & i = s; s - 1. \end{cases} \qquad (4.26)$$

The system (4.25) has a block structure

$$\left\| \begin{matrix} \hat{A} & B^T \\ B & 0 \end{matrix} \right\| \left\| \begin{matrix} X \\ \Lambda \end{matrix} \right\| = \left\| \begin{matrix} F \\ 0 \end{matrix} \right\|, \qquad (4.27)$$

where X is a vector incorporating the set of all α_k^i; Λ is a vector whose components are the Lagrange multipliers; and T is the transposition symbol.

From (4.27) we have

$$\hat{A}X + B^T\Lambda = F, \qquad BX = 0, \qquad (4.28)$$

whence, after eliminating X, we have formally

$$(B\hat{A}^{-1}B^T)\Lambda = B\hat{A}^{-1}F. \qquad (4.29)$$

Since $\hat{A}$ is positive definite its inverse $\hat{A}^{-1}$ exists and $B\hat{A}^{-1}B^T$ is positive definite. In fact,

$$(B\hat{A}^{-1}B^TW, W) = (\hat{A}^{-1}B^TW, B^TW) = (\hat{A}V, V) > 0,$$

where $V = \hat{A}^{-1}B^TW \neq 0$ for $W \neq 0$ (the vanishing of W for $B^TW = 0$ follows at once from (4.26) and (4.20)).

Using the block-diagonal structure of the matrix $\hat{A}$ we easily find that

$$B\hat{A}^{-1}B^T = \sum_{i=0}^{m} B_i A_i^{-1} B_i^T, \quad BA^{-1}F = \sum_{i=0}^{m} B_i A_i^{-1} F_i, \qquad (4.30)$$

where A_i and B_i are matrices with the elements $(A\omega_k^i, \omega_j^i)$ and (β_{ks}^i), respectively, and the F_i are vectors with the components $(f, \omega_0^i),\ldots,(f, \omega_{2N_i}^i)$.

As we see from (4.30) the inversion of the matrix $\hat{A}$ reduces to the inversion of its separate blocks, and this, as a rule, means the inversion of a matrix of not too high an order. Further, if m is not too large the order of the system (4.29) is small and the system, which has a positive definite symmetric matrix, is easily solved.

Finally, from the first equation in (4.28) we have

$$X_i = A_i^{-1}(F_i - B_i^T \Lambda), \tag{4.31}$$

where

$$X_i = (\alpha_0^i, \dots, \alpha_{2N_i}^i)^T, \qquad i = 0, 1, \dots, m.$$

Let us illustrate our current approach, using a typical diffusion problem in the theory of particle transport:

$$\left. \begin{array}{c} \dfrac{1}{r}\dfrac{d}{dr}rp\dfrac{d\phi}{dr} + q\phi = f, \\[2mm] \left.\dfrac{d\phi}{dr}\right|_{r=0} = \left.\dfrac{d\phi}{dr}\right|_{r=R} = 0, \end{array} \right\}$$

where $p(x)$, $q(x)$, and $f(x)$ are piecewise-constant nonnegative functions which we shall specify, as shown in Table 2.1, for the sake of concreteness.

Table 2.1

r	$p(r)$	$q(r)$	$f(r)$
$0 \le r < 12.7$	1.333	0.2	0
$12.7 \le r < 13$	0.3115	0.15	0
$13 \le r \le 15$	0.1282	0.015	1

In Table 2.2 we display the relative error of approximation in the form

$$E = \|\phi - \phi^0\|/\|\phi^0\|, \tag{4.32}$$

as a function of the number of basis functions, with the norm of the uniform metric, $\|\cdot\|_C$.

In Table 2.2, the n_i are the total numbers of basis functions in the ith interval of smoothness and $n = \sum_i n_i$.

Table 2.2 shows clearly the rapid decrease in E as the number of basis functions increases, testifying to the great potential of this approach in the implementation of the variational-difference methods for solving the problems of mathematical physics.

Table 2.2

n_1	n_2	n_3	n	E
2	3	3	8	0.138
3	5	5	13	0.141×10^{-1}
5	7	9	21	0.501×10^{-3}
6	9	11	26	0.782×10^{-4}

2.5 Variational-Difference Schemes for Two-Dimensional Equations of Elliptic Type

2.5.1 The Ritz Method

In this section we describe the way to obtain variational-difference schemes for (1.26), (1.27) under the following additional assumptions:

$$B_i(x) = 0 \quad \text{in} \quad D, \qquad i = 1, 2;$$

$$q(x) = 0 \quad \text{in} \quad D;$$

further, for any vector $\xi = (\xi_1, \xi_2)'$

$$\mu_0 \sum_{i=1}^{2} \xi_i^2 \leq \inf_{x \in D} \sum_{i,j=1}^{2} A_{ij}(x)\xi_i\xi_j \leq \sup_{x \in D} \sum_{i,j=1}^{2} A_{ij}(x)\xi_i\xi_j \leq \mu_1 \sum_{i=1}^{2} \xi_i^2 \qquad (5.1)$$

with some positive constants $\mu_0 \leq \mu_1$; the boundary ∂D of D is supposed to be piecewise-linear. In principle, the constraints on $q(x)$, μ_0, and μ_1 in the Ritz method could be weakened; for the sake of simplicity we will not do that. Our problem is thus to find the solution of the equation

$$-\sum_{i,j=1}^{2} \frac{\partial}{\partial x_i} A_{ij}(x) \frac{\partial u}{\partial x_j} = f \quad \text{in} \quad D \qquad (5.2)$$

with the boundary conditions

$$u = 0 \quad \text{on} \quad \partial D. \qquad (5.3)$$

As we have seen in Section 2.1 this is equivalent to finding a function which minimizes the quadratic functional

$$J(u) = \int_D \left[\sum_{i,j=1}^{2} A_{ij}(x) \frac{\partial u}{\partial x_i} \frac{\partial u}{\partial x_j} \right] dD - 2 \int_D uf \, dD \qquad (5.4)$$

over the space $\mathring{W}_2^1(D)$.

To solve the latter, we will use the Ritz method with a special kind of subspaces F_h. To this end, let us triangularize the region D, that is, cover it

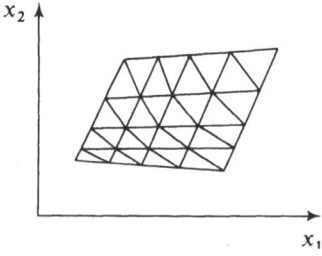

Figure 2.4

by a triangular net D_h (Figure 2.4). In each of the triangles we then consider an mth order polynomial in the variables x_1, x_2, of the following form:

$$g(x_1, x_2) = \sum_{i=0}^{m} \sum_{j_1 + j_2 = i} c_{j_1, j_2} x_1^{j_1} x_2^{j_2}. \tag{5.5}$$

In every triangle the coefficients of the polynomials are chosen in such a way, that the overall function becomes an element of $\mathring{W}_2^1(D)$, that is, the function must be continuous and zero on the boundary ∂D.

The method has been introduced by Courant [5] for the case $m = 1$, and also considered by Oganesyan [5] and others. The cases $m = 2, 3, 5$ have been solved in detail by Zlámal [5].

We will illustrate the method with the example $m = 1$, i.e., for a piecewise-linear function

$$g(x_1, x_2) = c_{0,0} + c_{1,0} x_1 + c_{0,1} x_2. \tag{5.6}$$

For a specific triangle the coefficients of this function are defined by means of given values $u(p_1), u(p_2)$, and $u(p_3)$ at the vertices p_1, p_2, and p_3 of the triangle. Performing this procedure in each triangle from the decomposition of the region D, while taking $u(p) = 0$ for $p \in \partial D$, we find that the resulting function is continuous in D and zero on the boundary ∂D.

For the quadratic interpolation $(m = 2)$

$$g(x_1, x_2) = c_{0,0} + c_{1,0} x_2 + c_{0,1} x_2 + c_{2,0} x_1^2 + c_{1,1} x_1 x_2 + c_{0,2} x_2^2, \tag{5.7}$$

a convenient way of making sure that the resulting function belongs to the space $\mathring{W}_2^1$ is as follows: the function u^h is specified at the triangle vertices p_1, p_2, and p_3, and at the points $p_{1,2}, p_{2,3}$, and $p_{3,1}$, which divide the segments $[p_1, p_2], [p_2, p_3]$, and $[p_3, p_1]$ in half (see Figure 2.5).

The continuity of u^h then follows from the simple fact that this function is uniquely defined on any triangle segment. For instance, $u^h(x_1, x_2)$ is uniquely defined on $[p_1, p_2]$ by specifying $u(p_1), u(p_{1,2})$, and $u(p_2)$.

If now the solution of the problem is sought in the form (5.7) in every triangular region Δ, then the difference equations are found in a routine way, using the Ritz method for the variational functional. It has been shown that if the solution $\hat{u}$ of equations (5.2) and (5.3) is from the class $C^{(3)}(D)$,

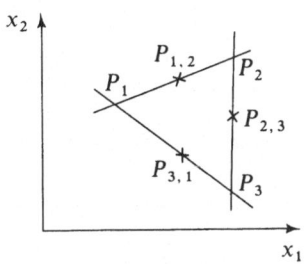

Figure 2.5

if the third derivatives of $\hat{u}(x)$ are bounded by M in modulus, and if in addition the minimal angle in the triangles which decompose the region D is bounded by a number $v_0 > 0$ from below, then the Ritz approximation error for the quadratic interpolation (5.7) can be estimated by

$$\|u^h - \hat{u}\|_{\mathring{W}_2^1} \leq Ch^2,$$

where $C = C_1 M/\sin v_0$ and C_1 are independent of the particular triangle (cf. Zlámal [5]).

If the smoothness assumptions on the solution of (5.2), (5.3) are taken in a weaker form ($\hat{u} \in C^{(2)}(D)$), then it is easy to prove that

$$\|u^h - \hat{u}\|_{\mathring{W}_2^1} \leq Ch, \qquad C > 0,$$

for the case of piecewise-linear approximations (cf. Oganesyan, Rivkind, and Rukhovets [5]).

This concludes the discussion of quadratic approximations on triangles, since what we have already said is sufficient for the derivation of complete algebraic systems; we will now concentrate on piecewise-linear approximations. In order to define uniquely a piecewise-linear function $g(x) \in \mathring{W}_2^1(D)$ in each triangle, it suffices (by what we have said earlier) to specify the values of $g(x)$ at the vertices of the triangles. For that, let us index all the inner vertices, using the notation $\{p_k\}_{k=1}^{N_h}$, and let us also denote by $D_{h,k}$ the union of all triangles with p_k the common vertex. Then the basis for the space F_h can be taken as the system of functions $\{\omega_k(x)\}_{k=1}^{N_h}$ defined by

(1) $\omega_k(p_j) = \delta_{k,j}$, where $\delta_{k,j}$ is the Kronecker symbol, and
(2) $\omega_k(x)$ is linear on each of the triangles, i.e., is represented by expression (5.6).

Thus ω_k can be viewed geometrically as a pyramid with the vertex at the point p_k and zero outside the region $D_{h,k}$ (Figure 2.6).

More specifically, let us assume that D is actually the unit square $\{x_1, x_2 : 0 < x_1, x_2 < 1\}$. Let us cover D with the usual uniform orthogonal net with the mesh size $h = 1/(N + 1)$, where N is a positive integer, and triangularize next the region D in the manner shown in Figure 2.7. The basis

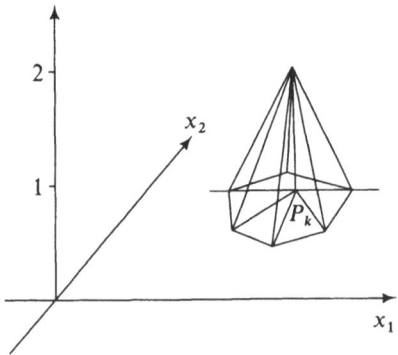

Figure 2.6

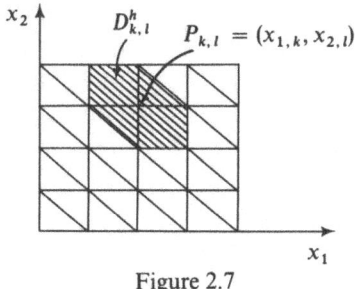

Figure 2.7

functions $\{\omega_k(x)\}_{k=1}^{N^2}$ will be denoted in this case by $\{\omega_{k,l}(x)\}_{k,l=1}^{N}$ (note $N_h = N^2$).

Consider the matrix A from the system of linear equations

$$A\alpha = g \qquad (5.8)$$

corresponding to the Ritz method. Here $\alpha = (\alpha_1, \ldots, \alpha_{N^2})'$ is a vector composed of the coefficients $\{\alpha_{N(k-1)+l} = \alpha_{k,l}\}_{k,l=1}^{N}$ from the decomposition

$$u^h(x) = \sum_{k,l=1}^{N} \alpha_{k,l} \omega_{k,l}(x), \qquad (5.9)$$

$g = (g_1, \ldots, g_{N^2})'$ is a vector with the components

$$g_{N(k-1)+l} = g_{k,l} = \int_{D_{k,l}} f\omega_{k,l}(x)\, dD, \qquad k,l = 1, \ldots, N, \qquad (5.10)$$

and for the elements of A we have the formulas

$$a_{N(k-1)+l, N(i-1)+j} = \int_D \sum_{s,t=1}^{2} A_{s,t}(x) \frac{\partial \omega_{k,l}}{\partial x_s} \frac{\partial \omega_{i,j}}{\partial x_t}\, dD, \qquad k,l,i,j = 1, \ldots, N. \qquad (5.11)$$

Let us use the notation $a_{k,l}^{i,j} = a_{N(k-1)+l, N(i-1)+j}$.

Considering the form of the functions $\{\omega_{k,l}(x)\}_{k,l=1}^{N}$ (see Figures 2.6 and 2.7), it is not difficult to show that

$$a_{k,l}^{i,j} = 0$$

provided at least one of the two following inequalities holds true:

$$|i - k| > 1, \qquad |j - l| > 1, \qquad k,l,i,j = 1, \ldots, N.$$

From this, one can show directly that A is a three-diagonal, block matrix of the form

$$A = \begin{Vmatrix} A_{11} & A_{12} & 0 & \cdots & 0 & 0 \\ A_{21} & A_{22} & A_{23} & \cdots & 0 & 0 \\ \hdotsfor{6} \\ 0 & 0 & 0 & \cdots & A_{N,N-1} & A_{NN} \end{Vmatrix}, \qquad (5.12)$$

where $A_{kk} = A_{k,k}^*$, $A_{k,k+1} = A_{k+1,k}^*$ $(k = 1, \ldots, N)$, and each of the matrices $A_{k,l}$ is itself a three-diagonal matrix of order N. Further analysis shows that the matrices $\{A_{k,k-1}\}_{k=2}^N$ are in fact two-diagonal matrices:

$$A_{k,k-1} = \begin{Vmatrix} a_{k,1}^{k-1,1} & a_{k,1}^{k-1,2} & 0 & \cdots & 0 & 0 \\ 0 & a_{k,2}^{k-1,2} & a_{k,2}^{k-1,3} & \cdots & 0 & 0 \\ \cdots\cdots\cdots\cdots\cdots\cdots\cdots\cdots\cdots\cdots \\ 0 & 0 & 0 & \cdots & 0 & a_{k,N}^{k-1,N} \end{Vmatrix}, \quad k = 2,\ldots,N.$$

$$(5.13)$$

Let us compute the elements $\{a_{k,l}^{i,j}\}$ of the matrix A for the particular case of (5.2), (5.3):

$$-\frac{\partial}{\partial x} p(x,y) \frac{\partial u}{\partial x} - \frac{\partial}{\partial y} q(x,y) \frac{\partial u}{\partial y} = f \quad \text{in} \quad D, \qquad u = 0 \quad \text{on} \quad \partial D. \quad (5.14)$$

In order to do that, we first represent $D_{k,l}^h$ as the union of the six triangles $\{D_{k,l,m}^h\}_{m=1}^6$, the indexing of which is clear from Figure 2.8. By a straightforward computation we obtain

$$\omega_{k,l}(x,y) = \begin{cases} 1 - \dfrac{1}{h}(x_k - x) - \dfrac{1}{h}(y_l - y), & \text{if} \quad x, y \in D_{k,l,1}^h, \\[2mm] 1 - \dfrac{1}{h}(x_k - x), & \text{if} \quad x, y \in D_{k,l,2}^h, \\[2mm] 1 + \dfrac{1}{h}(y_l - y), & \text{if} \quad x, y \in D_{k,l,3}^h, \\[2mm] 1 + \dfrac{1}{h}(x_k - x) + \dfrac{1}{h}(y_l - y), & \text{if} \quad x, y \in D_{k,l,4}^h, \\[2mm] 1 + \dfrac{1}{h}(x_k - x), & \text{if} \quad x, y \in D_{k,l,5}^h, \\[2mm] 1 - \dfrac{1}{h}(y_l - y), & \text{if} \quad x, y \in D_{k,l,6}^h. \end{cases} \quad (5.15)$$

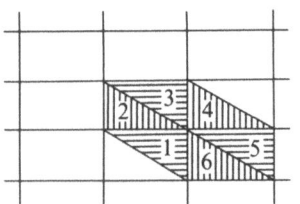

Figure 2.8

Since the matrix A is symmetric, and since the matrices $\{A_{k,k}\}_{k=1}^{N}$ are three-diagonal and the matrices $\{A_{k,k-1}\}_{k=2}^{N}$, $\{A_{k,k+1}\}_{k=1}^{N-1}$ are two-diagonal, it is enough to compute the elements

$$a_{k,l}^{k,l}, \; a_{k,l}^{k,l-1}, \; a_{k,l}^{k-1,l}, \; a_{k,l}^{k-1,l+1}, \qquad 1 \le k, l \le N.$$

From (5.11) and (5.15) we have (for simplicity we put $D_i = D_{k,l,i}^h$)

$$a_{k,l}^{k,l} = \int_{D_{k,l}^h} \left[p(x,y) \left(\frac{\partial \omega_{k,l}}{\partial x} \right)^2 + q(x,y) \left(\frac{\partial \omega_{k,l}}{\partial y} \right)^2 \right] dx \, dy$$

$$= \frac{1}{h^2} \left[\int_{D_1 \cup D_2} p(x,y) \, dx \, dy + \int_{D_4 \cup D_5} p(x,y) \, dx \, dy \right.$$

$$\left. + \int_{D_5 \cup D_4} q(x,y) \, dx \, dy + \int_{D_1 \cup D_6} q(x,y) \, dx \, dy \right];$$

$$a_{k,l}^{k,l-1} = \int_{D_{k,l}^h \cup D_{k,l-1}^h} \left[p(x,y) \frac{\partial \omega_{k,l}}{\partial x} \cdot \frac{\partial \omega_{k,l-1}}{\partial x} + q(x,y) \frac{\partial \omega_{k,l}}{\partial y} \frac{\partial \omega_{k,l-1}}{\partial y} \right] dx \, dy$$

$$= -\frac{1}{h^2} \left[\int_{D_1 \cup D_6} q(x,y) \, dx \, dy \right];$$

$$a_{k,l}^{k-1,l} = -\frac{1}{h^2} \left[\int_{D_1 \cup D_2} p(x,y) \, dx \, dy \right]; \qquad a_{k,l}^{k-1,l+1} = 0. \tag{5.16}$$

From this one can immediately show that the matrices $\{A_{k,k-1}\}_{k=2}^{N}$ are diagonal. Moreover, if we introduce a vector u with the components $u_{k,l} = \alpha_{k,l}$, where α is a vector from the system (5.8), then it is interesting to note that the system $Au = g$ has the representation

$$(A_1 + A_2)u = g, \tag{5.17}$$

where

$$(A_1 u)_{k,l} = -\tilde{P}_{k-1/2,l} u_{k-1,l} + (\tilde{P}_{k-1/2,l} + \tilde{P}_{k+1/2,l}) u_{k,l} - \tilde{P}_{k+1/2,l} u_{k+1,l} \tag{5.18}$$

and

$$(A_2 u)_{k,l} = -\tilde{Q}_{k,l-1/2} u_{k,l-1} + (\tilde{Q}_{k,l-1/2} + \tilde{Q}_{k,l+1/2}) u_{k,l} - \tilde{Q}_{k,l+1/2} u_{k,l+1}.$$

Here we have used the notation

$$\tilde{P}_{k \pm 1/2, l} = \frac{1}{h^2} \int_{D_{k,l}^h \cap D_{k \pm 1,l}^h} p(x,y) \, dx \, dy,$$

$$\tilde{Q}_{k,l \pm 1/2} = \frac{1}{h^2} \int_{D_{k,l}^h \cap D_{k,l \pm 1}^h} q(x,y) \, dx \, dy. \tag{5.19}$$

From (5.17)–(5.19) it is easy to see that the variational-difference scheme we have derived by the Ritz method is essentially identical to standard difference schemes, as far as the structure and distribution of the nonzero elements and their form is concerned. In particular, for constant $p(x, y)$ and $q(x, y)$, both difference- and variational-difference analogs of the differential operator are exactly the same (the analogs of the right-hand sides may differ). This fact allows one to solve system (5.17) by effective iteration methods, such as (among others) the splitting method and the sequential over-relaxation method.

2.5.2 The Galërkin Method

Since the basic features of solving two-dimensional problems have been already demonstrated with an example of the Ritz method, we will not dwell on the detailed constructions of the Galërkin variational-difference schemes. Let us only remark that for the nonzero coefficients $\{B_i(x)\}_{i=1}^2$ in (1.26), (1.27) the matrix $\tilde{A}$ of the Galërkin method with the basis functions (5.15) differs from the matrix A of the system of (5.17) by a certain matrix B ($B = \tilde{A} - A$). The elements $b_{k,l}^{i,j}$ of the latter matrix are given by

$$b_{k,l}^{i,j} = \int_{D_{k,l}^h \cap D_{i,j}^h} \left[B_1(x, y) \frac{\partial \omega_{ij}}{\partial x} \omega_{k,l} + B_2(x, y) \frac{\partial \omega_{ij}}{\partial y} \omega_{k,l} \right] dx \, dy$$

$$k, l, i, j = 1, \ldots, N. \qquad (5.20)$$

From this it can be seen that the matrices $\{A_{k,k-1}\}_{k=2}^N$ in the Galërkin method are not diagonal even for the equation

$$\sum_{i=1}^2 \left[-\frac{\partial^2 u}{\partial x_i^2} + B_i(x)u \right] = f \quad \text{in} \quad D, \quad u = 0 \quad \text{on} \quad \partial D. \qquad (5.21)$$

A more complicated matrix structure of the algebraic system calls for simpler subspaces F_h than those in the Ritz method. The same conclusion follows if we try to construct the Galërkin variational-difference schemes of more than first-order approximation.

Using the results from Section 2.4, the subspaces F_h can be constructed fairly easily for the regions which can be represented as finite unions of rectangles. Below we describe the structure of these spaces F_h with an example of a piecewise-linear approximation and construct a variational-difference scheme for a particular case.

Assume the region D is a union of r rectangles $\{D_i\}_{i=1}^r$, the sides of which are parallel to the coordinate lines. Let $\tilde{D}$ be the smallest rectangle containing the region D (see Figure 2.9, where $\tilde{D} = \{(x, y): a \leq x \leq b, c \leq y \leq d\}$. On the intervals $[a, b]$ and $[c, d]$ let us choose the nets $a = x_0 < x_1 < \cdots < x_{N+1} = b$ and $c = y_0 < y_1 < \cdots < y_{M+1} = d$, respectively, in such a way, that any boundary segment of any rectangle from the decomposition of D belongs to one of the lines

$$x \equiv x_k, \quad y \equiv y_l, \quad k = 0, 1, \ldots, N + 1; l = 0, 1, \ldots, M + 1. \quad (5.22)$$

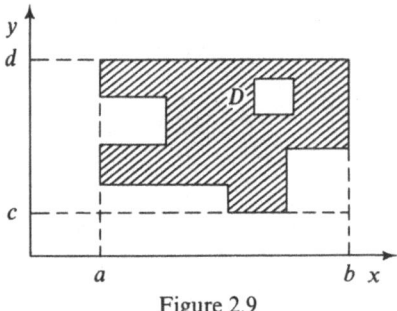

Figure 2.9

Finally, denote by D^h the net region consisting of the points (x_k, y_l) belonging to D $(k = 1, \ldots, N; l = 1, \ldots, M)$.

Let us now turn to constructing the subspace $F_h \in \mathring{W}_2^1(D)$. Introduce the functions

$$\omega_{x,k}(x) = \begin{cases} \dfrac{x - x_{k-1}}{x_k - x_{k-1}}, & \text{if } x \in [x_{k-1}, x_k], \\[2mm] \dfrac{x - x_{k+1}}{x_k - x_{k+1}}, & \text{if } x \in [x_k, x_{k+1}], \quad k = 1, \ldots, N, \\[2mm] 0, & \text{otherwise} \end{cases} \tag{5.23}$$

$$\omega_{y,l}(y) = \begin{cases} \dfrac{y - y_{l-1}}{y_l - y_{l-1}}, & \text{if } y \in [y_{l-1}, y_l], \\[2mm] \dfrac{y - y_{l+1}}{y_l - y_{l+1}}, & \text{if } y \in [y_l, y_{l+1}], \quad l = 1, \ldots, M, \tag{5.24} \\[2mm] 0 & \text{otherwise} \end{cases}$$

and the system of functions

$$\omega_{k,l}(x, y) = \omega_{x,k}(x)\omega_{y,l}(y), \qquad (x_k, y_l) \in D^h \tag{5.25}$$

Let us take for F_h the linear hull of the functions $(\omega_{k,l})$. Since the system $\{\omega_{k,l}\}$ is linearly independent, it can be clearly taken for the basis of the space F_h.

Next we will derive the linear equations corresponding to the Galërkin method in the subspace F_h with the basis (5.25) for the following simple problem:

$$-\frac{\partial^2 u}{\partial x^2} - \frac{\partial^2 u}{\partial y^2} + \frac{\partial u}{\partial x} = f \quad \text{in} \quad D \qquad u = 0 \quad \text{on} \quad \partial D. \tag{5.26}$$

In agreement with Section 2.1, if the approximation u^h is taken as

$$u^h = \sum_{(x_i, y_j) \in D^h} u_{i,j}\omega_{i,j}(x, y), \tag{5.27}$$

then the linear system is of the form

$$Au = g, \tag{5.28}$$

where the components of u and g are $\{u_{k,l}\}$ from (5.27), and $g_{k,l} = \int_D f\omega_{k,l}\,dD$, respectively; the elements of A are given by

$$a_{k,l}^{i,j} = \int_D \left[\frac{\partial\omega_{i,j}}{\partial x} \cdot \frac{\partial\omega_{k,l}}{\partial x} + \frac{\partial\omega_{i,j}}{\partial y} \cdot \frac{\partial\omega_{k,l}}{\partial y} + \frac{\partial\omega_{i,j}}{\partial x}\omega_{k,l}\right] dD$$

$$= \int_D \left[\omega_j\omega_l \frac{\partial\omega_i}{\partial x} \cdot \frac{\partial\omega_k}{\partial x} + \omega_i\omega_k \frac{\partial\omega_j}{\partial y} \cdot \frac{\partial\omega_l}{\partial y} + \omega_j \frac{\partial\omega_i}{\partial x}\omega_k\omega_l\right] dD, \quad (5.29)$$

where $\omega_i = \omega_i(x)$, $\omega_k = \omega_k(k)$, $\omega_j = \omega_j(y)$, and $\omega_l = \omega_l(y)$.

Similarly, as in the previous section, we can easily see that $a_{k,l}^{i,j} = 0$ if either one of the inequalities below is satisfied:

$$|i - k| > 1, \qquad |j - l| > 1.$$

Hence it follows that A is a three-diagonal block matrix of the form (5.12). We now give the final formulas for $a_{k,l}^{i,j}$, assuming for simplicity that the net is uniform with the mesh size h:

$$a_{k,l}^{k,l} = \frac{1}{h^2}\int_{x_{k-1}}^{x_{k+1}} dx \int_{y_{l-1}}^{y_{l+1}} dy[\omega_l^2(y) + \omega_k^2(x)] = \tfrac{8}{3},$$

$$a_{k,l}^{k,l-1} = \frac{1}{h^2}\int_{x_{k-1}}^{x_{k+1}} dx \int_{y_{l-1}}^{y_l} dy[\omega_{l-1}(y)\omega_l(y) - \omega_k^2(x)] = -\tfrac{1}{3}, \qquad (5.30a)$$

$$a_{k,l}^{k,l+1} = -\tfrac{1}{3},$$

$$a_{k,l}^{k-1,l} = \int_{x_{k-1}}^{x_k} dx \int_{y_{l-1}}^{y_{l+1}} dy\left[-\frac{\omega_l^2(y)}{h^2} + \frac{\omega_{k-1}(x)\omega_k(x)}{h^2} - \frac{1}{h}\omega_k(x)\omega_l^2(y)\right]$$

$$= -\frac{1}{3} - \frac{h}{3}, \qquad (5.30b)$$

$$a_{k,l}^{k-1,l-1} = \int_{x_{k-1}}^{x_k} dx \int_{y_{l-1}}^{y_l} dy\left[-\frac{1}{h^2}\omega_l(y)\omega_{l-1}(y)\right.$$

$$\left. - \frac{1}{h^2}\omega_k(x)\omega_{k-1}(x) - \omega_k(x)\omega_l(y)\omega_{l-1}(y)\right] = -\frac{1}{3} - \frac{h}{12},$$

$$a_{k,l}^{k-1,l+1} = \int_{x_{k-1}}^{x_k} dx \int_{y_l}^{y_{l+1}} dy\left[-\frac{\omega_{l+1}(y)\omega_l(y)}{h^2}\right.$$

$$\left. - \frac{\omega_{l+1}(x)\omega_{k+1}(x)}{h^2} - \frac{\omega_k(x)\omega_l(y)\omega_{l+1}(y)}{h}\right] = -\frac{1}{3} - \frac{h}{12},$$

$$\qquad (5.30c)$$

$$a_{k,l}^{k+1,l} = -\frac{1}{3} + \frac{h}{3}, \; a_{k,l}^{k+1,l-1} = -\frac{1}{3} + \frac{h}{12},$$

$$a_{k,l}^{k+1,l+1} = -\frac{1}{3} + \frac{h}{12}, \; (x_k, y_l) \in D^h.$$

Thus, it turns out that in the case $f(x, y) = f = \text{const}$ our variational-difference scheme is equivalent to a somewhat unusual difference scheme

$$\frac{8}{3} u_{k,l} - \frac{1}{3} u_{k,l-1} - \frac{1}{3} u_{k,l+1} - \left(\frac{1}{3} + \frac{h}{3}\right) u_{k-1,l} - \left(\frac{1}{3} + \frac{h}{12}\right) u_{k-1,l-1}$$

$$- \left(\frac{1}{3} + \frac{h}{12}\right) u_{k-1,l+1} - \left(\frac{1}{3} - \frac{h}{3}\right) u_{k+1,l} - \left(\frac{1}{3} - \frac{h}{12}\right) u_{k+1,l-1}$$

$$- \left(\frac{1}{3} - \frac{h}{12}\right) u_{k+1,l+1} = h^2 f. \tag{5.31}$$

The difference scheme can be easily obtained using three-point approximations of the second derivatives, provided the differential part in (5.26) is replaced in advance by the approximate expression on the net points D_h:

$$\frac{\partial^2 u}{\partial x^2} + \frac{\partial^2 u}{\partial y^2} - \frac{\partial u}{\partial x}\bigg|_{\substack{x=x_k \\ y=y_l}} \approx \frac{1}{6}\left[\sum_{i=k-1}^{k+1} \beta_{k-1} \left(\frac{\partial^2 u}{\partial y^2}\right)_{\substack{x=x_i \\ y=y_l}}\right.$$

$$\left. + \sum_{j=l-1}^{l+1} \beta_{l-j} \left(\frac{\partial^2 u}{\partial x^2} - \frac{\partial u}{\partial x}\right)_{\substack{x=x_k \\ y=y_j}}\right], \tag{5.32}$$

where

$$\beta_{-1} = \beta_1 = 1, \qquad \beta_0 = 4, \quad \text{and} \quad (x_k, y_l) \in D^h.$$

2.5.3 Methods for Constructing Subspaces

In constructing a variational-difference scheme of two-dimensional elliptic equations we need the two-dimensional subspaces F_h of $\overset{\circ}{W}{}^1_2(D)$. As we saw in Sections 2.5.1 and 2.5.2, the structure of the matrix of the variational-difference scheme depends essentially on the choice of the subspace

$$F_h \subset \overset{\circ}{W}{}^1_2(D)$$

and the method of triangulating the region D. In this section we shall consider a few very simple methods for triangulating a bounded two-dimensional region D with a smooth boundary S (cf. Oganesyan, Rivkind, and Rukhovets [5]).

We shall formulate the fundamental requirements to be satisfied by a net region D with a smooth boundary S which have no common inner points; the boundary S^h of D^h is piecewise-linear.

(1) There is a one-to-one mapping of the points of S and those of S^h, via the normals to S, and the distance between corresponding points does not exceed $\delta_1 h^2$, where $\delta_1 > 0$ and does not depend on h.

(2) The lengths of the sides and the areas of the triangles Δ_k that form the net region lie, respectively, within the limits $[l_1 h, l_2 h]$ and $[\gamma_1 h^2, \gamma_2 h^2]$, where the positive constants l_1, l_2, γ_1, and γ_2 do not depend on h.

It is easily seen that for a given region D various different net regions can be constructed, each satisfying the two conditions listed above; we may therefore impose additional constraints on the net region. We first recall that in the rectangular regions and in the regions composed of rectangles in Sections 2.5.1 and 2.5.2, the triangulation was carried out in a natural way on the basis of the rectangular nets. The problem of constructing net regions with a prescribed structure, having many domains in common with a triangulated rectangle, has been studied by Godunov and Prokopov [4]. With this in mind, we impose the following constraint on the structure of the net region D^h.

(3) There exists a continuous one-to-one mapping of the net region D^h on the region whose boundary consists of segments parallel to the coordinate axes, or forming a 45° angle with them. This mapping is linear on every triangle Δ_k and maps it into a right-angled triangle with legs of length h.

This condition implies that no more than eight triangles of the net region may have a common vertex.

One of the simplest methods for constructing such a triangulation is as follows (Matsokin [4]): we enclose the region D in a rectangle Π and cover Π with a square net having mesh size h. The vertices of this net, that lie near S, the boundary of D, are then shifted to S in such a way that the polygon S^h, of which they are the vertices, will approximate S satisfactorily, i.e., so that Condition (1) above is satisfied. Then we divide all the rectangles into triangles in such a way that Condition (2) is satisfied. Condition (3) is satisfied because of the way in which the triangulation was constructed. An example is shown in Figure 2.10.

It is not difficult to compute the actual values of the constants in Condition (2), which characterize the lengths of the sides and the areas of the triangles

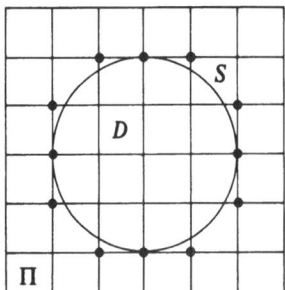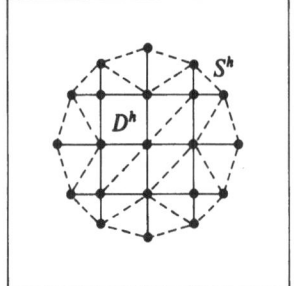

Figure 2.10

in the triangulation, independently of the configuration of the region D: $l_1 = 0.5$, $l_2 = \sqrt{18}/2$, $\gamma_1 = 0.125$, and $\gamma_2 = 1.125$. Of course, the constant δ_1 depends on the curvature of S.

The above process applies to multiply connected as well as singly connected domains, and is easily implemented.

For singly connected domains we may propose a process that constructs a net region which satisfies our three conditions and is mappable by a piecewise-linear transformation into a rectangular polygon (Matsokin [4], D'yakonov [5]).

For simplicity, we shall consider only the case when D is convex. The idea behind the construction process is simple. The region D is first mapped onto a rectangle by a continuous mapping with discontinuous first derivatives. Then the rectangle is triangulated in such a way that the sides of the triangles do not intersect the lines of discontinuity of the derivatives of the mapping. The inverse mapping yields a curvilinear trangulation of D which, on rectification, becomes the desired net region D^h.

The mapping onto the rectangle can be carried out quite simply: for instance, we may enclose D in a square and expand it along rays radiating from the intersection of the diagonals of the square. These diagonals will be lines of discontinuity of the derivatives of the mapping. In fact, this mapping onto the square is used only to derive the rule for constructing the net region. For this reason our current process is known as the method of fictitious mapping onto a rectangle. Figure 2.11 shows such a triangulation for an L-shaped region.

We shall now illustrate the usefulness of such net regions. Let F_h be a set of continuous functions, linear on every elementary triangle Δ_k of D^h and vanishing outside D^h. It is known that a system of variational-difference equations for the boundary problem

$$-\Delta u(x, y) = f(x, y), \qquad (x, y) \in D,$$

$$u(x, y) = 0, \qquad (x, y) \in S$$

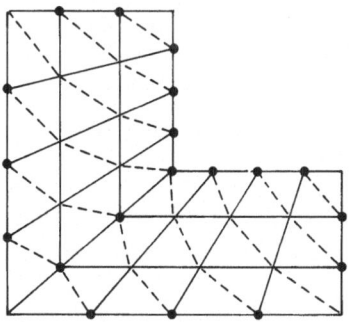

Figure 2.11

can be constructed by the Ritz method from the conditions for a minimum of the quadratic functional

$$J(v_h) = \int_{D^h} \left(\left| \frac{\partial v_h}{\partial x} \right|^2 + \left| \frac{\partial v_h}{\partial y} \right|^2 \right) dx \, dy - 2 \int_{D^h} f v_h \, dx \, dy$$

in the finite-dimensional space F_h.

Since the triangulation of the net region D^h is topologically equivalent to the simplest triangulation of a region consisting of rectangles, we may number the vertices of the triangles in accordance with the corresponding mapping and define the basis $\{\omega_{k,l}(x, y)\}$ in the space F_h, as we did in Section 2.5.2. Then the matrix structure of the system of algebraic linear equations obtained from the approximate variational equation is, in principle, no different from that of the matrix (5.8).

In conclusion, we note that the net region construction process described in this section can be generalized to three-dimensional regions (Matsokin [4]).

2.6 Variational Methods for Multi-Dimensional Problems

We will briefly discuss some possibilities for constructing the variational-difference schemes for problems with more than two independent variables.

2.6.1 Methods of Choosing the Subspaces

Consider a bounded region D in the space of the variables x, y, and z. Suppose the boundary ∂D is piecewise-linear. In this case there is a well-known method of constructing the subspaces $F_h \subset \mathring{W}_2^1(D)$; it is as follows: first, the region D is triangularized, i.e., D is covered by a nonintersecting finite system of three-edged pyramids Δ_k, such that $D = \bigcup_{k=1}^N \Delta_k$. Denote by h_k the length of the longest edge of the pyramid Δ_k. In analogy to the one dimensional and two-dimensional cases, we construct the sequence of spaces F_h of piecewise-polynomial functions, where the index $h = \max_{1 \le k \le N} h_k$. Let us illustrate this procedure with an example of piecewise-linear approximations and with the region $D = \{x, y, z : 0 \le x \le 1, 0 \le y \le 1, 0 \le z \le 1\}$.

Let us decompose the interval $[0, 1]$ by a uniform grid $0 = \xi_0 < \xi_1 < \cdots < \xi_{n+1} = 1$, $\xi_k = \Delta\xi \times k = k/(n + 1)$. Consider *elementary cubes*, with edges of equal length $\Delta\xi$, and with the vertices at the points $[x = \xi_k, y = \xi_l, z = \xi_m]$ $(k, l, m = 0, 1, \ldots, n + 1)$. This system of cubes covers D, and hence it is enough to triangularize the elementary cubes. There are many ways of doing this; one of them is shown in Figure 2.12.

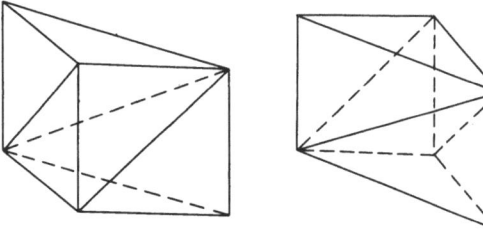

Figure 2.12

Here every elementary cube is taken as the union of six pyramids, the maximum edge length being $h = \sqrt{3}\Delta\xi = \sqrt{3}/(n + 1)$. Thus, the whole of D is covered by a total of $N = 6(n + 1)^3$ pyramids. The space F_h is now as follows: in each pyramid Δ_k consider a polynomial of the form

$$g_k(x, y, z) = \sum_{t=0}^{m} \sum_{i_1+i_2+i_3=t} C^k_{i_1, i_2, i_3} x^{i_1} y^{i_2} z^{i_3} \tag{6.1}$$

such that the function

$$g(x, y, z) = \begin{cases} g_1(x, y, z) & \text{if} \quad (x, y, z) \in \Delta_1, \\ \vdots \\ g_N(x, y, z), & \text{if} \quad (x, y, z) \in \Delta_N \end{cases} \tag{6.2}$$

belongs to the space $\mathring{W}^1_2(D)$; that is, its first derivatives are square integrable, the function itself being zero on the boundary. For this, it is enough that the piecewise-polynomial function g is continuous and is zero on ∂D.

In the case of piecewise-linear functions

$$g_k(x, y, z) = C^k_{0,0,0} + C^k_{1,0,0} x + C^k_{0,1,0} y + C^k_{0,0,1} z \tag{6.3}$$

the continuity of $g(x, y, z)$ is assured if, for instance, the coefficients of the functions g_k are specified by the function values of g at the vertices of the corresponding pyramid Δ_k. In other words, it is enough to require that g_k assumes strictly specified values at the vertices of the pyramid Δ_k.

The approach just described is sufficiently universal and can be applied to problems with more general boundaries. It leads, however, to very complex algorithms even for the simplest boundaries. Below we give another method of constructing the subspaces F_h which, in our opinion, is preferable for the multi-dimensional regions of a special kind.

Consider a region D in the p-dimensional Euclidean space, and assume D can be represented as a finite union of rectangles D_ν. Let $\tilde{D} = \{x_i : a_i \leq x_i \leq b_i, i = 1, \ldots, p\}$ be the smallest rectangle containing D (it is assumed that the edges of D_ν are parallel to the coordinate lines). For such i ($1 \leq i \leq p$), consider a subdivision of the interval $[a_i, b_i]$ of the x_i axis

$$a_i = x_{i,0} < x_{i,1} < \cdots < x_{i,N_i+1} = b_i, \qquad i = 1, \ldots, p,$$

and define the net D^h as the set of points $x_k = (x_{i,k_1}, x_{2,k_2}, \ldots, x_{p,k_p})$ belonging to D ($1 \leq k_i \leq N_i$ for $i = 1, \ldots, p$).

On each one-dimensional net introduce the family of one-dimensional basis functions

$$
g_{i,k_i}(x_i) = \begin{cases} \dfrac{x - x_{i,k_i-1}}{x_{i,k_i} - x_{i,k_i-1}}, & \text{if } x_i \in [x_{i,k_i-1}, x_{i,k_i}], \\[3mm] \dfrac{x - x_{i,k_i+1}}{x_{i,k_i} - x_{i,k_i+1}}, & \text{if } x_i \in [x_{i,k_i}, x_{i,k_i+1}], \\[3mm] 0, & \text{otherwise,} \end{cases} \tag{6.4}
$$

$$
k_i = 1, \ldots, N_i; \ 1 \le i \le p.
$$

Then the functions

$$
g_k(x) = \prod_{i=1}^{p} g_{i,k_i}(x), \qquad x_k \in D^h; \ k = (k_1, k_2, \ldots, k_p) \tag{6.5}
$$

can be taken for the basis of the space of piecewise-linear functions $F_h \subset \overset{\circ}{W}_2^1(D)$. The case when $p = 2$ was considered in detail in the previous section.

2.6.2 Coordinate-by-Coordinate Methods for Multi-Dimensional Problems

The approaches we have used above to construct the variational-difference schemes and the corresponding subspaces F_h become very complicated in practical implementations. In particular, this involves the choice of the basis functions (the first method in the previous section), the computation of the nonzero elements of the matrix characterizing the system, etc. These difficulties may be overcome by combining the variational approach and the ordinary difference approach into one single method. The system of net equations in this case will not be of the strictly difference or variational-difference types any more; in many cases, however, the mixture will allow one to relax the assumptions of one approach at the expense of borrowing from the other approach. Let us illustrate this with an example of a p-dimensional equation of elliptic type.

Consider the equation

$$
- \sum_{i=1}^{p} \frac{\partial}{\partial x_i} P_i(x) \frac{\partial u}{\partial x_i} = f \tag{6.6}
$$

in a bounded region D with the boundary ∂D. Suppose

$$
u = 0 \quad \text{on} \quad \partial D \tag{6.7}
$$

(the first boundary-value problem). If $p \geq 2$, and the region D is not one of those considered above or for which the sequence of subspaces $F_h \subset \mathring{W}_2^1(D)$ can be easily constructed, we propose the following method: it combines the difference and the variational approaches.

Define in the region D a rectangular net D^h by intersecting the region D with a set of hyperplanes parallel to the coordinate planes. Denote by $x_k = (x_{1,k_1}, x_{2,k_2}, \ldots, x_{p,k_p})$ the net points of D^h. The net equations are obtained from the approximations of the problem (6.6), (6.7) at every net point of D^h, while the approximations themselves are in turn derived with the help of the one-dimensional variational-difference schemes. The procedure in this case is as follows.

Assume we need to approximate problem (6.6), (6.7) at $x_k \in D^h$. The first thing to do is to rewrite the equation in the form

$$\sum_{i=1}^{p} \left[-\frac{\partial}{\partial x_i} P_i(x) \frac{\partial u}{\partial x_i} - f_i \right] = 0, \tag{6.8}$$

where $\sum_{i=1}^{p} f_i = f$. For each member of the sum (6.8) we then construct the variational difference scheme along the corresponding coordinate line x_i, parallel to the ith coordinate axis ($1 \leq i \leq p$). The next step is to add up the one-dimensional schemes. In other words, the net analog of (6.8) has the form

$$\sum_{i=1}^{p} (L_i^{(h)} u^h - f_i^h) = 0, \tag{6.9}$$

where $(L_i^{(h)} u^h - f_i^h)$ is the variational-difference analog of

$$-\frac{\partial}{\partial x_i} P_i(x) \frac{\partial u}{\partial x_i} - f_i$$

along segments parallel to the ith coordinate axis. Methods for constructing variational-difference schemes for one-dimensional operators were discussed in detail in Section 2.3.

Let us illustrate the proposed approach with an example of piecewise-linear approximations, when the region D is the p-dimensional unit box with a uniform net D_h ($h = 1/(N + 1)$). From Section 2.4 and equation (6.9) we have that the system of net equations for equation (6.8) is of the form

$$\frac{1}{h^2} \sum_{i=1}^{p} [-P_{k_i-1/2} u_{k_i-1}^h + (P_{k_i-1/2} + P_{k_i+1/2}) u_{k_i}^h - P_{k_i+1/2} u_{k_i+1}^h] - f_{k_i}^h = 0,$$

$$x_k \in D_h. \tag{6.10}$$

Here we have used the notation ($1 \leq i \leq p$):

$$u^h_{k_i+\alpha} = u^h(x_{k_1}, \ldots, x_{k_i-1}, x_{k_i+\alpha}, x_{k_i+1}, \ldots, x_{k_p}) \qquad \alpha = 0, \pm 1;$$

$$P_{k_i+\alpha} = \frac{1}{h} \int_{x_{k_i+\alpha-1/2}}^{x_{k_i+\alpha+1/2}} P(x_{k_i}, \ldots, x_{k_i-1}, x_i, x_{k_i+1}, \ldots, x_{k_p}) \, dx \qquad \alpha = \pm 1/2.$$

$$(6.11)$$

2.7 The Method of Fictive Domains

Consider the elliptic differential equation

$$Lu \equiv - \sum_{i,j=1}^{n} \frac{\partial}{\partial x_i} \left(a_{ij}(x) \frac{\partial u}{\partial x_j} \right) + c(x)u = f(x), \qquad x = (x_1, \ldots, x_n) \in D_1,$$

$$(7.1)$$

defined on a bounded domain D_1 with the (first) boundary condition

$$u(x) = 0, \qquad x \in \partial D_1. \tag{7.2}$$

We shall suppose that the coefficients and the solution of the problem (7.1), (7.2) are sufficiently smooth, that $a_{ij}(x) = a_{ji}(x)$, $c(x) \geq 0$, and that

$$\inf_{x \in D_1} \sum_{i,j=1}^{n} a_{ij}(x)\xi_i\xi_j \geq \mu \sum_{i=1}^{n} \xi_i^2 \tag{7.3}$$

with a positive constant μ independent of the arbitrary vector $\xi = (\xi_1, \ldots, \xi_n)$.

The difficulties in establishing a program for a numerical solution of the boundary problem (7.1), (7.2) by either a difference or variational-difference scheme depend, in many respects, on the geometry of the domain D_1. It is, therefore, worth while to construct implementations for a wide class of domains rather than for concrete ones. One possible method is to replace the problem (7.1), (7.2) by one that is in some definite sense close to the original but defined on a simpler domain, say on a parallelepiped. This is called the method of fictive domains (cf. Saul'ev [4], Kopchenov [4], Konovalov [4], Rukhovets [4]).

It is founded on the fact, well known to physicists, that in a medium with a relatively high diffusion coefficient the variations in the density of the diffusant are small. We try to extend the initial domain D_1 to a parallelepiped and extend the definition of the coefficients a_{ij} in (7.1) to sufficiently large values in the extended domain. In prescribing the boundary conditions (7.2) on the boundary of the parallelepiped we may expect that the solution of the new problem will differ by little from zero in the extended domain and will closely agree with the solution of (7.1), (7.2) in the initial domain D_1.

We illustrate this idea by a simple example, from a paper by Saul'ev [4], using the one-dimensional boundary problem

$$\frac{d^2u}{dx^2} = -2, \qquad 0 < x < 0.5, \qquad u(0) = u(0.5) = 0, \tag{7.4}$$

which has the exact solution $u(x) = x(0.5 - x)$. We replace (7.4) by the problem

$$\frac{d}{dx}\left(a(x)\frac{dv}{dx}\right) = f(x), \qquad 0 < x < 1, \qquad v(0) = v(1) = 0,$$

$$\left(a\frac{dv}{dx}\right)(0.5 - 0) = \left(a\frac{dv}{dx}\right)(0.5 + 0),$$

$$a(x) = \begin{cases} 1, & 0 < x < 0.5, \\ 1/\varepsilon^2, & 0.5 < x < 1, \end{cases} \tag{7.5}$$

$$f(x) = \begin{cases} -2, & 0 < x < 0.5, \\ 0, & 0.5 < x < 1. \end{cases}$$

The exact solution of (7.5) has the form

$$v(x) = \begin{cases} x\left(\dfrac{1 + 2\varepsilon^2}{1 + \varepsilon^2}\,0.5 - x\right), & 0 \le x \le 0.5, \\[3mm] \dfrac{\varepsilon^2}{2(1 + \varepsilon^2)}\,(1 - x), & 0.5 \le x \le 1. \end{cases}$$

It is clear that

$$\lim_{\varepsilon \to 0} v(x) = u(x), \qquad x \in [0; 0.5],$$

$$\lim_{\varepsilon \to 0} v(x) = 0, \qquad x \in (0.5; 1).$$

Therefore, for sufficiently small ε the domain $0.5 < x \le 1$ may be regarded as fictitious. A similar phenomenon appears in the multi-dimensional case as well.

Let us now formalize the method of fictitious domains for the first boundary problem (7.1), (7.2). We denote by D_2 the extension of the domain D_1 to the parallelepiped D, and by S the common portion of the boundaries of D_1 and D_2. In D we consider the equation

$$L_\varepsilon v \equiv -\sum_{i,j=1}^{n} \frac{\partial}{\partial x_i}\left(A_{ij}(x)\frac{\partial v}{\partial x_j}\right) + Cv = F(x), \tag{7.6}$$

where

$$A_{ij} = \begin{cases} a_{ij}(x), & x \in D_1, \\ 0, & x \in D_2; i \ne j, \\ \varepsilon^{-2}, & x \in D_2; i = j, \end{cases}$$

$$C(x) = \begin{cases} c(x), & x \in D_1, \\ 0, & x \in D_2, \end{cases}$$

$$F(x) = \begin{cases} f(x), & x \in D_1, \\ 0, & x \in D_2 \end{cases}$$

Note that (7.6) need not be homogeneous in D_2. In solving the problem of the torsion of a homogeneous cylindrical rod, Kononvalov [4] chose this equation to be homogeneous throughout D, which allowed him to give the problem a definite physical meaning. We pose for it the boundary problem

$$v(x) = 0, \qquad x \in \partial D, \tag{7.7}$$

$$[v(x)]|_S = 0, \qquad \left[\sum_{i,j=1}^{n} A_{ij} \cos(v, x_i) \frac{\partial v}{\partial x_j}\right]\Bigg|_S = 0. \tag{7.8}$$

Here v is the normal to the boundary S; $[\ ]|_S$ denotes a jump of the enclosed function on the surface S.

We assume that the solution $u(x)$ of (7.1), (7.2) vanishes in the domain D_2 and we estimate the difference $\omega(x) = u(x) - v(x)$. It is easily seen that the function $w(x)$ satisfies the boundary conditions

$$L_\varepsilon \omega(x) = 0, \qquad x \in D; x \notin S, \tag{7.9}$$

$$\omega(x) = 0, \qquad x \in \partial D, \tag{7.10}$$

$$[\omega(x)]|_S = 0, \qquad \left[\sum_{i,j=1}^{n} A_{ij}(x)\cos(v, x_i) \frac{\partial \omega}{\partial x_i}\right]\Bigg|_S = \phi(x), \tag{7.11}$$

where

$$\phi(x) = \sum_{i,j=1}^{n} a_{ij} \cos(v, x_i) \frac{\partial u}{\partial x_i}, \qquad x \in S.$$

Multiplying (7.9) by $\omega(x)$ and integrating over D, subject to (7.10) and (7.11), we arrive at the identity

$$\int_{D_1} \left(\sum_{i,j=1}^{n} a_{ij} \frac{\partial \omega}{\partial x_i} \frac{\partial \omega}{\partial x_j} + C\omega^2\right) dx + \frac{1}{\varepsilon^2} \int_{D_2} \sum_{i=1}^{n} \left(\frac{\partial \omega}{\partial x_i}\right)^2 dx$$

$$= -\int_S \phi(x)\omega(x)\, ds. \tag{7.12}$$

Discarding the first term on the left-hand side (cf. (7.3)) and applying the Cauchy–Schwartz inequality to the right-hand side of (7.12) we find that

$$\frac{1}{\varepsilon^2} \int_{D_2} \sum_{i=1}^{n} \left(\frac{\partial \omega}{\partial x_i}\right)^2 dx \leq \sqrt{\int_S \phi^2(x)\, ds} \sqrt{\int_S \omega^2(x)\, ds}. \tag{7.13}$$

To estimate the right-hand side of this equation we use first the relation

$$\int_S \omega^2(x)\, ds \leq C_1 \left(\delta \int_{\omega_\delta} \omega^2(x)\, dx + \frac{1}{\delta} \int_{\omega_\delta} \sum_{i=1}^{n} \left(\frac{\partial \omega}{\partial x_i}\right)^2 dx\right), \tag{7.14}$$

which holds in the boundary-band ω_δ of width $0 < \delta \leq \delta_0$, where δ_0 does not depend on the function $\omega(x)$ (cf. Oganesyan, Rivkind, and Rukhovets [5]), and also the Friedrichs inequality

$$\int_D \omega^2(x)\, dx \leq C_2 \int_D \sum_{i=1}^{n} \left(\frac{\partial \omega}{\partial x_i}\right)^2 dx, \tag{7.15}$$

which is valid for functions $w(x)$ vanishing on a portion of the boundary of D.

We set $\delta = \delta_0$ in (7.14) and carry out the integration on the right-hand side; we find that

$$\int_S \omega^2(x)\, ds \leq C_3\left(\int_{D_2} \omega^2(x)\, dx + \int_{D_2} \sum_{i=1}^n \left(\frac{\partial \omega}{\partial x_i}\right)^2 dx\right). \qquad (7.16)$$

Since $\omega(x) = 0$ for $x \in \partial D$, we may apply the inequality (7.15) to the first term on the right-hand side of (7.16). We find that

$$\int_S \omega^2(x)\, ds \leq C_4 \int_{D_2} \sum_{i=1}^n \left(\frac{\partial \omega}{\partial x_i}\right)^2 dx. \qquad (7.17)$$

Taking account of (7.17), we find that (7.13) yields the estimate

$$\left(\int_{D_2} \sum_{i=1}^n \left(\frac{\partial \omega}{\partial x_i}\right)^2 dx\right)^{1/2} \leq C_4 \varepsilon^2, \qquad (7.18)$$

and from (7.15) we have

$$\int_{D_2} \omega^2(x)\, dx \leq C_5 \varepsilon^2. \qquad (7.19)$$

Similarly, using the ellipticity condition (7.3) and the nonnegativity of $f(x)$, we obtain from the identity (7.12) the inequality

$$\mu \int_{D_1} \sum_{i=1}^n \left(\frac{\partial \omega}{\partial x_i}\right)^2 dx \leq \sqrt{\int_S \phi^2\, ds} \sqrt{\int_S \omega^2(x)\, ds} \leq C_4 \varepsilon^2. \qquad (7.20)$$

Moreover, using the generalized Friedrichs inequality

$$\int_{D_1} \omega^2(x)\, dx \leq C_6\left(\int_S \omega^2(x)\, ds + \int_{D_1} \sum_{i=1}^n \left(\frac{\partial \omega}{\partial x_i}\right)^2 dx\right) \qquad (7.21)$$

and the estimates (7.17), (7.18), and (7.20) we arrive at the inequality

$$\int_{D_1} \omega^2(x)\, dx \leq C_7 \varepsilon^2. \qquad (7.22)$$

Thus we have shown that the solution $v(x)$ of the problem (7.6)–(7.8) approximates the solution $u(x)$ of the problem (7.1), (7.2) in $W_2^1(D_1)$ to within ε, i.e.,

$$\|u - v\|_{W_2^1(D_1)} \leq C_8 \varepsilon, \qquad (7.23)$$

where the constant C_8 is independent of ε.

A more detailed analysis (Kopchenov [4]) leads to the following estimate

$$\|u - v\|_{C(D_1)} \leq C_9 \varepsilon^2. \qquad (7.24)$$

which is optimal in ε.

We may now state the following scheme for approximating the solution of the first boundary problem (7.1), (7.2) in an arbitrary bounded domain D_1: we enclose D_1 in the smallest parallelepiped D. We choose ε so that the solution of the problem (7.6)–(7.8) shall approximate the solution of the initial problem with the necessary precision. Then we solve the problem (7.6)–(7.8) by the difference method, to within the necessary precision.

The convergence of the difference solution to the exact solution has been studied for this problem, by, for example, Rivkind [4].

In concluding this section, we formulate the method of fictive domains as it applies to the solution of the third boundary problem:

$$- \sum_{i,j=1}^{n} \frac{\partial}{\partial x_i} \left(a_{ij}(x) \frac{\partial u}{\partial x_j} \right) + \sum_{i=1}^{n} b_i(x) \frac{\partial u}{\partial x_i} + c(x)u = f(x), \qquad (7.25)$$

$$x = (x_1, \ldots, x_n) \in D_1,$$

$$\sum_{i,j=1}^{n} a_{ij}(x)\cos(v, x_i) \frac{\partial u}{\partial x_j} + \sigma(x)u = 0, \qquad x \in \partial D_1. \qquad (7.26)$$

Here $\sigma(x)$ is a function sufficiently smooth and nonnegative on ∂D_1.

We extend the domain D_1 to a parallelepiped $D \supset \bar{D}_1$. We assume that there exists a sphere of radius $\rho > 0$, lying within the domain D_1, such that a tangent lying within D_1 may be drawn to it from any point on the boundary ∂D_1. In the band near the boundary of D_1, of width $\varepsilon \leq \rho$ we define the function $\sigma_\varepsilon(x)$ by the formula

$$\sigma_\varepsilon(x) = \frac{2}{\varepsilon} \sigma(x_\tau) \left(1 + \frac{\tau}{\varepsilon} \right), \qquad -\varepsilon \leq \tau \leq 0; x_{\tau i} = x_i - \tau \cos(v, x_i),$$

where $|\tau|$ is the distance along the normal from the point x to ∂D_1. In the rest of D we set $\sigma_\varepsilon(x) = 0$.

We consider the equation in D

$$- \sum_{i,j=1}^{n} \frac{\partial}{\partial x_i} \left(A_{ij}(x) \frac{\partial v}{\partial x_j} \right) + \sum_{i=1}^{n} B_i(x) \frac{\partial v}{\partial x_i} + C(x)v + \sigma_\varepsilon(x)v = F(x), \quad (7.27)$$

where

$$A_{ij}(x) = \begin{cases} a_{ij}(x), & x \in D_1, \\ 0, & x \in D_2; i \neq j, \\ \varepsilon, & x \in D_2; i = j; \end{cases}$$

$$B_i(x) = \begin{cases} b_i(x), & x \in D_1, \\ 0, & x \in D_2; \end{cases}$$

$$C(x) = \begin{cases} c(x), & x \in D_1, \\ 0, & x \in D_2; \end{cases}$$

$$F(x) = \begin{cases} f(x), & x \in D_1, \\ 0, & x \in D_2. \end{cases}$$

For (7.27) we pose the boundary problem

$$v(x) = 0, \qquad x \in \partial D, \tag{7.28}$$

and the compatibility conditions

$$[v(x)]|_{\partial D_1} = 0, \qquad \left[\sum_{i,j=1}^{n} A_{ij}(x)\cos(v, x_i) \frac{\partial v}{\partial x_j} \right]\Bigg|_{\partial D_1} = 0. \tag{7.29}$$

If the coefficients in the problem (7.25), (7.26) are sufficiently smooth, we may prove the following estimate:

$$\|u - v\|_{W_2^1(D_1)}^2 \le \varepsilon C_{10} \|f\|_{L_2(D_1)}^2, \tag{7.30}$$

where the constant C_{10} does not depend on the choice of $\varepsilon > 0$.

The passage from the third boundary problem (7.25), (7.26) to the first boundary problem (7.27)–(7.29) may be negotiated as follows: we suppose that as $\varepsilon \to 0$ the solution of the latter problem converges to some function $\tilde{u}(x)$ in the norm of the space $W_2^1(D)$. The solution $v(x)$ of this problem satisfies the integral identity

$$\int_{D_1} \sum_{i,j=1}^{n} a_{ij} \frac{\partial v}{\partial x_i} \frac{\partial \phi}{\partial x_j} \, dx + \varepsilon \int_{D_1} \sum_{i,j=1}^{n} \frac{\partial v}{\partial x_i} \frac{\partial \phi}{\partial x_i} \, dx$$

$$+ \int_{D_1} C(x) v \phi \, dx + \int_{D_1} \sigma_\varepsilon v \phi \, dx = \int_{D_1} f \phi \, dx \tag{7.31}$$

for arbitrary $\phi \in \mathring{W}_2^1(D_1)$ (for simplicity we set $b_i(x) = 0$). Letting $\varepsilon \to 0$ in (7.31), and noting the form of $\sigma_\varepsilon(x)$, we obtain the identity (cf. Rukhovets [4])

$$\int_{D_1} \sum_{i,j=1}^{n} a_{ij} \frac{\partial \tilde{u}}{\partial x_i} \frac{\partial \phi}{\partial x_j} \, dx + \int_{D_1} C \tilde{u} \phi \, dx + \int_{D_1} \sigma \tilde{u} \phi \, ds = \int_{D_1} f \phi \, dx \tag{7.32}$$

which holds for all $\phi \in W_2^1(D_1)$. But since the solution of (7.25), (7.26) satisfies the integral identity (7.32), the function $u(x)$ and $\tilde{u}(x)$ must coincide in D_1.

CHAPTER 3

Interpolation of Net Functions

The interpolation problem for a function of a continuous argument, given discrete data, is closely related to the problem of constructing variational-difference schemes and continuous representations of solutions of difference problems. For, as a rule, in order to obtain the difference equations, one has to discretize the operator and the solution of the problem by taking suitable projections. At the same time the solution of the differential problem usually represents an approximate solution of the original problem on a discrete set of points. Let us assume that the difference problem has been solved and that we know the approximate solution of the problem. The procedure now involves an interpolation from the results obtained onto the whole region on which the solution of the original problem is sought. An interpretation of this kind requires that certain conditions are taken into account: namely, if the solution of the difference equations has been obtained with a certain accuracy, then the accuracy of the interpolation must be at least as high. If we possess additional information regarding the error of the approximate solution, then the interpolation may be implemented, not from the exact data, but from those involving possible error at the net points. *A priori* information about the smoothness of solutions may then also allow one to improve the accuracy of the approximate solution in some cases, using one or another difference method. Apart from this, the interpolation problem also has its own merits.

As a rule, the interpolation algorithms for the exact data, specified on a discrete set of points, are based on Lagrange interpolation polynomials. Also, the interpolated function $\phi(x)$ is *a priori* assumed to be suitably differentiable.

Another problem closely related to interpolation arises when the function values of ϕ at the net points x_k are known only up to a certain error, the maximum magnitude of which is *a priori* known at each of the points. In this case, we want to construct a curve which would be in some sense the best approximation of the function, the latter being specified at the net points up to random errors. Problems of this kind are usually solved by the least-squares method.

The theory of interpolation has been enriched recently by what is known as *spline interpretation methods*. Usually a spline is a piecewise-polynomial function defined in the region D, i.e., a function for which there exists a decomposition of D into subregions in each of which the function is a polynomial of some degree m. Also the function, as a rule, is continuous in D, together with its derivatives to order $m - 1$, and the mth derivatives are square integrable. In practice, the most often used polynomials are cubics.

Spline interpolation methods are discussed in detail in papers by Holladey [14], Walsh, Alberg, and Nilson [14], Schoenberg [14], Anselon and Laurent [14], Zavyalov [14]. Generalizations are found in Birkhoff, Schultz, and Varga [15], Varga [1], and others.

3.1 Interpolation of Functions of One Variable

3.1.1 Interpolation of Functions of One Variable by Cubic Splines

Consider the net $a = x_0 < x_1 < \cdots < x_n = b$ on an interval $[a, b]$ of the real line. Assume we are given the values $\{f_k\}_{k=0}^n$ of a function $f(x)$, with $[a, b]$ as its domain of definition. The piecewise-cubic interpolation problem for this case is formulated as follows. Find a function $g(x)$ on $[a, b]$ satisfying the four conditions below:

(1) $g(x) \in C^{(2)}(a, b)$, that is, g is continuous along with its derivatives, up to and including the second order.

(2) On each of the intervals $[x_{k-1}, x_k]$, $g(x)$ is identical to a cubic polynomial of the form

$$g(x) \equiv g_k(x) = \sum_{l=0}^{3} a_l^{(k)}(x_k - x)^l, \qquad k = 1, \ldots, n. \tag{1.1}$$

(3) At the net points x_k ($k = 0, 1, \ldots, n$),

$$g(x_k) = f_k, \qquad k = 0, 1, \ldots, n. \tag{1.2}$$

(4) The boundary conditions

$$g''(a) = g''(b) = 0. \tag{1.3}$$

are satisfied by $g(x)$.

The advantages of this type of interpolation will be understood later, after we establish the certain natural extremal property of the function $g(x)$.

We will now prove that the piecewise-cubic interpolation function $g(x)$ is uniquely defined by the above requirements (1)–(4).

Since the second derivative of $g(x)$ is continuous and linear on every interval of the net $[x_{i-1}, x_1]$ $(i = 1, \ldots, n)$, we may write for $x_{i-1} \leq x \leq x_i$:

$$g''(x) = m_{i-1} \frac{x_i - x}{h_i} + m_i \frac{x - x_{i-1}}{h_i}, \tag{1.4}$$

where $h_i = x_i - x_{i-1}$, $m_k = g''(x_k)$. We integrate twice on both sides of (1.4), obtaining

$$g(x) = m_{i-1} \frac{(x_i - x)^3}{6h_i} + m_i \frac{(x - x_{i-1})^3}{6h_i} + A_i \frac{x_i - x}{h_i} + B_i \frac{x - x_{i-1}}{h_i}, \tag{1.5}$$

where A_i and B_i are constants of integration, determined by the conditions $g(x_{i-1}) = f_{i-1}$, $g(x_i) = f_i$. Setting $x = x_i$ and $x = x_{i-1}$ in (1.5), we find

$$m_i \frac{h_i^2}{6} + B_i = f_i$$

$$m_{i-1} \frac{h_i^2}{6} + A_i = f_{i-1}.$$

Finally, we have

$$g(x) = m_{i-1} \frac{(x_i - x)^3}{6h_i} + m_i \frac{(x - x_{i-1})^3}{6h_i}$$

$$+ \left(f_{i-1} - \frac{m_{i-1} h_i^2}{6} \right) \frac{x_i - x}{h_i} + \left(f_i - \frac{m_i h_i^2}{6} \right) \frac{x - x_{i-1}}{h_i}, \tag{1.6}$$

$$g'(x) = -m_{i-1} \frac{(x_i - x)^2}{2h_i} + m_i \frac{(x - x_{i-1})^2}{2h_i} + \frac{f_i - f_{i-1}}{h_i} - \frac{m_i - m_{i-1}}{6} h_i. \tag{1.7}$$

From (1.7) we find one-sided limits for the derivatives at the points $x_1, x_2, \ldots, x_{n-1}$:

$$g'(x_i - 0) = \frac{h_i}{6} m_{i-1} + \frac{h_i}{3} m_i + \frac{f_i - f_{i-1}}{h_i},$$

$$g'(x_i + 0) = -\frac{h_{i+1}}{3} m_i - \frac{h_{i+1}}{6} m_{i+1} + \frac{f_{i+1} - f_i}{h_{i+1}}.$$

By Condition (1) the functions $g''(x)$ and $g'(x)$ are continuous on $[a, b]$. Since $g'(x)$ is continuous at the points $x_1, x_2, \ldots, x_{n-1}$ we have $n - 1$ equations

$$\frac{h_i}{6} m_{i-1} + \frac{h_i + h_{i+1}}{3} m_i + \frac{h_{i+1}}{6} m_{i+1} = \frac{f_{i+1} - f_i}{h_{i+1}} - \frac{f_i - f_{i-1}}{h_i}. \tag{1.8}$$

If we supplement these equations with the equations $m_0 = m_n = 0$ [from (1.3)] we have a linear algebraic system of equations for the unknown $m_1, m_2, \ldots, m_{n-1}$:

$$Am = Hf. \tag{1.9}$$

The square matrix A has the form

$$A = \begin{Vmatrix} \dfrac{h_1 + h_2}{3} & \dfrac{h_2}{6} & 0 & \cdots & 0 & 0 \\[2ex] \dfrac{h_2}{6} & \dfrac{h_2 + h_3}{3} & \dfrac{h_3}{6} & \cdots & 0 & 0 \\[2ex] 0 & \dfrac{h_3}{6} & \dfrac{h_3 + h_4}{3} & \cdots & 0 & 0 \\[1ex] \cdots & \cdots & \cdots & \cdots & \cdots & \cdots \\[1ex] 0 & 0 & 0 & \cdots & \dfrac{h_{n-1}}{6} & \dfrac{h_n + h_{n-1}}{3} \end{Vmatrix}; \tag{1.10}$$

the vectors m and f and the rectangular matrix H are as follows:

$$m = \begin{Vmatrix} m_1 \\ m_2 \\ \vdots \\ m_{n-1} \end{Vmatrix}, \qquad f = \begin{Vmatrix} f_1 \\ f_2 \\ \vdots \\ f_n \end{Vmatrix},$$

$$H = \begin{Vmatrix} \dfrac{1}{h_1} & \left(-\dfrac{1}{h_1} - \dfrac{1}{h_2}\right) & \dfrac{1}{h_2} & \cdots & 0 & 0 \\[2ex] 0 & \dfrac{1}{h_2} & \left(-\dfrac{1}{h_2} - \dfrac{1}{h_3}\right) & \cdots & 0 & 0 \\[1ex] \cdots & \cdots & \cdots & \cdots & \cdots & \cdots \\[1ex] 0 & 0 & 0 & \cdots & \left(-\dfrac{1}{h_{n-1}} - \dfrac{1}{h_n}\right) & \dfrac{1}{h_n} \end{Vmatrix}.$$

$$\tag{1.11}$$

The matrix A is symmetric, with a strong diagonal predominance. By Hershgorin's theorem on the localization of eigenvalues, it is positive definite, and, of course, nonsingular. The coefficients $m_1, m_2, \ldots, m_{n-1}$ are uniquely defined by (1.9); therefore the spline functions $g(x)$ are uniquely determined by (1.6) and the problem of finding the piecewise-cubic functions $g(x)$ has a unique solution.

Cubic spline functions have a very important property underlying their great effectiveness for spline interpolation. To see this, we consider the class

$W_2^2[a, b]$ of functions having mean-square-summable second derivatives on the interval $[a, b]$. We try to find an interpolation function

$$u \in W_2^2[a, b], \qquad u(x_k) = f_k, \qquad k = 0, 1, \ldots, n, \qquad (1.12)$$

which will minimize the functional

$$\Phi(u) = \int_a^b [u''(x)]^2 \, dx \qquad (1.13)$$

over the class $W_2^2[a, b]$. We assert that the minimum value is realized on the piecewise-cubic spline function that we have just constructed. In fact, let us consider the quantity

$$\Phi(u - g) = \int_a^b [u'' - g'']^2 \, dx. \qquad (1.14)$$

Intregrating by parts, and using the properties of the functions u and $g \in W_3^2$, we have

$$\Phi(u - g) = \Phi(u) - \Phi(g) - 2 \left[(u' - g')g'' \Big|_{x=a}^{x=b} - \int_a^b (u' - g')g''' \, dx \right]$$

$$= \Phi(u) - \Phi(g) - 2 \sum_{k=1}^n \int_{x_{k-1}}^{x_k} (u' - g')g''' \, dx.$$

But $g''' = c_k = \text{const}$ on the interval $[x_{k-1}, x_k]$, so that

$$\Phi(u - g) = \Phi(u) - \Phi(g) - 2 \sum_{k=1}^n c_k(u - g) \Big|_{x=x_{k-1}}^{x=x_k} = \Phi(u) - \Phi(g).$$

Hence, and from (1.14) it follows that

$$\Phi(g) = \Phi(u) - \Phi(u - g) \leq \Phi(u) \qquad (1.15)$$

for all functions

$$u \in W_2^2, \qquad u(x_k) = f_k, \qquad k = 0, 1, \ldots, n.$$

Thus, the minimum of the functional (1.13) is attained on the piecewise-cubic function $g(x)$. It is not hard to show that there are no other minimal points of the functional.

On the basis of (1.12), (1.13) another (equivalent) definition of a piecewise-cubic spline function can be given: it is that function belonging to $W_2^2[a, b]$ which assumes given values on the net points and minimizes the functional (1.13). This property of the spline function is interesting because the functional $\Phi(u)$ may be interpreted as an analog of the potential energy of an elastic rod fixed at the points (x_k, f_k) of the plane, and the minimum of this energy is attained on the cubic splines.

In this subsection we have limited ourselves to the consideration of cubic splines satisfying the boundary conditions (1.3), which represent the conditions of "free suspension" of the interpolating curves at the points a and b.

In practice, however, the slopes of the interpolating curves are often known at the boundary points. Then we naturally apply the conditions

$$g'(a) = f'_0, \qquad g'(b) = f'_n. \tag{1.16a}$$

If we know the curvatures at a and b, the natural conditions are

$$g''(a) = f''_0, \qquad g''(b) = f''_n. \tag{1.16b}$$

If we know *a priori* that the interpolating function is periodic, with period $b - a$, we should apply the conditions

$$g'(a) = g'(b), \qquad g''(a) = g''(b). \tag{1.16c}$$

Here, of course, $f_0 = f_n$.

We ask how the system of linear algebraic equations (1.10) changes in the presence of these boundary conditions. In the simple case, (1.3), we adjoin the equations $m_0 = m_n = 0$ to the system (1.8). Taking account of (1.7), the conditions (1.16a) lead to the equations:

$$\tfrac{2}{3}m_0 + \tfrac{1}{3}m_1 = \frac{2}{h_1}\left(\frac{f_1 - f_0}{h_1} - f'_0\right), \qquad \tfrac{1}{3}m_{n-1} + \tfrac{2}{3}m_n = \frac{2}{h_n}\left(f'_n - \frac{f_n - f_{n-1}}{h_n}\right).$$

The conditions (1.16b) lead to the equations:

$$m_0 = f''_0, \qquad m_n = f''_n,$$

and, finally, the condition that the spline is periodic leads to the equations:

$$m_0 = m_n,$$

$$\frac{h_n}{6}m_{n-1} + \frac{h_n + h_1}{3}m_n = \frac{f_1 - f_n}{h_1} - \frac{f_n - f_{n-1}}{h_n}.$$

These equations are to be adjoined to the system (1.8). It is, of course, possible to combine various conditions at the points a and b.

We should note that the splines subjected to various types of boundary conditions all minimize the functional (1.13) over the subclass of functions in this space that satisfy the given boundary conditions, but not over the whole of $W_2^2(a, b)$.

3.1.2 Piecewise-Cubic Interpolation with Smoothing

We now consider the problem of a smooth construction of a function defined on a net

$$a = x_0 < x_1 < \cdots < x_n = b.$$

Here, however, the values $\tilde{f}_k$ of the function at the net points are assumed to be perturbed by errors. In this case it makes no sense to construct an inter-polation function that agrees precisely with the given function at the net points. Rather, we want to construct a function that behaves more smoothly near the given values, than the function we are interpolating. We shall call such a function a *smoothing* function, rather than an interpolator.

We require that the desired smoothing function $g(x)$ minimize the functional

$$\Phi_1(u) = \int_a^b [u'']^2 \, dx + \sum_{k=0}^n p_k [u(x_k) - \tilde{f}_k]^2 \tag{1.17}$$

over the class $W_2^2[a, b]$. Here the p_k are positive numbers. The functional $\Phi_1(u)$ combines the interpolation conditions to extend the curve in the neighborhood of the given values, and the minimization of the bending of the curve. The greater the weights p_k, the more influence the interpolation conditions have in the functional, and the nearer the smoothing function comes to the given values.

We shall show that the solution of the variational problem (1.17) is a cubic spline, that is, a function satisfying Conditions (1), (2), and (4) of the preceding section. Let $u_0 \in W_2^2[a, b]$ be a solution. We construct the spline $g(x)$ so that $g(x_k) = u_0(x_k)$ $(k = 0, 1, \ldots, n)$. The second term in (1.17) is the same for both $g(x)$ and $u_0(x)$; therefore

$$\int_a^b [u_0'']^2 \, dx \le \int_a^b [g'']^2 \, dx. \tag{1.18}$$

But, as we saw in the preceding section, $g(x)$ is the unique interpolator of $u_0(x)$ that minimizes $\int_a^b [u'']^2 \, dx$. Therefore, $u_0 \equiv g$.

Thus, we seek the minimum of the functional $\Phi_1(u)$ only over the class of cubic splines. Since a cubic spline is uniquely defined by the set of values $\{\mu_k\}_{k=0}^n$, that it assumes at the net points, $\{x_k\}_{k=0}^n$, the minimization of $\Phi_1(u)$ reduces to the search for the minimum of a function of the variables μ_0, $\mu_1, \ldots, \mu_n$.

We already know that $g''(x)$ is a piecewise-linear function and that

$$g''(x) = m_{k-1} \frac{x_k - x}{h_k} + m_k \frac{x - x_{k-1}}{h_k}, \tag{1.19}$$

where $x \in [x_{k-1}, x_k]$, $m_k = g''(x_k^0)$, $k = 1, \ldots, n-1$, $m_0 = m_n = 0$. Therefore

$$\Phi_1(g) = \sum_{k=1}^n \int_{x_{k-1}}^{x_k} \left[m_{k-1} \frac{x_k - x}{h_k} + m_k \frac{x - x_{k-1}}{h_k} \right]^2 dx + \sum_{k=0}^n p_k (\mu_k - \tilde{f}_k)^2. \tag{1.20}$$

If we integrate (1.20) we obtain

$$\sum_{k=1}^n \int_{x_{k-1}}^{x_k} \left[m_{k-1} \frac{x_k - x}{h_k} + m_k \frac{x - x_{k-1}}{h_k} \right]^2 dx$$

$$= \sum_{k=1}^n m_k \left[m_{k-1} \frac{h_k}{6} + \frac{h_k + h_{k+1}}{3} m_k + \frac{h_{k+1}}{6} m_{k+1} \right] = (Am, m). \tag{1.21}$$

Here A is the known matrix (1.10). So

$$\Phi_1(g) = (Am, m) + \sum_{k=0}^{n} p_k(\mu_k - \tilde{f}_k)^2. \tag{1.22}$$

In view of (1.9) m is linear in $\mu = (\mu_0, \mu_1, \ldots, \mu_n)$, and therefore $\Phi_1(g)$ is a positive definite form over μ. Its extremum can only be a minimum, for which the necessary condition is

$$\frac{\partial \Phi_1}{\partial \mu_s} \equiv \frac{\partial}{\partial \mu_s}(Am, m) + 2p_s(\mu_s - \tilde{f}_s) = 0, \qquad s = 0, 1, \ldots, n.$$

But the matrix A does not depend on μ. Therefore, by (1.9),

$$\frac{\partial}{\partial \mu_s}(Am, m) = 2\left(\frac{\partial(Am)}{\partial \mu_s}, m\right) = 2\left(\frac{\partial(H\mu)}{\partial \mu_s}, m\right) = 2\left(\frac{\partial \mu}{\partial \mu_s}, H^*m\right) = 2(H^*m)_s.$$

Here H is defined by (1.11). It follows that the vector form of the minimum condition is

$$H^*m + P\mu = P\tilde{f}, \tag{1.23}$$

where $\tilde{f} = (\tilde{f}_0, \tilde{f}_1, \ldots, \tilde{f}_n)$, and P is the diagonal matrix

$$P = \begin{Vmatrix} p_0 & 0 & \cdots & 0 \\ 0 & p_1 & \cdots & 0 \\ \multicolumn{4}{c}{\cdots\cdots\cdots\cdots\cdots} \\ 0 & 0 & \cdots & p_n \end{Vmatrix}. \tag{1.24}$$

Multiplying (1.23) on the left by HP^{-1} we find that

$$HP^{-1}H^*m + H\mu = H\tilde{f},$$

or, finally, using (1.9),

$$(A + HP^{-1}H^*)m = H\tilde{f}. \tag{1.25}$$

The matrix of the system (1.25) is pentadiagonal, symmetric, and positive definite. We could solve (1.25) by Gauss's method; then, since the vector m is determined, we must find the vector of net point values of the smoothing spline by a formula easily derived from (1.23):

$$\mu = \tilde{f} - P^{-1}H^*m. \tag{1.26}$$

This formula, with (1.6), determines the spline $g(x)$.

3.1.3 Smooth Construction

We now discuss another process for the smooth construction of a net function, differing to some extent from the method of splines, but also very effective in its algorithm, which was proposed by Ryabenkii [14].

We shall describe a very quick process for interpolating a function belonging to an arbitrary smoothness class C^p. Let $x_1 < x_2 < \cdots < x_{n-1} < x_n$ be a fixed net for which the values $f_1, f_2, \ldots, f_n$ of f are known at the net points. We assume that n is sufficiently large, and we choose an integer $p \ll n$. We first construct an interpolation of class C^p on the inverval (x_1, x_2). We denote by $P_0(x)$ a Lagrange polynomial of degree not exceeding p, coinciding with the given function values at the net points $x_1, x_2, \ldots, x_{p+1}$. We denote by $P_1(x)$ the Lagrange polynomial of the same degree (not exceeding p) that takes on the functional values at the points $x_2, x_3, \ldots, x_{p+2}$. Using these, we construct a polynomial $Q_1(x)$ of degree not exceeding $2p + 1$ and satisfying the conditions

$$\frac{d^k Q_1}{dx^k}\bigg|_{x=x_1} = \frac{d^k P_0}{dx^k}\bigg|_{x=x_1}, \qquad \frac{d^k Q_1}{dx^k}\bigg|_{x=x_2} = \frac{d^k P_1}{dx^k}\bigg|_{x=x_2}, \qquad k = 0, 1, \ldots, p.$$

$$(1.27)$$

It is clear that the polynomial $Q_1(x)$ exists and is uniquely defined by the conditions (1.27).

Let the interpolating polynomial $g(x)$ coincide with $Q_1(x)$ on the interval (x_1, x_2). As an example, we set $p = 1$. Then on (x_1, x_2) the polynomial $Q_1(x)$ is a cubic of the form

$$Q_1(x) = a_0 + a_1(x - x_1) + a_2(x - x_1)^2 + a_3(x - x_1)^3;$$

its coefficients are computed from the formulas

$$a_0 = f_1,$$

$$a_1 = \frac{f_2 - f_1}{x_2 - x_1},$$

$$a_2 = -\frac{1}{x_2 - x_1}\left(\frac{f_3 - f_2}{x_3 - x_2} - \frac{f_2 - f_1}{x_2 - x_1}\right), \qquad (1.28)$$

$$a_3 = \frac{1}{(x_2 - x_1)^2}\left(\frac{f_3 - f_2}{x_3 - x_2} - \frac{f_2 - f_1}{x_2 - x_1}\right).$$

The interpolator $g(x)$ is computed on (x_2, x_3) in exactly the same way, taking x_2 as the starting point. It is clear that the construction of $g(x)$ can be carried through up to the interval (x_{n-p-1}, x_{n-p}). The so-constructed interpolator $g(x)$ is a piecewise-polynomial function belonging to C^p; the polynomials on the net intervals are of degree $2p + 1$.

If we set $p = 2$, which corresponds to the smoothness of the cubic splines, we shall be dealing with fifth-degree polynomials. This degree is higher than two, but nevertheless we need not solve a set of linear algebraic equations to determine $g(x)$.

The construction of interpolating functions of several variables differs little from the one-dimensional case, to which, in fact, it reduces.

Smooth completions have excellent approximating properties. In fact, given an interpolated function of several variables $f \in C^q$, and a completion of it $g(x) \in C^p$, where $p \geq q$, the function itself and its derivatives satisfy the inequalities

$$\|D^k f - D^k g\|_C \leq C(p) h^{q-|k|} \sup_x \max_{|\alpha|=q} |D^\alpha f(x)|, \qquad (1.29)$$

where the multi-index $k = (k_1, k_2, \ldots, k_r)$ is such that

$$|k| = \sum_{i=1}^r k_i < q, \qquad D^k = \frac{\partial^{k_1 + \cdots + k_r}}{\partial x^{k_1} \cdots \partial x^{k_r}},$$

$C(p)$ depends only on p; it is independent of both the mesh size h and the function f.

Smooth completions have yet another property. If we construct a basis function, i.e., a function having the value 1 on some net point and 0 on all others, it is finite for all p and the "radius" of the support does not exceed p. If, however, we construct a basis consisting of piecewise-cubic splines equal to 1 at one point and equal to 0 at all others, we will in general not have a finite basis. This fact makes the smooth completions highly suitable for variational problems, which are solved by the method of finite elements.

3.1.4 The Convergence of Spline Functions

In this section we shall illustrate by some simple examples the technique of deriving estimates for the rate of convergence of cubic splines and their derivatives. For simplicity, we limit ourselves to the boundary conditions (1.16a) and, in order not to complicate the formulas, we use uniform nets. These limitations play no essential role in the demonstrations.

We introduce a needed concept. Suppose that in the interval $[a, b]$ we are given a continuous function $\phi(x)$. Then the quantity

$$\omega(h, \phi) = \sup_{x', x'' \in [a, b]; |x' - x| \leq h} |\phi(x') - \phi(x'')|, \qquad (1.30)$$

representing the maximum variation of $\phi(x)$ on an interval of length h within $[a, b]$, is called the *modulus of continuity* of $\phi(x)$.

Now suppose given on $[a, b]$ a twice differentiable function $f(x)$, which we are to interpolate by cubic splines on a net with mesh size h. As we know from Section 3.1.1, the second derivatives m_i of the spline at the net points satisfy a system of the type of (1.9)

$$Am = Hf.$$

We transform this system as follows: we divide both sides by h and subtract from both sides the vector AHf/h^2. Then we have the equation

$$\frac{1}{h} Am - \frac{1}{h^2} AHf = \left(I - \frac{1}{h} A \right) \frac{Hf}{h}. \qquad (1.31)$$

Let

$$d = \frac{1}{h} Hf \quad \text{and} \quad B = \frac{1}{h} A.$$

Then

$$Bm - Bd = (I - B) d. \tag{1.32}$$

It is clear that the jth component d_j of the vector d has the form

$$d_j = \frac{(f_{j+1} - f_j)/h - (f_j - f_{j-1})/h}{h}. \tag{1.33}$$

Therefore, the equation

$$d_j = f''(\xi_j) \tag{1.34}$$

holds at some point ξ_j of the interval (x_{j-1}, x_{j+1}). From the form of the boundary condition (1.16a) it is clear that

$$d_0 = f''(\xi_0), \qquad d_n = f''(\xi_n). \tag{1.35}$$

If we consider that the sum of the coefficients in every row of the matrix B is equal to unity $(\frac{1}{6} + \frac{2}{6} + \frac{1}{6})$, we see that

$$\|(I - B) d\| \leq \omega(h, f''). \tag{1.36}$$

It is known that the eigenvalues of a matrix lie in the union of discs in the complex plane with centers equal to the diagonal elements of the matrix, and radii equal to the sum of the absolute values of the off-diagonal elements of each row (or column); this assertion is known as Gershgorin's lemma. If we apply it to the matrix B^{-1} we see that

$$\|B^{-1}\| \leq 3. \tag{1.37}$$

Therefore,

$$\|m - d\| \leq \|B^{-1}\| \, \|(I - B) d\| \leq 3\omega(h, f''). \tag{1.38}$$

Clearly,

$$|f''(x_j) - d_j| \leq \omega(h, f''). \tag{1.39}$$

Then,

$$|f''(x_j) - m_j| \leq 4\omega(h, f''). \tag{1.40}$$

Since the second derivative $g''(x)$ of the spline $g(x)$ is piecewise-linear,

$$|f''(x) - g''(x)| \leq 5\omega(h, f''). \tag{1.41}$$

Since $g(x_j) = f(x_j)$, we know by Rolle's theorem that in each interval (x_{j-1}, x_j) there exists a point η_j for which $f'(\eta_j) = g'(\eta_j)$. Consequently,

$$|f'(x) - g'(x)| = \left| \int_{\eta_j}^{x} [f''(x) - g''(x)] \, dx \right| \leq 5h\omega(h, f''). \tag{1.42}$$

Repeating the integration, we find that

$$|f(x) - g(x)| \le \tfrac{5}{2}h^2\omega(h, f''). \qquad (1.43)$$

Thus we have obtained estimates of the rate of convergence of the spline itself and of its first and second derivatives. We note that these estimates are valid for the boundary conditions (1.16b) and (1.16c).

As the smoothness of the function $f(x)$ increases the estimates improve. However, the use of cubic splines cannot yield a rate of convergence better than $O(h^4)$ unless, of course, the function $f(x)$ is a cubic polynomial.

3.2 Interpolation of Functions of Two or More Variables

The problem of two-dimensional interpolation by the use of piecewise-bicubic functions has been studied by many authors (cf. Alberg, Nilson, and Walsh [14], Zav'yalov [14], and others). We here briefly examine the following typical problem.

Let $D = \{x, y: a \le x \le b; c \le y \le d\}$ be some rectangle in the (x, y) plane. We construct in D the net

$$D_h = \{x_k, y_l: a = x_0 < x_1 < \cdots < x_n = b; c = y_0 < y_1 < \cdots < y_m = d\}.$$

Then the piecewise-bicubic interpolation of a function $f(x, y)$ given at the points of D_h consists in the construction of a function $g(x, y)$ satisfying the following conditions:

(1)
$$g(x, y) \in C^2(D); \qquad (2.1)$$

(2) in each cell of the net $g(x, y)$ is a bicubic polynomial of the form

$$g(x, y) = g_{k, l}(x, y) = \sum_{i, j=0}^{3} a_{i, j}^{k, l}(x_k - x)^i(y_l - y)^j; \qquad (2.2)$$

(3) on the net D_h itself $g(x, y)$ takes on the given values

$$g(x_k, y_l) = f_{kl}, \qquad k = 0, \ldots, n; l = 0, \ldots, m; \qquad (2.3)$$

(4)
$$\left. \frac{\partial^2 g}{\partial v^2} \right|_\Gamma = 0 \qquad (2.4)$$

(where v is the outward normal to the boundary Γ of the region D).

In principle, we construct such a function just as we did in the one-dimensional case. We recall that in order to calculate a one-dimensional spline at an arbitrary point, using the simple formulas (1.6), we need to know the values of the function itself and its second derivatives at the net points. To find these second derivatives we must solve a linear algebraic equation system with a triply diagonal matrix. We ask what preliminary computations must be made in order to find explicit formulas for a function at an arbitrary point in the two-dimensional case.

Let us look first at the one-dimensional cubic spline interpolation on the net lines $y = y_j$ $(j = 0, 1, \ldots, m)$. For this interpolation we are to solve $m + 1$ linear algebraic systems of the type of (1.9). We obtain the values of the function $g_{xx}(x, y)$ on D_h. Next we solve the $n + 1$ problems of spline interpolation on the lines $x = x_i$ $(i = 0, 1, \ldots, n)$ and we find the values of $g_{yy}(x, y)$ on D_h. Now suppose we are required to compute the values of $g(x, y)$ at an arbitrary point (x, y). Suppose

$$x_{i-1} \leq x \leq x_i, \qquad y_{j-1} \leq y \leq y_j.$$

We obtain the values of $g(x, y)$ at the points (x_i, y), (x_{i-1}, y) via formulas analogous to (1.6):

$$g(x_{i-1}, y) = N_{i-1,j-1} \frac{(y_j - y)^3}{6\tau_j} + N_{i-1,j} \frac{(y - y_{j-1})^3}{6\tau_j}$$

$$+ \left(f_{i-1,j-1} - \frac{N_{i-1,j-1}\tau_j^2}{6} \right) \frac{y_j - y}{\tau_j}$$

$$+ \left(f_{i-1,j} - \frac{N_{i-1,j}\tau_j^2}{6} \right) \frac{y - y_{j-1}}{\tau_j}, \tag{2.5}$$

$$g(x_i, y) = N_{i,j-1} \frac{(y_j - y)^3}{6\tau_j} + N_{i,j} \frac{(y - y_{j-1})^3}{6\tau_j}$$

$$+ \left(f_{i,j-1} - \frac{N_{i,j-1}\tau_j^2}{6} \right) \frac{y_j - y}{\tau_j} + \left(f_{i,j} - \frac{N_{i,j}\tau_j^2}{6} \right) \frac{y - y_{j-1}}{\tau_j}. \tag{2.6}$$

Here, and later,

$$N_{ij} = g_{yy}(x_i, y_j), \qquad M_{ij} = g_{xx}(x_i, y_j),$$

$$h_i = x_i - x_{i-1}, \qquad \tau_j = y_j - y_{j-1}.$$

If the values of $g_{xx}(x_{i-1}, y)$ and $g_{xx}(x_i, y)$ are known, formulas like (1.6) enable us to find the value of $g(x, y)$. The function $g_{xx}(x, y)$ is piecewise-cubic in y. We solve $m + 1$ one-dimensional problems on the lines $y = y_j$ for the function $g_{xx}(x, y)$, the net point values of which are already given. The result is the function $g_{xxyy}(x, y)$ on the net D_h. We write $K_{ij} = g_{xxyy}(x_i, y_j)$. Then

$$g_{xx}(x_{i-1}, y) = K_{i-1,j-1} \frac{(y_j - y)^3}{6\tau_j} + K_{i-1,j} \frac{(y - y_{j-1})^3}{6\tau_j}$$

$$+ \left(M_{i-1,j-1} - \frac{K_{i-1,j-1}\tau_j^2}{6} \right) \frac{y_i - y}{\tau_j} + \left(M_{i-1,j} - \frac{K_{i-1,j}\tau_j^2}{6} \right) \frac{y - y_{j-1}}{\tau_j}, \tag{2.7}$$

$$g_{xx}(x_i, y) = K_{i, j-1} \frac{(y_j - y)^3}{6\tau_j} + K_{i, j} \frac{(y - y_{j-1})^3}{6\tau_j}$$

$$+ \left(M_{i, j} - \frac{K_{i, j-1} \tau_j^2}{6} \right) \frac{y_j - y}{\tau_j} + \left(M_{i, j} - \frac{K_{ij} \tau_j^2}{6} \right) \frac{y - y_{j-1}}{\tau_j}. \quad (2.8)$$

Using (1.6) again, we find

$$g(x, y) = g_{xx}(x_{i-1}, y) \frac{(x_i - x)^3}{6h_i} + g_{xx}(x_i, y) \frac{(x - x_{i-1})^3}{6h_i}$$

$$+ \left(g(x_{i-1}, y) - \frac{g_{xx}(x_{i-1}, y)h_i^2}{6} \right) \frac{x_i - x}{h_i}$$

$$+ \left(g(x_i, y) - \frac{g_{xx}(x_i, y)h_i^2}{6} \right) \frac{x - x_{i-1}}{h_i}. \quad (2.9)$$

If we sum up what we have said, we can estimate the amount of computation needed to determine a bicubic spline $g(x, y)$. Before calculating the value of $g(x, y)$ at the points that interest us, we must solve $(n + 1) + (m + 1) + (m + 1) = 2m + n + 3$ linear algebraic systems of the type of (1.9) and determine N_{ij}, M_{ij}, K_{ij} $(i = 0, \ldots, n; j = 0, \ldots, m)$. Then, to compute $g(x, y)$ at one point of the region we must carry out five calculations of the type of (1.6), namely (2.5), (2.6), (2.7), (2.8), and (2.9). These formulas, in chain, can be used to obtain an explicit polynomial expression for $g(x, y)$ in every cell of the net. But to store the massif of the coefficients of the polynomial in the memory of an electronic computer requires 16 nm words; our algorithm requires only 4 nm words of storage, while the volume of calculation is increased by a factor of four.

We may interchange the roles of the variables in our algorithm, and then we must solve $2n + m + 3$ linear problems. The overall result is not changed, nor is the total volume of computations.

The algorithm we have described is easily generalized to a many-dimensional parallelepiped.

3.3 An *r*-Smooth Approximation to a Function of Several Variables

We consider one of the possible approaches to the completion of a function of several variables, as prescribed with errors at the points of a chaotic net (cf. Tsetsokho, Belonosov, and Belonosova [14]). Our exposition relates primarily to the description of the algorithms involved, but the theoretical aspects are not emphasized and are heuristically expounded. The method is based on the well-known concept of the resolution of the identity.

A sequence (ϕ_i) of functions belonging to the class $C^r(V)$ (with $V \subset R^m$ an open set) is said to be a resolution of the identity on the set $\Omega \subset V$ if

$$0 \le \phi_i(x) \le 1 \quad \text{and} \quad \sum_i \phi_i(x) = 1, \quad \text{for all} \quad x \in \Omega.$$

It is usually required that each ϕ_i vanish outside some given open set $V_i \subset V$. The ensemble of the V_i covers Ω. The resolution of the identity (ϕ_i) is said to be subordinate to this covering.

For applications we want locally finite coverings by standardized sets, m-dimensional intervals or spheres. The centers and diameters of these sets are to be chosen so that they correspond to the properties of the approximating function $f : \Omega \to R$. We shall consider coverings by m-dimensional cubes and will construct the subordinate resolutions of the identity in the following way.

We denote by $\xi^{(i)}$ the center of the cube V_i and by $\delta^{(i)}$ the length of its edge. We prescribe a (standard) function $\psi : R^m \to R$ of class $C^r(R^m)$ such that

$$\psi(x) > 0 \quad \text{if} \quad x \in (-\tfrac{1}{2}, \tfrac{1}{2}) \times \cdots \times (-\tfrac{1}{2}, \tfrac{1}{2}) = V_0,$$

$$\psi(x) = 0 \quad \text{if} \quad x \notin V_0.$$

We may, for instance, put

$$\psi(x) = \prod_{i=1}^{m} \cos^{r+1}(\pi x_i) \quad \text{or} \quad \psi(x) = \prod_{i=1}^{m} (\tfrac{1}{4} - x_1^2)^{r+1},$$

and so on, for $x = (x_1, \ldots, x_m) \in V_0$. Now set

$$\psi_i(x) = \begin{cases} \psi\left(\dfrac{x - \xi^{(i)}}{\delta^{(i)}}\right), & x \in V_i, \\ 0, & x \notin V_i, \end{cases}$$

for each value of the index i. The functions

$$\phi_i(x) = \frac{\psi_i(x)}{\sum_j \psi_j(x)}$$

belong to $C^r(V)$, where $V = \bigcup_i V_i$, and constitute a resolution of the identity on Ω, subordinated to the covering (V_i).

Let us suppose for the function $f : \Omega \to R$, a sequence $\{f_i : V_i \to R\}$ of functions of class $C^r(V)$ is known and is such that every f_i is "near" f on V_i; suppose that in the uniform norm we have for all i

$$\|(f - f_i)|_{V_i}\| \le \varepsilon.$$

Then the function $\tilde{f}$ defined by the equation

$$\tilde{f}(x) = \sum_i f_i(x)\phi_i(x), \qquad x \in \Omega, \tag{3.1}$$

will be "near" to f on Ω. In fact, because of the identity

$$f(x) = \sum_i f(x)\phi_i(x), \qquad x \in \Omega,$$

we find that for all $x \in \Omega$

$$|f(x) - \tilde{f}(x)| = \left| \sum_i (f(x) - f_i(x))\phi_i(x) \right|$$

$$\leq \max_i \sup_{x \in V_i} |f(x) - f_i(x)| \sum_i \phi_i(x) \leq \varepsilon,$$

i.e.,

$$\|f - \tilde{f}\| \leq \varepsilon.$$

We now use the resolution of the identity for the completion of the function f: if f is defined at the net points of some net $(x^{(j)} \in \Omega)$, then for every i in some set of net points $(x^{(j)})$ near the centers $\xi^{(i)}$ of the cubes V_i we must construct local approximations $f_i(x)$ and then paste together all the local approximations according to the formula (3.1). From now on, we shall suppose that the net points $x^{(i)}$ coincide with the centers $\xi^{(i)}$ of the cubes V_i and that the local approximations are constant functions equal to $f(x^{(i)})$.

Let us now estimate the approximation error in a special case:

$$f(x) = x, \qquad x \in R;$$

the net is uniform and infinite, with net points $x_k = kh$ $(k = 0, \pm 1, \pm 2, \dots)$; the centers $\xi^{(k)}$ of the intervals V_k coincide with the net points, and $\delta^{(k)} = (2N + 1)h$, where N is some nonnegative integer. We suppose that the resolution of the identity $\{\phi_k\}$ belongs to $C^r(R)$, $r \geq 3$, and that $\phi_0(x)$ is even.

Then the difference

$$\sigma(x) = x - \sum_k x_k \phi_k(x)$$

is an odd function, of period h, vanishing on the integer and half-integer net points. Therefore, we need consider only the cases:

(1) $0 \leq x \leq h/4;$

(2) $h/4 \leq x \leq h/2.$

Since $\phi_k(x) = \phi_0(x - x_k)$ and ϕ_0 is even, we have in Case 1 that

$$\tilde{x} = \sum_k x_k \phi_k(x) = \sum_k x_k \phi_0(x - x_k) = \sum_k x_k[\phi_0(x_k - x) - \phi_0(x_k)].$$

By Taylor's expansion we have

$$\tilde{x} = -x \sum_k x_k \phi_0'(x_k) + \frac{x^2}{2} \sum_k x_k \phi_0''(x_k) - \frac{x^3}{6} \sum_k x_k \phi_0'''(x_k - \theta_k x).$$

The second term on the right vanishes and therefore

$$\tilde{x} = -x \sum_k x_k \phi_0'(x_k) - \frac{x^3}{6} \sum_k x_k \phi_0'''(x_k - \theta_k x). \qquad (3.2)$$

If we apply Abel's transformation to the sum

$$\bar{x} = -x \sum_k x_k \phi_0'(x_k)$$

we find

$$\bar{x} = xh \sum_{k=-N}^{N} \sum_{r=-N}^{k} \phi_0'(rh) + xx_N \sum_{r=-N}^{N} \phi_0'(rh) = xh \sum_{k=-N}^{N-1} \sum_{r=-N}^{k} \phi_0'(rh).$$

Since

$$h \sum_{r=-N}^{k} \phi_0'(rh) = \int_{-(N+1/2)h}^{(k+1/2)h} \phi_0'(x)\, dx + \sigma_1(k) = \phi_0\left(x_k + \frac{h}{2}\right) + \sigma_1(k),$$

where

$$|\sigma_1(k)| \le \left(\max_x |\phi_0'''(x)|\right) \frac{h^3}{24} (k + N + 1), \qquad (3.3)$$

we have

$$\bar{x} = x \sum_k \phi_0\left(x_k + \frac{h}{2}\right) + x \sum_{k=-N}^{N-1} \sigma_1(x) = x + x \sum_{k=-N}^{N-1} \sigma_1(k). \qquad (3.4)$$

From (3.2)–(3.4) we obtain the estimate

$$|\sigma(x)| \le \left(\max_x |\phi_0'''(x)|\right)\left[\frac{x^3 h}{6} \sum_{k=-N}^{N} |k| + \frac{xh^3}{24} \sum_{k=-N}^{N-1} (k + N + 1)\right]$$

$$\le \left(\max_x |\phi_0'''(x)|\right)h^4 \tfrac{3}{2} N(N + \tfrac{5}{9}) = C \frac{h^4 3(N + \tfrac{5}{9})N}{2^7(2N + 1)^3 h^3} < C \frac{h}{341N}. \qquad (3.5)$$

Here C is the maximum value of the modulus of the third derivative of the function $x(y) = \phi_0(y(2N + 1)h)$. It is an increasing function of N.

In Case 2 we come to the same conclusion.

The usefulness of this estimate lies in the fact that it offers one of several criteria for choosing the initial function ψ used in the construction of the resolution of the identity: namely, we want to make the third derivative of ψ as small as possible.

A similar estimate can be obtained for a linear function and for several variables.

We now state the fundamental problem considered in this section.

Let the function $f : \Omega \to R$ be given on an irregular net $(x^{(i)})$ $(i = 1, 2, \ldots, n)$, to within an error ε (in the uniform norm). We are to find a function $\tilde{f}$ of class C^r such that

$$\|f - \tilde{f}\| = \max_{i=1,\ldots,n} |f(x^{(i)}) - \tilde{f}(x^{(i)})| \le \varepsilon, \qquad (3.6)$$

and to find a smoothing condition (which we here formulate heuristically) as a requirement that the derivatives of f, of a certain order, should not oscillate.

We repeatedly apply the method we have just described, namely completion using a sequence of resolutions of the identity. While describing the algorithm, we shall point out the relationships that in our opinion guarantee (at least, in the heuristic sense described above) that the smoothing conditions will be satisfied.

Let L_0 denote the operator (of the smoothing approximation) which acts from the space of net functions $f : x^{(i)} \to R$ to the space of differentiable functions $C'(V)$ according to the formula

$$L_0(f)(x) = \sum_{i=1}^{n} f(x^{(i)})\phi_i(x), \qquad (3.7)$$

where (ϕ_i) is a resolution of the identity on Ω of the form described above; the centers of the supports of the functions ϕ_i are the net points $x^{(i)}$.

We first estimate the distance of f from $L_0(f)$. We have

$$\|f - L_0(f)\| = \max_{j=1,\ldots,n} \left| f(x^{(j)}) - \sum_{i=1}^{n} f(x^{(i)})\phi_i(x^{(j)}) \right|$$

$$= \max_{j=1,\ldots,n} \left| f(x^{(j)})(1 - \phi_j(x^{(j)})) - \sum_{\substack{i=1 \\ i \neq j}}^{n} f(x^{(i)})\phi_i(x^{(j)}) \right|$$

$$\leq \|f\| \max \left| 1 - q_j(x^{(j)}) + \sum_{\substack{i=1 \\ i \neq j}}^{n} \phi_i(x^{(j)}) \right|$$

$$\leq 2\|f\| \left[1 - \min_{j=1,\ldots,n} \phi_j(x^{(j)}) \right]. \qquad (3.8)$$

It follows that a single application of (3.7) satisfies the approximation condition if we choose a resolution of the identity such that

$$\lambda = \min_{j=1,\ldots,n} \phi_j(x^{(j)}) \qquad (3.9)$$

is close enough to 1. But λ is close to 1 if either: (a) V_i contains only a few net points; or (b) the functions ϕ_j are highly dome-shaped, which, as can be seen from (3.5), worsens the approximating power of the operator L_0.

This argument indicates that a single application of L_0 will not solve our problem, and we proceed as follows.

Let $\{\phi_i^{(k)}\}_{k=1,2,\ldots}$ be a sequence of resolutions of the identity, supposed for definiteness to be subordinate to one and the same covering of Ω, and let $(L_0^{(k)})$ be the corresponding set of operators of the smoothing approximation, as defined by (3.7) for every $(\phi_i^{(k)})$. We shall seek an $\tilde{f}$ of the form

$$\tilde{f} = \sum_{k=1}^{k_0} L_0^{(k)}(f^{(k-1)}) \qquad (3.10)$$

for some k_0, where $\{f^{(k)}\}$ is a sequence of net functions defined as follows:

$$f^{(k)}(x^{(j)}) = \begin{cases} f(x^{(j)}), & k = 0, \\ f^{(k-1)}(x^{(j)}) - L_0^{(k)}(f^{(k-1)})(x^{(j)}), & k > 0, \end{cases}$$

$$j = 1, 2, \ldots, n. \qquad (3.11)$$

Summing these equations for $k = 0, 1, \ldots, k_0$ we have

$$f = \tilde{f} + f^{(k_0)},$$

so that to fulfill the approximation condition we must have $\|f^{(k_0)}\| \leq \varepsilon$. This condition, however, can always be satisfied by a choice of the sequence $\{\phi_i^{(k)}\}$ of resolutions of the identity. If, for instance, we have for some k_1

$$\lambda^{(k)} = \min_{j=1,\ldots,n} \phi_j^{(k)}(x^{(j)}) \geq \sigma > \tfrac{1}{2}$$

for all $k \geq k_1$, then by (3.8)

$$\|E - L_0^{(k)}\| \leq 2(1 - \sigma) < 1,$$

and therefore $\|f^{(k)}\| \xrightarrow[k \to \infty]{} 0$.

If we choose our sequence of resolutions of the identity $\{\phi_i^{(k)}\}$ to make the $\lambda^{(k)}$ as small as possible under the condition that the sequence $\{\|f^{(k)}\|\}$ decreases monotonically with some preset speed $v > 1$ (i.e.,

$$\|f^{(k)}\|/\|f^{(k+1)}\| \geq v)$$

we can secure as well the fulfillment of the smoothing condition (since then the nonsmooth contribution to the sum (3.10) will be brought in by terms $f^{(k)}$ having small norms).

For practical purposes it is reasonable to construct the sequence $\{\phi_i^{(k)}\}$ as follows: if $\psi^{(k)}(x)$ is the standard function for the resolution $(\phi_i^{(k)})$, the standard function $\psi^{(k+1)}(x)$ for the resolution $(\phi_i^{(k+1)})$ is taken as

$$\psi^{(k+1)}(x) = \psi^{(k)}(x)g^{(k)}(x),$$

where the multiplier $g^{(k)}(x)$ may have either the form

$$g^{(k)}(x) = \exp[-\mu_k(x_1^2 + x_2^2 + \cdots + x_m^2)], \qquad \mu_k \geq 0,$$

or the form

$$g^{(k)}(x) = [1 + \mu_k(x_1^2 + x_2^2 + \cdots + x_m^2)]^{-1}, \qquad \mu_k \geq 0,$$

and so on.

We set $\mu_k = 0$ if $\|f^{(k-1)}\|/\|f^{(k)}\| \geq v$; otherwise, $\mu_k = \mu > 0$.

Thus our algorithm contains three controlling parameters:

(1) the quantity $\lambda^{(0)} = \lambda$, defined by (3.9);
(2) the rate $v > 1$ of decrease of the residuals $\|f^{(k)}\|$;
(3) the parameter μ that controls the rate of growth of the "delta-formedness" of the resolutions of the identity.

Clearly, by optimizing over these parameters we can minimize functionals that define one or another concept of the "oscillability" of the derivatives of the function $\tilde{f}$.

Note. If we make up our sequence of resolutions of the identity from identical resolutions (ϕ_i), we have by (3.10), (3.11)

$$\tilde{f}(x) = \sum_{i=1}^{n} \left(\sum_{k=1}^{k_0} f^{(k-1)}(x^{(i)})\phi_i(x) \right).$$

One might hope to limit the selection of the resolutions of the identity to a single set, and find the solution $\tilde{f}$ of a given problem in the form

$$\tilde{f}(x) = \sum_{i=1}^{n} c_i \phi_i(x), \tag{3.12}$$

defining the coefficients c_i by the conditions

$$\left\| f - \sum_{i=1}^{n} c_i \phi_i(x) \right\| \le \varepsilon \tag{3.13}$$

with the understanding that the smoothing condition is also to be fulfilled.

This scheme misses the essence of the problem; the $f(x^i)$ are subject to error. In fact, if

$$\det \|\phi_i(x^{(j)})\| \ne 0, \tag{3.14}$$

condition (3.13) can be satisfied with $\varepsilon = 0$, but then the errors in the $f(x^{(i)})$ will be passed on to the extension (3.12). Rather, it is desirable to satisfy (3.13) as far as possible, and for this purpose to impose the condition

$$\max_{i=1,\dots,n} |f(x^{(i)}) - c_i| \le E, \qquad E > 0, \tag{3.15}$$

which will bound the norm $\| f - c \|$. But, we have no criterion for the choice of E; if we prescribe it arbitrarily (we want it to be small), we arrive at a generally insoluble problem (aside from the fact that (3.14) is not always satisfied).

The algorithm we have described above arose from these conflicts among the requirements for approximation and smoothing, and the condition (3.15).

3.4 Elements of the General Theory of Splines

We have shown that the cubic splines arise in the minimization of the quadratic functional

$$\int_a^b [u''(x)]^2 \, dx,$$

which is an analog of the deformation energy of an elastic rod. In fact, spline theory developed along two basic directions.

In the first, twofold differentiability was replaced by the requirement for a differential operator L, of general form and even with discontinuous coefficients, and for the minimization of the functional

$$\int_a^b [Lu(x)]^2 \, dx.$$

This approach made spline theory flexible, and at the same time prepared it for the approximation of functions with minimal energy arising in concrete physical problems.

The second approach abandoned the very simple task of dealing with curves fixed at net points, and dealt with other linear functionals arising more naturally from concrete problems. Such splines are those of even degree, constructed by integral methods from a function, trigonometric splines, and those constructed by other currently popular methods. All this, of course, was accompanied by the development of many-dimensional spline theory. All these approximations go together with smoothings.

A *general definition* of interpolating and smoothing splines (cf. Atteia [14]): let X, Y be two Hilbert spaces, and let $T: X \to Y$ be a linear bounded operator from X to Y. Suppose a system of linearly independent bounded linear functionals given on X, k_i $(i = 1, \ldots, n)$. An *interpolating spline* is an element $\sigma \in X$ satisfying the two conditions

(1) $$k_i(\sigma) = (k_i, \sigma)_X = r_i, \qquad i = 1, 2, \ldots, n;$$

(2) $$(T\sigma, T\sigma)_Y = \|T\sigma\|_Y^2 = \min. \tag{4.1}$$

Here the symbols $(,)_X, (,)_Y$ denote the scalar products in the spaces X and Y, respectively; the r_i $(i = \overline{1, n})$ are numbers given in advance. The value of k_i on the vector σ is replaced by a scalar product in accordance with the well-known Riesz theorem on the representation of linear functionals on Hilbert spaces. The corresponding *smoothing spline* is an element $\sigma_\alpha \in X$ which yields the minimum value of the functional

$$\Phi_\alpha(u) = \alpha \|Tu\|_Y^2 + \sum_{i=1}^n [(k_i, u) - r_i]^2, \qquad \alpha > 0. \tag{4.2}$$

General existence and uniqueness theorems have been formulated for spline functions defined according to (4.1) and (4.2) (cf. Atteia [14]; Laurent, Anselon [14]). These are currently applied for the construction of many different forms of spline approximation. There exists a general algorithm for the construction of interpolating and smoothing splines. It has been shown that in the end the problems (4.1) and (4.2) reduce to the solution of algebraic systems with symmetric positive definite matrices.

We now present a convergence theorem of a quite general form (cf. Vasilenko [14]). This theorem can be applied both in the classical case, the convergence of splines on rectangular nets, and in the more complicated cases, for the construction of splines on condensing chaotic nets or to analyze the convergence of splines constructed on arbitrary systems of functionals.

First, we consider the convergence of interpolating splines. Let X be a Hilbert space, ϕ^* an element of X that we want to approximate. We consider a (generally nonorthogonal) basis in X consisting of the vectors $k_1, k_2, \ldots$ $k_n, \ldots$. If $T: X \to Y$ is a bounded linear operator acting from X to another Hilbert space Y, we shall seek the approximating element as the spline σ_n:

$$(\sigma_n, k_i)_X = (\phi^*, k_i)_X = r_i, \qquad i = 1, \ldots, n,$$
$$\|T\sigma_n\|_Y^2 = \min. \tag{4.3}$$

If we now suppose satisfied some requirements guaranteeing the existence and uniqueness of σ_n for some $n = n_0$, the splines σ_n will exist for all $n \geq n_0$. We ask now about the convergence of the σ_n to the exact function ϕ^*. If the requirements of the theorem on existence and uniqueness are met, and if the kernel of the operator T has a finite dimension, then the σ_n converge to ϕ^*, and, what is more important in practice, the sequence $T\sigma_n$ converges to $T\phi^*$ in the Y-norm.

The general convergence theorem can be fruitfully applied to estimate convergence in terms of decrease in the net step (cf. Vasilenko [14]). Applied, for instance, to splines consisting of the abutments of monomials of degree $2n - 1$, it yields on the interval $[a, b]$ the following estimate for the approximation of $\phi^* \in W_2^n$ and its derivatives by the spline σ_N:

$$\|\sigma_N^{(i)} - \phi^{*(i)}\|_{C[a, b]} = o(h^{n-i-1/2}), \qquad i = 0, 1, \ldots, n - 1,$$
$$\|\sigma_N^{(n)} - \phi^{*(n)}\|_{L_2[a, b]} = o(1). \tag{4.4}$$

(The symbol $o(h^\alpha)$ indicates that the error tends to 0 faster than h^α.)

The general convergence theorem can also be applied in cases where the use of previously known methods has yielded no result. An illustration is the convergence of splines formed on a randomly distributed set of interpolation nodes in a plane region. The splines are constructed by minimizing the functional

$$\Phi(u) = \int_\Omega \left[\left(\frac{\partial^2 u}{\partial x^2}\right)^2 + 2\left(\frac{\partial^2 u}{\partial x \, \partial y}\right)^2 + \left(\frac{\partial^2 u}{\partial y^2}\right)^2 \right] d\Omega, \tag{4.5}$$

when the function u takes given values on a fixed set of nodes. If $\phi^* \in W_2^2(\Omega)$ the general theorem yields the estimate

$$\|\sigma_N - \phi^*\|_{C[\Omega_\delta]} = o(h^{3/2}). \tag{4.6}$$

Here σ_N is our spline; h is a parameter characterizing the condensation of the random net, defined as the parameter of an ε-net formed of interpolation nodes in Ω; and Ω_δ is an arbitrary δ-interior of Ω, i.e., a set of points in Ω, all further than δ from the boundary of Ω.

We have briefly surveyed the convergence theory of interpolating splines. The problem of constructing such splines is that of minimizing a quadratic functional given linear constraints, i.e., a problem of conditional minimization. The problem of constructing smoothing splines is one of unconditional minimization, which is often easier. A smoothing spline σ_α converges to an interpolating spline σ in the norm of X as $\alpha \to 0$.

We must admit that the algorithm for the construction of both types of spline, which seems universal as seen from the theoretical point of view, is in practice difficult to apply even in two-dimensional problems. It is, therefore, worthwhile to adopt approximate construction methods. We change the problem of finding an interpolating spline σ into that of finding a smoothing spline σ_α defined by minimizing the functional

$$\Phi(u) = \alpha \| Tu \|_X^2 + \sum_{i=1}^{n} [(k_i, u) - r_i]^2, \qquad \alpha > 0, \tag{4.7}$$

for small α over the whole space X. We consider a finite-dimensional subspace $E_k \subset X$, whose elements approximate those of X to whatever precision suits us, and we look for a minimum of $\Phi_\alpha(u)$ on $E_{k,t}$. The minimum is called a *spline on the subspace* (cf. Vasilenko [14]).

This approach is successful in the following problem. Let Ω be a rectangular region containing a random distribution of nodes p_i $(i = 1, 2, \ldots, s)$. We are to construct in Ω a smooth interpolating function. To do so, we introduce in it a rectangular net Ω_h and associate with it a finite space E_k of smooth interpolants.

We have already seen how to construct these spaces. We may use either *bicubic splines*, or *Ryabenkov interpolations* of class C^2. On our space we solve the minimization problem for the functional

$$\Phi_\alpha(u) = \alpha \int_\Omega [u_{xx}^2 + 2u_{xy}^2 + u_{yy}^2] \, d\Omega + \sum_{i=1}^{s} [u(P_i) - r_i] \tag{4.8}$$

for small $\alpha > 0$, obtaining a linear algebraic system. If we succeed in finding a basis for E_k consisting of finite functions, the matrix of the system is said to be a *band matrix*. Such a basis might consist of cubic B-splines, or, more suitably in our opinion, might consist of the local basis functions considered in Section 3.1.3.

In concluding this section, we consider one more question of practical importance—namely how to choose the smoothing parameter. There are several criteria, of which we consider one: the so-called "residuals criterion," which is useful in a wide variety of problems (cf. Morozov [14]).

Suppose we are solving the smoothing problem, i.e., finding an element σ_α of the Hilbert space X that minimizes the functional

$$\Phi_\alpha(u) = \alpha \| Tu \|_Y^2 + \sum_{i=1} [(k_i, u) - r_i]^2 \equiv \alpha \| Tu \|_Y^2 + \| Ku - r \|^2. \tag{4.9}$$

Here $K : X \to R^n$ is an operator defined by the formula

$$Ku = \{(u, k_1), \ldots, (u, k_n)\} \tag{4.10}$$

and $r = (r_1, r_2, \ldots, r_n)$ is a given vector. The residuals criterion consists in choosing α by the condition

$$\| K\sigma_\alpha - r \| = \varepsilon, \tag{4.11}$$

where ε (a fixed quantity) is the "admissible level of the residuals." If we adopt the notation

$$\phi(\alpha) = \|K\sigma_\alpha - r\|, \qquad 0 < \alpha < +\infty, \tag{4.12}$$

the determination of α requires the solution of the nonlinear equation

$$\phi(\alpha) = \varepsilon.$$

Let us introduce the variable $\beta = 1/\alpha$ and consider the equation

$$\phi(\alpha) = \phi(1/\beta) = \psi(\beta) = \varepsilon \tag{4.13}$$

in β. We note that $\psi(\beta)$ is strictly monotone decreasing and that the equation $\psi(\beta) = \varepsilon$ has a unique root for $\varepsilon < \varepsilon_0$, where $\varepsilon_0 = \|Ke - r\|$ and e is the least-squares solution of the problem

$$\|Ke - r\| = \min_{u \in \text{Ker } T} \|Ku - r\|$$

over the kernel Ker T of the operator T. It is natural to apply Newton's method in solving the nonlinear problem (4.13). We raise both sides of (4.13) to some real power q,

$$\psi^q(\beta) = \varepsilon^q, \tag{4.14}$$

and with the help of q seek to maximize the rate of convergence of Newton's process. The admissible values of q, for which the process converges for any initial approximation, lie in the ranges $-1 \le q < 0, q > 0$; and the maximum convergence rate is obtained for $q = -1$. Thus we wish to solve the equation

$$f(\beta) = 1/\phi, (1/\beta) = \varepsilon^{-1}. \tag{4.15}$$

The Newton iteration proceeds according to the formula

$$\beta^{p+1} = \beta^p - \frac{f(\beta^p) - \varepsilon^{-1}}{f'(\beta^p)} = \beta^p - (\beta^p)^2 \frac{\phi(1/\beta^p)}{\phi'(1/\beta^p)} \cdot \frac{\varepsilon - \phi(1/\beta^p)}{\varepsilon}. \tag{4.16}$$

We ask how much computation must be done at each step of the iteration: to compute $\phi(1/\beta^p) = \phi(\alpha^p)$ we need to solve a spline-smoothing problem for the fixed parameter $\alpha = \alpha^p$, and this means solving a linear algebraic system. Also, we need to know $\phi'(\alpha^p)$ and this means solving the same linear algebraic system with a different right-hand side.

We note in conclusion that the development of smoothing theory in its general form is wholly equivalent to the development of a method for regularizing ill-posed problems (cf. Tikhonov [16]). We may suppose, for instance, that K is a singular matrix and that the interpolation spline u is a normal solution of the simultaneous system

$$Ku = f, \qquad \|u\| = \min. \tag{4.17}$$

The solution of this system (the interpolation spline) may be approximated as $\alpha \to 0$ by the solution u_α of the regularized system

$$\alpha \|u_\alpha\|^2 + \|Ku_\alpha - f\|^2 = \min,$$
$$(\alpha I + K^*K)u_\alpha = K^*f$$

(4.18)

i.e., by a smoothing spline.

CHAPTER 4

Methods for Solving Stationary Problems of Mathematical Physics

Numerical methods for stationary problems of mathematical physics constitute a more or less self-contained subject, even though many stationary problems can be treated as asymptotic cases, $(t \to \infty)$, for nonstationary problems. In solving stationary problems by asymptotic methods, we do not pay any attention to the transient behavior, since it is of no interest whatsoever. In the nonstationary case, however, the transient behavior has a physical meaning. Generally speaking, this is the point where the differences between the two classes of problems start and end. Let us illustrate this with an example.

Consider the problem

$$A\phi = f,$$

where

$$A > 0, \quad \phi \in \Phi \quad \text{and} \quad f \in F.$$

Replace this problem by the nonstationary one:

$$\frac{\partial \psi}{\partial t} + A\psi = f,$$

$$\psi = 0 \quad \text{for} \quad t = 0.$$

(In what follows, we will not indicate the domains of the operators.) Solutions of these problems will be sought in the form

$$\phi = \sum_n \phi_n u_n, \quad \psi = \sum_n \psi_n u_n,$$

where

$$f = \sum_n f_n u_n, \quad A u_n = \lambda_n u_n, \quad A^* u_n^* = \lambda_n u_n^*,$$

$\phi_n = (\phi, u_n^*)$, $\psi_n = (\psi, u_n^*)$, $f_n = (f, u_n^*)$, $\{u_n\}$ and $\{u_n^*\}$ biorthogonal bases. In a standard manner we obtain the following relations for the Fourier coefficients:

$$\lambda_n \phi_n = f_n,$$

and

$$\frac{d\psi_n}{\partial t} + \lambda_n \psi_n = f_n, \qquad \psi_n(0) = 0.$$

Solving these equations, we obtain

$$\phi = \sum_n \frac{f_n}{\lambda_n} u_n, \qquad \psi = \sum_n \frac{f_n}{\lambda_n} (1 - e^{-\lambda_n t}) u_n.$$

Suppose that the spectrum of the operator A is real, and that, in fact, $\lambda_n > 0$ ($n = 1, 2, \ldots$). It follows that

$$\lim_{t \to \infty} \psi = \phi.$$

If the operator in the stationary case has an arbitrary spectrum, such a simple and transparent relationship among the solutions may fail to exist.

Clearly, the nonstationary problem for ψ can be solved using difference methods with respect to t. For example,

$$\frac{\psi^{j+1} - \psi^j}{\tau} + A\psi^j = f,$$

or

$$\psi^{j+1} = \psi^j - \tau(A\psi^j - f).$$

As for the stationary problem, the solution is given by

$$\lim_{j \to \infty} \psi^j = \phi,$$

assuming certain relation between τ and $\beta(A)$.

The parameter τ may or may not depend on j. In any case, from the point of view of solving the stationary problem, it is convenient to interpret the index j as denoting the iteration rather than time.

There is another peculiarity: in the nonstationary case the parameter τ should be small to guarantee the accuracy of the solution; in the stationary case the optimal iteration parameter τ is chosen to minimize the number of iterations, and may be large.

A survey of iterative methods of linear algebra can be found in the monographs by Faddeev and Faddeeva [8], Forsythe and Moler [8], Wilkinson [8], Householder [3], Voevodin [8], Bakhvalov [8], Marchuk and Lebedev [17], Varga [3], Marchuk and Kuznetsov [8], Samarskii and Nikolaev [8], and others.

4.1 General Concepts of Iteration Theory

Throughout the rest of the book we shall suppose that the operator A is a square matrix. Thus, we are supposing that the initial problem has already been reduced to the solution of a linear algebraic equation system. Except in Section 4.5 we shall suppose that the matrix A is nonsingular, and we shall suppose everywhere that it and all vectors we encounter are real. Thus, we are to solve the system

$$A\phi = f, \tag{1.1}$$

where A is a matrix and ϕ and f are vectors.

Most iteration methods used for the solution of linear systems can be described by the general formula

$$B_j \frac{\phi^{j+1} - \phi^j}{\tau_j} = -(A\phi^j - f), \tag{1.2}$$

where $\{B_j\}$ is a sequence of nonsingular matrices and $\{\tau_j\}$ is a sequence of real parameters. If we write $H_j = \tau_j B_j^{-1}$ we may write (1.2) in the equivalent form

$$\phi^{j+1} = \phi^j - H_j(A\phi^j - f). \tag{1.3}$$

The vectors $\xi^j = A\phi^j - f$ are called the (vector) *residuals* of the iteration method (1.2), and the vectors $\psi^j = \phi^j - \phi^*$, where $\phi^* = A^{-1}f$ is the exact solution of (1.1), are called the *error vectors*. Subtracting ϕ^* from both sides of (1.3) and writing $f = A\phi^*$, we arrive at an equation for the sequence of error vectors:

$$\psi^{j+1} = T_j\psi^j, \tag{1.4}$$

where

$$T_j = E - H_jA \tag{1.5}$$

is called the *operator of the jth step* of the iteration process (1.2). If we multiply (1.4) by the matrix A, we arrive at an equation for the sequence of residuals:

$$\xi^{j+1} = (E - AH_j)\xi^j. \tag{1.6}$$

We shall say that (1.2) converges if the sequence $\{\phi^j\}$ converges to the exact solution ϕ^* of (1.1) for all initial vectors; otherwise, we say it diverges. Clearly, (1.2) converges if and only if the sequences $\{\xi^j\}$ and $\{\psi^j\}$ converge to the null vector for all ψ^0 and ξ^0 $(A\psi^0 = \xi^0)$.

The iteration method (1.2) is said to be *stationary* if H_j does not depend on the number of iterations (i.e., T_j is constant); otherwise, it is *nonstationary*. As a special case we introduce the class of cyclic iteration methods, which may be either stationary or nonstationary.

We shall say that an iteration method is *cyclic* if for all $j \geq 0$, and for some fixed $s \geq 1$, $H_j = H_{j+s}$. It is easy to see that if we collapse every s successive iterations into one iteration, we obtain a stationary process of the form

$$\phi^{j+1} = \phi^j - \tilde{H}(A\phi^j - f), \tag{1.7}$$

where $\tilde{H}$ is defined by the equation

$$E - \tilde{H}A = \prod_{i=0}^{s-1} (E - H_i A). \tag{1.8}$$

In its original form, on the other hand, a cyclic process is nonstationary.

One of our main aims in the study of iteration processes is optimization, i.e., the choice of a sequence $\{H_j\}$ of matrices from some given class that will yield the most effective computational method. The specific function that we want to minimize is the total number of arithmetic and logical operations that must be performed in order to solve the problem with a given accuracy, $\varepsilon > 0$. A complete solution of this problem is attainable only in special circumstances; in practice, we minimize instead some function

$$W(\{H_j\}, \varepsilon), \tag{1.9}$$

which majorizes our specific function and in some definite sense approximates it. If we assume that for all H_j in the given class a given iteration process (1.2) requires the same number W_0 of arithmetic and logical operations, then W is usefully defined as

$$W = W_0 \cdot N(\{H_j\}, \varepsilon), \tag{1.10}$$

where $N(\{H_j\}, \varepsilon)$ is the number of iterations of the process (1.2) that must be performed in order to reduce some norm of the initial error ψ^0 by a factor of $1/\varepsilon$. In other words, $N(\{H_j\}, \varepsilon)$ is the number of iterations of (1.2) that must be performed to achieve an accuracy of ε when the norm of the initial error vector ψ^0 has the value 1. In the concrete examples that we give later we shall show how to define the functional N and minimize it.

4.2 Some Iterative Methods and Their Optimization

4.2.1 The Simplest Iteration Method

Suppose that the matrix A of the system

$$A\phi = f \tag{2.1}$$

is symmetric and positive definite, and that the bounds of its spectrum $\beta = \beta(A) \geq \alpha = \alpha(A) > 0$ are known. We apply the iteration process

$$\phi^{j+1} = \phi^j - \tau(A\phi^j - f) \tag{2.2}$$

with some initial vector ϕ^0 and a real parameter τ. The error vectors $\psi^j = \phi^j - \phi^*$ in this method satisfy the equations

$$\psi^{j+1} = T_\tau \psi^j, \tag{2.3}$$

where

$$T_\tau = E - \tau A \tag{2.4}$$

is the operator for a single step.

We denote by $\{\lambda_n\}$ the set of eigenvalues of A, and by $\{u_n\}$ the complete orthonormal system of the eigenvectors of A. Then, resolving the vectors $\{\psi^j\}$ on the orthogonal basis $\{u_n\}$ we have:

$$\psi^j = \sum_n \psi_n^j u_n, \qquad \psi_n^j = (\psi^j, u_n),$$

and using (2.3) we find that

$$\psi_n^{j+1} = (1 - \tau\lambda_n)\psi_n^j.$$

It follows that as $j \to \infty$ and for arbitrary ψ^0 the coefficients ψ_n^j converge to zero if and only if

$$|1 - \tau\lambda_n| < 1,$$

and that the Euclidean norm

$$\|\psi^j\|_2 = (\psi^j, \psi^j)^{1/2} = \left[\sum_n (\psi_n^j)^2 \right]^{1/2}$$

of the vector ψ^j converges to zero if and only if

$$q(\tau) = \beta(T_\tau) = \max_n |1 - \tau\lambda_n| < 1. \tag{2.5}$$

Let us look at this quantity $q(\tau)$. If we consider the system of inequalities

$$-1 < 1 - \tau\lambda_n < 1, \qquad \lambda_n \in [\alpha, \beta],$$

which is equivalent to (2.5), we conclude that (2.5) will be satisfied only if

$$0 < \tau < \min_n \frac{2}{\lambda_n} = \frac{2}{\beta}, \tag{2.6}$$

i.e., if $\tau \in (0, 2/\beta)$. Thus, to define the range of values of τ for which the process (2.2) converges we need only know the magnitude of β or an upper bound for it.

Let us consider now the optimization of the process (2.2). We begin with the inequality

$$\|\psi^j\|_2 \le \|T_\tau^j\|_2 \|\psi^0\|_2 \tag{2.7}$$

Since T_τ is symmetric,

$$\|T_\tau^j\|_2 = [\beta(T_\tau)]^j = [q(\tau)]^j, \tag{2.8}$$

and we see that to decrease $\|\psi^0\|_2$ by a factor of $1/\varepsilon$ ($\varepsilon < 1$), we need perform only N iterations, where N is defined by the equation

$$\|T_\tau^N\| = \varepsilon. \tag{2.9}$$

Using (2.8) we find that

$$N = \left[\frac{\ln \varepsilon}{\ln[q(\tau)]}\right] + 1. \tag{2.10}$$

Since the number W_0 of arithmetic and logical operations performed in a single iteration of (2.2) is independent of τ, the functional W in (1.10) is, in this case, defined by

$$W = W_0\left(\left[\frac{\ln \varepsilon}{\ln[q(\tau)]}\right] + 1\right). \tag{2.11}$$

(We assume that $\tau \in (0, 2/\beta)$.) It is easy to see now that the optimization problem (to minimize W with respect to τ) reduces to the minimization of $q(\tau)$ with respect to variation of τ over the range $(0, 2/\beta)$.

A simple analysis shows that

$$q(\tau) = \max\{|1 - \tau\alpha|, |1 - \tau\beta|\}, \tag{2.12}$$

and the optimal value of τ is found by solving the equation

$$1 - \tau_{\text{opt}}\alpha = -(1 - \tau_{\text{opt}}\beta)$$

and is computed by the formula

$$\tau_{\text{opt}} = \frac{2}{\beta + \alpha}. \tag{2.13}$$

Using this value of τ in (2.12) we find

$$q_{\text{opt}} = q(\tau_{\text{opt}}) = \frac{\beta - \alpha}{\beta + \alpha} = \frac{p - 1}{p + 1}, \tag{2.14}$$

where

$$p \equiv p(A) = \frac{\beta}{\alpha} \tag{2.15}$$

is called the *conditioning number* of the positive definite matrix A.

We now introduce an auxiliary quantity

$$R(T_\tau) = -\ln q(\tau), \tag{2.16}$$

which we shall call the *asymptotic rate of convergence* of the method (2.2). Clearly, $R^{-1}(T_\tau)$ is equal to the number of iterations that suffice, and for arbitrary ψ^0 are necessary, to decrease $\|\psi^0\|_2$ by a factor of e, where $e = 2.7\ldots$ is the base of the natural system of logarithms. The asymptotic rate of convergence provides a useful criterion for comparing different methods when the total number of arithmetic and logical operations per iteration is left out of account. With respect to our auxiliary quantity, equation (2.11) takes the form

$$W = W_0 \left[\frac{|\ln \varepsilon|}{R(T_r)} + 1 \right]. \tag{2.17}$$

In concluding, we consider ill-conditioned matrices, i.e., those for which $p \gg 1$. Recognizing that for $p \gg 1$

$$q_{opt} \approx 1 - 2/p, \tag{2.18}$$

we find that

$$R(T_r) \approx 2/p \tag{2.19}$$

and therefore

$$W \approx W_0 |\ln \varepsilon| \frac{p}{2}. \tag{2.20}$$

A similar analysis turns out to be extremely useful in making qualitative comparisons of different methods as applied to ill-conditioned matrices.

4.2.2 Convergence and Optimization of Stationary Iterative Methods

We have shown that a necessary and sufficient condition for the convergence of the iteration method (2.2), given an arbitrary initial approximation ϕ^0, is that

$$\beta(T) = \max_n |\lambda_n(T)| < 1, \tag{2.21}$$

where $\lambda_n(T)$ is an eigenvalue of the step operator T. We shall now show that (2.21) is a necessary and sufficient condition for the convergence of the stationary method

$$\phi^{j+1} = \phi^j - H(A\phi^j - f)$$

with the transition operator (step operator)

$$T = E - HA,$$

which is a constant matrix.

Let

$$J = \begin{Vmatrix} J_1 & 0 & \cdots & 0 \\ 0 & J_2 & \cdots & 0 \\ \multicolumn{4}{c}{\dotfill} \\ 0 & 0 & \cdots & J_k \end{Vmatrix}$$

be the normal Jordan form of the matrix T (i.e., $T = SJS^{-1}$, where the columns of S are the eigenvectors and root vectors of T) with the Jordan cells

$$J_i = \left\| \begin{array}{cccccc} \lambda_i & 1 & 0 & \cdots & 0 & 0 \\ 0 & \lambda_i & 1 & \cdots & 0 & 0 \\ 0 & 0 & \lambda_i & \cdots & 0 & 0 \\ \multicolumn{6}{c}{\dotfill} \\ 0 & 0 & 0 & \cdots & \lambda_i & 1 \\ 0 & 0 & 0 & \cdots & 0 & \lambda_i \end{array} \right\|$$

of order $k_i \geq 1$, corresponding to the eigenvalues $\lambda_i = \lambda_i(T)$ $(i = 1, \ldots, l)$. Then, since the error vectors ψ^j are given by

$$\psi^j = T^j \psi^0 = SJ^jS^{-1}\psi^0,$$

it is not difficult to see that our stationary iteration method will converge if and only if the matrices T^j (and therefore the matrices J^j) converge to the null matrix as $j \to \infty$.

Therefore, to prove our assertion about the necessary and sufficient conditions for the convergence of a stationary iteration process, we need only show that the matrices J_i^j converge to the null matrix as $j \to \infty$ if and only if $|\lambda_i| < 1$. We denote by $\alpha_{s,t}^{(j)}$ the elements of the matrix J_i^j. Then by an immediate multiplying-out of the matrix, it is not difficult to show that for $j \geq k_i - 1$

$$\alpha_{s,t}^{(j)} = \begin{cases} 0, & \text{if } s > t, \\ \lambda_i^j, & \text{if } s = t, \\ C_j^{t-s}\lambda_i^{j-t+s}, & \text{if } s < t, \end{cases}$$

where the

$$C_j^{t-s} = \frac{j(j-1)\cdots(j-t+s+1)}{(t-s)!}$$

are binomial coefficients. In other words,

$$J_i^j = \left\| \begin{array}{cccc} \lambda_i^j & C_j^1\lambda_i^{j-1} & \cdots & C_j^{k_i-1}\lambda_i^{j-k_i+1} \\ 0 & \lambda_i^j & \cdots & C_j^{k_i-2}\lambda_i^{j-k_i+2} \\ \multicolumn{4}{c}{\dotfill} \\ 0 & 0 & \cdots & \lambda_i^j \end{array} \right\|$$

for all $j \geq k_i - 1$ and $1 \leq i \leq l$. By some elementary calculations we can easily show that the $\alpha_{s,t}^{(j)}$ tend to zero for all $1 \leq s, t \leq k_i$ as $j \to \infty$, if and only if, $|\lambda_i| < 1$. Then the J_i^j tend to the null matrix. This proves our assertion. It has been shown that this result is provable for an arbitrary stationary iteration method.

We are concerned with a general approach to the optimization of stationary iteration methods as to their asymptotic behavior. We shall need a well-known fact from functional analysis, which in the matrix case is formulated as follows.

For any square matrix T and any matrix norm $\|\cdot\|$,

$$\lim_{k \to \infty} \|T^k\|^{1/k} = \beta(T). \tag{2.22}$$

Let us now suppose that the matrix H_τ of the stationary iteration process

$$\phi^{j+1} = \phi^j - H_\tau(A\phi^j - f) \tag{2.23}$$

depends on the parameters $\tau_1, \tau_2, \ldots, \tau_s$. Then if each step of the iteration process requires the same number of arithmetic and logical steps, independently of the values of the parameters, our preceding arguments (cf. Section 4.1) show that the optimization of (2.23) is equivalent to the minimization over $\tau_1, \ldots, \tau_s$ of the function

$$W = W_0 \cdot N(H_\tau, \varepsilon), \tag{2.24}$$

where $N(H_\tau, \varepsilon)$ is the number of iterations needed to reduce some norm of the error vector ψ^0 by $1/\varepsilon$ times.

Since

$$\|\psi^j\| \le \|T_\tau^j\| \|\psi^0\|, \tag{2.25}$$

where $T_\tau = E - H_\tau A$, and $\|\cdot\|$ is a norm, the equation defining the dependence of N on H_τ and ε has the form

$$\|T_\tau^N\| = \varepsilon. \tag{2.26}$$

It is clear that an exact solution of this equation can be obtained only under exceptional circumstances, as, for instance, in Section 4.2.1. In concrete situations, we must employ auxiliary assumptions when posing the question of an approximate solution of (2.26).

One possible approach to the determination of the dependence of N on H_τ and ε is via the asymptotic method: namely, if we assume $\varepsilon \ll 1$, and therefore $N \gg 1$, the theorem formulated above implies that

$$\|T_\tau^N\| = [\|T_\tau^N\|^{1/N}]^N \approx [\beta(T_\tau)]^N,$$

$$N(H_\tau, \varepsilon) \approx \frac{|\ln \varepsilon|}{R(T_\tau)}, \tag{2.27}$$

where $R(T_\tau) = -\ln \beta(T_\tau)$ is the asymptotic rate of convergence of the process (2.23). Then the function W is given by

$$W = W_0 \frac{|\ln \varepsilon|}{R(T_\tau)}. \tag{2.28}$$

We note that (2.17) is a particular case of (2.28).

Now (2.28) implies that, under our assumptions, we can optimize the process (2.23) by maximizing $R(T_\tau)$ [i.e., minimizing $\beta(T_\tau)$] with respect to those parameters $\tau_1, \ldots, \tau_s$ for which $\beta(T_\tau) < 1$.

Let us pause for a moment to consider the cyclic iteration methods having period $s \geq 1$. As we saw in Section 4.1, a cyclic iteration process can be reduced to a stationary process (1.7) with the transition operator

$$\tilde{T}_\tau = \prod_{i=0}^{s-1} (E - H_i A). \tag{2.29}$$

If we assume that each application of (1.7) requires sW_0 arithmetic and logical operations, where W_0 is the analogous number for a single step of the cyclic process, we find

$$W = W_0 \frac{|\ln \varepsilon|}{R(\tilde{T}_\tau)}, \tag{2.30}$$

where

$$R(\tilde{T}_\tau) = -\frac{1}{s} \ln \beta(\tilde{T}_\tau).$$

It follows that the optimization of a cyclic iteration process is accomplished by minimizing $\beta(\tilde{T}_\tau)$.

4.2.3 The Successive Over-Relaxation Method

Young and Frankel have devised a method which is popular in many applications. This is the so-called method of successive over-relaxation. The idea is as follows.

Consider a linear algebraic system

$$A\phi = f \tag{2.31}$$

with the block-structured tridiagonal matrix

$$A = \left\|\begin{array}{ccccc} E_1 & -S_1 & \cdots & 0 & 0 \\ -R_2 & E_2 & \cdots & 0 & 0 \\ \cdots & \cdots & \cdots & \cdots & \cdots \\ 0 & 0 & \cdots & E_{k-1} & -S_{k-1} \\ 0 & 0 & \cdots & -R_k & E_k \end{array}\right\|, \tag{2.32}$$

where the E_i are unit matrices of order n_i and R_i, S_i are matrices of order $n_i \times n_{i-1}$, $n_i \times n_{i+1}$, respectively. Matrices of this type often arise in the reduction of elliptic equations to systems of finite-difference or variational-difference equations.

We represent the matrix A in the form

$$A = E - R - S, \tag{2.33}$$

where

$$R = \begin{Vmatrix} 0 & 0 & \cdots & 0 & 0 \\ R_2 & 0 & \cdots & 0 & 0 \\ \cdots\cdots\cdots\cdots\cdots\cdots \\ 0 & 0 & \cdots & 0 & 0 \\ 0 & 0 & \cdots & R_k & 0 \end{Vmatrix}, \quad S = \begin{Vmatrix} 0 & S_1 & \cdots & 0 & 0 \\ 0 & 0 & \cdots & 0 & 0 \\ \cdots\cdots\cdots\cdots\cdots \\ 0 & 0 & \cdots & 0 & S_{k-1} \\ 0 & \cdot & \cdots & 0 & 0 \end{Vmatrix}$$

and E is a unit matrix. If we introduce the matrix

$$B = R + S, \tag{2.34}$$

the system (2.31) can be written in the form

$$\phi = B\phi + f. \tag{2.35}$$

We shall further suppose that the iteration process

$$\phi^{j+1} = B\phi^j + f \tag{2.36}$$

converges (i.e., $\beta(B) < 1$), that the eigenvalues of B are real, and that B has a complete set of eigenvectors.

We rewrite (2.31) as a system of matrix equations

$$\Phi_l - R_l\Phi_{l-1} - S_l\Phi_{l+1} = F_l, \qquad l = 1, \ldots, k, \tag{2.37}$$

under the assumption that

$$R_1 = S_k = 0 \quad \text{and} \quad \Phi_0 = \Phi_{k+1} = 0,$$

where Φ_l and F_l are the vector components of the vectors ϕ and f. Then the method of successive over-relaxation is defined by the formulas

$$\Phi_l^{j+1} = \Phi_l^j - \tau(\Phi_l^j - R_l\Phi_{l-1}^{j+1} - S_l\Phi_{l+1}^j - F_l), \qquad l = 1, \ldots, k, \tag{2.38}$$

where τ is a real parameter. We may also write these relations in the form

$$\Phi_l^{j+1} = \tau\Phi_l^{j+1/2} + (1 - \tau)\Phi_l^j,$$
$$\Phi_l^{j+1/2} = R_l\Phi_{l-1}^{j+1} + S_l\Phi_{l+1}^j + F_l, \qquad l = 1, \ldots, k.$$

From all these formulas we see that for every j the computations are accomplished sequentially, using the first k components. We set Φ_0^{j+1} and Φ_{k+1}^j equal to zero.

We now go from (2.38) to the equations for the error vectors $\Psi_l^j = \Phi_l^j - \Phi_l^*$, where $\{\Phi_l^*\}$ is the solution of (2.37)

$$\Psi_l^{j+1} = \Psi_l^j - \tau(\Psi_l^j - R_l\Psi_{j-1}^{l+1} - S_l\Psi_{l+1}^j), \qquad l = 1, \ldots, k. \tag{2.39}$$

It is clear that the transition operator of (2.37) is defined by the equation

$$T_\tau = (E - \tau R)^{-1}[(1 - \tau)E + \tau S] \tag{2.40}$$

and depends on a parameter τ. The spectral problem

$$T_\tau \Psi = \lambda(T_\tau)\Psi \tag{2.41}$$

is easily transformed into

$$\lambda(T_\tau)\Psi_l = \Psi_l - \tau[\Psi_l - \lambda(T_\tau)R_l\Psi_{l-1} - S_l\Psi_{l+1}], \qquad l = 1, \ldots, k. \tag{2.42}$$

We shall seek its solution in the form

$$\Psi_l = [\lambda(T_\tau)]^{1/2}w_l, \tag{2.43}$$

where the $\{w_l\}_{l=1}^k$ are the vector components of the eigenvector w of the spectral problem

$$Bw = \lambda(B)w, \tag{2.44}$$

and $w_0 = w_{k+1} = 0$. Using (2.43) in (2.42) and taking account of (2.44), we find

$$\lambda^{1/2}(T_\tau)[\lambda(T_\tau) - \tau\lambda^{1/2}(T_\tau)\lambda(B) + \tau - 1]w_l = 0, \qquad l = 1, \ldots, k. \tag{2.45}$$

We are interested in the convergence of the successive over-relaxation method; from this point of view the zero values of $\lambda(T_\tau)$ are of no interest. Also, at least one of the $\{w_l\}$ differs from zero. Therefore, we may derive from (2.45) the equation

$$\lambda(T_\tau) - \tau\lambda^{1/2}(T_\tau)\lambda(B) + \tau - 1 = 0,$$

which connects the eigenvalues of T_τ with those of B. This equation yields

$$\lambda^{1/2}(T_\tau) = \frac{\tau\lambda(B)}{2} \pm \sqrt{\frac{\tau^2\lambda^2(B)}{4} - \tau + 1}. \tag{2.46}$$

Before we analyze this expression, we note that if $\lambda(B)$ is an eigenvalue of B, then so is its negative, $-\lambda(B)$. For, if $\lambda(B)w = Bw$, where w is an eigenvector of B with the vector components $\{w_p\}_{p=1}^k$, then the vector $\tilde{w}$ with the vector components $\tilde{w}_p = (-1)^p w_p$ is also an eigenvector of B, corresponding to the eigenvalue $-\lambda(B)$, i.e.,

$$-\lambda(B)\tilde{w} = B\tilde{w}. \tag{2.47}$$

This, together with (2.46), means that in the analysis of $\lambda^{1/2}(T_\tau)$ we may restrict ourselves to the case $\lambda(B) \geq 0$.

Let us now investigate the value of $\lambda^{1/2}(T_\tau)$ as a function of τ, assuming that $\lambda(B) < 1$ is a fixed nonnegative eigenvalue of B. We first consider the case $\tau \leq 0$. Some simple calculations will show that for $\tau \leq 0$ the inequality

$$|\lambda^{1/2}(T_\tau)| = \left|\frac{\tau\lambda(B)}{2} \pm \sqrt{\frac{\tau^2\lambda^2(B)}{4} - \tau + 1}\right| < 1 \tag{2.48}$$

has no solution, and therefore the requirement that τ be positive is necessary for the convergence of the successive over-relaxation process.

We shall assume that τ is positive. Then, remembering that the convergence of the process is defined only for the eigenvalue of T_τ with maximum modulus, it is not difficult to see that we need only consider the formula

$$\lambda^{1/2}(T_\tau) = \frac{\tau\lambda(B)}{2} + \sqrt{\frac{\tau^2\lambda^2(B)}{4} - \tau + 1},$$

where we take the positive value of the radical when the expression within it is positive.

Moreover, for values of τ in the interval $[\tau_1, \tau_2]$, where

$$\tau_{1,2} = \frac{2}{\lambda^2(B)}[1 \pm \sqrt{1 - \lambda^2(B)}],$$

the radicand in (2.48) is nonpositive; for these values we have

$$|\lambda(T_\tau)| = |\lambda^{1/2}(T_\tau)|^2 = \left(\frac{\tau\lambda(B)}{2}\right)^2 - \left(\frac{\tau^2\lambda^2(B)}{4} - \tau + 1\right) = \tau - 1.$$

For $\tau \notin [\tau_1, \tau_2]$, the radicand in (2.48) is always positive and $\lambda^{1/2}(T_\tau)$ is nonnegative. Then it is not difficult to show that for $\tau \in [0, \tau_1]$ the inequality

$$\lambda^{1/2}(T_\tau) \geq 1$$

has no solution; whereas for $\tau \geq \tau_2$ the inequality

$$\lambda^{1/2}(T_\tau) \geq \frac{\tau\lambda(B)}{2} \geq \frac{\tau_2\lambda(B)}{2} = \frac{1}{\lambda(B)}[1 + \sqrt{1 - \lambda^2(B)}] > 1.$$

is satisfied. From all this we draw the following conclusion: given the assumptions we have made, the condition

$$\tau \in (0, 2)$$

is necessary and sufficient for the convergence of the successive overrelaxation process (2.38).

We consider the quantity

$$\lambda^{1/2}(T_\tau) = \begin{cases} \dfrac{\tau\lambda(B)}{2} + \sqrt{\dfrac{\tau^2\lambda^2(B)}{4} - \tau + 1}, & \tau \leq \tau_1, \\[2mm] \sqrt{\tau - 1}, & \tau \geq \tau_1 \end{cases}$$

as a function of the parameter τ. Since for the values of $\tau \in [0, \tau_1]$ we have

$$\frac{\partial}{\partial\tau}[\lambda^{1/2}(T_\tau)] = \frac{\lambda(B)\lambda^{1/2}(T_\tau) - 1}{2\sqrt{\dfrac{\tau^2\lambda^2(B)}{4} - \tau + 1}} < 0$$

and $\lambda^{1/2}(T_\tau) = 1$ for $\tau = 0$, the graph of $\|\lambda(T_\tau)\|$, for various $\lambda(B)$, has the form shown in Figure 4.1.

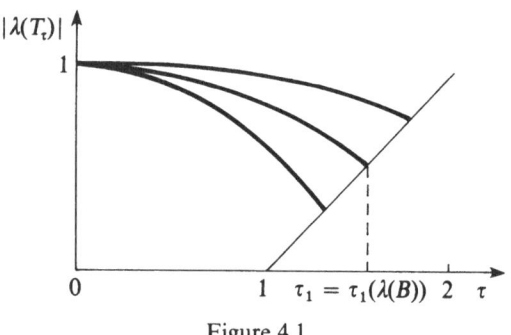

Figure 4.1

Since

$$\frac{\partial}{\partial \lambda(B)}[\lambda^{1/2}(T_\tau)] = \begin{cases} \dfrac{\tau \lambda^{1/2}(T_\tau)}{2\sqrt{\dfrac{\tau^2 \lambda^2(B)}{4} - \tau + 1}} > 0, & \text{for} \quad \tau < \tau_1, \\[6mm] 0, & \text{for} \quad \tau \geq \tau_1, \end{cases}$$

we have for $\tau \in (0, 2)$

$$\beta(T_\tau) = \begin{cases} \dfrac{\tau \beta(B)}{2} + \sqrt{\dfrac{\tau^2 \beta^2(B)}{4} - \tau + 1}, & \text{for} \quad \tau < \tau_1, \\[4mm] \tau - 1, & \text{for} \quad \tau \geq \tau_1. \end{cases}$$

It follows immediately from the last formula that the minimum of $\beta(T_\tau)$ is attained for $\tau = \tau_1$. Thus the value τ_{opt} that maximizes the asymptotic rate of convergence $R(T_\tau)$ of the successive over-relaxation process is to be found as

$$\tau_{\text{opt}} = \frac{2}{\beta^2(B)}[1 - \sqrt{1 - \beta^2(B)}] = \frac{2}{1 + \sqrt{1 + \beta^2(B)}}, \qquad (2.49)$$

and then

$$R(T_{\tau_{\text{opt}}}) = -\ln(\tau_{\text{opt}} - 1) = -\ln \frac{1 - \sqrt{1 - \beta^2(B)}}{1 + \sqrt{1 - \beta^2(B)}}. \qquad (2.50)$$

Let us now look at (2.49) and (2.50) when the matrices are ill-conditioned. (We suppose A to be symmetric.) In other words, we assume $p(A) = \beta(A)/\alpha(A) \gg 1$. We already know that if $\beta(B)$ is an eigenvalue of a matrix B of the type of (2.34) then so is $-\beta(B)$; therefore, $B = E - A$ implies that

$$\beta(B) = 1 - \alpha(A),$$

$$-\beta(B) = 1 - \beta(A).$$

It follows that $\alpha(A) = 2 - \beta(A)$; the condition $p(A) \gg 1$ implies that

$$\alpha(A) \approx 2/p,$$

$$\beta(B) \approx 1 - (2/p).$$

If we put these values into (2.49) and (2.50) we obtain the asymptotic relations

$$\tau_{\text{opt}} \approx 2 - \frac{4}{\sqrt{p}}, \qquad R(T_{\tau_{\text{opt}}}) \approx \frac{4}{\sqrt{p}},$$

$$\beta(T_{\tau_{\text{opt}}}) \approx 1 - \frac{4}{\sqrt{p}}, \qquad W \approx W_0 \frac{|\ln \varepsilon|}{4} \sqrt{p}. \tag{2.51}$$

Thus, for ill-conditioned matrices, given an optimal choice of the parameters, the successive over-relaxation method converges $2\sqrt{p}$ times as fast as the process (2.2). We must note, however, that for any specific number of iterations (especially in the first stages of the process) the over-relaxation process with optimal parameters converges more slowly than the third equality in (2.51) would suggest. The reason is that second-order cells appear in the Jordan form of the transition matrix $T_{\tau_{\text{opt}}}$ (cf. Young [10], Thie [10]).

4.2.4 The Chebyshev Iteration Method

We consider a linear algebraic system

$$A\phi = f \tag{2.52}$$

with a symmetric positive definite matrix A. We adopt the iterative process

$$\phi^{j+1} = \phi^j - \tau_j(A\phi^j - f), \tag{2.53}$$

to solve this equation. This process is sometimes called the *Richardson method of the first kind*. We optimize the process by choosing a sequence of parameters $\{\tau_j\}$ yielding the speediest convergence of the ϕ^j to the exact solution of (2.52).

A cyclic iteration process (2.53) with period $s \geq 1$ can be written (cf. Section 4.1) as a stationary process of the form

$$\phi^{j+(i/s)} = \phi^{j+[(i-1)/s]} - \tau_i(A\phi^{j+[(i-1)/s]} - f), \qquad i = 1, \ldots, s, \tag{2.54}$$

with the transition operator

$$\tilde{T}_\tau = \prod_{i=1}^{s} (E - \tau_i A). \tag{2.55}$$

Since A is symmetric,

$$\beta(\tilde{T}_\tau) = \|\tilde{T}_\tau\|_2 = \max_n \left| \prod_{i=1}^{s} (1 - \tau_i \lambda_n(A)) \right| \tag{2.56}$$

and, using the results of Section 4.2.2, the optimization of (2.53) reduces to the minimization of $\beta(\tilde{T}_r)$ with respect to the parameters $\tau_1, \ldots, \tau_s$. For $s = 1$, the exact solution was given in Section 4.2.1. There is no exact solution for $s > 1$, since a mere knowledge of the spectral bounds of A is insufficient, and finding all the eigenvalues of A is a more complicated task than the solution of (2.52) by simple iterative methods. Therefore, we minimize the function

$$q_s(\tau) = \max_{\alpha \leq \lambda \leq \beta} \left| \prod_{i=1}^{s} (1 - \tau_i \lambda) \right|, \tag{2.57}$$

where $0 < \alpha = \alpha(A) \leq \beta = \beta(A)$, which majorizes $\beta(\tilde{T}_r)$ and approximates it well enough.

We also consider the problem of constructing a polynomial $\tilde{P}_s(\lambda)$ of degree s which will be a solution of the problem

$$\min_{P_s(\lambda) \in Q_s} \max_{\alpha \leq \lambda \leq \alpha} |P_s(\lambda)| = \max_{\alpha \leq \lambda \leq \beta} |\tilde{P}_s(\lambda)|, \tag{2.58}$$

where Q_s is the set of all $P_s(\lambda)$ for which $P_s(0) = 1$. A solution of this last problem has been found by Markov, by the use of Chebyshev polynomials:

$$\tilde{P}_s(\lambda) = \frac{T_s\left(\dfrac{\beta + \alpha - 2\lambda}{\beta - \alpha}\right)}{T_s\left(\dfrac{\beta + \alpha}{\beta - \alpha}\right)} \tag{2.59}$$

where

$$T_s\left(\frac{\beta + \alpha - 2\lambda}{\beta - \alpha}\right) = r_s \prod_{i=1}^{s} (\lambda_i - \lambda); \tag{2.60}$$

and r_s is a constant. We have from this

$$\tilde{P}_s(\lambda) = \frac{\prod_{i=1}^{s} (\lambda_i - \lambda)}{\prod_{i=1}^{s} \lambda_i} = \prod_{i=1}^{s} \left(1 - \frac{1}{\lambda_i} \lambda\right). \tag{2.61}$$

and

$$\max_{\alpha \leq \lambda \leq \beta} |\tilde{P}_s(\lambda)| = \frac{1}{T_s\left(\dfrac{\beta + \alpha}{\beta - \alpha}\right)} \max_{\alpha \leq \lambda \leq \beta} \left| T_s\left(\frac{\beta + \alpha - 2\lambda}{\beta - \alpha}\right) \right|$$

$$= \frac{1}{T_s\left(\dfrac{\beta + \alpha}{\beta - \alpha}\right)} \max_{-1 \leq t \leq 1} |T_s(t)| = \frac{1}{T_s\left(\dfrac{\beta + \alpha}{\beta - \alpha}\right)}. \tag{2.62}$$

Let us now return to the optimization of (2.53), i.e., to the solution of the extremal problem

$$q_s(\tau_{opt}) = \min_{\tau} q_s(\tau), \tag{2.63}$$

where τ denotes the sequence of parameters $\tau_1, \ldots, \tau_s$. Let V_s denote the set of polynomials $P_s(\lambda)$ having the form

$$P_s(\lambda) = \prod_{i=1}^{s} (1 - a_i \lambda), \qquad (2.64)$$

we can formulate (2.63) as follows: find the polynomial $\hat{P}_s(\lambda)$ that solves the problem

$$\min_{P_s(\lambda) \in V_s} \max_{\alpha \le \lambda \le \beta} |P_s(\lambda)| = \max_{\alpha \le \lambda \le \beta} |\hat{P}(\lambda)|. \qquad (2.65)$$

Since $V_s \subset Q_s$,

$$\max_{\alpha \le \lambda \le \beta} |\tilde{P}_s(\lambda)| \le \max_{\alpha \le \lambda \le \beta} |\hat{P}_s(\lambda)|. \qquad (2.66)$$

If the solution of (2.58) is a polynomial $\tilde{P}_s(\lambda)$ of the form (2.64), i.e., if it belongs to V_s, then in (2.66) we have a strict equality. We conclude that for $\hat{P}_s(\lambda)$ we may take $\tilde{P}_s(\lambda)$. Finally, we note that the choice

$$\hat{P}_s(\lambda) = \tilde{P}_s(\lambda) \qquad (2.67)$$

is unique, since the solution of (2.58) is unique. (This has been proved, for example, in a paper by Flanders and Shortley [8].)

Let $k_s = (\sigma_1, \sigma_2, \ldots, \sigma_s)$ denote a permutation of the integers, of order s, with $1 \le \sigma_i \le s$ and $\sigma_i \ne \sigma_k$ for $i \ne k$ $(i, k = 1, 2, \ldots, s)$. Then it follows from (2.62) that for the optimal iteration process (2.54)

$$\tau_i = \frac{1}{\lambda_{\sigma_i}}, \qquad i = 1, \ldots, s, \qquad (2.68)$$

where

$$\lambda_i = \tfrac{1}{2}[\beta + \alpha - (\beta - \alpha)x_i], \qquad i = 1, \ldots, s,$$

and the x_i are roots of the polynomial $T_s(x)$. The process (2.54), with parameters chosen according to (2.68), will be called the *Chebyshev iteration method*.

Let us estimate its rate of convergence.

We shall show that the period s can be chosen in accordance with the condition that s iterations of the process (2.53) (which amounts to one iteration of (2.54)) reduce the initial error by a factor of $1/\varepsilon$. Begin with the equation

$$\varepsilon = \frac{1}{T_s\left(\dfrac{\beta + \alpha}{\beta - \alpha}\right)}.$$

If we introduce the notation

$$t_0 = \frac{\beta + \alpha}{\beta - \alpha}, \qquad \gamma = t_0 - \sqrt{t_0^2 - 1}$$

and use the relationship

$$t_0 - \sqrt{t_0^2 - 1} = \frac{1}{t_0 + \sqrt{t_0^2 - 1}},$$

then a simple transformation yields

$$\varepsilon = \frac{2\gamma^s}{\gamma^{2s} + 1}.$$

If we take this as a quadratic equation in γ^s and remember that $\gamma < 1$, we find

$$\gamma^s = \frac{1 - \sqrt{1 - \varepsilon^2}}{\varepsilon} = \frac{\varepsilon}{1 + \sqrt{1 - \varepsilon^2}}$$

and therefore

$$s = \frac{\ln \dfrac{\varepsilon}{1 + \sqrt{1 - \varepsilon^2}}}{\ln \gamma}.$$

When $\varepsilon \ll 1$ and $p(A) \gg 1$, this expression becomes

$$s \approx \frac{|\ln \varepsilon| \sqrt{p}}{2},$$

whence, we find

$$R(\tilde{T}_{\tau_{opt}}) \approx \frac{2}{\sqrt{p}},$$

$$W \approx W_0 |\ln \varepsilon| \frac{\sqrt{p}}{2} \qquad (2.69)$$

We conclude that the Chebyshev method is asymptotically $\sqrt{p}$ times as fast at the simplest one-step method, and twice as slow as the successive relaxation method.

It follows from what we have said, that to use the Chebyshev method we need to know the spectral bounds α and β of the matrix A. This is an urgent requirement, since for values of $x \notin [-1, 1]$ the Chebyshev polynomials increase rapidly, and errors in determining the spectral bounds can lead to slow convergence of the process.

As a rule, we determine the upper spectral bound of A by using Gershgorin's theorem. The determination of the lower bound is a highly complex affair. In many cases an *a priori* estimate can be made, but as a rule we must use an auxiliary iteration process—for instance the Lyusternik method described in Chapter 1, or the Lanczos minimized iteration method.

Another important problem in the use of the Chebyshev method is the correct ordering of the parameters. An arbitrary ordering may lead to instability in the numerical implementation of the process. Much attention has been paid to this problem, but it has been solved only recently, in papers by Lebedev and Finogenov [9], Lebedev [8], and Samarskii [3]. We shall describe two simple algorithms for ordering the parameters.

Let $s = 2^r$, where r is some positive integer, and let $n_{2^{r-1}} = (\sigma_1, \sigma_2, \ldots, \sigma_{s/2})$ be an ordering of the numbers x_i determined for $s = 2^{r-1}$ by the algorithm to be described. We now require that if $s = 2^r$ the ordering of the x_i shall be as follows:

$$n_{2^r} = (\sigma_1, s + 1 - \sigma_1, \sigma_2, s + 1 - \sigma_2, \ldots, \sigma_{s/2}, s + 1 - \sigma_{s/2}).$$

Thus, for instance, if $s = 16$ we have

$$n_{16} = (1, 16, 8, 9, 4, 13, 5, 12, 2, 15, 7, 10, 3, 14, 6, 11).$$

A second algorithm is as follows: let $s = 3^r$, where r is a positive integer, and let $n_{3^{r-1}} = (\sigma_1, \sigma_2, \ldots, \sigma_{3^{r-1}})$ be an ordering determined for $s = 3^{r-1}$ in accordance with the second algorithm. Then for $s = 3^r$ we require that the ordering be as follows:

$$n_{3^r} = (\sigma_1, 2 \cdot 3^{r-1} + \sigma_1, 2 \cdot 3^{r-1} + 1 - \sigma_1, \ldots, 2 \cdot 3^{r-1} + 1 - \sigma_{s/3}).$$

Thus, for instance, for $s = 9$, the second algorithm yields

$$n_9 = (1, 7, 6, 3, 9, 4, 2, 8, 5).$$

These algorithms distribute the transition operators $T_i = E - \tau_i A$ with large norms rather uniformly among the operators that diminish the norm of the error.

The iteration method (2.53) was described as optimal for a given s. Another class of iterative processes—stable infinitely extensible optimal methods of type (2.53) with Chebyshev parameters—allows us to extend the process (2.53) after s iterations, in such a way that it is stable and for some values of $j = j_k$ ($j_k \to \infty$, $j_k > s$), is again optimal. We shall explain the algorithm for constructing the parameters of such a method, in a particular case. Since

$$\cos 3x = \cos x(2 \cos 2x - 1),$$

we have, for the polynomials $T_s(t)$ and $T_{3s}(t)$, the following relationship

$$T_{3s}(t) = T_s(t)(2T_{2s}(t) - 1).$$

This shows that the set of roots of $T_{3s}(t)$ consists of the set (2.61)—the roots of $T_s(t)$—and the set of roots of the polynomial $2T_{2s}(t) - 1$:

$$\tilde{x}_i = \cos \frac{2i - 1}{6s} \pi, \qquad 2i \neq 1 \ (\text{mod } 3); \ 1 \leq i \leq 3s.$$

Therefore, if we first perform s iterations in which the parameters of (2.68) are used in (2.53), and then extend the process, using in (2.68) for the x_i (the λ_i are expressed in terms of the x_i) the parameters x_i shuffled by our permutation algorithms, we shall have an optimal method for $j = 3s$. By repeating this process we obtain an infinite sequence of the x_i, for which the method (2.53) is optimal for $j = 3^r s$.

We present some formulas defining the order for using the x_i when $s = 2$. We begin by setting $x_1 = -2^{-1/2}$, $x_2 = -x_1$. Suppose that we have constructed the order of use x_i ($i = 1, 2, \ldots, 2 \cdot 3^{r-1}$). We construct the segment from $(2 \cdot 3^{r-1} + 1, \ldots, 2 \cdot 3^r)$ as follows: using the permutation n_{3r-1} we form the quantities

$$t_{j-1} = \sin \frac{2(\sigma_j + [\sigma_{j/2}]) - 1}{4 \cdot 3^r} \pi, \qquad j = 1, \ldots, 3^{r-1}.$$

Then

$$x_{2 \cdot 3^{r-1} + 4j+1} = -t_j, \qquad x_{2 \cdot 3^{r-1} + 4j+2} = t_j,$$

$$x_{2 \cdot 3^{r-1} + 4j+3} = -\sqrt{1 - t_j^2}, \qquad x_{2 \cdot 3^{r-1} + 4j+4} = \sqrt{1 - t_j^2},$$

$$j = 0, \ldots, 3^{r-1} - 1.$$

Now, using the permutation n_{3r}, we construct the x_i ($i = 2 \cdot 3^r + 1$, $\ldots, 2 \cdot 3^{r+1}$), and so on.

Analysis of the Chebyshev iteration method (2.53) shows that in optimizing it we have nowhere used the property (2.56); there remains the minimization of the function $q_s(\tau)$, which depends only on the spectral bounds and on the fact that the eigenvalues of A are real. In view of this, when we optimize the method (2.53), we need require only that the eigenvalues of A are positive and its eigenvectors complete; we may drop the requirement that A be symmetric.

We are now able to construct a theory of optimization for iterative processes of the form

$$\phi^{j+1} = \phi^j - \tau_j H(A\phi^j - f) \tag{2.70}$$

under the assumption that all the eigenvectors of A are real and lie in the interval $[\alpha, \beta]$, where $0 < \alpha = \alpha(HA) \le \beta = \beta(HA)$.

We show that the listed requirements with respect to the spectrum of the matrix HA are satisfied if the matrices H and A are symmetric and positive definite. To do this, we introduce the notion of a positive quadratic root of a symmetric positive definite matrix D. Let $\{v_n\}$ be a complete orthonormal system of the eigenvectors of D corresponding to its positive eigenvalues $\{d_n\}$. Then there exists an orthogonal matrix P having as columns the vectors $\{v_k\}$, such that

$$D = PD_0 P^*,$$

where D_0 is a diagonal matrix with the eigenvalues $\{d_n\}$ of the matrix D as its diagonal. We define the matrix $D_0^{1/2}$ as the diagonal matrix with the quantities $\{d_n^{1/2}\}$ as its diagonal elements, and satisfying the equation

$$D_0^{1/2} D_0^{1/2} = D_0.$$

Clearly $D_0^{1/2}$ is uniquely defined by D_0. Now we define the matrix $D^{1/2}$ by the equation

$$D^{1/2} = PD_0^{1/2}P*.$$

It is easily seen that the matrix $D^{1/2}$ is symmetric and positive definite (since it is symmetric and all its eigenvalues $\{d_n^{1/2}\}$ are positive), and that it satisfies the equation

$$D^{1/2}D^{1/2} = [D^{1/2}]^2 = D.$$

On the basis of what we have said so far, we see that the matrix HA is similar to the symmetric positive definite matrix $S = A^{1/2}HA^{1/2}$;

$$A^{1/2}[HA][A^{1/2}]^{-1} = A^{1/2}HA^{1/2}.$$

It follows that all the eigenvalues of the nonsingular matrix HA are real and positive.

We now show that under our assumptions the matrix HA possesses a complete system of eigenvectors. This allows us to reach a solution either by cyclic or by infinitely extensible Chebyshev processes. We denote by $\{v_n\}$ a complete orthonormal system of eigenvectors of S corresponding to its positive eigenvalues $\{\mu_n\}$:

$$Sv_n = \mu_n v_n.$$

We multiply this equation by $[A^{1/2}]^{-1}$ and write $w_n = [A^{1/2}]^{-1}v_n$. This yields

$$HAw_n = \mu_n w_n.$$

Thus, the $\{\mu_n\}$ are eigenvalues of HA and the $\{w_n\}$ are the corresponding eigenvectors. The system $\{v_n\}$ forms a basis in the initial vector space and $\{w_n\}$ is derived from it by a nonsingular transformation. Therefore, the $\{w_n\}$ also form a basis and the matrix HA has a complete system of eigenvectors.

Let us look at another important case, in which the eigenvalues of HA are real and positive. We assume that A has the form

$$A = \begin{Vmatrix} E_1 & -S_1 \\ -R_2 & E_2 \end{Vmatrix}, \tag{2.71}$$

where E_1 and E_2 are unit matrices of order n_1 and n_2, respectively, and S_1 and R_2 are matrices of order $n_1 \times n_2$ and $n_2 \times n_1$. We suppose, moreover, that all the eigenvalues of the matrix

$$B = \begin{Vmatrix} 0 & S_1 \\ R_2 & 0 \end{Vmatrix}$$

are real and less than 1 in the absolute value ($|\lambda(B)| < 1$), and that the matrix H is defined by the equation

$$H = \begin{Vmatrix} E_1 & 0 \\ -R_1 & E_2 \end{Vmatrix}^{-1} = \begin{Vmatrix} E_1 & 0 \\ R_2 & E_2 \end{Vmatrix}. \tag{2.72}$$

This choice of H corresponds to the selection of the successive over-relaxation process (2.38) with parameter $\tau = 1$ (cf. Section 2.4.3). An arbitrary eigenvalue of the matrix

$$HA = \begin{Vmatrix} E_1 & -S_1 \\ 0 & E_2 - R_2 S_1 \end{Vmatrix}$$

is either equal to 1 or equal to one of the eigenvalues of the matrix $E_2 - R_2 S_1$: these are real and positive. To verify this last statement we need only observe: (a) that an arbitrary eigenvalue of $R_2 S_1$ is also an eigenvalue of the matrix

$$B^2 = \begin{Vmatrix} S_1 R_2 & 0 \\ 0 & R_2 S_1 \end{Vmatrix}$$

(the eigenvalues of B^2 are positive, since they are the squares of the real eigenvalues of B); and (b) that the inequality

$$\lambda(R_2 S_1) = \lambda(B^2) = \lambda^2(B) < 1.$$

is satisfied. Thus, all the eigenvalues of HA are real and positive; this means we may apply the optimization method (2.70) by using the theory developed above for constructing the Chebyshev parameters. Note that if we also assume that the matrix $A(R_2 = S_1^*)$ is symmetric, HA has a complete system of eigenvectors.

In conclusion, we estimate the rate of convergence of the method (2.70) with Chebyshev parameters, when the matrices A and H are defined by (2.71) and (2.72), via the ratio

$$p = p(A) = \frac{\beta(A)}{\alpha(A)}$$

under the assumption that $p \gg 1$. Using the results of Section 4.2.3 and the specific form of HA, it is easy to see that

$$\alpha(HA) = 1 - \beta(B^2) = 1 - \beta^2(B)$$
$$= 1 - (1 - \alpha(A))^2 = \alpha(A)(2 - \alpha(A)) \approx 2\alpha(A), \qquad \beta(HA) = 1$$

and therefore

$$p(HA) = \frac{\beta(HA)}{\alpha(HA)} \approx \frac{1}{2\alpha(A)} \approx \frac{1}{4}\frac{\beta(A)}{\alpha(A)} = \tfrac{1}{4}p.$$

Hence, by (2.69) we have for the process (2.70)

$$R(\tilde{T}_{\tau_{opt}}) \approx \frac{4}{\sqrt{p}},$$

$$W \approx W_0 |\ln \varepsilon| \frac{\sqrt{p}}{4}.$$

(2.73)

With this, we have shown that the asymptotic rate of convergence of the variant Chebyshev iteration method that we are now considering is equal to the asymptotic rate of convergence of the successive over-relaxation method with the parameter τ_{opt}.

4.2.5 Comparison of the Convergence Rates of Various Iteration Methods for a System of Finite-Difference Equations

In the preceding section we found some estimates for the rate of convergence of a number of iteration methods using the conditioning number of the matrix A. We now consider the application of these methods to the solution of systems of finite-difference equations approximating two-dimensional elliptic equations.

Let the system of finite-difference equations be written in the form

$$\phi_{k,l} = a_{k,l}\phi_{k-1,l} + b_{k,l}\phi_{k,l-1} + c_{k,l}\phi_{k+1,l} + d_{k,l}\phi_{k,l+1} + f_{k,l}, \quad (2.74)$$

where (k, l) run over some set Q of indices.

We shall suppose that to each net point of the region, on which the approximation problem is defined, there corresponds only one value of $\phi_{k,l}$ and only one equation, i.e., the number of unknowns and the number of equations both equal the number of net points. Then, to pass to the matrix formulation of the system of finite-difference equations we first number the net points and, correspondingly, number the components $\{\phi_{k,l}\}$ and the finite-difference equations in the same sequential order.

To illustrate this process, let us consider the system (2.74) when the net region is a square net and the domain of definition of the equation is the unit square. Then in (2.74) $k, l = 1, \ldots, m$ and the order of the matrix A of the corresponding system

$$A\phi = f \quad (2.75)$$

is $n = m^2$.

Our initial numbering will be done by the so-called "chess numbering," where all the components are divided into two groups: the first group contains those components with even values of $k + l$, and the second group contains those with odd values. We number the components in the first group first, then number those in the second. In the specific case when $m = 3$ the numbering is as shown in Figure 4.2.

The corresponding matrix $A = A_1$ of the system (2.75) is defined by the equation (valid for all m)

$$A_1 = \begin{Vmatrix} E_1 & -S_1 \\ -R_2 & E_2 \end{Vmatrix}, \quad (2.76)$$

where the order of the matrix E_1 is equal to the number of components in the first group and the order of E_2 is equal to the number in the second group.

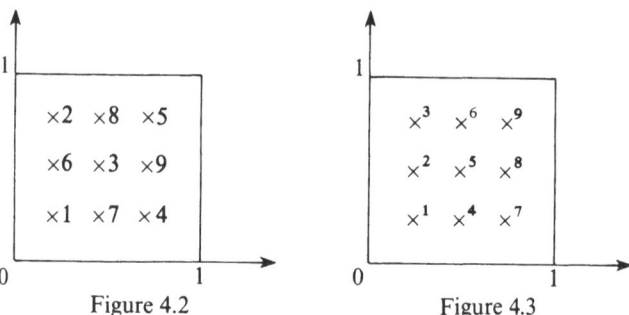

Figure 4.2   Figure 4.3

The second numbering process for $m = 3$ is shown in Figure 4.3.

If we assemble in one group all components having the same value of k we obtain three groups of components. The corresponding matrix $A = A_2$ of the system (2.75) has the form ($m = 3$)

$$A_2 = \begin{Vmatrix} A_{11} & A_{12} & 0 \\ A_{21} & A_{22} & A_{23} \\ 0 & A_{32} & A_{33} \end{Vmatrix} \tag{2.77}$$

If we multiply the system (2.75) with such a matrix A by the matrix D^{-1}, where

$$D = \begin{Vmatrix} A_{11} & 0 & 0 \\ 0 & A_{22} & 0 \\ 0 & 0 & A_{33} \end{Vmatrix} \tag{2.78}$$

is a block-diagonal matrix, we obtain the system

$$A_3 \phi \equiv \begin{Vmatrix} E_1 & -S_1 & 0 \\ -R_2 & E_2 & -S_2 \\ 0 & -R_3 & E_3 \end{Vmatrix} \phi = F, \tag{2.79}$$

where

$$F = D^{-1}f,$$
$$S_l = -A_{ll}^{-1}A_{l,l+1}, \qquad l = 1, 2, 3.$$
$$R_l = -A_{ll}^{-1}A_{l,l-1},$$

For arbitrary $k = m$ the matrix A_3 has the form (2.32).

If the operators of the original elliptic differential equation are symmetric and positive definite and if the approximation was made either by integral relationships or by variational methods, the matrices A will have a complete system of eigenvectors and positive eigenvalues. In exceptional cases (e.g., rectangular domains and constant coefficients) the spectral bounds can be computed exactly, as was done in Chapter 1 for the Laplacian difference operator in the square, or they may be estimated (sometimes very roughly),

Table 4.1

Method	Asymptotic convergence rate $R(T)$
Simplest iteration method	$2/p \approx \pi^2 h^2/2$
Chebyshev iteration	$2/\sqrt{p} \approx \pi h$
Successive over-relaxation	$4/\sqrt{p} \approx 2\pi h$
Chebyshev method per (2.70)–(2.72)	$4/\sqrt{p} \approx 2\pi h$

or they may be computed by some or other iterative process (Lyusternik, Lanczos, etc.).

As an example, let us look at the approximation of the Dirichlet problem for a Poisson equation in the unit square by means of ordinary five-point difference equations on a square net with step $h \ll 1$. We limit ourselves to the case in which the matrix A of the system (2.75) has the form (2.76). Then, as we may see by straightforward calculation

$$\alpha(A) = 2 \sin^2 \frac{\pi h}{2} \approx \frac{\pi^2 h^2}{2},$$

$$\beta(A) = 2 - \alpha(A) \approx 2 - \frac{\pi^2 h^2}{2}, \qquad (2.80)$$

$$p = p(A) = \frac{\beta(A)}{\alpha(A)} \approx \frac{4}{\pi^2 h^2}$$

Table 4.1 displays the asymptotic convergence rates for several different methods, bearing in mind the fact that for $h \ll 1$, the matrix A is ill-conditioned.

4.3 Nonstationary Iteration Methods

In this section we consider nonstationary iteration methods based on the sequential minimization of some quadratic functional. This section is based primarily on the work of Kuznetsov [8], [11].

4.3.1 Convergence Theorems

Suppose given a system of linear algebraic equations

$$A\phi = f \qquad (3.1)$$

and a quadratic functional

$$J(\phi) = (D(\phi - \phi^*), \phi - \phi^*), \qquad (3.2)$$

where $\phi^* = A^{-1}f$ is the exact solution of (3.1) and D is a symmetric positive definite matrix.

Since $J(\phi) > 0$ for arbitrary $\phi \neq \phi^*$ and $J(\phi^*) = 0$, the solving of (3.1) is equivalent to the minimization of (3.2), i.e., to the determination of a vector ϕ^* that minimizes $J(\phi)$. If $D = A^*A$, the functional

$$J(\phi) = (A\phi - f, A\phi - f) = \|A\phi - f\|_2^2 \tag{3.3}$$

is called the *functional of the square residual*.

If we assume that A is a symmetric positive definite matrix and $D = A$, then

$$J(\phi) = J_1(\phi) + (A\phi^*, \phi^*),$$

where

$$J_1(\phi) = (A\phi, \phi) - 2(\phi, f). \tag{3.4}$$

Thus, up to a constant $(A\phi^*, \phi^*) = (f, \phi^*)$, $J(\phi)$ coincides with the known variational functional $J_1(\phi)$. Note that the functionals (3.3) and (3.4) do not depend on the solution vector ϕ^*.

We may now formulate new principles for the optimization of iteration methods; we shall call them variational principles. We consider the iteration method

$$\phi^{j+1} = \phi^j - H_\tau(A\phi^j - f), \tag{3.5}$$

where the matrix H_τ depends on the parameters $\tau_1, \ldots, \tau_s$. We assume that for some values $\tau_1, \ldots, \tau_s$ in a set Q the method (3.5) converges, and that the sequence $\{\phi^j\}$ realizes a sequential minimization of the functional (3.2). In the preceding section we chose the parameters $\tau_1, \ldots, \tau_s$ to minimize the spectral radius $\beta(T_\tau)$ of the transition operator $T_\tau = E - H_\tau A$. In this section we will consider methods in which the parameters $\tau_1, \ldots, \tau_s$ are chosen to maximize, at each step, the minimization of the functional (3.2).

Since the matrix D is positive definite, the equation

$$\|\psi\|_D = (D\psi, \psi)^{1/2} \tag{3.6}$$

defines a norm in the space of the error vectors. The corresponding equation

$$\|B\|_D = \sup_{\psi \neq 0} \frac{\|B\psi\|_D}{\|\psi\|_D} \tag{3.7}$$

defines a norm of the matrix B. (These are often called D-norms.) We assumed earlier that for $(\tau_1, \ldots, \tau_s) \in Q$ the functional (3.2), with some matrix D, has been sequentially minimized. This means that for $\psi^j = \phi^j - \phi^*$,

$$\|T_\tau \psi^j\|_D < \|\psi^j\|_D. \tag{3.8}$$

Since for a stationary iteration method the matrix T_τ is a constant, it follows that for an arbitrary vector $\psi^j \neq 0$

$$\|T_\tau z^j\|_D < 1,$$

where $z^j = \psi^j/\|\psi^j\|_D$. Furthermore, since the set $v = \{z: \|z\|_D = 1\}$ [the notation $\{z: P(z)\}$ where P is a predicate means the set of z for which $P(z)$ is true] is closed and bounded, the supremum

$$\sup_{\|z\|_D = 1} \|T_\tau z\|_D = \|T_\tau\|_D$$

is attained at some vector z_0. From this and (3.8) we have

$$\|T_\tau\|_D = \|T_\tau z_0\|_D < 1. \tag{3.9}$$

Thus, we have shown that if $(\tau_1, \ldots, \tau_s) \in Q$,

$$\|T_\tau\|_D < 1.$$

We now formulate the nonstationary method corresponding to (3.5):

$$\phi^{j+1} = \phi^j - H_j(A\phi^j - f), \tag{3.10}$$

where $H_j = H(\tau_1^{(j)}, \ldots, \tau_s^{(j)})$ and the parameters $\{\tau_i^{(j)}\}$ satisfy the equation

$$J(\phi^j - H_j(A\phi^j - f)) = \inf_{\tau_1, \ldots, \tau_s} J(\phi^j - H_\tau(A\phi^j - f)). \tag{3.11}$$

If the equation

$$\|T_\tau\|_D = \beta(T_\tau),$$

holds for arbitrary $(\tau_1, \ldots, \tau_s) \in Q$ for which (3.5) converges, then the asymptotic convergence rate of the corresponding nonstationary iteration method is generally not less than the convergence rate with the optimal parameters.

As we proved in Section 4.2.2,

$$R(T_\tau) = -\ln \beta(T_\tau). \tag{3.12}$$

On the other hand,

$$\left[\frac{\|\prod_{j=1}^k T_j \psi^0\|_D}{\|\psi^0\|_D}\right]^{1/k} \leq \|\tilde{T}\|_D \leq \|T_\tau\|_D,$$

where $\tilde{T}$ is the nonlinear step operator of method (3.10), (3.11). Hence, we find that

$$R(\tilde{T}) \geq R(T_\tau). \tag{3.13}$$

Since this inequality holds for all $\tau_1, \ldots, \tau_s \in Q$, our assertion is proved.

It now follows that if the optimization parameters are chosen to minimize the D-norm of the transition operator T_τ, the variational optimization yields the highest convergence rate, at an arbitrary iteration and asymptotically.

4.3.2 The Method of Minimizing the Residuals

We set

$$D = A^*A \quad \text{and} \quad H = \tau E \tag{3.14}$$

and assume that $A = 0$.

We showed in Section 4.2.1 that in this case the iterative process

$$\phi^{j+1} = \phi^j - \tau(A\phi^j - f) \tag{3.15}$$

converges if $\tau \in (0, 2/\beta(A))$. A very simple spectral analysis shows that

$$\|E - \tau A\|_D = \|E - \tau A\|_2 = \beta(E - \tau A). \tag{3.16}$$

Let us look at the corresponding nonstationary process, called the *method of minimum residuals*:

$$\phi^{j+1} = \phi^j - \tau_j(A\phi^j - f), \tag{3.17}$$

where

$$\tau_j = \frac{(A\xi^j, \xi^j)}{(A\xi^j, A\xi^j)} = \frac{(A\xi^j, \xi^j)}{\|A\xi^j\|_2^2} \tag{3.18}$$

and $\xi^j = A\phi^j - f$ is the residuals vector. It follows from (3.17) that

$$\|(E - \tau_j A)\xi^j\|_D^2 = \|(E - \tau_j A)\xi^j\|_2^2 = \inf_\tau J(\phi^j - \tau\xi^j)$$

$$= \inf_\tau \|(E - \tau A)\xi^j\|_2^2 = \inf_\tau \{\|\xi^j\|_2^2 - 2\tau(A\xi^j, \xi^j) + \tau^2\|A\xi^j\|_2^2\}.$$

Clearly, τ_j is to be found from the equation

$$\frac{\partial}{\partial\tau} \{\|\xi^j\|_2^2 - 2\tau(A\xi^j, \xi^j) + \tau^2\|A\xi^j\|_2^2\} = 0.$$

Since (3.15) converges for $\tau \in (0, 2/\beta(A))$, the assertion we arrived at in Section 4.2 implies that the method of minimum residuals converges, and

$$\|\tilde{T}\| \leq \frac{\beta - \alpha}{\beta + \alpha}. \tag{3.19}$$

It is known (see, for example, Faddeev and Faddeeva [8], Kuznetsov [8]) that the exact equality holds in (3.19).

From the relationship

$$\|\xi^{j+1}\|_2^2 = \|\xi^j\|_2^2 - \frac{(A\xi^j, \xi^j)^2}{\|A\xi^j\|_2^2}$$

governing the sequential norms of the residuals in the process (3.17), (3.18), it is easily seen that for the norms of the residuals in a space of real vectors to form a monotone decreasing sequence, it is sufficient that the matrix A be positive definite over that space. Symmetry is not required (Kuznetsov [11]). This fact provides the basis for a theorem on the convergence of nonstationary iteration methods and, in particular, for a proof that the method of minimum residuals converges for systems with positive definite nonsymmetric matrices. This problem has received detailed treatment in a monograph of Marchuk and Kuznetsov [8]. We should note that the original exposition of the method of minimum residuals is due to Krasnosel'skii and Krein [11].

The method of minimum residuals has a peculiarity that is important in practice: namely, the initial iterations converge much more rapidly than the asymptotic rate of convergence (Kantorovich [11]). The error in the method of steepest descent is asymptotically a linear combination of two eigenvectors of the matrix A, corresponding to the eigenvalues $\alpha(A)$ and $\beta(A)$ (cf. Forsythe and Motzkin [11]). A similar situation occurs in the method of minimum residuals. Thus, the asymptotic property of an iterative process—to exhibit the rate of convergence at its worst—appears also in nonstationary methods.

In order to accelerate the convergence of the minimum residuals method, it is worthwhile to make occasional use of a single iteration of a two-step minimum residuals method

$$\phi^{j+1} = \phi^j - \tau_j(A\phi^j - f) - \gamma_j A(A\phi^j - f), \tag{3.20}$$

where τ_j and γ_j are the solutions of the system

$$\frac{\partial}{\partial \tau_j} \|\xi^j - \tau_j A\xi^j - \gamma_j A^2\xi^j\|_2^2 = 0,$$

$$\frac{\partial}{\partial \gamma_j} \|\xi^j - \tau_j A\xi^j - \gamma_j A^2\xi^j\|_2^2 = 0. \tag{3.21}$$

4.3.3 The Conjugate Gradient Method

We define a D-norm in the space of the initial vectors, and a subspace G_s with the basis $\{g_i\}_{i=1}^s$; then the problem of finding the best approximation to the solution $\phi^* = A^{-1}f$ of the system $A\phi = f$ on the manifold

$$U_s^0 = \phi^0 + G_s = \{\phi : \phi = \phi^0 + \psi, \psi \in G_s\}$$

may be formulated as follows: we are required to find a vector $\hat{\psi} \in G_s$ such that

$$\|\phi^* - (\phi^0 + \hat{\psi})\|_D = \min_{\psi \in G_s} \|\phi^* - (\phi^0 + \psi)\|_D$$

$$= \min_{\alpha_1, \alpha_2, \ldots, \alpha_s} \left\| \phi^* - \phi^0 - \sum_{i=1}^s \alpha_i g_i \right\|_D. \tag{3.22}$$

The system of equations defining the coefficients $\{\alpha_i^*\}$ in the resolution

$$\hat{\psi} = \sum_{i=1}^s \alpha_i^* g_i$$

has the form

$$B\alpha = F,$$

where $B = (b_{ij})$ is a matrix of order s with the elements

$$b_{ij} = (g_i, g_j)_D, \qquad i, j = 1, \ldots, s,$$

and $F = (F_1, \ldots, F_s)$ is a vector with the components

$$F_i = (\phi^* - \phi^0, g_i)_D, \qquad i = 1, \ldots, s.$$

We may now see that the simplest practical case occurs when

$$(Dg_i, g_j) = \delta_{ij} \|g_i\|_D^2,$$

where δ_{ij} is the Kronecker delta, i.e., when the $\{g_i\}$ form a D-orthogonal basis for the space G_s. If this condition is satisfied

$$\alpha_i^* = \frac{(\phi^* - \phi^0, g_i)_D}{\|g_i\|_D^2} = \frac{(D(\phi^* - \phi^0), g_i)}{\|g_i\|_D^2}, \qquad i = 1, \ldots, s. \tag{3.23}$$

If the vector $D\phi^*$ is known, the process (3.23) exists. For this, it is sufficient that either $D = A$ if $A = A^* > 0$, or that $D = A^*A$ for arbitrary A.

Let us specify the variational problem (3.22) in order to study one class of methods. We assume that the subspace G_s is the linear hull of a system of independent vectors

$$\{A^i(\phi^0 - \phi^*)\}_{i=1}^s = \{A^{i-1}(A\phi^0 - f)\}_{i=1}^s,$$

and the matrix A is symmetric and positive definite. We have already seen that if we can find an A-orthogonal basis $\{g_i\}_{i=1}^s$ for the subspace G_s, the desired approximation to the vector ϕ^* can be found by the formulas

$$\hat{\phi} = \phi^0 + \sum_{i=1}^s \alpha_i g_i,$$

$$\alpha_i = \frac{(\phi^* - \phi^0, g_i)_A}{(g_i, g_i)_A} = -\frac{(A\phi^0 - f, g_i)}{(g_i, Ag_i)}, \qquad i = 1, \ldots, s. \tag{3.24}$$

This process can be rewritten as

$$\phi^k = \phi^{k-1} - \alpha_k g_k,$$

$$\alpha_k = \frac{(\xi^{k-1}, g_k)}{(Ag_k, g_k)}, \qquad k = 1, \ldots, s, \tag{3.25}$$

where $\xi^k = (A\phi^k - f)$ is the residuals vector and $\hat{\phi} = \phi^s$.

The best-known method for constructing a basis in spaces of the type G_s is the Schmidt process. However, for high-order matrices A, this process demands for its implementation a large number of arithmetic operations and a large computer memory. When A is symmetric but not positive definite, the Lanczos minimal iteration method is effective for the construction of an A^2-orthogonal basis in G_s (here $D = A^2$). This method is described in detail in a monograph by Faddeev and Faddeeva [8]. The most economical of the known processes for A-orthogonalization of the vectors $\{A^{i-1}(A\phi^0 - f)\}_{i=1}^s$

for a symmetric positive definite matrix is known as the *conjugate gradient method*. The equations are as follows:

$$g_k = \begin{cases} \xi^0, & \text{if } k = 1, \\ \xi^{k-1} - b_k g_{k-1}, & \text{if } k > 1, \end{cases}$$

$$b_k = \frac{(A\xi^{k-1}, g_{k-1})}{(Ag_{k-1}, g_{k-1})}, \tag{3.26}$$

$$\phi^k = \phi^{k-1} - \alpha_k g_k,$$

$$\alpha_k = \frac{(\xi^{k-1}, g_k)}{(Ag_k, g_k)}, \qquad k = 1, \ldots, s,$$

where the $\{\xi^k = A\phi^k - f\}_{i=1}^s$ are the residuals vectors.

We shall show that the vectors $\{g_k\}_{k=1}^s$ constructed by this process form an A-orthogonal basis for the space G_s if the vectors $\{A^{k-1}\xi^0\}_{k=1}^s$ are linearly independent. First, however, we show that all the vectors are different from zero. In fact, if we carry out the sequential elimination in (3.26) we find that

$$g_k = \xi^0 + \sum_{i=1}^{k-1} \beta_{ki} A^{i-1} \xi^0, \qquad k = 1, \ldots, s \tag{3.27}$$

for some set of coefficients $\{\beta_{ki}\}$. Therefore, the assumption that $g_k = 0$ contradicts the linear independence of the vectors $\{A^{k-1}\xi^0\}_{k=1}^s$.

We must now show that the vectors $\{g_k\}_{k=1}^s$ are A-orthogonal. For $k = 1, 2$ the following equations are verifiable by immediate testing:

$$(Ag_k, g_j) = 0, \qquad j = 1, \ldots, k - 1,$$

$$(\xi^k, g_j) = 0, \qquad j = 1, \ldots, k, \tag{3.28}$$

$$\alpha_j > 0, \qquad j = 1, \ldots, k.$$

We assume that they hold for some $k \geq 2$, and prove that they then hold for $k + 1$. For this we need the equations

$$Ag_j = \frac{1}{\alpha_j} [g_j + b_j g_{j-1} - g_{j+1} - b_{j+1} g_j]$$

$$= \varepsilon_j g_{j+1} + \beta_j g_j + \gamma_j g_{j-1}, \qquad j = 1, \ldots, k - 1 \tag{3.29}$$

where $g_0 = 0$. These we derive from equations obtained from (3.26) and from the conditions

$$\xi^j = \xi^{j-1} - \alpha_j Ag_j, \qquad g_{j+1} = \xi^j - b_{j+1} g_j, \qquad j = 1, \ldots, k - 1.$$

Our assumptions imply that

$$(Ag_{k+1}, g_j) = (A\xi^k - b_{k+1} Ag_k, g_j) = (A\xi^k, g_j) - b_{k+1}(Ag_k, g_j)$$

$$= (\xi^k, Ag_j) = (\xi^k, \varepsilon_j g_{j+1} + \beta_j g_j + \gamma_j g_{j-1}) = 0,$$

$$j = 1, \ldots, k - 1$$

and, by construction, $(Ag_{k+1}, g_k) = 0$. Consequently,

$$(Ag_{k+1}, g_j) = 0 \tag{3.30}$$

for all $j \le k$. Furthermore, since $(\xi^{k+1}, g_{k+1}) = 0$, it follows from our construction, and from the assumptions (3.28) and (3.30), that

$$(\xi^{k+1}, g_j) = (\xi^k, g_j) - \alpha_{k+1}(Ag_{k+1}, g_j) = 0 \tag{3.31}$$

for all $1 \le j \le k + 1$. Combining (3.30) and (3.31), and remembering the inequality

$$\alpha_{k+1} = \frac{(\xi^k, g_{k+1})}{(Ag_{k+1}, g_{k+1})} = \frac{(\xi^k, \xi^k)}{(Ag_{k+1}, g_{k+1})} > 0,$$

we find that the equations (3.28) hold for $k + 1$. Extending this induction up to $k = s$, we arrive at the conclusion that the vector system $\{g_k\}$ is A-orthogonal. Therefore, the conjugate gradient method solves the variational problem (3.22).

In conclusion, we discuss the degenerate case, when the system of vectors $\{A^{i-1}\xi^0\}_{i=1}^k$ is linearly independent for some $k > 1$, but is not for $k + 1$, i.e.,

$$-\sum_{j=0}^k C_j A^j \xi^0 \equiv \sum_{j=0}^k C_j A^{j+1}(\phi^* - \phi^0) = 0 \tag{3.32}$$

for some set of coefficients $\{C_j\}_{j=0}^k$ containing at least one nonzero element. The coefficient C_0 is nonzero, else multiplication of (3.32) by A^{-1} would yield

$$\sum_{j=0}^k C_j A^j(\phi^* - \phi^0) = 0,$$

which contradicts the linear independence of the vector system $\{A^j(\phi^* - \phi^0)\}_{j=1}^k$. We have from (3.32)

$$\phi^* - \phi^0 = -\frac{1}{C_0} A^{-1} \sum_{j=1}^k C_j A^{j+1}(\phi^* - \phi^0) = \sum_{j=1}^k \frac{C_j}{C_0} A^{j-1} \xi^0,$$

which implies that $\phi^* - \phi^0 \in G_s$. From this and (3.28) it follows that the approximation $\hat{\psi}$ is equal to the vector $\phi^* - \phi^0$ $(\phi^* = \phi^0 + \hat{\psi} = \phi^k)$. In other words, in the degenerate case, the conjugate gradient method yields an exact solution of the system (3.1) at the kth step.

This, like other orthogonalization methods, can be widely applied to accelerate the convergence of stationary iteration methods. For instance, in the iterative process

$$\phi^{k+1} = \phi^k - B(A\phi^k - f), \qquad k = 1, 2, \ldots, \tag{3.33}$$

with symmetric positive definite matrices A and B the acceleration formulas derived by the conjugate gradient method have the following form:

$$g_k = \begin{cases} B\zeta^0, & \text{if } k = 1, \\ B\zeta^{k-1} - b_k g_{k-1}, & \text{if } k > 1, \end{cases}$$

$$b_k = \frac{(AB\zeta^{k-1}, g_{k-1})}{(Ag_{k-1}, g_{k-1})}, \tag{3.34}$$

$$\phi^k = \phi^{k-1} - \alpha_k g_k,$$

$$\alpha^k = \frac{(\zeta^{k-1}, g_k)}{(Ag_k, g_k)}, \qquad k = 1, 2, \ldots, s.$$

The conjugate gradient method is theoretically direct, since for $s > n$, where n is the order of the matrix A, the vector system $\{A^i \zeta^0\}_{i=0}^{s-1}$ is linearly dependent; therefore, for some $k \leq s$ the process ends with an exact solution. On the other hand, the implementation on an electronic computer for high-order matrices usually results after a few tens of iterations in a numerical instability of the orthogonalization algorithm (because of nonlinearity), and the real-word process no longer represents the method we try to implement. All this means that we may think of the conjugate gradient method as an s-step nonstationary iterative process, i.e., after s iterations we stop and renew our initial approximation. With this understanding, we can estimate the rate of convergence via the convergence rate of a cyclic Chebyshev iterative process. In fact, for all $s \geq 1$,

$$\min_{\alpha_1, \ldots, \alpha_s} \frac{\|\phi^* - \phi^0 - \sum_{i=1}^s \alpha_i A^{i-1} \zeta^0\|_A^2}{\|\phi^* - \phi^0\|_A^2} \leq \|\tilde{T}_s\|_A$$

$$\leq \left\| E - \sum_{i=1}^s \beta_i A^i \right\|_2^2$$

$$= \left[\beta\left(E - \sum_{i=1}^s \beta_i A^i \right) \right]^2, \tag{3.35}$$

where T_s is the operator transforming the error vector after s steps of the conjugate gradient method, and the $\{\beta_i\}_{i=1}^s$ are arbitrary real numbers. In particular, if we set

$$E - \sum_{i=1}^s \beta_i A^i = \prod_{i=1}^s (E - \tau_i A) \tag{3.36}$$

and use the estimate (3.12) we conclude that the s-step cyclic conjugate gradient method converges about as fast as the s-step cyclic Chebyshev method, so that the same estimates of the convergence rate apply. Let us note two important peculiarities of the conjugate gradient method that emerge in the solution of concrete computational tasks. First, the implementation of a single step in the conjugate gradient method requires (sometimes significantly) more arithmetic and logical operations than does the Cheby-shev method. Second, in practice (particularly in the first steps) it minimizes the A-norm of the error vector much more rapidly than our estimate would indicate. As an illustration we write out the relations arising from (3.22) for the conjugate gradient method (this is the process for defining the subspace G_s (cf. Il'in [8])):

$$(A\psi^k, \psi^k) \leq \max_{m \leq \lambda \leq M} |\lambda P_k^2(\lambda)| \cdot (\psi^0, \psi^0), \tag{3.37}$$

where ψ^k is the error vector in the kth step, ψ^0 is the initial error vector, and $P_k(\lambda)$ is a polynomial in λ of degree k, such that $P_k(0) = 1$. Then, if we set $M = 1$, which we may always do by normalizing the matrix A, and $m = 0$ (again without loss of generality), and if we set

$$P_k(\lambda) = (-1)^k \frac{\cos[(2k + 1)\arccos\sqrt{\lambda}]}{(2k + 1)\sqrt{\lambda}}, \tag{3.38}$$

we obtain the following estimate for the conjugate gradient method:

$$(A\psi^k, \psi^k) \leq \frac{(\psi^0, \psi^0)}{(2k + 1)^2}. \tag{3.39}$$

We note that because $m = 0$ the estimate (3.39) is valid even for singular matrices.

4.4 The Splitting-Up Method

The *alternating direction method*, introduced by Douglas, Peaceman, and Rachford has had wide application to the stationary problems of mathematical physics. There are now many versions of the method and many implementation schemes for it. The method itself is founded on specialized relaxation processes that allow the reduction of a complex process to a sequence of simple ones. We shall refer to all these as *splitting-up methods*, We include the classical method of *alternating directions* (cf. Section 9.2).

As an illustration, we choose the example of a system of linear algebraic equations

$$A\phi = f \tag{4.1}$$

with the assumption that

$$A = A_1 + A_2, \tag{4.2}$$

where A_1 and A_2 are positive definite matrices. We showed in Section 4.2 that if we replace the iterative process

$$\phi^{j+1} = \phi^j - \tau(A\phi^j - f) \tag{4.3}$$

by the over-relaxation process

$$B(\phi^{j+1} - \phi^j) = -(A\phi^j - f) \tag{4.4}$$

where the matrix B depends on the parameter, we can significantly increase the convergence rate in practice without increasing the number of operations per iteration. The shortcoming of this method lay in the stringency of the constraints on the form of the matrix A and on its spectral properties.

Let us consider another class of methods, in which the matrix $B = B_\tau$ is defined by the equation

$$B_\tau = \frac{1}{2\tau}(E + \tau A), \tag{4.5}$$

and the matrix A is assumed to be symmetric and positive definite. Then we have for the error vectors $\psi^j = \phi^j - \phi^*$

$$(E + \tau A)\psi^{j-1} = (E - \tau A)\psi^j$$

or, what is the same thing,

$$\psi^{j+1} = T_\tau \psi^j,$$

where

$$T_\tau = (E + \tau A)^{-1}(E - \tau A) = (E - \tau A)(E + \tau A)^{-1} \tag{4.6}$$

is the transition operator of the process (4.4). Using the symmetry and positive definiteness of the matrix A, we easily see that the operator T_τ is defined for all $\tau > 0$, is symmetric, and satisfies the norm equation

$$\|T_\tau\|_2 = \beta(T_\tau) = \max_{\lambda_n(A)} \left| \frac{1 - \tau\lambda_n(A)}{1 + \tau\lambda_n(A)} \right|. \tag{4.7}$$

Here

$$0 < \alpha = \alpha(A) \leq \lambda_n(A) \leq \beta = \beta(A). \tag{4.8}$$

We now introduce for investigation the function

$$q(\tau) = \max_{\lambda_n(A)} |g(\tau, \lambda_n(A))|, \tag{4.9}$$

where

$$g(\tau, \lambda) = \frac{1 - \tau\lambda}{1 + \tau\lambda}. \tag{4.10}$$

For $\tau \leq 0$, the function $q(\tau) \geq 1$, and the process (4.4) diverges. A necessary condition for its convergence is $\tau > 0$. Since if $\tau, \lambda > 0$

$$g'_\lambda(\tau, \lambda) = -\frac{2\tau}{(1 + \tau\lambda)^2} < 0,$$

and $\lambda_n(A) \in [\alpha, \beta]$ then

$$q(\tau) = \max_{\lambda_n(A)} |g(\tau, \lambda)| = \max\{|g(\tau, \alpha)|, |g(\tau, \beta)|\}$$

$$= \max\left\{\left|\frac{1 - \tau\alpha}{1 + \tau\alpha}\right|, \left|\frac{1 - \tau\beta}{1 + \tau\beta}\right|\right\} < 1. \quad (4.11)$$

It follows that (4.4) converges for all $\tau > 0$.

The method (4.4) is a stationary iterative process. Therefore, to optimize it (cf. Section 4.2.2) we minimize the quantity $\beta(T_\tau)$ with respect to τ, i.e., we solve the extremal problem

$$q(\tau_{opt}) = \min_{\tau > 0} q(\tau). \quad (4.12)$$

If we remember that

$$[g(\tau, \lambda)]'_\tau = -\frac{2\lambda}{(1 + \tau\lambda^2)} < 0 \quad (4.13)$$

for all $\tau, \lambda > 0$, an argument like the one we used in Section 4.2.2 for the simplest iteration process will show that τ_{opt} is a root of the equation

$$\frac{1 - \tau\alpha}{1 + \tau\lambda} = -\frac{1 - \tau\beta}{1 + \tau\beta} \quad (4.14)$$

and is explicitly computed as

$$\tau_{opt} = \frac{1}{\sqrt{\alpha\beta}}. \quad (4.15)$$

In fact, if $\tau = \tau_{opt}$

$$\frac{1 - \tau_{opt}\alpha}{1 + \tau_{opt}\alpha} = -\frac{1 - \tau_{opt}\beta}{1 + \tau_{opt}\beta} = \frac{1 - \sqrt{\alpha/\beta}}{1 + \sqrt{\alpha/\beta}} > 0$$

and therefore

$$q(\tau_{opt}) = \frac{1 - \sqrt{\alpha/\beta}}{1 + \sqrt{\alpha/\beta}}. \quad (4.16)$$

If $\tau < \tau_{opt}$, (4.13) implies that

$$q(\tau) = \frac{1 - \tau\alpha}{1 + \tau\alpha} > q(\tau_{opt}).$$

Finally, if $\tau > \tau_{opt}$

$$q(\tau) = -\frac{1 - \tau\beta}{1 + \tau\beta} > q(\tau_{opt}).$$

Let us estimate the asymptotic rate of convergence of the process (4.4), (4.5), namely

$$R(T_{\tau_{opt}}) = -\ln q(\tau_{opt}) \quad (4.17)$$

for an ill-conditioned matrix, i.e., when $p = p(A) \gg 1$. By (4.16) we have

$$q(\tau_{opt}) \approx 1 - \frac{2}{\sqrt{p}},$$

$$R(T_{\tau_{opt}}) \approx \frac{2}{\sqrt{p}}.$$

(4.18)

It follows that the asymptotic convergence rate of our current method is less by a factor of two, at least, than that of the optimal successive over-relaxation method. While admitting the merits of our process, we must not overlook its defect, which is contained in the fact that the solution of the system

$$Bz^{j+1} = -\xi^j$$

$(z^{j+1} = \phi^{j+1} - \phi^j)$ may require, at each step, at least as many arithmetic operations as the solution of the original equations. This brings us to the problem of replacing the operator $B = B_\tau$ by some "nearby" operator such that the rate of convergence is conserved and the number of operations demanded per step is comparable to that of the successive over-relaxation process. In a very large number of cases we can succeed by setting

$$B = \frac{1}{2\tau}(E + \tau A_1)(E + \tau A_2),$$

(4.19)

a choice which we shall examine later. This scheme and others like it have been investigated in papers by Douglas, Peaceman, and Rachford [15], Samarskii and Marchuk [15], Yanenko [15], Dyakonov [15], and many others.

4.4.1 The Commutative Case

We assume that the matrices A_1 and A_2 in the decomposition (4.2) are symmetric and have a common complete system of orthonormal eigenvectors $\{u_n\}$, i.e.,

$$A_1 u_n = \lambda_n(A_1)u_n,$$

$$A_2 u_n = \lambda_n(A_2)u_n,$$

(4.20)

for all n, and that, furthermore, the system $\{u_n\}$ is an orthonormal basis for the original vector space. We resolve an arbitrary vector ϕ on the system $\{u_n\}$:

$$\phi = \sum_n \phi_n u_n, \qquad \phi_n = (\phi, u_n).$$

It is easily seen that

$$A_1 A_2 \phi = A_1\left(\sum_n \lambda_n(A_2)\phi_n u_n\right) = \sum_n \lambda_n(A_1)\lambda_n(A_2)\phi_n u_n$$

$$= A_2 \sum_n \lambda_n(A_1)\phi_n u_n = A_2 A_1 \phi.$$

Since the vector ϕ is arbitrary, this equation implies that the matrices A_1 and A_2 must commute: $A_1A_2 = A_2A_1$.

Now we resolve the error vectors $\psi^j = \phi^j - \phi^*$ of the splitting-up method:

$$(E + \tau A_1)(E + \tau A_2)(\phi^{j+1} - \phi^j) = -2\tau(A\phi^j - f) \tag{4.21}$$

on the system $\{u_n\}$:

$$\psi^j = \sum_n \psi_n^j u_n.$$

Using (4.21) we have

$$\psi_n^{j+1} = \frac{1 - \tau\lambda_{1,n}}{1 + \tau\lambda_{1,n}} \frac{1 - \tau\lambda_{2,n}}{1 + \tau\lambda_{2,n}} \psi_n^j, \tag{4.22}$$

where, for simplicity, we have used the notation

$$\lambda_{i,n} = \lambda_n(A_i) \ (i = 1, 2).$$

It follows from (4.22) (as, in fact, even from the commutativity of the matrices A_1 and A_2) that the transition operator

$$T_\tau = E - 2\tau(E + \tau A_2)^{-1}(E + \tau A_1)^{-1}A$$

$$= (E + \tau A_2)^{-1}(E + \tau A_1)^{-1}(E - \tau A_1)(E - \tau A_2) \tag{4.23}$$

of the process (4.21) has a complete orthonormal system of eigenvectors $\{u_n\}$ and that its eigenvalues are real and given by the formulas

$$\lambda_n(T_\tau) = \frac{1 - \tau\lambda_{1,n}}{1 + \tau\lambda_{1,n}} \frac{1 - \tau\lambda_{2,n}}{1 + \tau\lambda_{2,n}}$$

in terms of the eigenvalues of the matrices A_1 and A_2. It follows that the matrix T_τ is symmetric, and therefore, that

$$\|T_\tau\|_2 = \beta(T_\tau) = \max_n \left| \frac{1 - \tau\lambda_{1,n}}{1 + \tau\lambda_{1,n}} \frac{1 - \tau\lambda_{2,n}}{1 + \tau\lambda_{2,n}} \right|. \tag{4.24}$$

In keeping with the optimization theory for stationary iterative processes that we developed in Section 4.2.2, equation (4.24) implies that if we determine the parameter $\tau = \tau_{\text{opt}}$ by the minimization condition

$$\beta(T_{\tau_{\text{opt}}}) = \min_n \beta(T_\tau) \tag{4.25}$$

we have an optimum not only in the asymptotic case, but also for all $\varepsilon > 0$—provided that we require a reduction by a factor $1/\varepsilon$ in the Euclidean norm of the error vector. We note also that because $\lambda_{1,n}$ and $\lambda_{2,n}$ are positive the inequality $\beta(T_\tau) < 1$ holds only for $\tau > 0$, which is therefore a necessary condition for the convergence of the splitting-up method now under consideration.

Let us now solve the optimization problem (4.25). In the general case this is a very complex problem, as yet unsolved. Instead of minimizing $\beta(T_\tau)$ we must be content with minimizing

$$q^2(\tau) = \max_n \left| \frac{1 - \tau\lambda_{1,n}}{1 + \tau\lambda_{1,n}} \right| \max_n \left| \frac{1 - \tau\lambda_{2,n}}{1 + \tau\lambda_{2,n}} \right|, \tag{4.26}$$

which majorizes $\beta(T_\tau)$ from above and approximates it well enough, in many cases coinciding with it. We shall assume that

$$\tilde{\alpha} = \alpha(A_1) = \alpha(A_2), \qquad \tilde{\beta} = \beta(A_1) = \beta(A_2). \tag{4.27}$$

Then, since the maximum of the function

$$g(\tau, \lambda) = \frac{1 - \tau\lambda}{1 + \tau\lambda}$$

depends only on the boundaries of the interval $[\tilde{\alpha}, \tilde{\beta}]$ which, as shown at the outset of this section, contains both $\lambda_{1,n}$, $\lambda_{2,n}$, we find from (4.27) that

$$q^2(\tau) = \left[\max_{\tilde{\alpha} \leq \lambda \leq \tilde{\beta}} |g(\tau, \lambda)| \right]^2.$$

Accordingly, the optimization problem reduces to the extremal problem

$$q(\tau_{\text{opt}}) = \min_{\tau > 0} q(\tau), \tag{4.28}$$

which we have already studied. Its solution is given by (4.15):

$$\tau_{\text{opt}} = \frac{1}{\sqrt{\tilde{\alpha}\tilde{\beta}}}. \tag{4.29}$$

From this we have

$$\beta(T_{\tau_{\text{opt}}}) \leq q^2(\tau_{\text{opt}}) = \left[\frac{1 - \sqrt{\tilde{\alpha}/\tilde{\beta}}}{1 + \sqrt{\tilde{\alpha}/\tilde{\beta}}} \right]^2. \tag{4.30}$$

We know that

$$\begin{aligned}
\alpha &= \alpha(A) \geq \alpha(A_1) + \alpha(A_2) = 2\tilde{\alpha}, \\
\beta &= \beta(A) \leq \beta(A_1) + \beta(A_2) = 2\tilde{\beta}
\end{aligned} \tag{4.31}$$

and therefore

$$\frac{\tilde{\alpha}}{\tilde{\beta}} \leq \frac{\alpha}{\beta} < 1.$$

Using this inequality and the monotone decrease from 1 to 0 of the function

$$u(t) = \frac{1 - \sqrt{t}}{1 + \sqrt{t}}$$

on the interval $[0, 1]$, and using also the inequality (4.30), we arrive at the estimate

$$\beta(T_{\tau_{opt}}) \leq u^2\left(\frac{\tilde{\alpha}}{\tilde{\beta}}\right) \leq u^2\left(\frac{\alpha}{\beta}\right) = \left[\frac{\sqrt{p} - 1}{\sqrt{p} + 1}\right]^2,$$

where $p = \beta/\alpha$ is the conditioning number of the matrix A.

Finally, assuming $p \gg 1$, we have the estimates

$$\beta(T_{\tau_{opt}}) \leq 1 - \frac{4}{\sqrt{p}}, \tag{4.32}$$

$$R(T_{\tau_{opt}}) \geq \frac{4}{\sqrt{p}}.$$

Comparing this estimate with the data given in Table 4.1 (cf. Section 4.2.5) we conclude that our current method has no worse a convergence rate than the best of the stationary iterative processes—the successive overrelaxation method. We note further, that with its optimal parameters our splitting-up method converges at least twice as rapidly as the method (4.4), (4.5), and is at the same time much more easily realizable in practice, since one of the requirements of the splitting-up (4.2) is the simplicity of inversion of the matrices $E + \tau A_i$ $(i = 1, 2)$.

One of the basic problems of this method is optimization of a multi-parameter scheme of the form

$$(E + \tau_j A_1)(E + \tau_j A_2)(\phi^{j+1} - \phi^j) = -2\tau_j(A\phi^j - f). \tag{4.33}$$

This we accomplish as follows: for a given $s \geq 1$ we are to define a sequence of parameter values $\{\tau_i\}_{i=1}^s$ minimizing a norm or the spectral radius of the operator

$$T^{(s)} = \prod_{i=1}^{s} T_i, \tag{4.34}$$

where

$$T_i = (E + \tau_i A_2)^{-1}(E + \tau_i A_1)^{-1}(E - \tau_i A_1)(E - \tau_i A_2). \tag{4.35}$$

In the commutative case this is equivalent to finding the extremal

$$\max_n \prod_{i=1}^{s} \left|\frac{1 - \tau_i \lambda_{1,n}}{1 + \tau_i \lambda_{1,n}} \frac{1 - \tau_i \lambda_{2,n}}{1 + \tau_i \lambda_{2,n}}\right| = \min_{\tau_1, \ldots, \tau_s}. \tag{4.36}$$

As in the one-parameter case, this problem is replaced by the minimization of the function

$$q(\tau_1, \ldots, \tau_s) = \max_{\alpha \leq \lambda \leq \beta} \prod_{i=1}^{s} \left|\frac{1 - \tau_i \lambda}{1 + \tau_i \lambda}\right|, \tag{4.37}$$

where

$$\alpha = \min(\alpha(A_1), \alpha(A_2)), \qquad \beta = \max(\beta(A_1), \beta(A_2)).$$

Clearly,

$$\beta(T^{(s)}) \le q^2(\tau_1, \ldots, \tau_s) < 1 \tag{4.38}$$

for all $\tau_i > 0$ $(i = 1, \ldots, s)$.

The following minimization problem was first solved by Zolotarev [9]; as applied to the commutative splitting-up scheme, it has been solved by Peaceman and Rachford [15], Douglas and Rachford [15], Jordan [15], Wachspress [15], and others. The problem is to find

$$q(\tau_1^0, \ldots, \tau_s^0) = \min_{\tau_1, \ldots, \tau_s} q(\tau_1, \ldots, \tau_s), \tag{4.39}$$

With a special choice of the cycle length s and of the parameters $\tau = (\tau_1, \ldots, \tau_s)$ as approximate solutions of (4.39), we find for an ill-conditioned matrix A the estimate

$$\frac{C}{\ln p} \le R(T_\tau^{(s)}),$$

where $p = p(A)$ and C is a constant not depending on s or p. In other words, this method has the best convergence rate of all that we have examined so far.

Let us look at yet another approach to the optimization of the splitting-up method, limiting ourselves for simplicity to the one-parameter case (4.21). It follows from what we have developed above that for arbitrary $\tau > 0$ the splitting-up method reduces the value of each of the coefficients ψ_n^j in the expansion of the error vector ψ^j,

$$\psi^j = \sum_n \psi_n^j u_n,$$

at each iteration, and for the optimally chosen τ it yields the best uniform reduction over all n. Let us now assume that for those eigenvalues $\lambda_n \in [\tilde{m}, \Delta]$, where

$$\tilde{m} = \min(\alpha(A_1), \alpha(A_2)) < \Delta$$

and

$$\Delta \ll \tilde{M} = \max(\beta(A_1), \beta(A_2)),$$

the corresponding coefficients $\{\psi_n^0\}$ in the resolution of the initial error vector ψ^0 predominate significantly over the remaining coefficients. Then it would seem to be worthwhile to perform several iterations of the process (4.21) with the value

$$\tau = \frac{1}{\sqrt{\tilde{m}\Delta}}.$$

of the parameter τ. Such a choice yields the best uniform reduction of the selected predominant coefficients during the first few steps of the sequential iteration; after this, we use the value of τ given by (4.29).

In practical situations the dominant coefficients turn out to be those corresponding to the eigenvectors with smallest eigenvalues; with a certain amount of idealization, we may even suppose that the coefficients decrease monotonely with increasing λ_n. Physically this fact corresponds to the dominant influence of the large-scale processes (with small values of λ_n) versus the small-scale processes (large values of λ_n). It is clear that the actual choice of the value of Δ must depend on our specific a priori knowledge about the problem.

4.4.2 The Noncommutative Case

We now consider the splitting-up method

$$(E + \tau A_1)(E + \tau A_2)(\phi^{j+1} - \phi^j) = -2\tau(A\phi^j - f) \qquad (4.40)$$

with noncommutative and positive definite matrices A_1 and A_2. We begin by introducing the symmetric positive definite matrix

$$D_\tau = D_{1,\tau}^* D_{1,\tau} \qquad (4.41)$$

and the vectors

$$z^j = D_{1,\tau} \psi^j. \qquad (4.42)$$

Here

$$D_{1,\tau} = (E + \tau A_1)^{-1} A \qquad (4.43)$$

and the $\psi^j = \phi^j - \phi^*$ are the error vectors. Then

$$z^j = (E + \tau A_1)^{-1} \xi^j,$$

where the $\xi^j = A\phi^j - f$ are the residuals vectors.

Using standard transformations, we may easily show that

$$\psi^{j+1} = T_\tau \psi^j, \qquad (4.44)$$

where

$$T_\tau = E - 2\tau(E + \tau A_2)^{-1}(E + \tau A_1)^{-1} A \qquad (4.45)$$

is the transition operator of the process (4.40). Also

$$z^{j+1} = \tilde{T}_\tau z^j, \qquad (4.46)$$

where

$$\tilde{T}_\tau = T_{1,\tau} T_{2,\tau} \qquad (4.47)$$

and

$$T_{i,\tau} = (E - \tau A_i)(E + \tau A_i)^{-1}, \qquad i = 1, 2. \qquad (4.48)$$

In the derivation of (4.46)–(4.48) we used the relations

$$\tilde{T}_\tau = (E + \tau A_1)^{-1} A T_\tau A^{-1}(E + \tau A_1)$$
$$= (E + \tau A_1)^{-1}[(E + \tau A_1)(E + \tau A_2) - 2\tau A](E + \tau A_2)^{-1}$$
$$= (E + \tau A_1)^{-1}(E - \tau A_1)(E - \tau A_2)(E + \tau A_2)^{-1}$$
$$= (E - \tau A_1)(E + \tau A_1)^{-1}(E - \tau A_2)(E + \tau A_2)^{-1}.$$

Thus, for an arbitrary vector ψ ($z = D_{1,\tau}\psi$, $x = T_{2,\tau}z$) we have the formulas

$$\|T_\tau \psi\|_{D_\tau} = \|\tilde{T}_\tau z\|_2 = \frac{\|T_{1,\tau}x\|_2}{\|x\|_2} \frac{\|T_{2,\tau}z\|_2}{\|z\|_2} \|\psi\|_{D_\tau}, \qquad (4.49)$$

and

$$\|T_\tau\|_{D_\tau} = \|\tilde{T}_\tau\|_2 \le \|T_{1,\tau}\|_2 \|T_{2,\tau}\|_2. \qquad (4.50)$$

Let us study $\|T_{i,\tau}\|_2$ ($i = 1, 2$) on the assumption that the inequalities

$$(A_i u, u) \ge m(u, u),$$
$$(A_i u, A_i u) \le M(A_i u, u), \qquad i = 1, 2. \qquad (4.51)$$

hold for all u.

By definition

$$\|T_{i,\tau}\|_2^2 = \sup_{v \ne 0} \frac{\|T_{i,\tau}v\|_2^2}{\|v\|_2^2} = \sup_{u \ne 0} \frac{\|(E - \tau A_i)u\|_2^2}{\|(E + \tau A_i)u\|_2^2}$$

$$= \sup_{u \ne 0} \frac{(u, u) - 2\tau(A_i u, u) + \tau^2(A_i u, A_i u)}{(u, u) + 2\tau(A_i u, u) + \tau^2(A_i u, A_i u)}$$

$$= 1 - 4\tau \inf_{u \ne 0} \frac{(A_i u, u)}{(u, u) + 2\tau(A_i u, u) + \tau^2(A_i u, A_i u)}, \qquad (4.52)$$

where $u = (E + \tau A_i)^{-1}v$. It is clear that the fraction

$$\frac{\|T_{i,\tau}v\|_2}{\|v\|_2}$$

for all nonzero v and all i, either is greater than 1 for $\tau < 0$ or is less than 1 for $\tau > 0$. Hence, by (4.49), it follows that the positivity of τ is a necesary condition for the convergence of the splitting-up method (4.40). We shall assume from now on that $\tau > 0$.

We shall estimate $\|T_{i,\tau}\|_2$, using (4.51) and (4.52):

$$\|T_{i,\tau}\|_2^2 \le 1 - 4\tau \inf_{u \ne 0} \frac{(A_i u, u)}{(u, u) + 2\tau(A_i u, u) + \tau^2 M(A_i u, u)}$$

$$= 1 - 4\tau \inf_{\|w\|_2 = 1} \frac{(A_i w, w)}{1 + (2\tau + M\tau^2)(A_i w, w)} \le 1 - 4\tau \inf_{t \ge m} \frac{t}{1 + (2\tau + M\tau^2)t}$$

$$= 1 - 4\tau \frac{m}{1 + (2\tau + M\tau^2)m} = \frac{1 - 2m\tau + mM\tau^2}{1 + 2m\tau + mM\tau^2} < 1.$$

Here we have used the notation

$$w = \frac{u}{\|u\|_2}, \qquad t = (A_i\omega, \omega)$$

and the fact that the function

$$u(t) = \frac{t}{1 + at}$$

increases uniformly for $\tau \geq 0$ and all positive a. Thus we have shown that under the assumptions (4.51)

$$\|\tilde{T}_\tau\|_{D_\tau} \leq \|T_{1,\tau}\|_2 \|T_{2,\tau}\|_2 \leq \frac{1 - 2m\tau + mM\tau^2}{1 + 2m\tau + mM\tau^2} < 1$$

for all $\tau > 0$.

As an approximate optimization of (4.40) we minimize the function

$$q(\tau) = \frac{1 - 2m\tau + mM\tau^2}{1 + 2m\tau + mM\tau^2} = 1 - \frac{4m\tau}{1 + 2m\tau + mM\tau^2} \qquad (4.53)$$

with respect to τ. Since $q(0) = q(+\infty) = 1$, and since for all nonnegative τ the function $q(\tau)$ is infinitely differentiable and

$$0 \leq q(\tau) < 1,$$

the value $\tau = \tau_{\text{opt}}$ that minimizes $q(\tau)$ is the positive solution of the equation

$$\frac{dq(\tau)}{d\tau} \equiv -4m \frac{1 - mM\tau^2}{(1 + 2m\tau + mM\tau^2)^2} = 0.$$

This yields

$$\tau_{\text{opt}} = \frac{1}{\sqrt{mM}}, \qquad (4.54)$$

so that

$$q(\tau_{\text{opt}}) = \frac{1 - \sqrt{m/M}}{1 + \sqrt{m/M}}, \qquad (4.55)$$

$$R(T_{\tau_{\text{opt}}}) \geq -\ln - \ln \left[\frac{1 - \sqrt{m/M}}{1 + \sqrt{m/M}}\right]. \qquad (4.56)$$

For the analysis of this estimate we assume that the matrices A_1 and A_2 are symmetric, whence

$$\alpha(A_1 = \alpha(A_2) = \tfrac{1}{2}\alpha(A),$$

$$\beta(A_1) = \beta(A_2) = \tfrac{1}{2}\beta(A).$$

Then

$$m = \tfrac{1}{2}\alpha(A), \qquad M = \tfrac{1}{2}\beta(A)$$

and therefore

$$q(\tau_{opt}) = \frac{\sqrt{p} - 1}{\sqrt{p} + 1}.$$

Thus for an ill-conditioned system $[p = p(A) \gg 1]$ we have the estimate

$$R(T_{opt}) \geq \frac{2}{\sqrt{p}}. \qquad (4.57)$$

At the same time, an immediate spectral analysis like the one carried out in Section 4.4.1 shows that for $\tau = \tau_{opt}$,

$$\|T_{i,\tau_{opt}}\|_2 = \max_{m \leq \lambda \leq M} \left| \frac{1 - \tau_{opt}\lambda}{1 + \tau_{opt}\lambda} \right| = \frac{1 - \sqrt{m/M}}{1 + \sqrt{m/M}} \qquad (4.58)$$

and therefore

$$\beta(T_{\tau_{opt}}) \leq \left[\frac{1 - \sqrt{m/M}}{1 + \sqrt{m/M}} \right]^2 \leq \left[\frac{\sqrt{p} - 1}{\sqrt{p} + 1} \right]^2. \qquad (4.59)$$

For an ill-conditioned system we have the following estimates:

$$\beta(T_{\tau_{opt}}) \leq 1 - \frac{4}{\sqrt{p}},$$

$$R(T_{\tau_{opt}}) \geq \frac{4}{\sqrt{p}}, \qquad (4.60)$$

that is, the convergence of the splitting-up method (4.40) is no slower, for $\tau = \tau_{opt}$, than that of any stationary iterative process so far considered. The estimates (4.57) and (4.60) differ because the first method of constructing the estimate is more general, and the additional constraints on the properties of the matrices A_1 and A_2 allow us to improve the estimate substantially.

In conclusion, we look briefly at another important subclass of the splitting-up processes (4.40). Suppose that the matrix A of the system $A\phi = f$ is symmetric and positive definite, and

$$A = A_1 + A_2,$$

$$A_1 = A_2^*. \qquad (4.61)$$

We have supposed that the matrices A_1 and A_2 are real. Therefore, if (4.61) is assumed, the equations

$$(A_1\psi, \psi) = (A_2\psi, \psi) = \tfrac{1}{2}(A\psi, \psi).$$

hold in the space of real vectors for all ψ. It follows that if A is positive definite, A_1 and A_2 will be also. Then by what we have shown above, the splitting-up method (4.40) with such matrices converges for all $\tau > 0$.

The matrices A_1 and A_2 are very often constructed from the matrix A in the following way: first represent A in block form

$$A = \begin{Vmatrix} A_{11} & A_{12} & \cdots & A_{1k} \\ A_{21} & A_{22} & \cdots & A_{2k} \\ \cdots\cdots\cdots\cdots\cdots\cdots \\ A_{k1} & A_{k2} & \cdots & A_{kk} \end{Vmatrix}$$

with square diagonal submatrices A_{ii}, and for A_1 choose a block-triangular matrix of the form

$$A_1 = \begin{Vmatrix} \tfrac{1}{2}A_{11} & 0 & \cdots & 0 \\ A_{21} & \tfrac{1}{2}A_{22} & \cdots & 0 \\ \cdots\cdots\cdots\cdots\cdots\cdots \\ A_{k1} & A_{k2} & \cdots & \tfrac{1}{2}A_{kk} \end{Vmatrix}$$

Set A_2 equal to A_1^*. The simplest special case of this procedure arises by setting $A_{ii} = a_{ii}$. Then A_1 is a lower triangular matrix and A_2 is upper triangular. Iterative splitting-up methods with $A_1 = A_2^*$ have been studied by Samarskii [15] and Il'in [15].

4.4.3 Variational and Chebyshevian Optimization of Splitting-Up Methods

In the preceding subsection we proved that for $\gamma_j = 2\tau$ in the iterative splitting-up method

$$(E + \tau A_1)(E + \tau A_2)(\phi^{j+1} - \phi^j) = -\gamma_j(A\phi^j - f), \qquad (4.62)$$

the inequality

$$q_j = \frac{\|(E - \gamma_j H_\tau A)\psi^j\|_{D_\tau}}{\|\psi^j\|_{D_\tau}} \le \|T_\tau\|_{D_\tau} < 1 \qquad (4.63)$$

holds for all $\tau > 0$, with

$$H_\tau = (E + \tau A_2)^{-1}(E + \tau A_1)^{-1},$$

$$T_\tau = E - 2\tau H_\tau A$$

and ψ^j as the error vector. In accordance with the results obtained in Section 4.3.1 this means that the nonstationary iterative method (4.62) will converge if its parameters γ_j are solutions of the equations

$$\frac{d}{d\gamma_i} \|(E - \gamma_j H_\tau A)\psi^j\|_{D_\tau} = 0, \qquad (4.64)$$

that is, if they are chosen so as to minimize, at every step of the iteration, the functional

$$J(\phi) = (D_\tau(\phi^* - \phi), \phi^* - \phi) = \|\phi^* - \phi\|_{D_\tau}^2.$$

The transition operator $T_{\tau, \gamma}$ of our process is estimated by

$$\|T_{\tau, \gamma}\|_{D_\tau} \leq \|T_\tau\|_{D_\tau} < 1. \tag{4.65}$$

for all $\tau > 0$. Solving (4.64) we find for γ_j:

$$\gamma_j = \frac{(H_\tau A \psi^j, \psi^j)_{D_\tau}}{\|H_\tau A \psi^j\|_{D_\tau}^2} = \frac{((E + \tau A_1)^{-1} A H_\tau \xi^j, (E + \tau A_1)^{-1} \xi^j)}{\|(E + \tau A_1)^{-1} A H_\tau \xi^j\|_2^2}, \tag{4.66}$$

where the $\xi^j = A\phi^j - f$ are the residuals vectors.

Another method of optimizing the splitting-up process (4.40) presupposes that $H_\tau = H_\tau^*$. We know from the work done in this section that this situation arises either in the commutative case if $A_i = A_i^*$ ($i = 1, 2$), or in the case $A_1 = A_2^*$:

$$H_\tau = (E + \tau A_2)^{-1}(E + \tau A_1)^{-1} = (E + \tau A_1^*)^{-1}(E + \tau A_2)^{-1}$$
$$= [(E + \tau A_1)^{-1}]^*(E + \tau A_1)^{-1} = H_\tau^*.$$

Using the assumption and the results of Section 4.2.4, we conclude that the parameters $\{\gamma_j\}$ of the cyclic iteration method

$$\phi^{j+1} = \phi^j - \gamma_j H_\tau(A\phi^j - f) \tag{4.67}$$

with cycle length s, can be selected on the basis of the general theory of Chebyshevian iteration methods developed in Section 4.2.4. We must replace $\alpha = \alpha(A)$ and $\beta = \beta(A)$ by

$$m = \alpha(H_\tau A),$$

$$M = \beta(H_\tau A).$$

To estimate these quantities, or to compute them, we may use either our estimates of the norm of the operator $T = E - 2\tau H_\tau A$, or an auxiliary iterative method of the Lyusternik type.

As an illustration, let us estimate m and M in the commutative case, assuming that $\tau = \tau_{opt}$ (cf. (4.29)) and that A is ill-conditioned. Then the first of the inequalities (4.32) implies that

$$m = \alpha(H_\tau A) \geq \frac{1}{2\tau} \frac{4}{\sqrt{p(A)}}, \tag{4.68}$$

$$M = \beta(H_\tau A) \leq \frac{1}{2\tau}\left(2 - \frac{4}{\sqrt{p(A)}}\right).$$

From these we find that

$$p(HA_\tau) = \frac{\beta(H_\tau A)}{\alpha(H_\tau A)} \leq \tfrac{1}{2}(\sqrt{p(A)} - 2) \approx \frac{\sqrt{p(A)}}{2}. \tag{4.69}$$

Returning to the estimates (2.69) and taking account of the fact that $p(H_\tau A) \gg 1$ if $p(A) \gg 1$, we find that

$$R(T_{\tau,\gamma}) \approx \frac{2}{\sqrt{p(H_\tau A)}} \approx \frac{2\sqrt{2}}{\sqrt[4]{p(A)}}. \tag{4.70}$$

for the method (4.67) when $s \gg 1$. This means that the use of the Chebyshev method lets us increase the effectiveness of the splitting-up method by a factor of $\sqrt{p(A)/2}$.

Another method for increasing the convergence rate of the splitting-up method, when $A_1 = A_2^*$, is based on the conjugate gradient method. Since the matrix H_τ is symmetric and positive definite, we may immediately rewrite the formulas (3.34) as:

$$g_k = \begin{cases} H_\tau \xi^0, & k = 1, \\ H_\tau \xi^{k-1} - b_k g_{k-1}, & k > 1, \end{cases}$$

$$b_k = \frac{(AH_\tau \xi^{k-1}, g_{k-1})}{(Ag_{k-1}, g_{k-1})}, \tag{4.71}$$

$$\phi^k = \phi^{k-1} - \alpha_k g_k,$$

$$\alpha_k = \frac{(\xi^{k-1}, g_k)}{(Ag_k, g_k)}, \qquad k = 1, \ldots, s.$$

Note that the convergence rate of (4.71) is roughly the same as that of (4.67) for an arbitrary s-period choice of the parameters $\{\gamma_j\}$.

4.5 Iteration Methods for Systems with Singular Matrices

We consider a system of linear algebraic equations

$$A\phi = f \tag{5.1}$$

with a symmetric positive semidefinite matrix A. We denote by $\{u_n\}$ the system of orthonormal eigenvectors of A, corresponding to the eigenvalues $\{\lambda_n\}$. We denote by ker A the null space of A, i.e., the set of vectors ψ such that $A\psi = 0$; we assume that its dimension is m and that the null eigenvalues are represented by $\{\lambda_n\}_{n=1}^m$, corresponding to the vectors $\{u_n \in \text{ker } A\}$ for $n = 1, \ldots, m$.

We now resolve the vectors f and ϕ of the system (5.1) on the basis $\{u_n\}$:

$$f = \sum_n f_n u_n,$$

$$\phi = \sum_n \phi_n u_n.$$

Inserting this resolution into (5.1) we find that

$$\lambda_n \phi_n = f_n$$

and, in particular, that

$$\lambda_n \phi_n = f_n \quad \text{for} \quad n = 1, \ldots, m.$$

But this latter set of equations is possible only if $f_n = 0$ $(n = 1, \ldots, m)$. Therefore, the compatibility conditions for the system require that f be normal to the kernel space $(f \perp \ker A)$, or, what amounts to the same thing

$$(f, u_n) = 0, \quad n = 1, \ldots, m.$$

When the system (5.1) is inconsistent the solution is often taken to be that of the consistent system

$$A\phi = \tilde{f}, \tag{5.2}$$

where

$$\tilde{f} = \sum_{n > m} f_n u_n,$$

This is called the *generalized solution* of (5.1). As is easily seen, it minimizes the Euclidean norm of the residuals vector. If A is arbitrary, the vector ϕ^* is called the generalized solution of the system if it is a solution of the equation

$$\|A\phi^* - f\|_2 = \min_{\phi} \|A\phi - f\|_2.$$

(Cf. Voevodin [8], Faddeev, Faddeeva, and Kublanovskaya [8], and others.)

4.5.1 Consistent Systems

Suppose that the system (5.1) is consistent and that we apply to it the stationary iterative process

$$B(\phi^{j+1} - \phi^j) = -(A\phi^j - f) \tag{5.3}$$

with a symmetric positive definite matrix B. To investigate the conditions of convergence for this process, we introduce the vectors $\psi^j = \phi^j - \phi^*$, where ϕ^* is some fixed solution of (5.1), and then write the iterative process for the $\{\psi^j\}$ in the form

$$\psi^{j+1} = T\psi^j, \tag{5.4}$$

where

$$T = E - B^{-1}A \tag{5.5}$$

is the transition operator of the method (5.3). Now let $z^j = B^{1/2j}\psi^j$. If we multiply (5.4) by $B^{1/2}$ we have

$$z^{j+1} = \tilde{T}z^j, \tag{5.6}$$

where

$$\tilde{T} = E - B^{-1/2}AB^{-1/2} \tag{5.7}$$

is symmetric and the matrix $s = B^{-1/2}AB^{-1/2}$ is symmetric and positive semidefinite. (By $B^{-1/2}$ we mean $[B^{1/2}]^{-1}$.)

Let $\{v_n\}$ be the complete orthonormal system of eigenvectors of $\tilde{T}$, corresponding to the eigenvalues $\{\mu_n\}$ and let the vectors z^j be resolved on this system:

$$z^j = \sum_n z_n^j v_n. \tag{5.8}$$

If we insert (5.8) into (5.6) we find that

$$z_n^{j+1} = \mu_n z_n^j = \mu_n^{j+1} z_n^0.$$

It follows that the conditions

$$|\mu_n| \le 1,$$

$$\mu_n \ne -1 \tag{5.9}$$

are necessary for the convergence of the process (5.3) under arbitrary initial approximations ϕ^0. In fact, if $|\mu_n| > 1$ or $\mu_n = -1$, the sequence z_n^j respectively diverges or fails to converge to a specific value for all $z_n^0 \ne 0$.

The condition (5.9) is sufficient for the convergence of the sequence $\{\phi^j\}$ to some solution $\tilde{\phi}^*$ of the system (5.1) for all initial ϕ^0. In fact, the inequality $|\mu_n| < 1$ always implies the convergence of z_n^j to zero as $j \to \infty$; in the case $\mu_n = 1$,

$$z_n^{j+1} = z_n^j.$$

Let us now suppose that (5.9) is satisfied, that the multiplicity of the root $\mu_n = 1$ is s (we could prove that $s = m$), and that $\mu_1 = \mu_2 = \cdots = \mu_s = 1$. Then it is easily seen that

$$\lim_{j \to \infty} z^j = z^\infty = \sum_{n=1}^{s} z_n^0 v_n.$$

The equation $\tilde{T}z^\infty = z^\infty$ (since $\tilde{T}v_n = v_n$ for $n = 1, \ldots, s$) implies that

$$B^{-1/2}AB^{-1/2}z^\infty = 0,$$

and therefore

$$B^{-1/2}z^\infty \in \ker A. \tag{5.10}$$

On the other hand, we have

$$B^{-1/2}z^\infty = \lim_{j \to \infty} B^{-1/2}z^j = \lim_{j \to \infty} \psi^j = \psi^\infty. \tag{5.11}$$

Combining (5.10) and (5.11), we conclude that the sequence $\{\phi^j\}$ in (5.3) converges to a vector

$$\phi^\infty = \phi^* + \psi^\infty, \tag{5.12}$$

which, since $\psi^\infty \in \ker A$, is a solution of (5.1); i.e., $\phi^\infty = \tilde{\phi}^*$, and depends on the choice of the initial approximation ϕ^0.

Sufficient conditions for the convergence of general stationary and non-stationary iterative methods for solving systems with singular matrices have been studied in detail by Kuznetsov and are given in a monograph by Marchuk and Kuznetsov [8].

4.5.2 Inconsistent Systems

We shall consider two approaches to the solution of the system (5.1) when the equations are inconsistent. Let us first assume that the matrices B and A in (5.3) have a common complete orthonormal system of eigenvectors $\{u_n\}$, i.e.,

$$\begin{aligned} Bu_n &= v_n u_n, \\ Au_n &= \lambda_n u_n, \end{aligned} \tag{5.13}$$

and that (5.9) holds for (5.1). The latter condition implies, as we have shown, that (5.3) converges for the corresponding consistent systems. If we resolve the vectors ϕ^j and f on the basis $\{u_n\}$ we have

$$\phi^j = \sum_n \phi_n^j u_n,$$

$$f = \sum_n f_n u_n,$$

and we obtain the equations

$$\phi_n^{j+1} = \left(1 - \frac{\lambda_n}{v_n}\right)\phi_n^j + \frac{1}{v_n} f_n = \left(1 - \frac{\lambda_n}{v_n}\right)^{j+1}\phi_n^0 + \frac{1}{v_n}\sum_{k=0}^{j}\left(1 - \frac{\lambda_n}{v_n}\right)^k f_n,$$

whence, by (5.9), we have

$$\lim_{j \to \infty} \phi_n^j = \frac{1}{\lambda_n} f_n$$

for $\lambda_n \neq 0$ $(n > m)$, and

$$\phi_n^j = \frac{j}{v_n} f_n + \phi_n^0$$

for $\lambda_n = 0$ $(n \leq m)$. Thus, the sequence ϕ^j diverges if $f_n \neq 0$, for even one $n \leq m$. Let us formulate the sequence

$$z^j = \phi^j - j(\phi^{j+1} - \phi^j) \tag{5.14}$$

and resolve the vectors z^j on the basis $\{u_n\}$:

$$z^j = \sum_n z_n^j u_n.$$

We find that

$$\begin{cases} \phi_n^j - j\left[\dfrac{\lambda_n}{v_n}\left(1 - \dfrac{\lambda_n}{v_n}\right)^f \phi_n^0 + \dfrac{1}{v_n}\left(1 - \dfrac{\lambda_n}{v_n}\right)^f f_n\right], & n > m, \\[4mm] \phi_n^0, & n \le m, \end{cases}$$

and therefore

$$\lim_{j \to \infty} z_n^j = \begin{cases} \dfrac{1}{\lambda_n} f_n, & n > m, \\[4mm] \phi_n^0, & n \le m. \end{cases} \tag{5.15}$$

Thus, the vector

$$z^\infty = \lim_{j \to \infty} z^j \tag{5.16}$$

exists and is the solution of the consistent system (5.2).

Molchanov and Yakovlev [8] have studied iteration methods of the type of (5.3), (5.14) for solving systems of linear equations with symmetric and positive semidefinite matrices.

Another approach to the solution of inconsistent systems, as applied to the system (5.1) is as follows: suppose that the null space of the matrix A is one-dimensional ($m = 1$), that the matrix B of the iterative process (5.3) is symmetric and positive definite, and that the conditions (5.9) are satisfied. Then, as is evident, the null spaces of the matrices A and $B^{-1}A$ coincide, all the eigenvalues of the latter are real, and it has a complete system of eigenvectors (as was shown in Section 4.2.4).

Let $\{v_n\}$ denote the system of eigenvectors of $B^{-1}A$ and $\{v_n^*\}$ the eigenvectors of $(B^{-1}A)^* = AB^{-1}$. It is well known that if these systems are complete they form a biorthogonal basis in the initial vector space, i.e.,

$$(v_i, v_j^*) = \begin{cases} 0, & i \ne j, \\ 1, & i = j. \end{cases}$$

Suppose that $v_1 \in \ker(B^{-1}A) = \ker A$. We resolve the vectors x^j of the iterative process

$$\begin{aligned} x^{j+1} &= x^j - B^{-1}Ax^j, \\ x^0 &= B^{-1}f \end{aligned} \tag{5.17}$$

on the system $\{v_n\}$ and obtain the equations

$$x^j = \sum_n x_n^j x_n, \qquad x_n^j = (x^j, v_n^*).$$

Then, by (5.9)

$$x^\infty = x_1^0 v_1.$$

The assumption that $x_1^0 = (B^{-1}f, v_1^*) = 0$ implies that the system

$$B^{-1}A\phi = B^{-1}f$$

is consistent and that, therefore, the initial system is consistent also. But we assumed it inconsistent; therefore, $x_1^0 \neq 0$ and the vector $v^\infty = x_1^0 v_1 \in \ker A$. Orthogonalizing the vector f to the vector v^∞;

$$\tilde{f} = f - \frac{(f, v^\infty)}{(v^\infty, v^\infty)} v^\infty, \tag{5.18}$$

and we arrive at the system

$$A\phi = \tilde{f}, \tag{5.19}$$

which is consistent and equivalent to the original system in the sense that it has the same set of generalized solutions.

The above method was proposed and generalized to the case of non-symmetric matrices and null spaces of arbitrary dimension by Kuznetsov [8].

4.5.3 The Matrix Analog of the Method of Fictive Regions

Suppose we are given a system of linear algebraic equations

$$A_0 \phi_0 = f_0 \tag{5.20}$$

with a symmetric positive definite matrix A_0. We consider another system as well,

$$A\phi = f \tag{5.21}$$

with a matrix A of the form

$$A = \left\| \begin{array}{cc} A_0 & 0 \\ 0 & A_1 \end{array} \right\|, \tag{5.22}$$

where A_1 is symmetric and positive semidefinite,

$$\phi = \left\| \begin{array}{c} \phi_0 \\ \phi_1 \end{array} \right\| \quad \text{and} \quad f = \left\| \begin{array}{c} f_0 \\ 0 \end{array} \right\|.$$

It is clear that A is a symmetric positive semidefinite matrix and that given any solution

$$\phi = \left\| \begin{array}{c} \phi_0 \\ \phi_1 \end{array} \right\|$$

of the system (5.21) the vector ϕ_0 is a solution of (5.20).

To solve (5.21) we apply the stationary iterative process

$$\phi^{j+1} = \phi^j - H(A\phi^j - f) \tag{5.23}$$

with a symmetric positive definite matrix H and transition operator

$$T = E - HA, \tag{5.24}$$

whose eigenvalues satisfy (5.9). As we proved in Section 4.5.1, the convergence rate of this process, if (5.21) is consistent (and it always is), is defined by that eigenvalue of T having the largest absolute value different from 1. It can be shown (Kuznetsov and Matsokin [4]) that this value can be minimized by the choice

$$A_1 = 0$$

for any symmetric positive definite matrix H.

We may illustrate this method by solving a five-point equation system approximating the Dirichlet problem for a Poisson equation in an L-shaped region D (Figure 4.4). In this figure, the points at which the five-point equations are defined by boundary values are marked by crosses, and the points at which they are defined by null coefficients are marked with circles (these correspond to $A_1 = 0$).

Let $B = \tau H^{-1}$ be the five-point difference analog of the Laplace operator at all nodes of a rectangular net with step h and covering the square. The maximal eigenvalue of the matrix T (different from 1) satisfies the inequality

$$|\lambda_{\max}| < 1 - Ch, \tag{5.25}$$

where $h \ll 1$ and C is a constant independent of h. The proof of (5.25) is as follows: let $h = 1/(n + 1)$, $k = (n + 1)/2$. We number the nodes in the net-

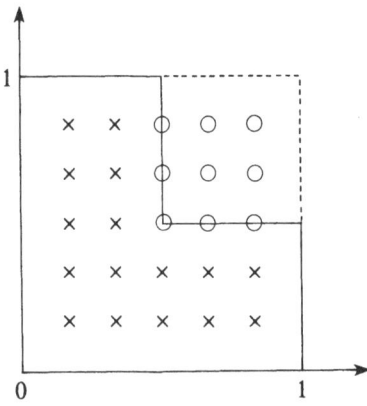

Figure 4.4

work so that those lying in the L-shaped region are numbered first and the rest afterward. Writing out the five-point equations

$$\frac{-u_{i-1j} + 2u_{ij} - u_{i+1j}}{h^2} + \frac{-u_{ij-1} + 2u_{ij} - u_{ij+1}}{h^2} = f_{ij} \qquad (5.26)$$

in sequence according to our numbering system and filling in the null coefficients, we arrive at the matrix problem

$$Au = f$$

of the form (5.21), (5.22) with a matrix

$$A = \left\| \begin{matrix} A_0 & 0 \\ 0 & 0 \end{matrix} \right\|$$

of order n^2. The matrix of the difference analog of the Laplace operator has the form

$$B = \left\| \begin{matrix} A_0 & -C \\ -C^T & B_1 \end{matrix} \right\|,$$

under the numbering system we have introduced, and B_1 is symmetric and positive definite.

The minimal nonzero eigenvalue of the matrix HA is given by the equation

$$\tau A\phi = \lambda B\phi, \qquad (5.27)$$

where $\phi = (\phi_0, \phi_1)^T$ is the corresponding eigenvector. This implies that

$$\phi_1 = B_1^{-1}C^T\phi_0. \qquad (5.28)$$

If we take the scalar product of (5.27) with ϕ and take account of (5.28) we have

$$\lambda = \tau \frac{(A\phi, \phi)}{(B\phi, \phi)} = \tau \frac{(A_0\phi_0, \phi_0)}{(A_0\phi_0, \phi_0) - (B_1^{-1}C^T\phi_0, C^T\phi_0)} \geq \tau, \qquad (5.29)$$

since $(B\phi, \phi) > 0$ and $(B^{-1}C^T\phi_0, C^T\phi_0) > 0$.

Now suppose that $\lambda = \rho$ is the maximal eigenvalue of the problem (5.27) and that ϕ is the corresponding eigenvector. We introduce the square nth order matrices

$$A_n = \frac{1}{h^2} \left\| \begin{matrix} 2 & -1 & 0 & \cdots & 0 & 0 & 0 \\ -1 & 2 & -1 & \cdots & 0 & 0 & 0 \\ \cdots\cdots\cdots\cdots\cdots\cdots\cdots\cdots\cdots\cdots \\ 0 & 0 & 0 & \cdots & -1 & 2 & -1 \\ 0 & 0 & 0 & \cdots & 0 & -1 & 2 \end{matrix} \right\|, \qquad \hat{A}_n = \left\| \begin{matrix} A_{k-1} & 0 \\ 0 & 0 \end{matrix} \right\|$$

and the vectors of length n formed from the components of the vector ϕ:

$$\phi^i = \begin{Vmatrix} \phi_{i1} \\ \phi_{i2} \\ \vdots \\ \phi_{in} \end{Vmatrix}, \quad i = 1, \ldots, n, \qquad \psi^j = \begin{Vmatrix} \phi_{1j} \\ \phi_{2j} \\ \vdots \\ \phi_{nj} \end{Vmatrix}, \quad j = 1, \ldots, n.$$

A scalar multiplication of (5.27) by ϕ yields the equation

$$\begin{aligned}
\rho &= \frac{(A\phi, \phi)}{(H^{-1}\phi, \phi)} = \tau \frac{(A\phi, \phi)}{(B\phi, \phi)} \\
&= \tau \frac{\sum_{i=1}^{k-1} [(A_n\phi^i, \phi^i) + (A_n\psi^i, \psi^i)] + \sum_{i=k}^{n} [(\hat{A}_n\phi^i, \phi^i) + (\hat{A}_n\psi^i, \psi^i)]}{\sum_{i=1}^{n} [(A_n\phi^i, \phi^i) + (A_n\psi^i, \psi^i)]}.
\end{aligned}$$

Note that if any one of the scalar products occurring in the denominator is equal to zero, the corresponding vector ϕ^i or ψ^i is also null. Since the inequality

$$\frac{\sum_{i=1}^{m} a_i}{\sum_{i=1}^{m} b_i} \leq \max_{1 \leq i \leq m} \frac{a_i}{b_i}$$

holds for all $a_i \geq 0$, $b_i > 0$, we find that

$$\rho \leq \tau \max\left\{1, \frac{(\hat{A}_n g, g)}{(A_n g, g)}\right\}, \tag{5.30}$$

where the nonnull vector g coincides with some one of the vectors ϕ^i, ψ^i $(i = 1, \ldots, n)$.

The quantity $(\hat{A}_n g, g)/(A_n g, g)$ does not exceed the spectral radius of the matrix $G = \hat{A}_n^{-1} A_n$. This matrix is easily computed:

$$G = \begin{Vmatrix} X & 0 \\ Y & 0 \end{Vmatrix},$$

where

$$X = (A_{k-1} - CA_k^{-1}C^T)^{-1} A_{k-1}, \qquad Y = A_k^{-1} C^T X,$$

and C is a $(k-1) \times k$-matrix with a single nonzero element in the lower left corner, equal to $1/h^2$.

It follows that the spectral radius of G coincides with the maximal eigenvalue of the problem

$$A_{k-1}v = \mu(A_{k-1} - CA_k^{-1}C^T)v.$$

If we compute the product $CA_k^{-1}C^T$ explicitly we see that the above problem can be reduced to the system

$$\frac{1}{k+1}\phi_{k-1} = \left(1 - \frac{1}{\mu}\right)\phi_1,$$

$$\frac{2}{k+1}\phi_{k-1} = \left(1 - \frac{1}{\mu}\right)\phi_2,$$

$$\cdots\cdots\cdots\cdots\cdots\cdots\cdots\cdots$$

$$\frac{k-1}{k+1}\phi_{k-1} = \left(1 - \frac{1}{\mu}\right)\phi_{k-1}.$$

We find immediately that

$$\mu = \frac{k+1}{2} = h^{-1}\frac{1+2h}{4}.$$

Taking account of (5.30) we conclude that

$$\rho \le \frac{\tau}{h}\frac{1+2h}{4}. \tag{5.31}$$

Finally, setting $\tau = 8h/(1+6h)$ we obtain the estimate (5.25) with $C \approx 8$.

4.6 Iterative Methods for Inaccurate Input Data

Consider a problem of linear algebra

$$A\phi = f, \tag{6.1}$$

where $A > 0$ and f are given.

Up to now we have assumed that the matrix A and the vector f are known without error, so that our method of solving equations like (6.1) tacitly presupposed an exact input data. In practice, however, we often have to deal with input data which are known only approximately: rather than (6.1) we have the equation

$$A^h\phi^h = f^h, \tag{6.2}$$

where the index h indicates that the given information depends on the approximation error or on various random errors. We will assume that the operator- and vector-related errors are known; more precisely, we will assume *a priori* knowledge of the following kind of estimates:

$$\|(A - A^h)\phi\| \le \xi(h), \qquad \|f - f^h\| \le \eta(h). \tag{6.3}$$

We now attempt to solve the problem of equation (6.1) via equation (6.2) and the *a priori* information (6.3). Since our results will trivially carry over for most of the iterative processes considered earlier, our description will thus be restricted to the algorithm based on the simple scheme. With this in mind, let us consider an iterative process of the form

$$[\phi^h]^{j+1} = [\phi^h]^j - \tau(A^h[\phi^h]^j - f^h), \qquad [\phi]^0 = \tau f^h, \tag{6.4}$$

where the parameter τ is assumed to satisfy

$$q = \|E - \tau A^h\| < 1. \tag{6.5}$$

This brings up the following question: in view of the estimates (6.3), for how long is the iterative process to be continued? It is natural to assume that for given errors the successive approximations should continue until the error of the iterative process becomes near to the error originating from the approximation. If the number of iterations is restricted so that these errors of different nature become approximately equal at the last iteration, then the error of the approximate solution will turn out unimprovable. Moreover, if the matrix A is ill-conditioned in addition (so that the inverse $[A^h]^{-1}$ may differ considerably from A^{-1}), the continuing iterations may in fact lead to significantly worse results rather than improvements. Thus we have the following problem: with the input errors given, find the optimal number of iterations for which all the errors of various nature match each other. This problem has been considered by Marchuk and Vasil'ev [16].

We argue as follows: consider the formal solutions of equations (6.1) and (6.2):

$$\phi = A^{-1}f; \qquad \phi^h = [A^h]^{-1}f^h \tag{6.6}$$

and write the identity

$$\phi^h - \phi = [A^h]^{-1}[f^h - f + (A - A^h)\phi]. \tag{6.7}$$

Hence

$$\|\phi^h - \phi\| \le \|[A^h]^{-1}\|[\|f - f^h\| + \|(A - A^h)\phi\|],$$

or, accounting for the *a priori* facts of (6.3),

$$\|\phi^h - \phi\| \le \|[A^h]^{-1}\|[\xi(h) + \eta(h)]. \tag{6.8}$$

Considering the iterative process (6.4), the equation below is not difficult to obtain:

$$\phi^h - [\phi^h]^{j+1} = (E - \tau A^h)^{j+2}[A^h]^{-1}f^h.$$

Hence

$$\|\phi^h - [\phi^h]^{j+1}\| \le q^{j+2}\|[A^h]^{-1}\| \, \|f^h\|. \tag{6.9}$$

But

$$\|\phi - [\phi^h]^{j+1}\| \le \|[\phi^h]^{j+1} - \phi^h\| + \|\phi^h - \phi\|,$$

by the triangle inequality, and thus we find

$$\|\phi - [\phi^h]^{j+1}\| \le q^{j+2}\|[A^h]^{-1}\| \, \|f^h\| + \|[A^h]^{-1}\|[\xi(h) + \eta(h)]. \tag{6.10}$$

The first summand on the right of (6.10) represents the error estimate of the iterative process, while the second term estimates the error due to the inaccurate input data. We want the two errors to be the same:

$$q^{j+2}\|[A^h]^{-1}\| \, \|f^h\| = \|[A^h]^{-1}\|[\xi(h) + \eta(h)], \tag{6.11}$$

Consequently, the index $j = j_0$ of the terminal iteration should satisfy

$$j_0 = \frac{1}{\ln q} \ln \frac{\xi(h) + \eta(h)}{\|f^h\|} - 2. \tag{6.12}$$

Since j_0 must be an integer, it is natural to take for j_0 the integer part of the expression above.

Let us note that the norm of the inverse operator does not enter equation (6.12). This simplifies considerably the computation of the optimal number of iterations.

We see that besides the *a priori* information $\xi(h)$, $\eta(h)$, and $\|f^h\|$, equation (6.12) also contains $q = \|E - \tau A^h\|$. This quantity can be found with the help of the largest eigenvalue of the operator T^*T, where $T = E - \tau A^h$; namely

$$q = \sqrt{\beta(T^*T)}.$$

The upper spectral boundary of T can be evaluated by the Lyusternik method described in Section 1.1.

Various approaches to the problems of linear algebra with ill-conditioned matrices and inaccurate input data have been considered by Tikhonov [16], Faddeeva [8], Lavrentiev [16], Voevodin [8], and others.

4.7 Direct Methods for Solving Finite-Difference Systems

A significant amoung of work has been done recently on direct methods for solving finite-difference equation systems. First of all, we should cite the Fourier methods, which, of course, had been applied earlier—but only in rare circumstances. This can be explained by the fact that the discrete Fourier method was inferior to other methods with respect to the number of arithmetic operations required for the solution of the problem: the bulk of its work went to finding the system of eigenfunctions and then to computing the coefficients in the Fourier series, and their sums.

An effective method for solving a system of finite-difference equations is the cyclic reduction process. This has been treated adequately in Buzbee, Golub, and Nilson [12]. In essence, it is an original modification of the Gaussian elimination algorithm and is a particular case of the factorization method.

4.7.1 The Fast Fourier Transform

The notion of the fast Fourier transform has been stated many times, but only recently has it been reduced (by Cooley and Tukey [13]) to an algorithm generalizing an algorithm of Good, and applied to yield a significant reduction in the number of operations required. This reduction has stimulated intense interest in the method.

Suppose that $f(k)$ is a function of a discrete argument, $k = 0, 1, \ldots, N - 1$. We shall represent it in the form of a finite Fourier series, i.e., as the sum:

$$f(k) = \sum_{n=0}^{N-1} A(n)W^{kn},$$

$$A(n) = \frac{1}{N} \sum_{k=0}^{N-1} f(k)W^{-kn}. \tag{7.1}$$

Here, following Cooley and Tukey, we have denoted by

$$W = \exp\left(\frac{2\pi i}{N}\right)$$

the principal Nth order root of unity.

By an *operation* we shall mean an addition followed by a multiplication in the complex arithmetic. Thus it follows from equations (7.1) that for a given $A(n)$ and W^{kn}, we need a total of N^2 operations for computing $f(k)$.

The essence of Cooley's and Tukey's idea consists of the following: if N is not a prime number, then the number of operations can be significantly reduced by representing equations (7.1) as a multiple series.

Consider the case $N = N_1 N_2$, where N_1, N_2 are integers. In order to write (7.1) as a double series, let us represent k and n in the form

$$k = k_1 N_1 + k_0; \; k_1 = 0, \ldots, N_2 - 1; \; k_0 = 0, \ldots, N_1 - 1;$$

$$n = n_1 N_2 + n_0; \; n_1 = 0, \ldots, N_1 - 1; \; n_0 = 0, \ldots, N_2 - 1. \tag{7.2}$$

Since

$$W^{k_1 N_1 n_1 N_2} = (W^N)^{k_1 n_1} = 1,$$

we have

$$W^{kn_1 N_2} = W^{k_0 n_1 N_2}$$

and

$$f(k) = f(k_1, k_0) = \sum_{n_0=0}^{N_2-1} \left[\left(\sum_{n_1=0}^{N_1-1} A(n_1, n_0)W^{k_0 n_1 N_2} \right) W^{kn_0} \right]. \tag{7.3}$$

Hence finding the sum of the series in (7.1) is equivalent to summing the double series (7.3), or, what is the same, summing in succession the series

$$A_1(n_0, k_0) = \sum_{n_1=0}^{N_1-1} A(n_1, n_0)W^{k_0 n_1 N_2}; \tag{7.4}$$

$$f(k_1, k_0) = \sum_{n_0=0}^{N_2-1} A_1(n_0, k_0)W^{(k_1 N_1 + k_0)n_0}. \tag{7.5}$$

But it follows from (7.2) and (7.4) that the computation of A_1 requires NN_1 operations. Having found A_1 we use (7.5) to find $f(k)$; this requires

NN_2 operations. In all, therefore, $N(N_1 + N_2)$ operations are required. The larger N, the greater the reduction in the number of operations.

Clearly, if N_2 is prime and N_1 composite, we may apply the same process to the sum (7.4), in which n_1 appears as a parameter, and again reduce the number of operations by representing N_1 as a product. In general, if $N = N_1 N_2, \ldots, N_m$, we replace N^2 operations by $N(N_1 + N_2 + \cdots + N_m)$ operations. The greatest reduction is attained when $N_i = 2$, 3, or 4. For instance, if $N = 256 = 2^8$, the number of operations is reduced by $256/(8 \times 2) = 16$ times; if $N = 243 = 3^5$, the reduction is $243/(5 \times 3) = 16.2$ times.

From the programing point of view, the most convenient case is that when $N_i = 2 \, (i = 1, 2, \ldots, m)$, though economical methods exist for other values of $N_i (N_i = 4, 8)$. In fact, suppose $N = 2^m$. To obtain the corresponding formulas we set $N_1 = 2, N_2 = 2^{m-1}$ and compute the sums (7.4), (7.5); then we continue the process. We have

$$k = \overline{k_{m-1} k_{m-2} \cdots k_1 k_0} \equiv k_{m-1} 2^{m-1} + k_{m-2} 2^{m-2} + \cdots + k_1 2 + k_0,$$

$$n = \overline{n_{m-1} n_{m-2} \cdots n_1 n_0} \equiv n_{m-1} 2^{m-1} + n_{m-2} 2^{m-2} + \cdots + n_1 2 + n_0,$$

where k_i, n_i are zero or one, we have

$$f(k_{m-1}, \ldots, k_0) = \sum_{n_0=0}^{1} \left\{ \sum_{n_1=0}^{1} \left[\cdots \sum_{n_{m-1}=0}^{1} (A(n_{m-1}, \ldots, n_0) \right. \right.$$

$$\left. \left. \times W^{k n_{m-1} 2^{m-1}}) \cdots W^{k n_1 2} \right] W^{k n_0} \right\}. \qquad (7.6)$$

Since

$$W^{k n_{m-1} 2^{m-1}} = W^{k_0 n_{m-1} 2^{m-1}}, \qquad W^{k n_{m-2} 2^{m-2}} = W^{\overline{k_1 k_0} n_{m-2} 2^{m-2}},$$

etc., finding the sum of (7.6) is equivalent to summing the m series below:

$$A_1(k_0, n_{m-2}, \ldots, n_0) = \sum_{n_{m-1}=0}^{1} A(n_{m-1}, \ldots, n_0) W^{k_0 n_{m-1} 2^{m-1}},$$

$$A_2(k_1, k_0, n_{m-3}, \ldots, n_0) = \sum_{n_{m-2}=0}^{1} A_1(k_0, n_{m-2}, \ldots, n_0) W^{\overline{k_1 k_0} n_{m-2} 2^{m-2}},$$

$$\cdots\cdots\cdots\cdots\cdots\cdots\cdots\cdots\cdots\cdots\cdots\cdots$$

$$A_m(k_{m-1}, \ldots, k_0) = \sum_{n_0=0}^{1} A_{m-1}(k_{m-2}, \ldots, k_0, n_0) W^{\overline{k_{m-1} \ldots k_0} n_0},$$

$$f(k) = A_m(k_{m-1}, \ldots, k_0). \qquad (7.7)$$

In conclusion, let us note that the fast Fourier transform is an effective tool in the correlation analysis of random samples $f(k) \, (k = 0, 1, \ldots, N-1)$.

Consider now the Dirichlet problem for the equation

$$-\Delta \phi + \mu \phi = f \quad \text{in} \quad D,$$
$$\phi = 0 \quad \text{on} \quad \partial D, \qquad (7.8)$$

Here μ is a given constant, and f a given sufficiently smooth function in the domain $D = \{0 \le x \le 1, 0 \le y \le 1\}$.

The corresponding difference problem is taken as follows:

$$\frac{-\phi_{k-1,l} - \phi_{k+1,l} - \phi_{k,l-1} - \phi_{k,l+1} + 4\phi_{k,l}}{h^2} + \mu\phi_{k,l} = f_{k,l} \quad \text{in} \quad D_h,$$

$$\phi_{k,l} = 0 \quad \text{on} \quad \partial D_h, \tag{7.9}$$

$$0 \le k \le \frac{1}{h} = N, \qquad 0 \le l \le \frac{1}{h} = N.$$

If $\mu \ge 0$, then (7.8), (7.9) has a unique solution. In case $\mu < 0$ one has to place additional requirements on μ and f in order to have the existence.

Suppose the solutions of (7.8) and (7.9) exist and are unique.

We introduce the notation:

$$\phi_l = \left\| \begin{matrix} \phi_{1,l} \\ \vdots \\ \phi_{N-1,l} \end{matrix} \right\|, \qquad f_l = \left\| \begin{matrix} f_{1,l} \\ \vdots \\ f_{N-1,l} \end{matrix} \right\|, \qquad l = 1, \dots, N-1,$$

$$A = \left\| \begin{matrix} 2 & -1 & 0 & \cdots & 0 & 0 & 0 \\ -1 & 2 & -1 & \cdots & 0 & 0 & 0 \\ \multicolumn{7}{c}{\cdots\cdots\cdots\cdots\cdots\cdots\cdots\cdots\cdots\cdots} \\ 0 & 0 & 0 & \cdots & -1 & 2 & -1 \\ 0 & 0 & 0 & \cdots & 0 & -1 & 2 \end{matrix} \right\|,$$

where A is a matrix of order $N - 1$. Let E denote the unit matrix of the same order.

We rewrite (7.9) in the form:

$$B\phi_1 - \phi_2 = h^2 f_1,$$
$$-\phi_{l-1} + B\phi_l - \phi_{l+1} = h^2 f_l, \qquad l = 2, \dots, N-2,$$
$$-\phi_{N-2} + B\phi_{N-1} = h^2 f_{N-1}, \tag{7.10}$$

where $B = A + (2 + \mu h^2)E$.

Observe that the matrices A and B have a common basis of eigenvectors, and that the solution of the complete eigenvalue problem

$$Au^{(m)} = \lambda_m(A)u^{(m)}$$

is of the form

$$\lambda_m(A) = 2\left(1 - \cos\frac{m\pi}{N}\right), \qquad u_k^{(m)} = \sqrt{\frac{2}{N}} \sin\frac{m\pi k}{N},$$

where $u_k^{(m)}$ are the components with index k of the eigenvector $u^{(m)}$ $(k = 1, 2, \ldots, N - 1; m = 1, 2, \ldots, N - 1)$. The factor $\sqrt{2/N}$ arises from the normalization condition

$$\|u^{(m)}\|^2 = \sum_{k=1}^{N-1} (u_k^{(m)})^2 = 1.$$

Since the vectors $u^{(m)}$ form an orthonormal basis in $(N - 1)$-dimensional space, the vectors ϕ_l and f_l $(l = 1, \ldots, N - 1)$ can be represented in the form

$$\phi_l = \sum_{m=1}^{N-1} \Phi_{m,l} u^{(m)},$$

$$f_l = \sum_{m=1}^{N-1} F_{m,l} u^{(m)}. \tag{7.11}$$

Inserting these expressions into (7.10) and multiplying both sides by the vector $u^{(m)}$, we obtain for any fixed m a system of equations with a tri-diagonal matrix:

$$\lambda_m \Phi_{m,1} - \Phi_{m,2} = F_{m,1},$$

$$-\Phi_{m,l-1} + \lambda_m \Phi_{m,l} - \Phi_{m,l+1} = F_{m,l}, \qquad l = 2, \ldots, N - 2, \tag{7.12}$$

$$-\Phi_{m,N-2} + \lambda_m \Phi_{m,N-1} = F_{m,N-1}.$$

Here $\lambda_m = \lambda_m(B) = \lambda_m(A) + 2 + \mu h^2$.

Thus, to solve the system (7.10) we need only compute the Fourier coefficients of the vectors f_l $N - 1$ times, solve the $N - 1$ system with tri-diagonal matrices of the type of (7.12) that define the Fourier coefficients of the vectors ϕ_l $(l = 1, \ldots, N - 1)$, and calculate the ϕ_l by (7.11). The expansion in a Fourier series can be accomplished by the fast Fourier transform. Then the formulas defining the Fourier coefficients $F_{m,l}$ of the vector f_l can be written in the following form:

$$F_{m,l} = \sqrt{\frac{2}{N}} \sum_{n=1}^{N-1} f_{n,l} \sin \frac{m\pi n}{N} = \sqrt{\frac{2}{N}} \sum_{n=0}^{2N-1} f_{n,l} \sin \frac{2m\pi n}{2N},$$

where $f_{0,l} = f_{N,l} = \cdots = f_{2N-1,l} = 0$. Let $\tilde{w}$ denote the value of the principal root of unity, of degree $M = 2N$; then

$$F_{m,l} = \sqrt{\frac{2}{N}} \, \mathrm{Im}\!\left(\sum_{n=0}^{M-1} f_{n,l} \tilde{w}^{nm} \right), \qquad m = 1, \ldots, N - 1,$$

and to compute the sums we may immediately apply the algorithm described above. We compute the vectors ϕ_l in a similar way.

This algorithm for the direct solution of the Helmholtz equation is applicable not only to the Dirichlet conditions, but also the von Neumann boundary conditions and to functions $\phi(x, y)$ that are periodic on the boundary of a square, for an equation of the form

$$a(y) \frac{\partial^2 \phi}{\partial x^2} + \frac{\partial}{\partial y}\left(b(y) \frac{\partial \phi}{\partial y}\right) - \mu(y)\phi = f(x, y).$$

Of course, in an electronic computer, where excess digits are discarded in the course of calculations and the results are rounded off, a significant decrease in precision may occur in the final result when a large number of operations are performed (cf. Kaneko and Liu [13], Segeth [13]).

4.7.2 The Cyclic Reduction Method

Let us re-examine the system (7.10), supposing now that $N = 2^{k+1}$. Recall that this is the most favorable case for the application of the fast Fourier transform for computing a finite Fourier series. Nevertheless, for $N = 2^{k+1}$, there is a direct method for solving (7.10) that compares, as to volume of arithmetic operations, with the algorithm based on Fourier expansions. This is called the cyclic reduction method; it requires no knowledge of the eigenvectors and values of the matrix B. It has recently become clear that there are certain advantages to this method.

The idea is that for even N we can extract from the system (7.10)

$$-\phi_{l-1} + B\phi_l - \phi_{l+1} = h^2 f_l, \qquad l = 1, \ldots, N - 1,$$

a similar system containing the ϕ_l only for even values of l.

We write out the following three matrix equations from (7.10):

$$-\phi_{l-2} + B\phi_{l-1} - \phi_l = h^2 f_{l-1},$$
$$-\phi_{l-1} + B\phi_l - \phi_{l+1} = h^2 f_l,$$
$$-\phi_l + B\phi_{l+1} - \phi_{l+2} = h^2 f_{l+1}$$

for even values of l. Multiplying both sides of the second equation by the matrix B and then adding all three equations, we find

$$-\phi_{l-2} + B^{(1)}\phi_l - \phi_{l+2} = h^2 f_l^{(1)}, \qquad l = 2, 4, \ldots, N - 2, \quad (7.13)$$

where

$$B^{(1)} = B^2 - 2E, \qquad f_l^{(1)} = f_{l-1} + Bf_l + f_{l+1}, \qquad l = 2, 4, \ldots, N - 2.$$

For simplicity of our notation, we set $\phi_0 = \phi_N = f_0 = f_N = 0$.

For all odd l we find the system

$$B\phi_l = h^2 f_l + \phi_{l-1} + \phi_{l+1}, \qquad l = 1, 3, \ldots, N - 1. \quad (7.14)$$

in which the ϕ_l for even values of l are assumed known.

This process for reducing the order of the system of matrix equations is known as *reduction*. Note that the system (7.13) can be solved by Fourier expansions, since the matrix $B^{(1)}$ has a basis of eigenvectors that is common to B and A (see, for example, Hockney [13]).

The system (7.13) contains $N/2 - 1$ matrix equations. Since $N = 2^{k+1}$, we may again reduce this system and obtain a system of the same form with $N/4 - 1$ matrix equations:

$$-\phi_{4(l-1)} + B^{(2)}\phi_{4l} - \phi_{4(l+1)} = h^2 f^{(2)}_{4l}, \qquad l = 1, 2, \ldots, N/4 - 1, \quad (7.15)$$

where

$$B^{(2)} = (B^{(1)})^2 - 2E,$$

$$f^{(2)}_{4l} = f^{(1)}_{4l-2} + f^{(1)}_{4l+2} + B^{(1)}f^{(1)}_{4l}.$$

For each even l not a multiple of 4, and for given ϕ_l, for which l is a multiple of four, we obtain the system of equations

$$B^{(1)}\phi_l = h^2 f^{(1)} + \phi_{l-2} + \phi_{l+2}, \qquad l = 2, 6, 10, \ldots, N - 2. \quad (7.16)$$

Applying the cyclic reduction k times we arrive at the equation

$$B^{(k)}\phi_{2^k} = h^2 f^{(k)}_{2^k}; \quad (7.17)$$

the remaining unknowns are given by successive solutions of the system

$$B^{(r)}\phi_l = h^2 f^{(r)}_l + \phi_{l-2^r} + \phi_{l+2^r}, \quad (7.18)$$

$$l = (2i + 1)2^r, \qquad i = 0, 1, \ldots, 2^{k-r} - 1; r = k - 1, k - 2, \ldots, 1, 0.$$

The matrices $B^{(r)}$ satisfy the equations

$$B^{(0)} = B, \qquad B^{(r+1)} = (B^{(r)})^2 - 2E, \qquad r = 0, 1, \ldots, k - 1, \quad (7.19)$$

and the vectors $f^{(r)}_l$ are defined by the formulas

$$f^{(0)}_l = f_l, \qquad l = 1, \ldots, N - 1;$$

$$f^{(r+1)}_l = f^{(r)}_{l-2^r} + f^{(r)}_{l+2^r} + B^{(r)}f^{(r)}_l, \quad (7.20)$$

$$l = j \cdot 2^{r+1}, \qquad j = 1, 2, \ldots, 2^{k-r} - 1; r = 0, 1, \ldots, k - 1.$$

The matrices $B^{(r)}$ are representable as products of 2^r-tri-diagonal matrices. In fact, consider the sequence

$$P_1(b) = b,$$

$$P_{2^{r+1}}(b) = (P_{2^r}(b))^2 - 2, \qquad r = 0, 1, \ldots$$

of polynomials. If $b = 2 \cos \phi$, then

$$P_{2^r}(b) = 2 \cos 2^r \phi.$$

Therefore, the quantities

$$b_i = 2 \cos \frac{(2i - 1)\pi}{2^{r+1}}, \qquad i = 1, 2, \ldots, 2^r$$

are roots of the polynomial $P_{2^r}(b)$ and the resolution we need has the form

$$B^{(r)} = \prod_{i=1}^{2^r} \left(B - 2 \cos \frac{(2i-1)\pi}{2^{r+1}} E \right). \tag{7.21}$$

Thus, there is no need to compute the $B^{(r)}$ explicitly, and to convert them into vectors we need only to fill out 2^r tracks for our tri-diagonal matrices (cf. Section 4.7.3).

Note that in the computations according to (7.20) the components of the vectors $f_i^{(r)}$ increase very rapidly, because of the rapid growth of the eigenvalues of the matrices $B^{(r)}$, and this leads to instability in the computations. In 1969, Buneman proposed a cyclic reduction algorithm that is stable from the computational point of view (cf. Buzbee, Golub, and Nilson [12]).

In this algorithm the vectors $f_i^{(r)}$ in (7.20) are represented in the form

$$h^2 f_i^{(r)} = q_i^{(r)} - B^{(r)} p_i^{(r)},$$
$$l = i \cdot 2^r, \qquad i = 1, \ldots, 2^{k+1-r} - 1. \tag{7.22}$$

Substituting (7.22) in (7.20) we obtain a formula for the successive computation of the vectors $p_i^{(r)}$ and $q_i^{(r)}$:

$$p_i^{(0)} = 0, \qquad q_i^{(0)} = h^2 f_i^{(0)}, \qquad l = 1, \ldots, 2^{k+1} - 1;$$
$$p_i^{(r+1)} = p_i^{(r)} + (B^{(r)})^{-1} (p_{i-2^r}^{(r)} + p_{i+2^r}^{(r)} - q_i^{(r)}),$$
$$q_i^{(r+1)} = q_{i-2^r}^{(r)} + q_{i+2^r}^{(r)} - 2p_i^{(r+1)}, \tag{7.23}$$
$$l = i \cdot 2^{r+1}, \qquad i = 1, \ldots, 2^{k-r} - 1.$$

Taking account of (7.22), we may rewrite the system of matrix equations (7.17), (7.18) as follows:

$$B^{(r)}(\phi_l + p_l^{(r)}) = q_l^{(r)} + \phi_{l-2^r} + \phi_{l+2^r},$$
$$l = (2i+1)2^r, \qquad i = 0, 1, \ldots, 2^{k-r} - 1; \tag{7.24}$$
$$r = k, k-1, \ldots, 1, 0.$$

The cyclic reduction method is applicable not only to the Helmholtz problem with Dirichlet conditions, but also to von Neumann boundary conditions and to the prescription of periodicity of the functions $\phi(x, y)$ on the boundary of a square for more complicated equations.

4.7.3 Factorization of Difference Equations

A large amount of work has been done recently on direct methods for solving finite-difference equations. The factorization method has proved to be effective for a broad class of one-dimensional equations.

As of now, an assessment of priorities in its development is not worth making, since we are dealing with a typical renaissance of excellent classical notions, with a re-ordering of the accents and emphasis on new areas, selected in accordance with the current state of affairs in numerical mathematics and technology.

The essence of the method can be demonstrated by a simple diffusion model:

$$p\frac{d^2\phi}{dx^2} - q\phi = -f(x)$$

$$\frac{d\phi}{dx} = 0 \quad \text{for} \quad x = 0,$$

$$\phi = 0 \quad \text{for} \quad x = 1,$$

where p and q are positive constants and f is a given continuous function of the argument x.

We are to find a continuous solution in the interval $0 \leq x \leq 1$.

We consider the operator

$$L = p\frac{d^2}{dx^2} - q$$

and represent it as the product of two operators:

$$L = p\left(\frac{d}{dx} + \beta\right)\left(\frac{d}{dx} - \alpha\right),$$

where α and β are functions defined by the identity

$$p\left(\frac{d}{dx} + \beta\right)\left(\frac{d}{dx} - \alpha\right) \equiv \frac{d^2}{dx^2} - q.$$

It is sufficient to require that

$$\alpha = \beta,$$

$$\frac{d\beta}{\alpha x} + \beta^2 = \frac{q}{p}.$$

We introduce a new function z by means of the formula

$$z = -\frac{d\phi}{dx} + \beta\phi.$$

Then the diffusion equation reduces to a first-order system

$$\frac{d\beta}{dx} + \beta^2 = \frac{q}{p},$$

$$\frac{dz}{dx} + \beta z = \frac{f}{p},$$

$$\frac{d\phi}{dx} - \beta\phi = -z.$$

We shall call this the factorizing system. Now bring in the boundary conditions:

$$\beta = 0 \quad \text{for} \quad x = 0,$$
$$z = 0 \quad \text{for} \quad x = 0,$$
$$\phi = 0 \quad \text{for} \quad x = 1.$$

It is immediately clear for this choice of the initial conditions in the Cauchy problem for our system, the boundary conditions in the diffusion problem will be fulfilled, i.e., $d\phi/dx = 0$ for $x = 0$ and $\phi = 0$ for $x = 1$. Thus, we have gone from the boundary problem for the diffusion equation to a Cauchy problem for three ordinary first-order differential equations, which are to be solved one after another. For stability of the computations, the calculation of β and z should be done for increasing values of x ,i.e., from left to right and of ϕ for decreasing values. The *factorization method* consists precisely in this reduction of a second-order boundary problem to a first-order Cauchy problem. It is easily generalized to the more complicated diffusion problem with piecewise-discontinuous functions p, q, and f, and to various boundary conditions.

For difference equations the factorization method (or, as it is usually called, the recursion method) is defined as follows: we write the initial difference equations in the form

$$-a_i\phi_{i-1} + p_i\phi_i - c_i\phi_{i+1} = f_i, \quad i = 1, 2, \ldots, n. \quad (7.25)$$

We shall suppose that the boundary conditions for the desired function (if written in discrete form they can be regarded as independent equations) can be written as follows:

$$a_1 = 0, c_n = 0, \quad p_i \geq a_i + c_i. \quad (7.26)$$

We write the original equation (7.25) in vector-matrix form:

$$A\Phi = F, \quad (7.27)$$

where $\Phi = (\phi_1, \phi_2, \ldots, \phi_n)^T$, $F = (f_1, f_2, \ldots, f_n)^T$ and then write (7.27) as

$$KS_1S_2\Phi = F, \quad (7.28)$$

in which

$$S_1 = \begin{Vmatrix} 1 & 0 & \cdots & 0 & 0 & 0 \\ -\alpha_2 & 1 & \cdots & 0 & 0 & 0 \\ \cdots & \cdots & \cdots & \cdots & \cdots & \cdots \\ 0 & 0 & \cdots & -\alpha_{n-1} & 1 & 0 \\ 0 & 0 & \cdots & 0 & -\alpha_n & 1 \end{Vmatrix},$$

$$S_2 = \begin{Vmatrix} 1 & -\xi_1 & 0 & \cdots & 0 & 0 \\ 0 & 1 & -\xi_2 & \cdots & 0 & 0 \\ \cdots & \cdots & \cdots & \cdots & \cdots & \cdots \\ 0 & 0 & 0 & \cdots & 1 & -\xi_{n-1} \\ 0 & 0 & 0 & \cdots & 0 & 1 \end{Vmatrix},$$

and K is a diagonal matrix acting as a coefficient.

Writing $S_2\Phi = Z$, $K^{-1} = \Gamma$ we obtain the system of equations

$$S_1 Z = \Gamma F, \qquad S_2\Phi = Z. \tag{7.29}$$

Let us write out the equations for Z and Φ in component form:

$$-\alpha_i Z_{i-1} + Z_i = \gamma_i f_i, \tag{7.30}$$

$$\phi_i - \xi_i \phi_{i+1} = Z_i. \tag{7.31}$$

To find the coefficients α_i, ξ_i, and γ_i we substitute into (7.30) the expressions for Z_{i-1} and Z_i given by (7.31):

$$-\alpha_i \phi_{i-1} + (1 + \alpha_i \xi_{i-1})\phi_i - \xi_i \phi_{i+1} = \gamma_i f_i.$$

Comparing this result with the original equation (7.25) we find that

$$-\alpha_i = \gamma_i a_i, \qquad \xi_i = \gamma_i c_i,$$

$$\gamma_i = (p_i - a_i \gamma_{i-1} c_{i-1})^{-1} \tag{7.32}$$

Finally, we write equations (7.30), (7.31) in the form

$$Z_i = \gamma_i(a_i Z_{i-1} + f_i),$$

$$\phi_i = \gamma_i c_i \phi_{i+1} + Z_i. \tag{7.33}$$

It is known (cf. Godunov and Ryaben'kii [3]) that the equations (7.9), (7.10) will be stable if the following conditions are satisfied: $a_k \geq 0$, $c_k \geq 0$, $p_k > 0$, and $a_k + c_k \leq p_k$, with strict inequality in the fourth condition for at least one value of k.

Note that the formulas for solving the difference equations (7.25) are usually written in the form

$$\beta_{i+1} = (p_i - a_i \beta_i)^{-1} c_i,$$

$$z_{i+1} = (p_i - a_i \beta_i)^{-1}(a_i z_i + f_i), \tag{7.34}$$

$$\phi_i = \beta_{i+1}\phi_{i+1} + z_{i+1}, \qquad i = 1, \ldots, n,$$

with the conditions

$$\beta_1 = z_1 = 0, \qquad \phi_n = z_{n+1}.$$

Recently, the factorization method (sometimes, as noted, called the 'recursion' method) has been generalized to the case of systems of ordinary first-order linear differential equations with arbitrary linear constraints; this includes, as a special case, multi-point and boundary conditions (cf. Fage [12]).

The algorithms (7.33), (7.34) are valid also when the functions ϕ_i and f_i are vectors and the coefficients a_i, c_i, and p_i are matrices. Then in (7.33) and (7.34) the order of the factors must be preserved.

The recursive formulas (7.33) and (7.34) can be used, with the conditions (7.26), to provide a solution.

The matrix factorization method is applicable to a two-dimensional elliptic difference equation, but is ineffective unless the number of conditional points is small for at least one of the variables.

Let us look at a simple and reasonably effective method for solving two- and three-dimensional elliptic difference equations—the Buleev incomplete factorization method. The idea is illustrated in the following example (cf. Buleev [12]).

Suppose that we have a two-dimensional difference equation

$$-a_{ik}\phi_{i-1,k} - c_{ik}\phi_{i+1,k} - b_{ik}\phi_{i,k-1} - d_{ik}\phi_{i,k+1} + p_{ik}\phi_{ik} = f_{ik}, \quad (7.35)$$

$$i = 1, 2, \ldots, m; k = 1, 2, \ldots, n,$$

$$a_{1k} = c_{mk} = b_{i1} = d_{in} = 0, \qquad p_{ik} \geq a_{ik} + c_{ik} + b_{ik} + d_{ik}. \quad (7.36)$$

We write it in vector–matrix form:

$$A\Phi = F,$$

where

$$\Phi = (\phi_1, \phi_2, \ldots, \phi_N)^T, \qquad F = (f_1, f_2, \ldots, f_N)^T, \qquad N = mn.$$

We add the vector $B\Phi$ to both sides and obtain the equation

$$(A + B)\Phi = F + B\Phi. \quad (7.37)$$

We choose B so that the matrix $A + B$ can be written as the product of two simple triangular matrices S_1 and S_2 and a diagonal matrix:

$$A + B = KS_1S_2$$

As we did with (7.27) we replace (7.37) by the system

$$S_1 Z = K^{-1}(F + B\Phi),$$
$$S_2 \Phi = Z, \quad (7.38)$$

and solve by the method of successive approximations.

We replace the original two-dimensional system (7.35) by the system

$$Z_{ik} = \alpha_{ik}Z_{i-1,k} + \beta_{ik}Z_{i,k-1} + \gamma_{ik}[f_{ik} + D_{ik}(\phi) - q_{ik}\phi_{ik}], \quad (7.39)$$
$$\phi_{ik} = \xi_{ik}\phi_{i+1,k} + \eta_{ik}\phi_{i,k+1} + Z_{ik}.$$

(The iteration index is omitted; $D_{ik}(\phi)$ means $(D\Phi)_{ik}$ where $D = KB + \text{diag } q_{ik}$; diag q_{ik} is a diagonal matrix with the elements q_{ik}.)

For the coefficients α_{ik}, β_{ik}, ξ_{ik}, η_{ik}, and γ_{ik} we have formulas similar to (7.32), (7.33):

$$\alpha_{ik} = \gamma_{ik}a_{ik}, \qquad \beta_{ik} = \gamma_{ik}b_{ik}, \qquad \xi_{ik} = \gamma_{ik}c_{ik}, \qquad \eta_{ik} = \gamma_{ik}d_{ik},$$

$$\gamma_{ik} = (p_{ik} - q_{ik} - a_{ik}c_{i-1,k}\gamma_{i-1,k} - b_{ik}d_{i,k-1}\gamma_{i,k-1})^{-1},$$

and for the iterative operator $D_{ik}(\phi)$ we have

$$D_{ik}(\phi) = a_{ik}\eta_{i-1,k}\phi_{i-1,k+1} + b_{ik}\xi_{i,k-1}\phi_{i+1,k-1}.$$

The coefficient q_{ik} is to be set equal to

$$\theta_{ik}(a_{ik}\eta_{i-1,k} + b_{ik}\xi_{i,k-1}), \qquad 0 \le \theta < 1.$$

To guarantee the convergence of (7.39) for $\theta > 0$ we adjoin the supplementary simple iteration

$$\tilde{\phi}_{ik}^{(l)} = \frac{1}{\gamma_{ik}p_{ik}}(\alpha_{ik}\phi_{i-1,k}^{(l)} + \beta_{ik}\phi_{i,k-1}^{(l)} + \xi_{ik}\phi_{i+1,k}^{(l)} + \eta_{ik}\phi_{i,k+1}^{(l)} + \gamma_{ik}f_{ik}), \quad (7.40)$$

and use the lth approximation of $\tilde{\phi}_{ik}$ in the equation for $Z_{ik}^{(l+1)}$ in order to compute the $(l + 1)$st approximation of ϕ_{ik}. If the problem is ill-conditioned, the square-bracketed expression in (7.39) should be replaced by the expression

$$f_{ik} + D_{ik}(\phi) - q_{ik}\phi_{ik} + \kappa_{ik}(a_{ik} + b_{ik})\phi_{ik},$$

with $0 \le \kappa_{ik} \le 1$, at all net points corresponding to the right-hand and upper boundaries of the region.

It is not difficult to show by induction that the coefficients $\alpha_{ik}, \beta_{ik}, \xi_{ik}$, and η_{ik} satisfy the conditions

$$\xi_{ik} + \delta_{ik} \le 1,$$

$$\alpha_{ik} + \beta_{ik} \le \frac{a_{ik} + b_{ik}}{c_{ik} + d_{ik} + (1 - \theta_{ik})(\alpha_{ik}\eta_{i-1,k} + \beta_{ik}\xi_{i,k-1}) + \kappa_{ik}(a_{ik} + b_{ik})}$$

$$(7.41)$$

The condition $\alpha_{in} + \beta_{in} \le 1$ can always be guaranteed by proper choice of θ_{ik} and κ_{ik}. Thus the spatial countable stability of (7.39) can always be secured.

The von Neumann problem is soluble by (7.39), (7.41) without strengthening of the initial function at any point, with $\kappa_{in} > 0$ in (7.41). After each iteration according to the scheme (7.39), (7.41) we compute the mean value of the approximation.

Let us look at yet another incomplete factorization scheme, defined by the equations

$$Z_{ik} = \alpha_{ik}Z_{i-1,k} + \gamma_{ik}[f_{ik} + D_{ik}(\phi) - q_{ik}\phi_{ik}], \qquad (7.42)$$

$$\phi_{ik} - \beta_{ik}\phi_{i,k-1} - \delta_{ik}\phi_{i,k+1} = \xi_{ik}\phi_{i+1,k} + Z_{ik}.$$

The coefficients $\alpha_{ik}, \xi_{ik}, \beta_{ik}, \delta_{ik}, \gamma_{ik}$, and $D_{ik}(\phi)$ are given by

$$\alpha_{ik} = \gamma_{ik}a_{ik}, \qquad \beta_{ik} = \gamma_{ik}b_{ik}, \qquad \xi_{ik} = \gamma_{ik}c_{ik}, \qquad \delta_{ik} = \gamma_{ik}d_{ik},$$

$$\gamma_{ik} = (p_{ik} - q_{ik} - a_{ik}c_{i-1,k}\gamma_{i-1,k})^{-1},$$

$$D_{ik}(\phi) = a_{ik}(\beta_{i-1,k}\phi_{i-1,k-1} + \delta_{i-1,k}\phi_{i-1,k-1}).$$

The coefficient q_{ik} is set equal to $\theta a_{ik}(\beta_{i-1,k} + \delta_{i-1,k})$, where $0 \le \theta \le 1$.

The square-bracketed expression in (7.41) is usefully replaced at points on the right-hand boundary of the region by the expression

$$f_{ik} + D_{ik}(\phi) - q_{ik}\phi_{ik} + \kappa_{ik}a_{ik}\phi_{ik}, \qquad 0 \le \kappa_{ik} \le 1.$$

In this scheme

$$\beta_{ik} + \delta_{ik} + \xi_{ik} + \gamma_{ik}q_{ik} + (1 - \theta)(\beta_{i-1,k} + \delta_{i-1,k})\alpha_{ik} + \kappa_{ik}\alpha_{ik} \le 1.$$

It is quite natural to ask how we can decrease the relative weight of the iterand expressions in our equations of the type of (7.41), i.e., how to obtain a satisfactory simple factorization of the difference operator $A + B$ that will be "near" the given operator A.

In other words, we seek a system of equations equivalent to the original two-dimensional equation (7.35), in the following form:

$$Z_{ik} = \alpha_{ik}Z_{i-1,k} + \mu_{ik}Z_{i-1,k-1} + \nu_{ik}Z_{i-1,k+1}$$
$$+ \gamma_{ik}[f_{ik} + D_{ik}(\phi) - s_{ik}\phi_{ik}], \quad (7.43)$$
$$\phi_{ik} - \beta_{ik}\phi_{i,k-1} - \delta_{ik}\phi_{i,k+1} = \xi_{ik}\phi_{i+1,k} + Z_{ik},$$

where the $\alpha_{ik}, \beta_{ik}, \delta_{ik}, \xi_{ik}, \mu_{ik}, \nu_{ik}$, and γ_{ik} are coefficients not yet known.

Since there are seven unknown coefficients and five equations, two of the coefficients may be chosen arbitrarily. We set

$$\mu_{ik} = \alpha_{ik}\beta_{i-1,k}, \qquad \nu_{ik} = \alpha_{ik}\delta_{i-1,k}. \quad (7.44)$$

Then the principal terms in the iteration operator $D_{ik}(\phi)$ vanish and the expressions for $\gamma_{ik}D_{ik}(\phi)$ and the coefficients $\alpha_{ik}, \beta_{ik}, \delta_{ik}, \xi_{ik}$, and γ_{ik} become

$$\alpha_{ik} = \gamma_{ik}e_{ik},$$
$$\xi_{ik} = \gamma_{ik}c_{ik},$$
$$\beta_{ik} = \gamma_{ik}b_{ik} + \alpha_{ik}\beta_{i-1,k}\xi_{i-1,k-1}, \quad (7.45)$$
$$\delta_{ik} = \gamma_{ik}d_{ik} + \alpha_{ik}\delta_{i-1,k}\xi_{i-1,k+1},$$
$$\gamma_{ik} = (p_{ik} - s_{ik} - e_{ik}\xi_{i-1,k})^{-1},$$
$$\gamma_{ik}D_{ik}(\phi) = \alpha_{ik}(\beta_{i-1,k}\beta_{i-1,k-1}\phi_{i-1,k-2} + \delta_{i-1,k}\delta_{i-1,k+1}\phi_{i-1,k+2}),$$
$$(7.46)$$

where

$$e_{ik} = a_{ik}(1 - \beta_{i-1,k}\delta_{i-1,k-1} - \delta_{i-1,k}\delta_{i-1,k+1})^{-1}. \quad (7.47)$$

Let the values of the coefficient $\gamma_{ik}s_{ik}$ be proportional to the sum of the coefficients in the operator $\gamma_{ik}D_{ik}(\phi)$:

$$\gamma_{ik}s_{ik} = \theta\alpha_{ik}(\beta_{i-1,k}\beta_{i-1,k-1} + \delta_{i-1,k}\delta_{i-1,k+1}), \qquad 0 \le \theta \le 1. \quad (7.48)$$

Taking account of (7.48) and (7.47) we calculate γ_{ik} by the formula

$$\gamma_{ik} = [p_{ik} - e_{ik}(\xi_{i-1,k} + \theta\beta_{i-1,k}\beta_{i-1,k-1} + \theta\delta_{i-1,k}\delta_{i-1,k+1})]^{-1}. \quad (7.49)$$

The coefficients in (7.43) are to be computed in the following order

$$e_{ik}, \quad \gamma_{ik}, \quad \alpha_{ik}, \quad \xi_{ik}, \quad \beta_{ik}, \quad \delta_{ik}.$$

We may assume that in (7.49) there exists an explicit trial for the vector

$$\gamma_i = (\gamma_{i1}, \gamma_{i2}, \ldots, \gamma_{in}).$$

Lastly, we re-write (7.43) in the following form:

$$
\begin{aligned}
Z_{ik} = {} & \alpha_{ik}(Z_{i-1,k} + \beta_{i-1,k}Z_{i-1,k-1} + \delta_{i-1,k}Z_{i-1,k+1}) \\
& + \gamma_{ik}f_{ik} + \alpha_{ik}[\beta_{i-1,k}\beta_{i-1,k-1}(\phi_{i-1,k-2} - \theta\phi_{ik}) \\
& + \delta_{i-1,k}\delta_{i-1,k+1}(\phi_{i-1,k+2} - \theta\phi_{ik})],
\end{aligned}
\tag{7.50}
$$

$$\phi_{ik} - \beta_{ik}\phi_{i,k-1} - \delta_{ik}\phi_{i,k+1} = \xi_{ik}\phi_{i+1,k} + Z_{ik}.$$

These equations are to be solved by the method of successive approximations. At the outset there is an explicit trial for the vector $Z_i = (Z_{i1}, Z_{i2}, \ldots, Z_{in})$ and an implicit trial in the reverse direction for the vector

$$\Phi_i = (\phi_{i1}, \phi_{i2}, \ldots, \phi_{in}).$$

At the points corresponding to the right-hand boundary, the accelerating correction, i.e., the square bracket in (7.43), is to be taken in the form

$$f_{ik} + D_{ik}(\phi) - \theta e_{ik}(\beta_{i-1,k}\beta_{i-1,k-1} + \delta_{i-1,k}\delta_{i-1,k-1})\phi_{ik} + \kappa_{ik}e_{ik}\phi_{ik},$$

Finally, we can show by induction that

$$\beta_{ik} + \delta_{ik} + \xi_{ik} + \gamma_{ik}q_{ik} + (1 - \theta)\lambda_{ik}\alpha_{ik} + \kappa_{ik}\alpha_{ik} \leq 1,$$

where

$$\lambda_{ik} = \beta_{i-1,k}\beta_{i-1,k-1} + \delta_{i-1,k}\delta_{i-1,k-1}.$$

The solution of Poisson's equation in the square for arbitrary boundary conditions shows that the iterative process (7.50) converges for values of θ in the interval $0 \leq \theta \leq 1$, and the best convergence is obtained for $\theta = 0.8$ to 0.9.

Table 4.2 displays the dependence of the norm of the iteration operator in (7.50), expressed via q, on the parameter θ for the Dirichlet problem

$$\frac{\partial^2 \phi}{\partial x^2} + \frac{\partial \phi^2}{\partial y^2} = -2(2 - x^2 - y^2),$$

$\phi = 0$ for $x = \pm 1$, $\phi = 0$ for $y = \pm 1$, for mesh sizes $\Delta x = \Delta y = \frac{1}{12}$ (529 countable nodes).

Table 4.2

θ	0	0.3	0.5	0.6	0.7	0.8	0.85	0.9	0.95	1.0
q	0.86	0.82	0.76	0.72	0.65	0.47	0.43	0.49	0.61	1.0

According to the results shown in Table 4.2, a single iteration in the above problem, per the scheme (7.50), with $\theta = 0.7$, 0.8, and 0.9, yields the same results, respectively, as 44, 77, and 72 single simple iterations.

Thus, for optimal values of the parameter θ the scheme (7.50), in our given Dirichlet problem, is just as effective as the known alternative direction methods with a constant value of the iteration parameter τ.

As in the scheme (7.42), it is useful to add a term $\kappa_{ik} \alpha_{ik} \phi_{ik}$ to the iterand expression $\gamma_{ik} D_{ik}(\phi) - \gamma_{ik} S_{ik} \phi_{ik}$ in (7.43), on the right-hand boundary of the region. Then we may show by induction that the coefficients in (7.43) satisfy the conditions

$$\beta_{ik} + \delta_{ik} + \xi_{ik} \leq 1,$$

$$\alpha_{ik} + \mu_{ik} + v_{ik} \leq \frac{a_{ik}(1 + \beta_{i-1,k} + \delta_{i-1,k})}{(1 - \sigma_{ik})(b_{ik} + d_{ik} + c_{ik}) + [(1 - \theta)\lambda_{ik} + \rho_{ik} + \kappa_{ik}]a_{ik}},$$

where

$$\lambda_{ik} = \beta_{i-1,k}\beta_{i-1,k-1} + \delta_{i-1,k}\delta_{i-1,k+1},$$

$$\sigma_{ik} = \beta_{i-1,k}\delta_{i-1,k-1} + \delta_{i-1,k}\beta_{i-1,k+1},$$

$$\rho_{ik} = \beta_{i-1,k}\xi_{i-1,k-1} + \delta_{i-1,k}\xi_{i-1,k+1}.$$

As applied to ill-conditioned problems either scheme (7.43) or (7.50) is somewhat slower than the alternating direction method.

The von Neumann problem is effectively solved by the scheme (7.50) with $\theta = 0.8$ to 0.9. *We must note, however, that this method is to some extent inferior to the alternating direction method.*

It is natural enough that the first iteration of the scheme (7.50) yields a result close to the exact solution if we set θ equal or nearly equal to 1. Therefore, in solving laborious problems it is reasonable to give θ such a value for the first iteration, or for the first few, before going to the optimum value.

Note that the arbitrariness of the shape of the two-dimensional region, and of the boundary conditions, or the original function ϕ, present no difficulty in the realization of the incomplete factorization method. It is only necessary that the boundary conditions, written in difference form, contain the original function within the limits of the standard five-point pattern, i.e., it must have a form like that in (7.35) and two nonzero peripheral coefficients in it must, if they are positive, not exceed in their sum, the coefficient p_{ik}.

As we have noted earlier, the incomplete factorization method is effective in the solution of the Dirichlet problem. Efforts to find good schemes for solving the von Neumann problem have led to improved incomplete factorization methods. In operator form, the method is as follows:

$$z_{ik} = \alpha_{ik} z_{i-1,k} + \gamma_{ik}[f_{ik} + D_{ik}(\phi) + E_{ik}(\phi)],$$

$$\phi_{ik} - \beta_{ik}\phi_{i,k-1} - \delta_{ik}\phi_{i,k+1} = \zeta_{ik}\phi_{i+1,k} + z_{ik}, \tag{7.51}$$

where $\alpha_{ik}, \beta_{ik}, \zeta_{ik}, \delta_{ik}, \gamma_{ik}$ are unknown coefficients and the operator E contains only the function ϕ in five-point form.

This is equivalent to the single equation

$$(1 + \alpha_{ik}\zeta_{i-1,k})\phi_{ik} - \alpha_{ik}\phi_{i-1,k} - \zeta_{ik}\phi_{i+1,k} - \beta_{ik}\phi_{i,k-1}$$
$$-\delta_{ik}\phi_{i,k+1} + \alpha_{ik}\beta_{i-1,k}\phi_{i-1,k-1} + \alpha_{ik}\delta_{i-1,k}\phi_{i-1,k+1}$$
$$-\gamma_{ik}D_{ik}(\phi) + \gamma_{ik}E_{ik}(\phi) = \gamma_{ik}f_{ik}. \tag{7.52}$$

Taken together, (7.52) and (7.35) imply that

$$\gamma_{ik}D_{ik}(\phi) = \alpha_{ik}\beta_{i-1,k}\phi_{i-1,k-1} + \alpha_{ik}\delta_{i-1,k}\phi_{i-1,k+1}. \tag{7.53}$$

We set

$$\gamma_{ik}E_{ik}(\phi) = \alpha_{ik}\beta_{i-1,k}(\kappa\phi_{i-1,k} + \eta\phi_{i,k-1} + \sigma_i\omega\phi_{i+1,k} - \theta_i\phi_{ik})$$
$$+ \alpha_{ik}\delta_{i-1,k}(\kappa\phi_{i-1,k} + \eta\phi_{i,k+1} - \sigma_i\omega\phi_{i+1,k} - \theta_i\phi_{ik}), \tag{7.54}$$

where κ, η, ω, and θ_i are undefined parameters† and

$$\sigma_i = \begin{cases} 1, & \text{for } i = 1, 2, \ldots, m - 1, \\ 0, & \text{for } i = m. \end{cases}$$

The terms in (7.54) that contain ϕ_{ik} serve to "strengthen" the principal diagonal of the factorizing matrix $(A + D - E)$, so that in the end the linear combination $\kappa\phi_{i-1,k} + \eta\phi_{i,k-1} + \sigma_i\omega\phi_{i+1,k}$ offsets the sum $\phi_{i-1,k-1} + \theta_i\phi_{i,k}$, and the linear combination $\kappa\phi_{i-1,k} + \eta\phi_{i,k+1} + \sigma_i\omega\phi_{i+1,k}$ offsets the sum $\phi_{i-1,k+1} + \theta_i\phi_{ik}$.

We require that the parameters satisfy the condition

$$1 + \theta_i = \kappa + \eta + \sigma_i\omega + \varepsilon, \tag{7.56}$$

where ε is a small positive quantity or zero.

By combining (7.52) and (7.35) while taking account of (7.54), we obtain the following system of relationships connecting the coefficients of the system (7.51) with those of (7.35):

$$\alpha_{ik} - \kappa\alpha_{ik}(\beta_{i-1,k} + \delta_{i-1,k}) = \gamma_{ik}a_{ik},$$
$$\beta_{ik} - \eta\alpha_{ik}\beta_{i-1,k} = \gamma_{ik}b_{ik},$$
$$\delta_{ik} - \eta\alpha_{ik}\delta_{i-1,k} = \gamma_{ik}d_{ik}, \tag{7.57}$$
$$\zeta_{ik} - \sigma_i\omega\alpha_{ik}(\beta_{i-1,k} + \delta_{i-1,k}) = \gamma_{ik}c_{ik}$$
$$1 + \alpha_{ik}\zeta_{i-1,k} - \theta_i\alpha_{ik}(\beta_{i-1,k} + \delta_{i-1,k}) = \gamma_{ik}p_{ik},$$

---

† In order not to break the pattern of the general formula for $E_{ik}(\phi)$ in the right-hand column through the introduction of the factor $\sigma_m = 0$, we may use the boundary condition on the right, i.e.,

$$-a_{m+1,k}\phi_{m,k} + \phi_{m+1,k} = f_{m+1,k} \tag{7.55}$$

as an independent equation. Then the functions $D_{ik}(\phi)$, $E_{ik}(\phi)$, and z_{ik} are not computed for $i = m + 1$. We begin the computation of (7.51) in the mth column, after eliminating the functions $\phi_{m+1,k}$ by means of (7.55).

or

$$\gamma_{ik} = [p_{ik} + (1 - \kappa\beta_{i-1,k} - \kappa\delta_{i-1,k})^{-1}a_{ik}$$
$$\times (\theta_i\beta_{i-1,k} + \theta_i\delta_{i-1,k} - \xi_{i-1,k})]^{-1}, \tag{7.58}$$

Finally, the iterative scheme (7.51), taking account of (7.53) and (7.54), may be written as

$$z_{ik}^1 = a_{ik}z_{i-1,k}^{(1)} + \gamma_{ik}f_{ik}$$
$$+ \alpha_{ik}[\beta_{i-1,k}(\phi_{i-1,k-1}^{(l-1)} - \eta\phi_{i,k-1}^{(l-1)}) + \delta_{i-1,k}(\phi_{i-1,k+1}^{(l-1)} - \eta\phi_{i,k+1}^{(l-1)})]$$
$$+ \alpha_{ik}(\beta_{i-1,k} + \delta_{i-1,k})(\theta_i\phi_{ik}^{(l-1)} - \delta_i\omega\phi_{i+1,k}^{(l-1)}),$$

$$\tag{7.59}$$

$$\phi_{ik}^{(l)} - \beta_{ik}\phi_{i,k-1}^{(l)} - \delta_{ik}\phi_{i,k+1}^{(l)} = \xi_{ik}\phi_{i+1,k}^{(l)} + z_{ik}^{(l)}. \tag{7.60}$$

The equation (7.70) is solved by the one-dimensional sweep method:

$$\rho_{ik} = (1 - \beta_{ik}\delta_{i,k-1}\rho_{i,k-1})^{-1} \tag{7.61}$$

$$\omega_{ik} = \rho_{ik}(\beta_{ik}\omega_{i,k-1} + \xi_{ik}\phi_{i+1,k}^{(l)} + z_{ik}^{(l)}),$$

$$\phi_{ik} = \rho_{ik}\delta_{ik}\phi_{i,k+1} + \omega_{ik}. \tag{7.62}$$

In the iterative scheme (7.51) we must, of course, approach a condition such that in the operator $D_{ik}(\phi) - E_{ik}(\phi)$ the sum of the coefficients is close to zero, and the sum of the moduli of the coefficients is as small as possible in relation to the coefficient a_{ik} in the original equation (7.35).

We shall say that the operator $(A + B)_{ik}$ is close to the operator A_{ik} in the approximation sense if the sum of the coefficients in B_{ik} is close to zero, and that it is close in the norm of the residuals if the sum of the moduli of the coefficients in B_{ik} is small compared to the value of the coefficient a_{ik} in A_{ik}.

In the scheme now under consideration the nearness of the operator $(A + B)_{ik}$ to A_{ik} in the approximation sense will guarantee the relation (7.56).

The optimal value of θ must be commensurable with the coefficients of $\phi_{i-1,k-1}$ and $\phi_{i-1,k+1}$, that is, it must have a value near 1. If we approach the condition in which the coefficients β_{ik}, δ_{ik}, and ξ_{ik} are approximately proportional to the coefficients b_{ik}, d_{ik}, and c_{ik}, then the parameters η and ω should, by (7.69), be related by the approximate equation

$$2\omega \approx \eta$$

An experimental solution of the diffusion and convective transport problems in a rectangular region has shown that the optimal values of the parameters κ, η, ω, and ε are

$$\kappa = 0.5, \ldots, 1, \quad \eta = 1, \quad \omega = 0, \ldots, 0.4, \quad \varepsilon = 0,$$

and $\kappa + \omega = 1.0, \ldots, 1.1$. For the Poisson equation the optimal choice of parameters is:

$$\kappa = \eta = 1, \quad \omega = 0.1, \quad \varepsilon = 0.$$

Let us investigate the spatial stability of the scheme (7.51). We shall suppose, for generality, that we have in (7.51)

$$
\begin{aligned}
\gamma_{ik} D_{ik}(\phi) &- \gamma_{ik}(\phi) \\
&= d_{ik}[\beta_{i-1,k-1} - \eta\phi_{i,k-1}) + \delta_{i-1,k}(\phi_{i-1,k-1} - \eta\phi_{i,k+1})] \\
&\quad + \alpha_{ik}(\beta_{i-1,k} + \delta_{i-1,k})(\theta_i\phi_{ik} - \kappa\phi_{i-1,k} - \sigma_i\omega\phi_{i+1,k}) \\
&\quad + s_{ik}\alpha_{ik}\phi_{ik}.
\end{aligned}
$$

Then, it is not difficult to show by induction that the coefficients α_{ik}, β_{ik}, δ_{ik}, and ξ_{ik} satisfy the conditions

$$
\beta_{ik} + \delta_{ik} + \xi_{ik} \le 1, \tag{7.63}
$$

$$
\alpha_{ik} = \frac{a_{ik}}{(1 - \kappa v_{i-1,k})(l_{ik} + d_{ik} + p_{ik}) + [(\eta + \sigma_i\omega + \varepsilon)v_{ik} + s_{ik}]a_{ik}} \tag{7.64}
$$

By a choice of the parameters κ, η, ω, ε, and s we can always guarantee the spatial countable stability of the scheme. Moreover, the conditions (7.63), (7.64) easily allow us to construct an iterative scheme meeting the earlier established constraints on the coefficients in (7.51).

CHAPTER 5

Methods for Solving Nonstationary Problems

The main object to be considered is the following evolution problem of mathematical physics:

$$\frac{\partial \phi}{\partial t} + A\phi = f \quad \text{in} \quad D \times D_t, \qquad \phi = g \quad \text{in} \quad D \quad \text{at} \quad t = 0,$$

where $A \geq 0$, and the solution ϕ as well as the functions f and g are assumed sufficiently smooth. In addition, the solution of the problem is further required to satisfy certain boundary conditions on ∂D (the boundary of the region D).†

5.1 Second-Order Approximation Difference Schemes with Time-Varying Operators

Consider the evolution equation

$$\frac{\partial \phi}{\partial t} + A\phi = 0 \quad \text{in} \quad D \times D_t, \qquad \phi = g \quad \text{in} \quad D \quad \text{at} \quad t = 0. \qquad (1.1)$$

The corresponding difference equation is taken in the form

$$\frac{\phi^{j+1} - \phi^j}{\tau} + A\frac{\phi^{j+1} + \phi^j}{2} = 0, \qquad \phi^0 = g. \qquad (1.2)$$

Provided the solution is sufficiently smooth it is not difficult to verify that (1.2) approximates problem (1.1) with second-order accuracy with respect

---

† Everywhere in this chapter (if there are no specific indications to the contrary) it is assumed that the operator A is reduced to a finite-difference one.

to τ. The difference scheme (1.2) is usually referred to as the *Crank–Nicholson* scheme. It is interesting to note that (1.2) is obtained by applying alternately the first-order accuracy schemes (explicit and implicit) which are considered on the intervals $t_j \leq t \leq t_{j+1/2}$ and $t_{j+1/2} \leq t \leq t_{j+1}$, correspondingly, the linear operator A being assumed independent of t:

$$\frac{\phi^{j+1/2} - \phi^j}{\tau/2} + A\phi^j = 0,$$

$$\frac{\phi^{j+1} - \phi^{j+1/2}}{\tau/2} + A\phi^{j+1} = 0. \tag{1.3}$$

The Crank–Nicholson scheme is obtained from this by eliminating the unknowns $\phi^{j+1/2}$.

Assume now that the operator A depends on time and consider equation (1.1) with A replaced by the approximating difference operator denoted by Λ.

$$\frac{\phi^{j+1} - \phi^j}{\tau} + \Lambda^j \frac{\phi^{j+1} + \phi^j}{2} = 0, \qquad \phi^0 = g, \tag{1.4}$$

$$(\Lambda^j \phi, \phi) \geq 0 \tag{1.5}$$

for any function from the subspace Φ.

Solving for ϕ^{j+1} in equation (1.4) we obtain

$$\phi^{j+1} = \left(E + \frac{\tau}{2}\Lambda^j\right)^{-1}\left(E - \frac{\tau}{2}\Lambda^j\right)\phi^j, \tag{1.6}$$

or

$$\phi^{j+1} = T^j\phi^j, \tag{1.7}$$

where T^j is the transition operator

$$T^j = \left(E + \frac{\tau\Lambda^j}{2}\right)^{-1}\left(E - \frac{\tau\Lambda^j}{2}\right). \tag{1.8}$$

For proof of countable stability we do not necessarily need to evaluate the norms of the transition operators T^j. Indeed, taking the inner product of equation (1.4) with $(\phi^{j+1} + \phi^j)/2$, we obtain

$$\frac{(\phi^{j+1}, \phi^{j+1}) - (\phi^j, \phi^j)}{2\tau} + \left(\Lambda^j \frac{\phi^{j+1} + \phi^j}{2}, \frac{\phi^{j+1} + \phi^j}{2}\right) = 0. \tag{1.9}$$

Since Λ^j is positive semidefinite by assumption [see equation (1.5)], we have

$$\|\phi^{j+1}\| \leq \|\phi^j\|, \tag{1.10}$$

i.e., stability.

By no means does this lessen the importance of norm estimates of transitions operators in the analysis of difference schemes. In the present case such an estimate can be obtained with the help of relation (1.10):

$$\|T^j\| \leq 1. \tag{1.11}$$

Using (1.8) and the relations $\Lambda^j \geq 0$, $\tau > 0$, we note that the above inequality is also an immediate consequence of the Kellogg lemma [see Section 1.1]. If Λ^j is skew symmetric, that is, if

$$(\Lambda^j \phi, \phi) = 0,$$

then

$$\|\phi^{j+1}\| = \|\phi^j\|, \tag{1.12}$$

or we have a strict equality in (1.10)

One can show, using a procedure similar to the one above, that in this case

$$\|T^j\| = 1. \tag{1.13}$$

Next, let us discuss the Crank–Nicholson approximation scheme for the case where A changes with time. Define the operators H and H_τ by

$$H\phi \equiv \frac{\partial \phi}{\partial t} + A\phi, \tag{1.14}$$

$$(H_\tau \phi)^j = \frac{(\phi)^{j+1} - (\phi)^j}{\tau} + \Lambda^j \frac{(\phi)^{j+1} + (\phi)^j}{2}, \tag{1.15}$$

where $(\phi)^j$ denotes the projection of the exact solution of (1.1) on the net D_τ. The following norm is convenient for evaluating the approximation of the operator H:

$$\|H_\tau \phi\|_{c_\tau} = \max_{t_j \in D_\tau} \|(H_\tau \phi)^j\| \tag{1.16}$$

where $\|\cdot\|$ is the norm on the net elements (for $t = t_j$). In order to estimate the norm (1.16) we first expand the solution of the original equation (1.1) into the Taylor series:

$$(\phi)^{j+1} = (\phi)^j + \tau(\phi_t)^j + \frac{\tau^2}{2}(\phi_{tt})^j + \cdots. \tag{1.17}$$

Note the immediate relations

$$\phi_t = -A\phi, \qquad \phi_{tt} = A^2\phi - A_t\phi, \tag{1.18}$$

where we use the convention $A_t \equiv \partial A/\partial t$. With the above in mind, the Taylor series (1.17) can be rewritten as

$$(\phi)^{j+1} = (\phi)^j - \tau A^j(\phi)^j + \frac{\tau^2}{2}[(A^j)^2(\phi)^j - A_t^j(\phi)^j] - \cdots. \tag{1.19}$$

Substituting (1.19) into (1.16), and noting (1.15) we obtain

$$\|f - H_\tau \phi\|_{c_\tau} = \max_{t_j} \|\Lambda^j(\phi)^j - A^j(\phi)^j$$
$$+ \frac{\tau}{2}\{(A^j)^2 - A_t^j - \Lambda^j A^j\}(\phi)^j + O(\tau^2)\|, \tag{1.20}$$

where f^j is the right-hand side of equation (1.4), here equal to zero.

If we take

$$\Lambda^j = A^j = A(t_j)$$

(1.21)

for the approximating operator Λ^j, then (1.20) implies

$$\|f - H_\tau \phi\|_{c_\tau} = \frac{\tau}{2} \max_{t_\tau \in D_\tau} \|A_t^j(\phi)^j\| + O(\tau^2),$$

and the approximation is first order. Note that in the particular case of a time-invariant A the approximation (1.21) guarantees second order with respect to τ on the class of sufficiently smooth functions.

Assume now that the approximating operator Λ^j is chosen in the form

$$\Lambda^j = A^j + \frac{\tau}{2} A_t^j.$$

(1.22)

In this case

$$\|f - H_\tau \phi\|_{c_\tau} = O(\tau^2).$$

Let us note that the second-order approximation of the Crank–Nicholson scheme is also obtained by choosing the operator Λ^j in the form

$$\Lambda^j = A^{j+1/2},$$

(1.23)

or

$$\Lambda^j = \tfrac{1}{2}(A^{j+1} + A^j),$$

(1.24)

The forms (1.22), (1.23), and (1.24) of the second-order approximation of the operator A are used in many applications, in particular for numerical integration of quasi-linear equations.

5.2 Nonhomogeneous Equations of the Evolution Type

The foregoing paragraph dealt with homogeneous equations. Now we will consider the nonhomogeneous equation

$$\frac{\partial \phi}{\partial t} + A\phi = f \quad \text{in} \quad D \times D_t,$$

$$\phi = g \quad \text{in} \quad D \quad \text{for} \quad t = 0.$$

(2.1)

Assuming the hypothesis discussed in Section 5.1, the difference approximation for (2.1) based on the Crank–Nicholson scheme has the form

$$\frac{\phi^{j+1} - \phi^j}{\tau} + \Lambda^j \frac{\phi^{j+1} + \phi^j}{2} = f^j, \qquad \phi^0 = g,$$

(2.2)

where

$$f^j = f(t_{j+1/2}).$$

It is easy to see that the difference problem of (2.2) is the second-order approximation in τ of (2.1). On each interval let us write the formal solution of (2.2) as

$$\phi^{j+1} = T^j \phi^j + \tau \left(E + \frac{\tau}{2} \Lambda^j \right)^{-1} f^j. \tag{2.3}$$

In the case of the homogeneous equation it was shown in Section 5.1 that for $\Lambda^j \geq 0$ the following estimate holds:

$$\| T^j \| \leq 1. \tag{2.4}$$

Naturally, this norm estimate is independent of the right-hand side f. From (2.3) it follows that

$$\| \phi^{j+1} \| \leq \| T^j \| \, \| \phi^j \| + \tau \left\| \left(E + \frac{\tau}{2} \Lambda^j \right)^{-1} \right\| \| f^j \|. \tag{2.5}$$

In order to establish stability we exploit the estimate of equation (1.25) from Chapter 1. Since $\tau > 0$ and

$$(\Lambda^j \phi^j, \phi^j) \geq 0, \tag{2.6}$$

we obtain

$$\left\| \left(E + \frac{\tau}{2} \Lambda^j \right)^{-1} \right\| \leq 1, \tag{2.7}$$

and hence inequality (2.5) becomes

$$\| \phi^{j+1} \| \leq \| \phi^j \| + \tau \| f^j \|, \tag{2.8}$$

using (2.4) and (2.7). Setting $\| \phi^0 \| = \| g \|$ and $\| f \| = \max_j \| f_j \|$, the recursive relations (2.8) yields

$$\| \phi^j \| \leq \| g \| + j\tau \| f \|, \qquad j\tau \leq \text{const.} \tag{2.9}$$

The stability of the difference scheme is thus established. Furthermore, relation (2.9) is an *a priori* norm estimate of the solution.

5.3 Splitting-Up Methods for Nonstationary Problems

Complicated problems of mathematical physics can often be reduced to those consisting of a chain of simpler problems which can be effectively solved with a computer. This kind of reduction is possible in the cases where the original positive semidefinite operator characterizing the problem can be decomposed into a sum of positive semidefinite operators with a simple structure. Such methods will be referred to as *splitting-up* methods. Splitting-up methods were initiated by Douglas, Peaceman and Rachford [15], and then further developed by Soviet mathematicians Bagrinovskii and Godunov [15], Yanenko [3, 15], Samarskii [3, 15], Dyakonov [3, 15], Saul'ev [3], Marchuk [15], and others.

Splitting-up methods were originally formulated and theoretically justified for simple problems with commuting positive definite operators. It has become clear that the various splitting-up methods as introduced by various authors for these kinds of problems are either essentially equivalent (the only difference being the actual realization of the scheme), or otherwise quite similar.

The circle of nontrivial applications of splitting-up methods has been considerably enlarged. At present the splitting-up methods have become a powerful tool for solving highly complicated problems of mathematical physics. The most complete theory has been given for the case when the operator corresponding to the original problem has a representation as a sum of two simpler operators. We will begin our exposition with this latter case. In our opinion the most important method for applications is the *component-by-component splitting-up* method, explained below. We hope that the reader takes notice of this while reading the present chapter.

Thus, consider the evolution equation

$$\frac{\partial \phi}{\partial t} + A\phi = f \quad \text{in} \quad D \times D_t,$$

$$\phi = g \quad \text{in} \quad D \quad \text{for} \quad t = 0,$$

(3.1)

where the operator $A \geq 0$ does not depend on time and has the representation

$$A = A_1 + A_2,$$ (3.2)

$$A_1 \geq 0, \qquad A_2 \geq 0.$$ (3.3)

Assume further that the solution of (3.1) is sufficiently smooth. Wherever it is necessary for a proof we will assume that (3.1) is already reduced to a difference form and therefore A, A_1, and A_2 are matrices.

5.3.1 The Stabilization Method

Let us consider first the approximate solution of equations (3.1)–(3.3) under the condition that $f = 0$:

$$\left(E + \frac{\tau}{2} A_1\right)\left(E + \frac{\tau}{2} A_2\right)\frac{\phi^{j+1} - \phi^j}{\tau} + A\phi^j = 0, \qquad \phi^0 = g. \quad (3.4)$$

It is not difficult to see that under sufficient smoothness assumptions equations (3.4) approximate the original problem expressed in equations (3.1)–(3.3) up to the second order of accuracy in τ. Indeed, with the help of some algebra, equations (3.4) can be rewritten as

$$\left(E + \frac{\tau^2}{4} A_1 A_2\right) \frac{\phi^{j+1} - \phi^j}{\tau} + A \frac{\phi^{j+1} + \phi^j}{2} = 0, \qquad \phi^0 = g. \qquad (3.5)$$

From this we see that the difference equation (3.5) coincides in approximation order with the Crank–Nicholson scheme

$$\frac{\phi^{j+1} - \phi^j}{\tau} + A \frac{\phi^{j+1} + \phi^j}{2} = 0, \qquad \phi^0 = g, \qquad (3.6)$$

provided that the solution is smooth enough, that is, the second-order approximation in τ.

Let us turn next to the stability analysis of (3.4). To this end, rewrite this equation in the form

$$\left(E + \frac{\tau}{2} A_1\right)\left(E + \frac{\tau}{2} A_2\right)\phi^{j+1} = \left(E - \frac{\tau}{2} A_1\right)\left(E - \frac{\tau}{2} A_2\right)\phi^j, \qquad \phi^0 = g.$$
$$(3.7)$$

Solving for ϕ^{j+1}, we obtain

$$\phi^{j+1} = \left(E + \frac{\tau}{2} A_2\right)^{-1}\left(E + \frac{\tau}{2} A_1\right)^{-1}\left(E - \frac{\tau}{2} A_1\right)\left(E - \frac{\tau}{2} A_2\right)\phi^j. \qquad (3.8)$$

For each unknown ϕ^j define a new unknown ψ^j by the relation

$$\psi^j = \left(E + \frac{\tau}{2} A_2\right)\phi^j. \qquad (3.9)$$

Thus the new variable ψ^j satisfies

$$\psi^{j+1} = T\psi^j \qquad (3.10)$$

where the transition operator T is given by

$$T = \left(E + \frac{\tau}{2} A_1\right)^{-1}\left(E - \frac{\tau}{2} A_1\right)\left(E - \frac{\tau}{2} A_2\right)\left(E + \frac{\tau}{2} A_2\right)^{-1}. \qquad (3.11)$$

With the help of (3.10) we obtain the following estimate in the energy norm:

$$\|\psi^{j+1}\| \le \|T\| \|\psi^j\|. \qquad (3.12)$$

Let us estimate the norm of T:

$$\|T\| \le \|T_1\| \|T_2\|,$$

where

$$T_\alpha = \left(E - \frac{\tau}{2} A_\alpha\right)\left(E + \frac{\tau}{2} A_\alpha\right)^{-1}, \qquad \alpha = 1, 2. \tag{3.13}$$

Here use was made of the relationship

$$\left(E + \frac{\tau}{2} A_\alpha\right)^{-1}\left(E - \frac{\tau}{2} A_\alpha\right) = \left(E - \frac{\tau}{2} A_\alpha\right)\left(E + \frac{\tau}{2} A_\alpha\right)^{-1},$$

which follows from the immediate identity

$$\left(E - \frac{\tau}{2} A_\alpha\right)^{-1}\left(E - \frac{\tau}{2} A_\alpha\right) = \left(E + \frac{\tau}{2} A_\alpha\right)\left(E + \frac{\tau}{2} A_\alpha\right)^{-1}. \tag{3.14}$$

Indeed, multiplying (3.14) by $[E + (\tau/2)A_\alpha]^{-1} \times [E - (\tau/2)A_\alpha]$ from the left and using the fact that the operators $[E - (\tau/2)A_\alpha]$ and $[E + (\tau/2)A_\alpha]$ commute (direct verification) gives the claimed property.

Thus the problem of stability has been reduced to estimating the norms of T_α.

Application of Kellogg's lemma for estimating $\|T_1\|$ and $\|T_2\|$ in (3.13) yields

$$\|T\| \leq 1 \tag{3.15}$$

and consequently

$$\|\psi^{j+1}\| \leq \|\psi^j\|. \tag{3.16}$$

Our final goal however is to establish the stability of the original difference problem (3.4). For that we rewrite relation (3.16) in the form

$$\left\|\left(E + \frac{\tau}{2} A_2\right)\phi^{j+1}\right\| \leq \left\|\left(E + \frac{\tau}{2} A_2\right)\phi^j\right\|. \tag{3.17}$$

Introducing the notation

$$\left\|\left(E + \frac{\tau}{2} A_2\right)\phi\right\| = (C_2 \phi, \phi)^{1/2} = \|\phi\|_{C_2}, \tag{3.18}$$

where

$$C_\alpha = \left(E + \frac{\tau}{2} A_\alpha^*\right)\left(E + \frac{\tau}{2} A_\alpha\right), \qquad \alpha = 1, 2,$$

it is easy to see that $C_\alpha > 0$ and that $\|\cdot\|_{C_2}$ is indeed a norm.

Hence it follows that in the given norm we have absolute stability:

$$\|\phi^{j+1}\|_{C_2} \leq \|\phi^j\|_{C_2}. \tag{3.19}$$

We thus conclude that if $A_1 \geq 0$ and $A_2 \geq 0$, and if the entries of these matrices are independent of time, then, assuming the solution of problem (3.1) is smooth enough, the difference scheme is absolutely stable and approximates the problem (3.1) with second-order accuracy in τ.

In conclusion, let us point out that the difference scheme of the stabilization method can be conveniently implemented on a computer. For this purpose the difference equation is put in the form

$$F^j = A\phi^j$$

$$\left(E + \frac{\tau}{2} A_1\right)\xi^{j+1/2} = -F^j,$$

$$\left(E + \frac{\tau}{2} A_2\right)\xi^{j+1} = \xi^{j+1/2}, \tag{3.20}$$

$$\phi^{j+1} = \phi^j + \tau\xi^{j+1}.$$

Here $\xi^{j+1/2}$ and ξ^{j+1} are certain auxiliary variables which facilitate the reduction of (3.4) to the sequence of simpler problems (3.20). Note that the first and last relations in (3.20) are explicit. It means that the operator inversion is only needed in the second and third equations which involve the simple operators A_1 and A_2.

Consider next the nonhomogeneous problem

$$\frac{\partial\phi}{\partial t} + A\phi = f, \qquad \phi = g \quad \text{for} \quad t = 0, \tag{3.21}$$

where $A = A_1 + A_2$, $A_1 \geq 0$, $A_2 \geq 0$. In this case the stabilization method scheme is written as follows:

$$\left(E + \frac{\tau}{2} A_1\right)\left(E + \frac{\tau}{2} A_2\right)\frac{\phi^{j+1} - \phi^j}{\tau} + A\phi^j = f^j, \qquad \phi^0 = g. \tag{3.22}$$

where

$$f^j = f(t_{j+1/2}). \tag{3.23}$$

It can be shown that under the assumption of (3.23) the difference problem (3.22) approximates the original problem (3.21) up to the second order in τ.

Let us investigate the stability of the difference scheme. To this end we write equation (3.22) in the form

$$\psi^{j+1} = T\psi^j + \tau\left(E + \frac{\tau}{2} A_1\right)^{-1} f^j, \tag{3.24}$$

where

$$\psi^j = \left(E + \frac{\tau}{2} A_2\right)\phi^j. \tag{3.25}$$

From (3.24) it follows that

$$\|\psi^{j+1}\| \leq \|T\| \|\psi^j\| + \tau \left\|\left(E + \frac{\tau}{2} A_1\right)^{-1}\right\| \|f^j\|. \tag{3.26}$$

Since it has already been established that for a homogeneous equation

$$\|T\| \le 1,$$

it follows that

$$\|\psi^{j+1}\| \le \|\psi^j\| + \tau \left\|\left(E + \frac{\tau}{2} A_1\right)^{-1}\right\| \|f^j\|. \tag{3.27}$$

A simple manipulation yields

$$\|f^j\| = \left\|\left(E + \frac{\tau}{2} A_2\right)^{-1}\left(E + \frac{\tau}{2} A_2\right)f^j\right\| \le \left\|\left(E + \frac{\tau}{2} A_2\right)^{-1}\right\|\left\|\left(E + \frac{\tau}{2} A_2\right)f^j\right\|. \tag{3.28}$$

Taking (3.18), (3.25), and (3.28) into account, (3.27) implies

$$\|\phi^{j+1}\|_{C_2} \le \|\phi^j\|_{C_2} + \tau \left\|\left(E + \frac{\tau}{2} A_1\right)^{-1}\right\|\left\|\left(E + \frac{\tau}{2} A_2\right)^{-1}\right\| \|f^j\|_{C_2}. \tag{3.29}$$

Furthermore, making use of the norm estimate

$$\left\|\left(E + \frac{\tau}{2} A_\alpha\right)^{-1}\right\| \le 1, \tag{3.30}$$

and the inequality $A_\alpha \ge 0$ [established in (1.25); see Chapter 1], we arrive at

$$\|\phi^{j+1}\|_{C_2} \le \|\phi^j\|_{C_2} + \tau \|f^j\|_{C_2}. \tag{3.31}$$

With the help of recursive relations we obtain the estimate

$$\|\phi^j\|_{C_2} \le \|g\|_{C_2} + j\tau \|f\|_{C_2}, \tag{3.32}$$

where

$$\|f\|_{C_2} = \max_j \|f^j\|_{C_2}. \tag{3.33}$$

Therefore, if the matrices A_1, A_2 are nonnegative and their entries are independent of time, then, under the assumption of sufficient smoothness of the solution ϕ of (3.1) and of the function f, the difference scheme (3.22) is absolutely stable and approximates the problem (3.1) with the second-order in τ.

In conclusion, let us point out once more that the proof given above holds if and only if the original operator A is time invariant.

5.3.2 The Predictor–Corrector Method

Let us discuss a splitting method known as the *predictor-corrector* method. The essence of the method is to decompose the whole interval $0 \le t \le T$ into subintervals $t_j \le t \le t_{j+1}$ and then to solve on each of these problem (3.1) in two ways as follows: first, using the first-order approximation scheme with a comparatively large "degree" of stability, we find an approximate solution at the time instant $t_{j+1/2} = t_j + \tau/2$. After that we write the second-order scheme (corrector) on the whole interval (t_j, t_{j+1}), the main feature being that for its construction we make use of the "coarse" solution at $t_{j+1/2}$ obtained by means of the predictor (the first step).

Consequently, the predictor–corrector scheme can be written in the form

$$\frac{\phi^{j+1/4} - \phi^j}{\tau/2} + A_1 \phi^{j+1/4} = 0,$$

$$\frac{\phi^{j+1/2} - \phi^{j+1/4}}{\tau/2} + A_2 \phi^{j+1/2} = 0, \tag{3.34}$$

$$\frac{\phi^{j+1} - \phi^j}{\tau} + A\phi^{j+1/2} = 0,$$

where we assume $\phi^0 = g$.

Let us look at the scheme in some detail. First of all the elimination of the auxiliary variable $\phi^{j+1/4}$ from the first two equations in (3.34) reduces this system to

$$\left(E + \frac{\tau}{2} A_1\right)\left(E + \frac{\tau}{2} A_2\right)\phi^{j+1/2} = \phi^j,$$

$$\frac{\phi^{j+1} - \phi^j}{\tau} + A\phi^{j+1/2} = 0. \tag{3.35}$$

Furthermore, eliminating $\phi^{j+1/2}$, we obtain

$$\frac{\phi^{j+1} - \phi^j}{\tau} + A\left(E + \frac{\tau}{2} A_2\right)^{-1}\left(E + \frac{\tau}{2} A_1\right)^{-1}\phi^j = 0, \qquad \phi^0 = g. \tag{3.36}$$

Let us now analyze the approximation problem. To this end let us rewrite (3.36) in the form

$$\left(E + \frac{\tau}{2} A_1\right)\left(E + \frac{\tau}{2} A_2\right)\frac{\phi^{j+1} - \phi^j}{\tau} + \Lambda\phi^j = 0,$$

where

$$\Lambda = \left(E + \frac{\tau}{2} A_1\right)\left(E + \frac{\tau}{2} A_2\right)A\left(E + \frac{\tau}{2} A_2\right)^{-1}\left(E + \frac{\tau}{2} A_1\right)^{-1}.$$

Expanding into a power series in τ and assuming that

$$\frac{\tau}{2}\|A_\alpha\| < 1, \qquad \alpha = 1, 2,$$

one easily obtains

$$\Lambda = A + O(\tau^2).$$

Using the estimate used for the stabilization method we conclude that the predictor–corrector method yields a second-order approximation in τ.

Consider next the stability of this method. First let us write (3.36) in the form

$$\left(E + \frac{\tau}{2} A_1\right)\left(E + \frac{\tau}{2} A_2\right)\frac{\Phi^{j+1} - \Phi^j}{\tau} + A\Phi^j = 0, \qquad (3.37)$$

where

$$\Phi^j = \left(E + \frac{\tau}{2} A_2\right)^{-1}\left(E + \frac{\tau}{2} A_1\right)^{-1}\phi^j. \qquad (3.38)$$

Since

$$\|\Phi^{j+1}\|_{C_2} \leq \|\Phi^j\|_{C_2}, \qquad (3.39)$$

the difference equation (3.37) is stable. Substituting (3.38) into (3.39) and making use of (3.18) we obtain

$$\left\|\left(E + \frac{\tau}{2} A_1\right)^{-1}\phi^{j+1}\right\| \leq \left\|\left(E + \frac{\tau}{2} A_1\right)^{-1}\phi^j\right\| \qquad (3.40)$$

or

$$\|\phi^{j+1}\|_{C_1^{-1}} \leq \|\phi^j\|_{C_1^{-1}}, \qquad (3.41)$$

where

$$C_1^{-1} = \left(E + \frac{\tau}{2} A_1^*\right)^{-1}\left(E + \frac{\tau}{2} A_1\right)^{-1}.$$

We have thus established stability in the metric of (3.41). Consequently, if $A_1 \geq 0$, $A_2 \geq 0$, and if the entries of these matrices are time invariant, then the difference scheme (3.34) is absolutely stable and yields an approximate solution of second-order accuracy in τ, provided (3.1) has a sufficiently smooth solution.

For the nonhomogeneous problem the predictor–corrector method is formulated as follows:

$$\frac{\phi^{j+1/4} - \phi^j}{\tau/2} + A_1\phi^{j+1/4} = f^j,$$

$$\frac{\phi^{j+1/2} - \phi^{j+1/4}}{\tau/2} + A_2\phi^{j+1/2} = 0, \qquad (3.42)$$

$$\frac{\phi^{j+1} - \phi^j}{\tau} + A\phi^{j+1/2} = f^j,$$

where

$$f^j = f(t_{j+1/2}).$$

Taking f^j as indicated it can be shown that (3.42) approximates the original problem with second-order accuracy in τ. The stability of (3.42) is established as follows. First, eliminate $\phi^{j+1/2}$ and $\phi^{j+1/4}$; the result is

$$\frac{\phi^{j+1} - \phi^j}{\tau} + A\left(E + \frac{\tau}{2} A_2\right)^{-1}\left(E + \frac{\tau}{2} A_1\right)^{-1}\left(\phi^j + \frac{\tau}{2} f^j\right) = f^j. \quad (3.43)$$

Introduce the notation

$$\psi^j = \left(E + \frac{\tau}{2} A_1\right)^{-1}\left(\phi^j + \frac{\tau}{2} f^j\right)$$

Using this, the relation (3.43) can be written as

$$\frac{\psi^{j+1} - \psi^j}{\tau} + \left(E + \frac{\tau}{2} A_1\right)^{-1} A\left(E + \frac{\tau}{2} A_2\right)^{-1} \psi^j$$

$$= \left(E + \frac{\tau}{2} A_1\right)^{-1}\left(\frac{f^j + f^{j+1}}{2}\right).$$

Hence

$$\psi^{j+1} = \left[E - \tau\left(E + \frac{\tau}{2} A_1\right)^{-1} A\left(E + \frac{\tau}{2} A_2\right)^{-1}\right]\psi^j$$

$$+ \tau\left(E + \frac{\tau}{2} A_1\right)^{-1}\left(\frac{f^j + f^{j+1}}{2}\right). \quad (3.44)$$

The following relation holds

$$E - \tau\left(E + \frac{\tau}{2} A_1\right)^{-1} A\left(E + \frac{\tau}{2} A_2\right)^{-1}$$

$$= \left(E + \frac{\tau}{2} A_1\right)^{-1}\left[\left(E + \frac{\tau}{2} A_1\right)\left(E + \frac{\tau}{2} A_2\right) - \tau A\right]\left(E + \frac{\tau}{2} A_2\right)^{-1}$$

$$= \left[\left(E + \frac{\tau}{2} A_1\right)^{-1}\left(E - \frac{\tau}{2} A_1\right)\right]\left[\left(E - \frac{\tau}{2} A_2\right)\left(E + \frac{\tau}{2} A_2\right)^{-1}\right].$$

Thus, according to Kellogg's lemma, and using equation (1.25) from Chapter 1, we obtain

$$\left\|\phi^j + \frac{\tau}{2} f^j\right\|_{C_{\bar{1}}^{1}} \leq \left\|g + \frac{\tau}{2} f^0\right\|_{C_{\bar{1}}^{1}} + \tau j\|f\|_{C_{\bar{1}}^{1}} \quad (3.45)$$

where $\|f\|_{C_{\bar{1}}^{1}} = \max_j \|f^j\|_{C_{\bar{1}}^{1}}$; that is, stability follows provided $0 \leq t_j \leq T$.

Hence, if the matrices $A_1 \geq 0$, $A_2 \geq 0$ are time invariant, then the difference scheme (3.42) is absolutely stable and provides a second-order approximation (in τ) of the solution, if only the right-hand side f in (3.1) along with the solutions are smooth enough.

In conclusion, we note that although the difference scheme (3.34) is absolutely stable, the corrector portion of it, taken by itself, may be absolutely unstable. Let us prove this. To simplify the computation, we consider the case when A is the difference analog of the two-dimensional Laplace operator and D is the unit square, on the boundary of which the solution vanishes.

In this case the corrector has the form

$$\frac{\phi_{k,l}^{j+1} - \phi_{k,l}^{j-1}}{2\tau} - \frac{\phi_{k+1,l}^{j} - 2\phi_{k,l}^{j} + \phi_{k-1,l}^{j}}{h^2} - \frac{\phi_{k,l+1}^{j} - 2\phi_{k,l}^{j} + \phi_{k,l-1}^{j}}{h^2} = 0$$

(since the predictor now takes no part in the computations, we will, for convenience, consider only integer values of j).

We consider a solution ϕ^j of this difference problem for which the inequality (3.16) in the definition of stability fails to hold. Such a solution will be sought in the form

$$\phi_{k,l}^{j} = \lambda^j \sin m\pi kh \cdot \sin p\pi lh,$$

where j on the left side is an index, and on the right, an exponent. Substituting this expression in the difference equation, we arrive at the following characteristic equation:

$$\lambda^2 + 8a_{mp}\lambda - 1 = 0,$$

where

$$a_{mp} = \frac{\tau}{h^2}\left(\sin^2 \frac{m\pi h}{2} + \sin^2 \frac{p\pi h}{2}\right).$$

Choosing

$$\lambda = -4a_{mp} - \sqrt{1 + 16a_{mp}^2},$$

for the value of λ, we find that as $\tau \to 0$ (since $\tau/h^2 = \text{const}$)

$$\|\phi^j\|/\|\phi^0\| = |\lambda|^j \to \infty,$$

that is, the scheme is absolutely unstable.

Thus, despite the fact that the difference scheme used as a corrector is absolutely unstable, the reserve of stability possessed by the predictor is sufficient for the absolute stability of the process as a whole.

5.3.3 The Component-by-Component Splitting-Up Method

The stabilization method and the predictor–corrector method are equivalent in accuracy and are absolutely stable provided $A_\alpha \geq 0$.

It is to be kept in mind, however, that we have assumed A_α to be time-invariant. This constraint has made it possible to analyze completely the stability, provided only the constructive assumption of positive semi-definiteness of the operators A_α. Unfortunately, in the time-dependent case the stability analysis of this kind can not be done in general. An exception is the component-by-component splitting-up method which we will formulate presently.

Consider equations (3.1)–(3.3) and let $A_1 \geq 0$, $A_2 \geq 0$. Approximate these matrices on $t_j \leq t \leq t_{j+1}$ in the form

$$\Lambda_\alpha^j = A_\alpha(t_{j+1/2}),$$

assuming that their entries are sufficiently smooth. The difference scheme below (suggested by Yanenko [15]) consists of a sequence of simple Crank–Nicholson schemes:†

$$\frac{\phi^{j+1/2} - \phi^j}{\tau} + \Lambda_1^j \frac{\phi^{j+1/2} + \phi^j}{2} = 0,$$

$$\frac{\phi^{j+1} - \phi^{j+1/2}}{\tau} + \Lambda_2^j \frac{\phi^{j+1} + \phi^{j+1/2}}{2} = 0. \qquad (3.46)$$

By eliminating the auxiliary functions $\phi^{j+1/2}$, the system of difference equations of (3.46) can be reduced to one single equation

$$\phi^{j+1} = T^j \phi^j, \qquad (3.47)$$

where

$$T^j = \left(E + \frac{\tau}{2}\Lambda_2^j\right)^{-1}\left(E - \frac{\tau}{2}\Lambda_2^j\right)\left(E + \frac{\tau}{2}\Lambda_1^j\right)^{-1}\left(E - \frac{\tau}{2}\Lambda_1^j\right). \quad (3.48)$$

At first, let us consider the approximation problem. To this end, let us expand the operator T^j into a power series in τ and assume that

$$\frac{\tau}{2} \|\Lambda_\alpha^j\| < 1.$$

Performing some simple manipulations we obtain

$$T^j = E - \tau\Lambda^j + \frac{\tau^2}{2}\left[(\Lambda_1^j)^2 + 2\Lambda_2^j\Lambda_1^j + (\Lambda_2^j)^2\right] - \cdots. \qquad (3.49)$$

If the operators Λ_α^j commute, that is, if $\Lambda_1^j\Lambda_2^j = \Lambda_2^j\Lambda_1^j$, then (3.49) can be written as

$$T^j = E - \tau\Lambda^j + \frac{\tau^2}{2}(\Lambda^j)^2 - \cdots. \qquad (3.50)$$

Hence if the matrices $A_1(t) \geq 0$, $A_2(t) \geq 0$, and the solution ϕ of equations (3.1)–(3.3) are smooth enough, then the difference scheme (3.46) is absolutely stable, as implied by the inequality $\|T^j\| \leq 1$ (from Kellogg's lemma).

---

† The theoretical foundations and some modifications of the scheme were given by the author in a paper at the Symposium on Numerical Solutions of Partial Differential Equations (SYNSPADE, 1970, USA).

Moreover, the scheme approximates the original equation [equation (3.1)] with the second- or first-order accuracy in τ, according to whether Λ_1^j and Λ_2^j commute or not.

The next step is to approximate the operators $A_1(t)$ and $A_2(t)$ on the interval $t_j \leq t \leq t_{j+1}$ the same way as in (3.46), and on the interval $t_{j-1} \leq t \leq t_{j+1}$ by taking

$$\Lambda_\alpha^j = A_\alpha(t_j).$$

Consider the following two systems of difference equations:

$$\frac{\phi^{j-1/2} - \phi^{j-1}}{\tau} + \Lambda_1^j \frac{\phi^{j-1/2} + \phi^{j-1}}{2} = 0,$$

$$\frac{\phi^j - \phi^{j-1/2}}{\tau} + \Lambda_2^j \frac{\phi^j + \phi^{j-1/2}}{2} = 0, \tag{3.51}$$

$$\frac{\phi^{j+1/2} - \phi^j}{\tau} + \Lambda_2^j \frac{\phi^{j+1/2} + \phi^j}{2} = 0,$$

$$\frac{\phi^{j+1} - \phi^{j+1/2}}{\tau} + \Lambda_1^j \frac{\phi^{j+1} + \phi^{j+1/2}}{2} = 0. \tag{3.52}$$

The computational cycle consists of a sequential application of schemes (3.51) and (3.52). In analogy with the above, it can be shown that the full computational cycle, using (3.51) and (3.52), gives us

$$\phi^{j+1} = T^j \phi^{j-1}, \tag{3.53}$$

where

$$T^j = \left(E + \frac{\tau}{2}\Lambda_1^j\right)^{-1}\left(E - \frac{\tau}{2}\Lambda_1^j\right)\left(E + \frac{\tau}{2}\Lambda_2^j\right)^{-1}$$

$$\times \left(E - \frac{\tau}{2}\Lambda_2^j\right)\left(E + \frac{\tau}{2}\Lambda_2^j\right)^{-1}\left(E - \frac{\tau}{2}\Lambda_2^j\right)\left(E + \frac{\tau}{2}\Lambda_1^j\right)^{-1}$$

$$\times \left(E - \frac{\tau}{2}\Lambda_1^j\right) = E - 2\tau\Lambda^j + \frac{(2\tau)^2}{2}(\Lambda^j)^2 - \cdots.$$

Let us now compare the transition operator T^j above with that corresponding to the following Crank–Nicholson scheme:

$$\frac{\phi^{j+1} - \phi^{j-1}}{2\tau} + \Lambda^j \frac{\phi^{j+1} + \phi^{j-1}}{2} = 0.$$

It then follows that with an accuracy up to τ^2 the two operators T^j (corresponding to the two-cycle splitting scheme and to the Crank–Nicholson scheme on the doubled interval) coincide, no matter whether the operators A_α commute or not. Thus, this approach avoids the considerable constraint of commutativity.

Next, let us turn to the problem of countable stability. Consider relation (3.47); from this

$$\|\phi^{j+1}\| \leq \|T^j\| \|\phi^j\|.$$

Since for $A_\alpha \geq 0$

$$\|T^j\| \leq 1$$

as we have seen above, we obtain the estimate

$$\|\phi^{j+1}\| \leq \|\phi^j\|. \tag{3.54}$$

From this one can immediately show that

$$\|\phi^j\| \leq \|g\|. \tag{3.55}$$

In the case of the two-cycle method one has the estimate of (3.54) at each step of the cycle. This means that the two-cycle method is absolutely stable. (Note that a similar approach of symmetrization has been suggested independently by Strang [7] for the alternating-directions method.)

Consequently, if $A_1(t) \geq 0$ and $A_2(t) \geq 0$, then, assuming the entries of the matrices as well as the solution ϕ of equations (3.1)–(3.3) are smooth enough, the difference system of (3.51), (3.52) is absolutely stable, and scheme (3.53) represents a second-order approximation (in τ) of equation (3.1).

Let us now seek the solution of the nonhomogeneous problem by means of the two-cycle, full splitting. With this in mind, consider the difference system of the form (3.51), (3.52), which can be written more conveniently as follows:

$$
\begin{aligned}
\left(E + \frac{\tau}{2} \Lambda_1^j\right)\phi^{j-1/2} &= \left(E - \frac{\tau}{2} \Lambda_1^j\right)\phi^{j-1}, \\
\left(E + \frac{\tau}{2} \Lambda_2^j\right)(\phi^j - \tau f^j) &= \left(E - \frac{\tau}{2} \Lambda_2^j\right)\phi^{j-1/2}, \\
\left(E + \frac{\tau}{2} \Lambda_2^j\right)\phi^{j+1/2} &= \left(E - \frac{\tau}{2} \Lambda_2^j\right)(\phi^j + \tau f^j), \\
\left(E + \frac{\tau}{2} \Lambda_1^j\right)\phi^{j+1} &= \left(E - \frac{\tau}{2} \Lambda_1^j\right)\phi^{j+1/2},
\end{aligned}
\tag{3.56}
$$

where $f^j = f(t_j)$. Solving for ϕ^{j+1}, we obtain

$$\phi^{j+1} = T^j \phi^{j-1} + 2\tau T_1^j T_2^j f^j, \tag{3.57}$$

where

$$T^j = T_1^j T_2^j T_2^j T_1^j \tag{3.58}$$

$$T_\alpha^j = \left(E + \frac{\tau}{2} \Lambda_\alpha^j\right)^{-1}\left(E - \frac{\tau}{2} \Lambda_\alpha^j\right). \tag{3.59}$$

Using a power-series expansion in the small parameter τ, expression (3.57) becomes

$$\phi^{j+1} = \left[E - 2\tau\Lambda^j + \frac{(2\tau)^2}{2} (\Lambda^j)^2 \right] \phi^{j-1} + 2\tau(E - \tau\Lambda^j)f^j + O(\tau^3), \quad (3.60)$$

which can further be written in the form

$$\frac{\phi^{j+1} - \phi^{j-1}}{2\tau} + \Lambda^j(E - \tau\Lambda^j)\phi^{j-1} = (E - \tau\Lambda^j)f^j + O(\tau^2). \quad (3.61)$$

Expanding into a Taylor series in the vicinity of t_{j-1}, we may eliminate ϕ^{j-1} from the last equation. Thus

$$\phi^j = \phi^{j-1} + \left(\frac{\partial\phi}{\partial t}\right)^{j-1} \tau + O(\tau^2). \quad (3.62)$$

With the help of the relation

$$\left(\frac{\partial\phi}{\partial t}\right)^{j-1} = -\Lambda^j\phi^{j-1} + f^j + O(\tau) \quad (3.63)$$

we further eliminate the derivative $\partial\phi/\partial t$. Hence substituting (3.63) into (3.62) we obtain

$$\phi^j = (E - \tau\Lambda^j)\phi^{j-1} + \tau f^j + O(\tau^2).$$

Thus

$$(E - \tau\Lambda^j)\phi^{j-1} = \phi^j - \tau f^j + O(\tau^2). \quad (3.64)$$

Using (3.64) in (3.61), we have finally

$$\frac{\phi^{j+1} - \phi^{j-1}}{2\tau} + \Lambda^j\phi^j = f^j + O(\tau^2). \quad (3.65)$$

Clearly, (3.65) is a second-order approximation (in τ) of the original equation (3.1) on the interval $t_{j-1} \le t \le t_{j+1}$. We have thus found a difference approximation of the nonhomogeneous evolution equation of second order by means of the two-cycle method.

The stability is established by elementary manipulations, if we use the energy norms. Indeed, from the norm estimate of (3.57)

$$\|\phi^{j+1}\| \le \|T^j\| \|\phi^{j-1}\| + 2\tau\|T_1^j\| \|T_2^j\| \|f^j\|. \quad (3.66)$$

It has been shown above that $\|T_\alpha^j\| \le 1$, and hence

$$\|T^j\| \le \|T_1^j\| \|T_2^j\| \|T_2^j\| \|T_1^j\| \le 1.$$

Therefore

$$\|\phi^{j+1}\| \le \|\phi^{j-1}\| + 2\tau\|f^j\|. \quad (3.67)$$

Using the recursive relation (3.63) we have

$$\|\phi^j\| \le \|g\| + \tau j \|f\|, \tag{3.68}$$

where

$$\|f\| = \max_j \|f^j\|.$$

The countable stability of the scheme on any finite time interval now follows from relation (3.68).

The equations (3.56) can also be equivalently written as

$$\left(E + \frac{\tau}{2}\Lambda_1^j\right)\phi^{j-2/3} = \left(E - \frac{\tau}{2}\Lambda_1^j\right)\phi^{j-1},$$

$$\left(E + \frac{\tau}{2}\Lambda_2^j\right)\phi^{j-1/3} = \left(E - \frac{\tau}{2}\Lambda_2^j\right)\phi^{j-2/3}.$$

$$\phi^{j+1/3} = \phi^{j-1/3} + 2\tau f^j, \tag{3.69}$$

$$\left(E + \frac{\tau}{2}\Lambda_2^j\right)\phi^{j+2/3} = \left(E - \frac{\tau}{2}\Lambda_2^j\right)\phi^{j+1/3},$$

$$\left(E + \frac{\tau}{2}\Lambda_1^j\right)\phi^{j+1} = \left(E - \frac{\tau}{2}\Lambda_1^j\right)\phi^{j+2/3}.$$

Eliminating the unknowns indexed by fractions we arrive at the explicit formula

$$\phi^{j+1} = T_1^j T_2^j T_2^j T_1^j \phi^{j-1} + 2\tau T_1^j T_2^j f^j, \tag{3.70}$$

which coincides with (3.57). Of the two forms (3.57) and (3.69), the latter is preferable in some cases.

In summary, if the matrices $A_1(t) \ge 0$, $A_2(t) \ge 0$, along with the function $f(t)$ and the solution ϕ are smooth enough, then the difference system (3.56) is absolutely stable on $0 \le t \le T$ and approximates the original equation with second-order accuracy in τ.

5.3.4 Some General Remarks

To begin with, let us compare the splittin-up methods of this section under the assumption that A, A_1, and A_2 are independent of time, $A_1 \ge 0$, $A_2 \ge 0$, and $A_1 A_2 = A_2 A_1$. It is, namely, this simple case which is fairly completely discussed in the literature. Let us assume that the splitting-up differential schemes are already formally resolved with respect to the solution sought at each step. It is not difficult to see that all the splitting-up schemes for the homogeneous ($f = 0$) evolution problem (3.1) are mutually equivalent and hence represent only various realization schemes. In particular, they can be written in the form

$$\phi^{j+1} = T\phi^j,$$

where

$$T = \left(E + \frac{\tau}{2}A_1\right)^{-1}\left(E - \frac{\tau}{2}A_1\right)\left(E + \frac{\tau}{2}A_2\right)^{-1}\left(E - \frac{\tau}{2}A_2\right).$$

The splitting-up schemes for nonhomogeneous evolution problems are equivalent only as far as the order of approximation accuracy is concerned. This means that the various splitting-up schemes which approximate the nonhomogeneous problem with the second-order of accuracy in τ lead to different results, the difference being within the limits of order $O(\tau^2)$.

The second remark concerns the case where $A_1 \geq 0$, $A_2 \geq 0$ are time invariant, but $A_1 A_2 \neq A_2 A_1$. As shown above, any one of the three splitting-up schemes (that is, the stabilization method, the predictor–corrector method, and the component-by-component method) can be used to obtain an approximate solution of the evolution problem. Although equivalent in order of accuracy, these methods differ considerably. Even in the homogeneous case the transition operators are quite different. It is still much too early for any recommendation as to the areas of the most effective applications of one scheme or another, since this problem has not yet been adequately analyzed. However, the fact that there are three independent techniques available enhances the confidence when attempting to solve complicated problems.

The third remark deals with the most general case where the operators $A_1 \geq 0$, $A_2 \geq 0$ may depend on time and may not commute. We recommend in this case the component-by-component splitting. In its two-cycle form it gives a second-order approximation of the solution.

5.4 Multi-Component Splitting

So far we have assumed that the original operator A is represented as a sum of two operators of simpler structure. As it is often the case, more complex physical problems require that A be decomposed into a large number of components. In general we have

$$A = \sum_{\alpha=1}^{n} A_\alpha, \tag{4.1}$$

where $A_\alpha \geq 0$. We will consider only the case $n > 2$, since the case $n = 2$ has been dealt with at length in Section 5.3.

First of all, it can be seen that the straightforward generalizations of the splitting-up methods considered above are not possible in general.

5.4.1 The Stabilization Method

Assuming the validity of equation (4.1), the stabilization method can be represented in the form

$$\prod_{\alpha=1}^{n}\left(E + \frac{\tau}{2}A_\alpha\right)\frac{\phi^{j+1} - \phi^j}{\tau} + A\phi^j = f^j, \qquad \phi^0 = g, \tag{4.2}$$

where

$$f^j = f(t_{j+1/2}).$$

The realization scheme of the algorithm is as follows:

$$F^j = -A\phi^j + f^j,$$

$$\left(E + \frac{\tau}{2}A_1\right)\xi^{j+(1/n)} = F^j,$$

$$\left(E + \frac{\tau}{2}A_2\right)\xi^{j+(2/n)} = \xi^{j+(1/n)}, \tag{4.3}$$

$$\dots\dots\dots\dots\dots\dots\dots\dots\dots$$

$$\left(E + \frac{\tau}{2}A_n\right)\xi^{j+1} = \xi^{j+(n-1)/n}$$

$$\phi^{j+1} = \phi^j + \tau\xi^{j+1}.$$

It is easy to check that the stabilization method is of second-order approximation in τ, provided the solution is sufficiently smooth. The countable stability is assured if

$$\|T\| < 1, \tag{4.4}$$

where the transition operator T is given by

$$T = E - \tau\prod_{\alpha=n}^{1}\left(E + \frac{\tau}{2}A_\alpha\right)^{-1}A. \tag{4.5}$$

Unfortunately, the condition $A_\alpha \geq 0$ does not imply the stability in any norm, contrary to what we have in the case $n = 2$.

To establish stability, one usually uses the following simple algorithmic approach. Put f^j equal to zero in (4.2) and solve for ϕ^{j+1} to obtain

$$\phi^{j+1} = T\phi^j. \tag{4.6}$$

Since T is assumed independent of time (or the index j), we can solve (4.6) with the initial condition

$$\phi^0 = g \tag{4.7}$$

and for a fixed parameter τ, which ensures the necessary approximation. If the norm $\|\phi^j\|$ does not increase, then $\|T\| < 1$, and therefore the condition of countable stability can be taken as satisfied. Having this, we may turn to the nonhomogeneous problem. Equation (4.2) can be rewritten as follows:

$$\phi^{j+1} = T\phi^j + \tau \prod_{\alpha=n}^{1} \left(E + \frac{\tau}{2} A_\alpha \right)^{-1} f^j. \tag{4.8}$$

Hence

$$\|\phi^{j+1}\| \le \|T\| \, \|\phi^j\| + \tau \prod_{\alpha=n}^{1} \left\| \left(E + \frac{\tau}{2} A_\alpha \right)^{-1} \right\| \|f^j\|,$$

or, as a consequence of the inequalities (4.4) and (1.25) from Chapter 1, we have

$$\|\phi^{j+1}\| \le \|\phi^j\| + \tau \|f^j\|. \tag{4.9}$$

Using the recursive relation we arrive at the condition of stability in the energy metric, namely

$$\|\phi\| \le \|g\| + \tau j \|f\|, \tag{4.10}$$

where

$$\|f\| = \max_j \|f^j\|. \tag{4.11}$$

Note that when solving the homogeneous equation (4.6) we have made use of the initial condition (4.7). This was not necessary at all. Any function could have been taken from the same space to which g belongs.

5.4.2 The Predictor–Corrector Method

In this case the splitting-up scheme becomes

$$\left(E + \frac{\tau}{2} A_1 \right) \phi^{j+(1/2n)} = \phi^j + \frac{\tau}{2} f^j$$

$$\left(E + \frac{\tau}{2} A_2 \right) \phi^{j+(2/2n)} = \phi^{j+(1/2n)},$$

$$\cdots\cdots\cdots\cdots\cdots\cdots\cdots\cdots\cdots\cdots\cdots\cdots \tag{4.12}$$

$$\left(E + \frac{\tau}{2} A_n \right) \phi^{j+1/2} = \phi^{j+[(n-1)/2n]},$$

$$\frac{\phi^{j+1} - \phi^j}{\tau} + A\phi^{j+1/2} = f^j,$$

where, again, it is assumed that $A_\alpha \geq 0$ and $f^j = f(t_{j+1/2})$. The above system can be reduced to a single equation of the form

$$\frac{\phi^{j+1} - \phi^j}{\tau} + A \prod_{\alpha=n}^{1} \left(E + \frac{\tau}{2} A_\alpha \right)^{-1} \left(\phi^j + \frac{\tau}{2} f^j \right) = f^j \qquad (4.13)$$

with

$$\phi^0 = g.$$

Provided there is sufficient smoothness, the predictor–corrector method is of second-order accuracy in τ. Let us rewrite (4.13) as follows:

$$\phi^{j+1} = T\phi^j + \frac{\tau}{2}(E + T)f^j, \qquad (4.14)$$

where the transition operator T is given by

$$T = E - \tau A \prod_{\alpha=n}^{1} \left(E + \frac{\tau}{2} A_\alpha \right)^{-1}. \qquad (4.15)$$

The requirement of countable stability eventually comes down to estimating the norm of T. Unfortunately, also in this case the constructive condition $A_\alpha \geq 0$ does not yield the proof of stability for the scheme. This remains an open problem.

In order to complete the analysis of the two schemes considered above, let us take a cimple case when the operators A_α are mutually commuting and have common bases. It turns out that this additional hypothesis along with $A_\alpha \geq 0$ are enough to prove stability of the considered schemes. Indeed, commutativity implies that the transition operators for the two schemes coincide. For the sake of simplicity consider the homogeneous problem (4.6), (4.7), and let us seek the solution in the spectral form

$$\phi^j = \sum_k \phi_k^j u_k, \qquad (4.16)$$

where u_k are the eigenfunctions of problem (1.7) in Chapter 1. $\phi_k^j = (\phi^j, u_k^*)$, where u_k^* are the eigenfunctions of the adjoint problem (1.7) in Chapter 1. Since $\{u_k\}$ is the common basis, we have

$$Au_k = \lambda_k u_k, \qquad A_\alpha u_k = \lambda_k^\alpha u_k, \qquad \lambda_k = \sum_{\alpha=1}^{n} \lambda_k^\alpha. \qquad (4.17)$$

Substituting the expansions of (4.16) and the corresponding representations for g into (4.6), (4.7), we obtain the following expressions for the Fourier coefficients:

$$\phi_k^{j+1} = T_k \phi_k^j, \qquad \phi_k^0 = g_k, \qquad (4.18)$$

where

$$T_k = 1 - \frac{\tau \lambda_k}{\prod\limits_{\alpha=1}^{n} \left(1 + \frac{\tau}{2} \lambda_k^\alpha \right)}. \qquad (4.19)$$

The formula for T_k above can be written as

$$T_k = \frac{\mu_k - \dfrac{\tau}{2}\lambda_k}{\mu_k + \dfrac{\tau}{2}\lambda_k}, \tag{4.20}$$

where μ_k are positive constants, provided $\lambda_k^{\alpha} \geq 0$. From (4.20) we have

$$|T_k| \leq 1, \tag{4.21}$$

which proves the claim, in agreement with Section 1.3.

The stabilization method and the predictor–corrector method for the n-component splitting can also be applied to the case where A depends on time. The stability analysis however becomes much more complex. Therefore, it is difficult to say to what extent we can justify the application of these schemes in general situations. In particular, this stimulated the present author to formulate a more or less universal approach for solving various complicated and sufficiently general problems, the basic idea being that of splitting. The two-cycle sequential splitting method is of this type.

5.4.3 The Component-by-Component Splitting-Up Method Based on the Elementary Schemes

Let us try to construct a second-order difference approximation of the problem so that it is absolutely stable. In accord with the assumptions regarding the component-by-component splitting let us suppose that

$$\Lambda^j = \sum_{\alpha=1}^{n} \Lambda_{\alpha}^j, \tag{4.22}$$

where all Λ_{α}^j are nonnegative definite operators, that is $\Lambda_{\alpha}^j \geq 0$. Consider the following system:

$$\left(E + \frac{\tau}{2}\Lambda_{\alpha}^j\right)\Phi^{j+(\alpha/n)} = \left(E - \frac{\tau}{2}\Lambda_{\alpha}^j\right)\Phi^{j+[(\alpha-1)/n]}, \qquad \alpha = 1, 2, \ldots, n. \tag{4.23}$$

Assume that $\Lambda_{\alpha}^j \geq 0$ commute and $\Lambda_{\alpha}^j = A_{\alpha}^{j+1/2}$ or $\Lambda_{\alpha}^j = \frac{1}{2}(A^{j+1} + A^j)$. Then (4.23) is an unconditionally stable, second-order approximation scheme. This can be seen quite easily using the Fourier method. In the noncommutative case, however, the scheme will only be of first-order accuracy in τ. Of more interest in applications is the following second-order approximation scheme suggested by Dyakonov [15]:

$$\Phi^{j+(\alpha/2n)} = \left(E - \frac{\tau}{2}\Lambda_{\alpha}^j\right)\Phi^{j+[(\alpha-1)/2n]}, \qquad \alpha = 1, 2, \ldots, n,$$

$$\left(E + \frac{\tau}{2}\Lambda_{2n-\alpha+1}^j\right)\Phi^{j+(\alpha/2n)} = \Phi^{j+[(\alpha-1)/2n]}, \qquad \alpha = n+1, n+2, \ldots, 2n.$$

$$\tag{4.24}$$

Let us try to find a special construction of the full splitting-up method based on (4.23) which would solve the Cauchy problem corresponding to nonnegative definite and noncommuting operators Λ_α^j, and which would be of second-order accuracy. In a sense, this would solve completely the splitting-up problem.

Note that (4.23) reduces to a single equation of the form

$$\Phi^{j+1} = \prod_{\alpha=1}^{n} \left(E + \frac{\tau}{2} \Lambda_\alpha^j \right)^{-1} \left(E - \frac{\tau}{2} \Lambda_\alpha^j \right) \Phi^j. \tag{4.25}$$

Using (4.25), we find the norm estimates

$$\|\Phi^{j+1}\| \leq \prod_{\alpha=1}^{n} \left\| \left(E + \frac{\tau}{2} \Lambda_\alpha^j \right)^{-1} \left(E - \frac{\tau}{2} \Lambda_\alpha^j \right) \right\| \|\Phi^j\|. \tag{4.26}$$

From Kellogg's lemma

$$\|\Phi^{j+1}\| \leq \|\Phi^j\| \leq \cdots \leq \|g\|. \tag{4.27}$$

If the operators are skew symmetric, that is $(\Lambda_\alpha^j \phi, \phi) = 0$, then

$$\|\Phi^{j+1}\| = \|\Phi^j\| = \cdots = \|g\|. \tag{4.28}$$

Thus we have absolute stability for the scheme.

To determine the order of approximation, let us expand the expression below in term of powers of τ (assuming $(\tau/2)\|\Lambda_\alpha\| < 1$):

$$T^j = \prod_{\alpha=1}^{n} \left(E + \frac{\tau}{2} \Lambda_\alpha^j \right)^{-1} \left(E - \frac{\tau}{2} \Lambda_\alpha^j \right).$$

Since

$$T^j = \prod_{\alpha=1}^{n} T_\alpha^j,$$

and since T_α^j can be expanded into the series

$$T_\alpha^j = E - \tau\Lambda_\alpha^j + \frac{\tau^2}{2} (\Lambda_\alpha^j)^2 \cdots, \tag{4.29}$$

we have that

$$T^j = E - \tau\Lambda^j + \frac{\tau^2}{2} \left[(\Lambda^j)^2 + \sum_{\alpha=1}^{n} \sum_{\beta=\alpha+1}^{n} (\Lambda_\alpha^j \Lambda_\beta^j - \Lambda_\beta^j \Lambda_\alpha^j) \right] + O(\tau^3). \tag{4.30}$$

In the case where the operators Λ_α^j commute, the expression under the double sum is zero and (4.30) becomes

$$T^j = E - \tau\Lambda^j + \frac{\tau^2}{2} (\Lambda^j)^2 + O(\tau^3). \tag{4.31}$$

Comparing (4.31) with (1.13) and (1.22)–(1.24), we conclude that in this particular case scheme (4.23) approximates with the second order in τ. If the operators Λ_α^j do not commute, then the splitting-up scheme turns out to be only of first-order accuracy in τ. To obtain the second order in this latter case it is necessary to use the scheme

$$\Phi^j = \prod_{\alpha=1}^{n} T_\alpha^j \Phi^{j-1}, \qquad \Phi^{j+1} = \prod_{\alpha=n}^{1} T_\alpha^j \Phi^j, \tag{4.32}$$

rather than (4.23). In the language of algorithms, this says that the system of equations (4.23) is solved first on the interval $t_{j-1} \le t \le t_j$ for $\alpha = 1, 2, \ldots, n$, and then again on the interval $t_j \le t \le t_{j+1}$, but with the indices α taken in the reversed order ($\alpha = n, n-1, \ldots, 1$):

$$\left(E + \frac{\tau}{2} \Lambda_\alpha^j\right)\Phi^{j+(\alpha/n)-1} = \left(E - \frac{\tau}{2} \Lambda_\alpha^j\right)\Phi^{j+[(\alpha-1)/n]-1}, \qquad \alpha = 1, 2, \ldots, n,$$

(4.33)

$$\left(E + \frac{\tau}{2} \Lambda_\alpha^j\right)\Phi^{j+1-[(\alpha-1)/n]} = \left(E - \frac{\tau}{2} \Lambda_\alpha\right)\Phi^{j+1-(\alpha/n)}, \qquad \alpha = n, n-1, \ldots, 1.$$

For the full cycle (4.33) we have clearly

$$\Phi^{j+1} = T^j \Phi^{j-1},$$

where

$$T^j = \prod_{\alpha=1}^{n} T_\alpha^j \prod_{\alpha=n}^{1} T_\alpha^j = E - 2\tau\Lambda^j + \frac{(2\tau)^2}{2} (\Lambda^j)^2 + O(\tau^3).$$

Thus, if we take for Λ_α^j one of the analogs introduced in (1.22)–(1.24), scheme (4.33) becomes a second-order approximation on the interval $t_{j-1} \le t \le t_{j+1}$.

Finally let us note that the difference system (4.33) is absolutely stable, provided $\Lambda_\alpha^j \ge 0$. In a sense, we have thus obtained an optimal multi-component splitting-up algorithm.

Consider next the nonhomogeneous equation

$$\frac{\partial\phi}{\partial t} + A\phi = f,$$

(4.34)

$$\phi = g \quad \text{for} \quad t = 0,$$

where $A(t) \ge 0$ and $A = \sum_{\alpha=1}^{n} A_\alpha$, $A_\alpha(t) \ge 0$. The splitting-up scheme on the interval $t_{j-1} \le t \le t_{j+1}$ is as follows:

$$\left(E + \frac{\tau}{2} \Lambda_1^j\right)\phi^{j-[(n-1)/n]} = \left(E - \frac{\tau}{2} \Lambda_1^j\right)\phi^{j-1},$$

$$\cdots\cdots\cdots\cdots\cdots\cdots\cdots\cdots\cdots$$

$$\left(E + \frac{\tau}{2} \Lambda_n^j\right)(\phi^j - \tau f^j) = \left(E - \frac{\tau}{2} \Lambda_n^j\right)\phi^{j-(1/n)},$$

(4.35)

$$\left(E + \frac{\tau}{2} \Lambda_n^j\right)\phi^{j+(1/n)} = \left(E - \frac{\tau}{2} \Lambda_n^j\right)(\phi^j + \tau f^j),$$

$$\cdots\cdots\cdots\cdots\cdots\cdots\cdots\cdots\cdots$$

$$\left(E + \frac{\tau}{2} \Lambda_1^j\right)\phi^{j+1} = \left(E - \frac{\tau}{2} \Lambda_1^j\right)\phi^{j+[(n-1)/n]},$$

where

$$\Lambda_\alpha^j = A_\alpha(t_j).$$

It is not difficult to see that this scheme approximates with second order in τ and is absolutely stable, provided ϕ is sufficiently smooth.

The n-component system of equations (4.35) can be rewritten (as in the case $\alpha = 2$) in the following equivalent form:

$$\left(E + \frac{\tau}{2} \Lambda_\alpha \right) \Phi^{j-[(n+1-\alpha)/(n+1)]} = \left(E - \frac{\tau}{2} \Lambda_\alpha \right) \Phi^{j-[(n+1-\alpha+1)/(n+1)]}$$

$$\alpha = 1, 2, \ldots, n,$$

$$\Phi^{j+[1/(n+1)]} = \Phi^{j-[1/(n+1)]} + 2\tau f^j, \tag{4.36}$$

$$\left(E + \frac{\tau}{2} \Lambda_{n-\alpha+2} \right) \Phi^{j+[\alpha/(n+1)]} = \left(E - \frac{\tau}{2} \Lambda_{n-\alpha+2} \right) \Phi^{j+[(\alpha-1)/(n+1)]}$$

$$\alpha = 2, 3, \ldots, n+1.$$

Let us now discuss the splitting-up method for the backward implicit difference approximations. To this end, consider the problem

$$\frac{\partial \phi}{\partial t} + A\phi = 0 \quad \text{in} \quad D \times D_t,$$

$$\phi = g \quad \text{in} \quad D \quad \text{for} \quad t = 0. \tag{4.37}$$

(For methodological reasons, we consider a homogeneous boundary condition.)

Suppose that $A = \sum_{\alpha=1}^n A_\alpha$, and the $A_\alpha \geq 0$ are time invariant. Take the splitting algorithm in the form

$$\frac{\phi^{j+(1/n)} - \phi^j}{\tau} + A_1 \phi^{j+(1/n)} = 0,$$

$$\cdots\cdots\cdots\cdots\cdots\cdots\cdots\cdots\cdots \tag{4.38}$$

$$\frac{\phi^{j+1} - \phi^{j+[(n-1)/n]}}{\tau} + A_n \phi^{j+1} = 0.$$

We will show that such an algorithm is absolutely stable. Indeed, consider the equation

$$\frac{\phi^{j+(\alpha/n)} - \phi^{j+[(\alpha-1)/n]}}{\tau} + A_\alpha \phi^{j+(\alpha/n)} = 0. \tag{4.39}$$

Take the scalar product of this equation with $\phi^{j+(\alpha/n)}$. Then

$$(\phi^{j+(\alpha/n)} - \phi^{j+[(\alpha-1)/n]}, \phi^{j+(\alpha/n)}) + \tau(A_\alpha \phi^{j+(\alpha/n)}, \phi^{j+(\alpha/n)}) = 0.$$

A_α being positive semidefinite, we have further

$$(\phi^{j+(\alpha/n)} - \phi^{j+[(\alpha-1)/n]}, \phi^{j+(\alpha/n)}) \leq 0$$

or

$$(\phi^{j+(\alpha/n)}, \phi^{j+(\alpha/n)}) \leq (\phi^{j+(\alpha/n)}, \phi^{j+[(\alpha-1)/n]}).$$

But since

$$(\phi^{j+(\alpha/n)}, \phi^{j+[(\alpha-1)/n]}) \leq \tfrac{1}{2}[(\phi^{j+(\alpha/n)}, \phi^{j+(\alpha/n)}) + (\phi^{j+[(\alpha-1)/n]}, \phi^{j+[(\alpha-1)/n]})],$$

the following inequality holds:

$$\|\phi^{j+(\alpha/n)}\|^2 \leq \|\phi^{j+[(\alpha-1)/n]}\|^2, \qquad \alpha = 1, 2, \ldots, n.$$

Hence

$$\|\phi^{j+1}\| \leq \|\phi^j\|. \tag{4.40}$$

In other words, under the assumptions postulated, the computations based on the splitting-up scheme (4.38) will be absolutely stable.

It is a simple matter to verify that system (4.38) represents a first-order approximation in τ of the original problem.

Next, consider the nonhomogeneous problem:

$$\frac{\partial \phi}{\partial t} + A\phi = f \quad \text{in} \quad D \times D_t,$$

$$\phi = g \quad \text{in} \quad D \quad \text{for} \quad t = 0. \tag{4.41}$$

The splitting-up scheme can be taken in the form

$$\frac{\phi^{j+(1/n)} - \phi^j}{\tau} + A_1 \phi^{j+(1/n)} = 0,$$

$$\cdots\cdots\cdots\cdots\cdots\cdots\cdots\cdots\cdots\cdots \tag{4.42}$$

$$\frac{\phi^{j+1} - \phi^{j+[(n-1)/n]}}{\tau} + A_n \phi^{j+1} = f^j.$$

This type of splitting-up scheme approximates the original nonhomogeneous equation up to the first order of accuracy in τ.

Stability is shown as follows: take the scalar products of each of equations (4.42) with $\phi^{j+(1/n)}, \ldots, \phi^{j+1}$ correspondingly. Similarly as before, we obtain

$$\|\phi^{j+(\alpha/n)}\| \leq \|\phi^{j+(\alpha-1/n)}\|, \qquad \alpha = 1, 2, \ldots, n - 1. \tag{4.43}$$

Accounting for $A_n \geq 0$, we have further

$$(\phi^{j+1}, \phi^{j+1}) \leq (\phi^{j+[(n-1)/n]}, \phi^{j+1}) + \tau(f^j, \phi^{j+1}).$$

From the Cauchy–Schwartz inequality

$$|(\phi^{j+[(n-1)/n]}, \phi^{j+1})| \leq \|\phi^{j+[(n-1)/n]}\| \|\phi^{j+1}\|,$$

$$|(f^j, \phi^{j+1})| \leq \|f^j\| \|\phi^{j+1}\|.$$

Hence

$$\|\phi^{j+1}\|^2 \leq \|\phi^{j+[(n-1)/n]}\| \|\phi^{j+1}\| + \tau \|f^j\| \|\phi^{j+1}\|,$$

or, dividing by $\|\phi^{j+1}\|$,

$$\|\phi^{j+1}\| \leq \|\phi^{j+[(n-1)/n]}\| + \tau \|f^j\|.$$

The elimination of fractional indices yields

$$\|\phi^{j+1}\| \leq \|\phi^j\| + \tau\|f^j\|. \qquad (4.44)$$

Since

$$\|\phi^0\| = \|g\|,$$

the elimination of intermediate values of the solution gives

$$\|\phi^{j+1}\| \leq \|g\| + \tau j\|f\|, \qquad (4.45)$$

where

$$\|f\| = \max_j \|f^j\|.$$

The absolute stability of the difference scheme for any instant of the time interval $0 \leq t_j \leq T$ follows from this.

The algorithm above can be generalized to cover the case where the operators A depends on time. This is done by taking a suitable difference approximation of this operator on every subinterval $t_j \leq t \leq t_{j+1}$.

5.4.4 Splitting-Up of Quasi-Linear Problems

Consider the evolution problem

$$\frac{\partial \phi}{\partial t} + A(t, \phi)\phi = 0 \quad \text{in} \quad D \times D_t,$$

$$\phi = g \quad \text{in} \quad D \quad \text{for} \quad t = 0, \qquad (4.46)$$

where the operator A depends both on time and the solution of the problem. Suppose that this operator is a sum of nonnegative operators, that is,

$$A(t, \phi) = \sum_{\alpha=1}^{n} A_\alpha(t, \phi), \qquad (4.47)$$

$A_\alpha(t, \phi) \geq 0$ and is sufficiently smooth. Suppose further than the solution ϕ is also a sufficiently smooth function of time. Take the interval $t_{j-1} \leq t \leq t_{j+1}$ and consider the following splitting-up scheme:

$$\frac{\phi^{j+(1/n)-1} - \phi^{j-1}}{\tau} + A_1^j \frac{\phi^{j+(1/n)-1} + \phi^{j-1}}{2} = 0,$$

$$\cdots\cdots\cdots\cdots\cdots\cdots\cdots\cdots\cdots\cdots\cdots\cdots\cdots$$

$$\frac{\phi^j - \phi^{j-(1/n)}}{\tau} + A_n^j \frac{\phi^j + \phi^{j-(1/n)}}{2} = 0,$$

$$\frac{\phi^{j+(1/n)} - \phi^j}{\tau} + A_n^j \frac{\phi^{j+(1/n)} + \phi^j}{2} = 0, \qquad (4.48)$$

$$\cdots\cdots\cdots\cdots\cdots\cdots\cdots\cdots\cdots\cdots\cdots\cdots\cdots$$

$$\frac{\phi^{j+1} - \phi^{j+[(n-1)/n]}}{\tau} + A_1^j \frac{\phi^{j+1} + \phi^{j+[(n-1)/n]}}{2} = 0,$$

where

$$A^j = A(t_j, \tilde{\phi}^j),$$

$$\tilde{\phi}^j = \phi^{j-1} - \tau A^{j-1}(t_{j-1}, \phi^{j-1})\phi^{j-1}, \tag{4.49}$$

$$\tau = t_j - t_{j-1}.$$

By methods similar to those used above for the linear operators which depend on the time variable only, one can easily prove that the splitting-up scheme (4.48) and (4.49) is a second-order approximation in τ and is absolutely stable. The splitting-up method for the nonhomogeneous quasi-linear equations is derived the same way. This opens up a large variety of possibilities for applying component-by-component splitting to the nonstationary quasi-linear problems of hydrodynamics, meteorology, oceanology, and other significant fields.

5.5 General Approach to Component-by-Component Splitting

Many problems of mathematical physics can often be solved by splitting the original differential (integral, integro-differential) equations into simpler ones and then reducing them further to difference forms by means of the algorithms given in this chapter. This technique is closely related to the *weak approximation* of the original equations by equations with a simpler structure; it has been described by Samarskii [3, 15], Yanenko [3], Demidov [15], Dyakonov [15], Lebedev [15], and subsequently expanded upon by many authors. This problem also will be the subject of our present discussion.

Assume that the following equations describe a certain problem of mathematical physics:

$$\frac{\partial \phi}{dt} + A\phi = 0 \quad \text{in} \quad D \times D_t,$$

$$\phi = g \quad \text{in} \quad D \quad \text{for} \quad t = 0. \tag{5.1}$$

Assume further that

$$A = \sum_{\alpha=1}^{n} A_\alpha, \tag{5.2}$$

with $A_\alpha \geq 0$. The solution ϕ and the function f are assumed to be sufficiently smooth. On each interval $\Theta_j = \{t_j \leq t \leq t_{j+1}\}$ let us now represent (5.1) in the form

$$\frac{\partial \phi_\alpha}{\partial t} + A_\alpha \phi_\alpha = 0 \quad \text{in} \quad D \times \Theta_j,$$

$$\phi_\alpha^j = \phi_{\alpha-1}^{j+1} \quad \text{in} \quad D \quad \text{for} \quad t = t_j, \qquad \alpha = 1, 2, \ldots, n. \tag{5.3}$$

We have used the notation

$$\phi_0^{j+1} = \phi^j, \qquad \phi_n^{j+1} = \phi^{j+1}. \tag{5.4}$$

As has been shown earlier, by applying the Crank–Nicholson scheme to each of the equations, we arrive at the system of difference equations

$$\frac{\phi^{j+(\alpha/n)} - \phi^{j+[(\alpha-1)/n]}}{\tau} + A_\alpha \frac{\phi^{j+(\alpha/n)} + \phi^{j+[(\alpha-1)/n]}}{2} = 0, \tag{5.5}$$

$$\alpha = 1, 2, \ldots, n,$$

where

$$\phi^{j+(\alpha/n)} = \phi_\alpha^{j+1}; \qquad \phi^{j+1} = \phi_n^{j+1}. \tag{5.6}$$

Suppose that each of the operators A_α is itself represented as

$$A_\alpha = \sum_{\beta=1}^{m_\alpha} A_{\alpha\beta}, \tag{5.7}$$

where $A_{\alpha\beta} \geq 0$. At this point one may question the usefulness of splitting the operator A twice in a row. Is it now more straightforward to split the operator into $A_{\alpha\beta}$ right away at the outset? Let us note that disregarding the formal equivalence, it is useful in many cases to decompose first the complicated problem of mathematical physics into simpler problems and then reduce them later in an independent fashion into even simpler ones.

Consider system (5.3), and let us split it further into more elementary problems using (5.7):

$$\frac{\phi_\alpha^{j+(\beta/m_\alpha)} - \phi_\alpha^{j+[(\beta-1)/m_\alpha]}}{\tau} + A_{\alpha\beta} \frac{\phi_\alpha^{j+(\beta/m_\alpha)} + \phi_\alpha^{j+[(\beta-1)/m_\alpha]}}{2} = 0. \tag{5.8}$$

$$\alpha = 1, 2, \ldots, n; \beta = 1, 2, \ldots, m_\alpha,$$

where

$$\phi_1^j = \phi^j; \qquad \phi_\alpha^j = \phi_{\alpha-1}^{j+1}, \qquad \alpha > 1; \qquad \phi_n^{j+1} = \phi^{j+1}.$$

It is not difficult to see that system (5.8) approximates the original problem (5.1) with second-order accuracy in τ, provided the operators $A_{\alpha\beta}$ commute. For proof we rearrange first the components of A and write

$$A = \sum_{\alpha=1}^{n} \sum_{\beta=1}^{m_\alpha} A_{\alpha\beta} = \sum_{\gamma=1}^{p} A_\gamma.$$

Then we obtain

$$\frac{\phi^{j+(\gamma/p)} - \phi^{j+[(\gamma-1)/p]}}{\tau} + A_\gamma \frac{\phi^{j+(\gamma/p)} + \phi^{j+[(\gamma-1)/p]}}{2} = 0, \qquad \gamma = 1, 2, \ldots, p, \tag{5.9}$$

which represent a second-order approximation τ of problem (5.1), as we know from Section 4.4. This result remains true even if $A_{\alpha\beta}$ depend on time, in which case it is necessary to construct a second-order approximation of the operators $A_{\alpha\beta} = \Lambda^j_{\alpha\beta}$ on each of the intervals $t_j \leq t \leq t_{j+1}$. If $\Lambda^j_{\alpha\beta}$ do not commute, then using the two-cycle procedure described in Section 4.4, we obtain a second-order scheme on each $t_{j-1} \leq t \leq t_{j+1}$.

To summarize, the evolution problem of the type of (5.1) with $A_\alpha \geq 0$ can be considered [upon reduction to the particular evolution problems (5.3)] as a set of new independent evolution problems; if at least one of the elementary evolution problems is reduced to the first-order difference scheme then the original problem (5.1) is also approximated with first-order accuracy in τ. If every elementary problem is approximated with a second-order accuracy, then in the framework of the two-cycle procedure in α and β we arrive at a second-order approximation of (5.1). Note that if the operators $A_{\alpha\beta}$ do not commute, we may still obtain a first-order approximation of (5.1) without any reference to the two-cycle procedure. Indeed, in the non-commutative case the original problem becomes

$$\frac{\partial \phi}{\partial t} + \sum_{\alpha=1}^{n} A_\alpha \phi = 0 \quad \text{in} \quad D \times \Theta_j,$$

$$\phi = \phi^i \quad \text{in} \quad D \quad \text{for} \quad t = t_j. \tag{5.10}$$

The above is reduced to the system

$$\frac{\partial \phi_\alpha}{\partial t} + A_\alpha \phi_\alpha = 0, \qquad \phi^j_\alpha = \phi^{j+1}_{\alpha-1}, \qquad \alpha = 1, 2, \ldots, n. \tag{5.11}$$

Let $A_\alpha = \sum_{\beta=1}^{m_\alpha} A_{\alpha\beta}$. For each of problems (5.11) we use the two-cycle method:

$$\frac{\phi^{j+(\beta/2m_\alpha)}_\alpha - \phi^{j+[(\beta-1)/2m_\alpha]}_\alpha}{\tau/2} + A_{\alpha\beta} \frac{\phi^{j+(\beta/2m_\alpha)} + \phi^{j+[(\beta-1)/2m_\alpha]}}{2} = 0,$$

$$\beta = 1, 2, \ldots, m_\alpha, \tag{5.12}$$

$$\frac{\phi^{j+(\beta/2m_\alpha)}_\alpha - \phi^{j+[(\beta-1)/2m_\alpha]}_\alpha}{\tau/2} + A_{\alpha, 2m_\alpha+1-\beta} \frac{\phi^{j+(\beta/2m_\alpha)} + \phi^{j+[(\beta-1)/2m_\alpha]}}{2} = 0.$$

$$\beta = m_\alpha + 1, m_\alpha + 2, \ldots, 2m_\alpha.$$

The initial conditions for (5.12) are correspondingly

$$\phi^j_1 = \phi^j, \qquad \phi^j_\alpha = \phi^{j+1}_{\alpha-1}, \qquad \alpha = 2, \ldots, n. \tag{5.13}$$

It is easy to verify that problem (5.12) approximates any of problems (5.11) with an accuracy up to τ^2 on the interval $t_j \leq t \leq t_{j+1}$.

In order that the overall algorithm yields a solution with an accuracy of τ^2 it is necessary to rearrange the basic cycles in addition. Thus, instead of (5.11) we have to take

$$\frac{\partial \phi_\alpha}{\partial t} + A_\alpha \phi_\alpha = 0, \qquad \alpha = 1, 2, \ldots, n,$$

$$\phi^{j-1}_1 = \phi^{j-1}, \qquad \phi^{j-1}_\alpha = \phi^j_{\alpha-1}, \qquad \alpha > 1, \qquad \phi^j = \phi^j_n \tag{5.14}$$

on $t_{j-1} \le t \le t_j$, and

$$\frac{\partial \phi_\alpha}{\partial t} + A_{n-\alpha+1}\phi_\alpha = 0, \qquad \alpha = 1, 2, \ldots, n,$$

$$\phi_1^j = \phi^j, \qquad \phi_\alpha^j = \phi_{\alpha-1}^{j+1}, \qquad \alpha > 1, \qquad \phi^{j+1} = \phi_n^{j+1} \tag{5.15}$$

on $t_j \le t \le t_{j+1}$. We also assume that each of problems (5.14) and (5.15) is solved by means of the two-cycle method of the form of (5.12). Note that under the condition $A_{\alpha\beta} \ge 0$ the component-by-component splitting-up method is absolutely stable.

In conclusion, let us describe the general splitting-up scheme for the nonhomogeneous equation

$$\frac{\partial \phi}{\partial t} + \sum_{\alpha=1}^{n} A_\alpha \phi = f,$$

$$\phi = g \quad \text{for} \quad t = 0 \tag{5.16}$$

on the interval $t_{j-1} \le t \le t_{j+1}$, based on the two-cycle method. Consider the weak approximation schemes in the differential form.

Let

$$\frac{\partial \phi_\alpha}{\partial t} + A_\alpha \phi_\alpha = 0, \qquad \alpha = 1, 2, \ldots, n-1,$$

$$\frac{\partial \phi_n}{\partial t} + A_n \phi_n = f + \frac{\tau}{2} A_n f \tag{5.17}$$

on $t_{j-1} \le t \le t_j$, and let

$$\frac{\partial \phi_{n+1}}{\partial t} + A_n \phi_{n+1} = f - \frac{\tau}{2} A_n f,$$

$$\frac{\partial \phi_{n+\alpha}}{\partial t} + A_{n-\alpha+1}\phi_{n+\alpha} = 0, \qquad \alpha = 2, 3, \ldots, n \tag{5.18}$$

on $t_j \le t \le t_{j+1}$. Assume that

$$\phi_1(t_{j-1}) = \phi(t_{j-1}), \qquad \phi_{\alpha+1}(t_{j-1}) = \phi_\alpha(t_j), \qquad \alpha = 1, 2, \ldots, n \tag{5.19}$$

and

$$\phi_{\alpha+1}(t_j) = \phi_\alpha(t_{j+1}), \qquad \alpha = n+1, n+2, \ldots, 2n. \tag{5.20}$$

Use now the Crank–Nicholson scheme to solve equations (5.17)–(5.19) on $t_{j-1} \le t \le t_{j+1}$, letting $f = f^j$. We obtain system (4.36).

Together with system (5.17), (5.18), let us consider the following system:

$$\frac{\partial \phi_1}{\partial t} + A_1 \phi_1 = 0,$$

$$\cdots\cdots\cdots\cdots\cdots \tag{5.21}$$

$$\frac{\partial \phi_n}{\partial t} + A_n \phi_n = 0$$

on the interval $t_{j-1} \le t \le t_j$,

$$\frac{\partial \phi_{n+1}}{\partial t} = f \tag{5.22}$$

on $t_{j-1} \leq t \leq t_{j+1}$, and

$$\frac{\partial \phi_{n+2}}{\partial t} + A_n \phi_{n+2} = 0,$$

$$\dots\dots\dots\dots\dots\dots\dots \tag{5.23}$$

$$\frac{\partial \phi_{2n+1}}{\partial t} + A_1 \phi_{2n+1} = 0$$

on the intervals $t_j \leq t \leq t_{j+1}$. The initial conditions for (5.21) are

$$\phi_1(t_{j-1}) = \phi(t_{j-1}), \qquad \phi_\alpha(t_{j-1}) = \phi_{\alpha-1}(t_j), \qquad \alpha = 2, 3, \dots, n, \tag{5.24}$$

for (5.22)

$$\phi_{n+1}(t_{j-1}) = \phi_n(t_j) \tag{5.25}$$

and for (5.23)

$$\phi_\alpha(t_j) = \phi_{\alpha-1}(t_{j+1}), \qquad \alpha = n + 2, n + 3, \dots, 2n + 1. \tag{5.26}$$

The approximation and stability of the schemes obtained guarantee convergence (see Section 1.4).

5.6 Methods of Solving Equations of the Hyperbolic Type

A large class of problems of mathematical physics is related to the equations of hyperbolic type; the origins of the corresponding numerical methods are found in the fundamental contribution by Courant, Friedrichs, and Lewy [7]. A sizable amount of research was later undertaken by Lady-zhenskaya [2], Godunov [2], Samarskii [3], Konovalov [15], and others. Recently a number of effective algorithms for hyperbolic-type problems have been constructed and applied in the theory of oscillations, theory of elasticity, etc., for the case of multi-dimensional regions. For smooth operators, solutions and input data, these algorithms are based on special splitting-up algorithms which will be discussed presently.

5.6.1 The Stabilization Method

Consider the problem

$$\frac{\partial^2 \phi}{\partial t^2} + A\phi = f \quad \text{in} \quad D \times D_t, \tag{6.1}$$

$$\phi = p, \frac{\partial \phi}{\partial t} = q \quad \text{in} \quad D \quad \text{for} \quad t = 0.$$

We will assume that A is a time invariant finite-difference operator and that the functions p and q have properties which guarantee a sufficient smoothness of the solutions. Assume further that the operator A is positive definite; that is,

$$(A\phi, \phi) \geq \gamma^2(\phi, \phi). \tag{6.2}$$

Let us recall that for positive definite operators A we have

$$\gamma^2 = \alpha\left(\frac{A^* + A}{2}\right),$$

where α is the smallest eigenvalue of the operator $(A + A^*)/2$.

Consider the difference approximation of equation (6.1) in the form

$$\frac{\phi^{j+1} - 2\phi^j + \phi^{j-1}}{\tau^2} + A\phi^j = f^j. \tag{6.3}$$

It is not difficult to show that on smooth solutions the difference scheme (6.3) approximates the original equation (6.1) with an accuracy of τ^2. Let us now complement equation (6.3) with the initial data. In order not to destroy the approximation order, consider the initial conditions in the following form

$$\phi^0 = p, \qquad \phi^1 = \left(E - \frac{\tau^2}{2}A\right)p + \tau q + \frac{\tau^2}{2}f^0. \tag{6.4}$$

The second of the above relations has been obtained by expanding the solution of (6.1) into a Taylor series around $t = 0$, and subsequently eliminating the derivatives using the equation and the initial conditions from (6.1).

The problem specification (6.3) and (6.4) is thus complete. At this point it is necessary to investigate the stability of scheme (6.3). To do this, we will use the spectral method.

Let the eigenfunctions u_n and u_n^*, and the real eigenvalues $\lambda_n > 0$ correspond to the spectral problems

$$Au = \lambda u, \qquad A^*u^* = \lambda u^*. \tag{6.5}$$

Suppose further that $\{u_n\}$ form a basis. The solution of the equation will be sought in the form

$$\phi^j = \sum_n \phi_n^j u_n, \tag{6.6}$$

where

$$\phi_n^j = (\phi^j, u_n^*).$$

Substituting the Fourier series (6.6) into (6.3) and subsequently taking the scalar product with u_n^*, we obtain the following equations for the Fourier coefficients:

$$\frac{\phi_n^{j+1} - 2\phi_n^j + \phi_n^{j-1}}{\tau^2} + \lambda_n \phi_n^j = f_n^j. \tag{6.7}$$

The general solution of the homogeneous equation which corresponds to (6.7) will be sought as a power function:

$$\phi_n^j = \eta_n^j. \tag{6.8}$$

(Let us emphasize that the letter j in the left side of the above relation stands for an index, while on the right it indicates the power.) We substitute next (6.8) into (6.7) and take $f_n^j = 0$; the result is the characteristic equation for η_n:

$$\eta_n^2 - 2\left(1 - \frac{\tau^2 \lambda_n}{2}\right)\eta_n + 1 = 0. \tag{6.9}$$

If we assume

$$\left|1 - \frac{\tau^2 \lambda_n}{2}\right| \leq 1, \tag{6.10}$$

it is easy to see that the roots of equation (6.9) are complex conjugate (in the case of the strong inequality) with the amplitude one:

$$|\eta_n| = 1. \tag{6.11}$$

From (6.10) we have

$$\tau^2 \leq \frac{4}{\lambda_n}, \qquad n = 1, 2, \ldots . \tag{6.12}$$

Clearly, (6.12) holds true for all λ_n, if τ is such that

$$\tau \leq \frac{2}{\sqrt{\beta(A)}}, \tag{6.13}$$

where $\beta(A)$ is an upper bound of the spectrum of A. For symmetric operators $\beta(A) = \|A\|$ and, consequently,

$$\tau \leq \frac{2}{\sqrt{\|A\|}}. \tag{6.14}$$

Let us now turn to the implicit difference schemes. Consider

$$\frac{\phi^{j+1} - 2\phi^j + \phi^{j-1}}{\tau^2} + A\frac{\phi^{j+1} + \phi^{j-1}}{2} = f^j. \tag{6.15}$$

The above scheme is of second-order accuracy in τ, and together with (6.4) it approximates problem 6.1) up to the second power in τ. The characteristic equation for (6.15) has the form

$$\eta_n^2 - \frac{2}{1 + \frac{\tau^2 \lambda_n}{2}}\eta_n + 1 = 0, \tag{6.16}$$

and consequently

$$\eta_n = \frac{1}{1 + \frac{\tau^2 \lambda_n}{2}} \pm \sqrt{\left(\frac{1}{1 + \frac{\tau^2 \lambda_n}{2}}\right)^2 - 1}. \tag{6.17}$$

Hence we can see that for any τ an n

$$|\eta_n| = 1. \tag{6.18}$$

Scheme (6.15) is absolutely stable (see Richtmyer and Morton [3]). Consider now the case where

$$A = \sum_{\alpha=1}^{n} A_\alpha, \tag{6.19}$$

with all A_α nonnegative. In order to obtain an approximate solution of (6.1) we exploit in this case the following difference approximation:

$$B \frac{\phi^{j+1} - 2\phi^j + \phi^{j-1}}{\tau^2} + A\phi^j = f^j, \tag{6.20}$$

where

$$B = \prod_{\alpha=1}^{n} \left(E + \frac{\tau^2}{2} A_\alpha\right). \tag{6.21}$$

From (6.20) and (6.21) we have that (6.20) is a second-order approximation of (6.1). Since (6.20) can be written as

$$\frac{\phi^{j+1} - 2\phi^j + \phi^{j-1}}{\tau^2} + B^{-1}A\phi^j = B^{-1}f^j, \tag{6.22}$$

the Fourier analysis will imply the stability of (6.20), (6.21), provided

$$\tau \leq \frac{2}{\sqrt{\beta(B^{-1}A)}}. \tag{6.23}$$

In this fashion the problem of choosing the parameter τ satisfying the stability condition has been reduced to computation of the largest eigenvalue for the problem

$$Au = \lambda Bu, \tag{6.24}$$

under the assumption that all eigenvalues of $B^{-1}A$ are positive. The problem can be solved by the Lyusternik iterative process.

Let us form the realization of the difference scheme corresponding to equation (6.20):

$$\left(E + \frac{\tau^2}{2} A_1\right)\xi^{j+(1/n)} = -A\phi^j + f^j$$

$$\left(E + \frac{\tau^2}{2} A_2\right)\xi^{j+(2/n)} = \xi^{j+(1/n)},$$

$$\cdots\cdots\cdots\cdots\cdots\cdots\cdots\cdots\cdots\cdots\cdots\cdots$$

$$\left(E + \frac{\tau^2}{2} A_n\right)\xi^{j+1} = \xi^{j+[(n-1)/n]},$$

$$\phi^{j+1} = 2\phi^j - \phi^{j-1} + \tau^2\xi^{j+1}.$$

(6.25)

This problem is solved sequentially for $j = 1, 2, \ldots$ using the initial data of (6.4). Scheme (6.20) is a splitting-up scheme.

5.6.2 Reduction of the Wave Equation to an Evolution Problem

Computer-oriented constructions of absolutely stable second-order difference approximations for hyperbolic equations have led eventually to the necessity of developing special splitting-up methods similar to those for evolution problems.

The basic idea behind the formal reduction of a hyperbolic problem to an evolution problem will be illustrated on a simple example of membrane oscillations, with periodic boundary conditions relative the square $D = \{0 \le x \le 1, 0 \le y \le 1\}$:

$$\frac{\partial^2 \phi}{\partial t^2} = \frac{\partial}{\partial x} a^2 \frac{\partial \phi}{\partial x} + \frac{\partial}{\partial y} a^2 \frac{\partial \phi}{\partial y} \quad \text{in} \quad D \times D_t,$$

$$\phi = p, \quad \frac{\partial \phi}{\partial t} = q \quad \text{in} \quad D \quad \text{for} \quad t = 0,$$

(6.26)

where $a^2 = a^2(x, y)$ is the square of the propagation velocity of the disturbances, and $p = p(x, y)$ and $q = q(x, y)$ are given functions. Periodic solutions are known to be sufficiently smooth for our purposes in all arguments x, y, and t.

First of all let us rewrite the wave equation (6.26) as a system of equations. There results

$$\frac{\partial u}{\partial t} - a \frac{\partial \phi}{\partial x} = 0,$$

(6.27)

$$\frac{\partial v}{\partial t} - a \frac{\partial \phi}{\partial y} = 0 \quad \text{in} \quad D \times D_t,$$

$$\frac{\partial \phi}{\partial t} - \left(\frac{\partial au}{\partial x} + \frac{\partial av}{\partial y}\right) = 0.$$

Let us take the following initial data for the functions u, v, and ϕ:

$$u = u^0(x, y), \qquad v = v^0(x, y), \qquad \phi = p(x, y) \quad \text{for} \quad t = 0. \quad (6.28)$$

The functions u^0, v^0 must satisfy the smoothness conditions and also satisfy the relation

$$\frac{\partial au^0}{\partial x} + \frac{\partial av^0}{\partial y} = q(x, y). \tag{6.29}$$

Introduce next the matrix A and the vector ϕ:

$$A = \left\| \begin{array}{ccc} 0 & 0 & -a\dfrac{\partial}{\partial x} \\[2ex] 0 & 0 & -a\dfrac{\partial}{\partial y} \\[2ex] -\dfrac{\partial}{\partial x}a & -\dfrac{\partial}{\partial y}a & 0 \end{array} \right\|, \qquad \phi = \left| \begin{array}{c} u \\ v \\ \phi \end{array} \right|.$$

With this notation we can rewrite system (6.27) and the initial data of (6.28) as follows:

$$\frac{\partial \phi}{\partial t} + A\phi = 0 \quad \text{in} \quad D \times D_t,$$

$$\phi = \phi^0 \quad \text{in} \quad D \quad \text{for} \quad t = 0, \tag{6.30}$$

where ϕ^0 has the components u^0, v^0, and ϕ^0.

In order to investigate the properties of the symmetric operator A, let us form the functional

$$(A\phi, \phi) = -\int_D \left[\frac{\partial}{\partial x}(au\phi) + \frac{\partial}{\partial y}(av\phi) \right] dD = -\int_S au_n\phi \, dS. \tag{6.31}$$

Here u_n is the component of the vector $\mathbf{u} = u\mathbf{i} + v\mathbf{j}$, which is orthogonal to the surface S. Because of the periodicity of a, ϕ (and also the derivatives of the solution), the values of u_n at the boundary points of the square D, which are symmetric relative to its center, are equal in magnitude and of opposite sign. Hence the surface integral in (6.31) becomes zero and we obtain the condition

$$(A\phi, \phi) = 0. \tag{6.32}$$

Note that if we require that the membrane be fixed in a frame ($\phi = 0$) rather than the periodicity conditions, then even in this case (6.32) holds true, as can be seen from (6.31). This remark is correct for an arbitrary region D. Condition (6.32) guarantees the uniqueness of the solution for the problem.

Indeed, taking the scalar product of (6.30) with ϕ and exploiting relation (6.32), we obtain

$$\frac{d}{dt} \|\phi\|^2 = 0, \tag{6.33}$$

where

$$\|\phi\| = \left\{ \int_D (u^2 + v^2 + \phi^2) \, dD \right\}^{1/2}.$$

Assume that

$$u^0 = 0, \qquad v^0 = 0, \qquad \phi^0 = 0.$$

Then

$$\|\phi^0\| = 0. \tag{6.34}$$

Solving equation (6.33) with the initial conditions of (6.34), we obtain

$$\|\phi\| = \|\phi^0\| = 0.$$

This means that

$$u = v = \phi = 0$$

at all times, which proves uniqueness.

Let us now turn to the problem of formulating the splitting-up method for problem (6.30). To this end let us introduce the following matrices:

$$A_1 = \left\| \begin{array}{ccc} 0 & 0 & -a\dfrac{\partial}{\partial x} \\[2mm] 0 & 0 & 0 \\[2mm] -\dfrac{\partial}{\partial x}a & 0 & 0 \end{array} \right\|, \qquad A_2 = \left\| \begin{array}{ccc} 0 & 0 & 0 \\[2mm] 0 & 0 & -a\dfrac{\partial}{\partial y} \\[2mm] 0 & -\dfrac{\partial}{\partial y}a & 0 \end{array} \right\|.$$

Clearly

$$A = A_1 + A_2. \tag{6.35}$$

In analogy with the foregoing one can show that

$$(A_1\phi, \phi) = 0, \qquad (A_2\phi, \phi) = 0. \tag{6.36}$$

Therefore, on each interval $t_j \le t \le t_{j+1}$ problem (6.30) can be solved using one of the splitting-up methods discussed in Section 5.3: either the stabilization method, or the predictor–corrector method, or, finally, the component-by-component splitting-up method. Note at this point that if we consider a multi-dimensional wave equation rather than two dimensions, then it would be desirable to use the component-by-component splitting, since it leads to an absolutely stable, second-order difference approximation under the minimal requirements regarding the definiteness of equation (6.36)-type operators.

Let us now perform the reduction of problem (6.30) on each of the intervals $t_j \leq t \leq t_{j+1}$, for example using the component-by-component splitting-up method. We have

$$
\frac{\phi^{j+1/2} - \phi^j}{\tau} + A_1 \frac{\phi^{j+1/2} + \phi^j}{2} = 0,
$$
$$
\frac{\phi^{j+1} - \phi^{j+1/2}}{\tau} + A_2 \frac{\phi^{j+1} + \phi^{j+1/2}}{2} = 0.
$$
(6.37)

Going back to the scalar form, these equations become

$$
\frac{u^{j+1/2} - u^j}{\tau} = a \frac{\partial}{\partial x} \left(\frac{\phi^{j+1/2} + \phi^j}{2} \right),
$$
$$
\frac{v^{j+1/2} - v^j}{\tau} = 0,
$$
$$
\frac{\phi^{j+1/2} - \phi^j}{\tau} = \frac{\partial}{\partial x} \left(a \frac{u^{j+1/2} + u^j}{2} \right)
$$
(6.38)

and

$$
\frac{u^{j+1} - u^{j+1/2}}{\tau} = 0,
$$
$$
\frac{v^{j+1} - v^{j+1/2}}{\tau} = a \frac{\partial}{\partial y} \left(\frac{\phi^{j+1} + \phi^{j+1/2}}{2} \right),
$$
$$
\frac{\phi^{j+1} - \phi^{j+1/2}}{\tau} = \frac{\partial}{\partial y} \left(a \frac{v^{j+1} + v^{j+1/2}}{2} \right).
$$
(6.39)

Taking into account that $v^{j+1/2} = v^j$ and $u^{j+1/2} = u^{j+1}$, system (6.38) can be somewhat simplified by writing

$$
\frac{u^{j+1} - u^j}{\tau} = a \frac{\partial}{\partial x} \left(\frac{\phi^{j+1/2} + \phi^j}{2} \right),
$$
(6.40)

$$
\frac{\phi^{j+1/2} - \phi^j}{\tau} = \frac{\partial}{\partial x} \left(a \frac{u^{j+1} + u^j}{2} \right),
$$
$$
\frac{v^{j+1} - v^j}{\tau} = a \frac{\partial}{\partial y} \left(\frac{\phi^{j+1} + \phi^{j+1/2}}{2} \right),
$$
$$
\frac{\phi^{j+1} - \phi^{j+1/2}}{\tau} = \frac{\partial}{\partial y} \left(a \frac{v^{j+1} + v^j}{2} \right).
$$
(6.41)

System (6.40) is solved for u^{j+1} and $\phi^{j+1/2}$, and (6.41) for v^{j+1} and ϕ^{j+1}.

Difference approximations with respect to x and y will be chosen so as to obtain eventually the absolutely stable schemes for u, v, and ϕ, which will be of second-order accuracy and which will preserve conditions (6.36) for the finite-difference representations of A_1, A_2. Put

$$\frac{u_{k,l}^{j+1} - u_{k,l}^{j}}{\tau} = \frac{a_{k,l}}{h}\left[\left(\frac{\phi^{j+1/2} + \phi^{j}}{2}\right)_{k,l} - \left(\frac{\phi^{j+1/2} + \phi^{j}}{2}\right)_{k-1,l}\right],$$

$$\frac{\phi_{k,l}^{j+1/2} - \phi_{k,l}^{j}}{\tau} = \frac{1}{h}\left[a_{k+1,l}\left(\frac{u^{j+1} + u^{j}}{2}\right)_{k+1,l} - a_{k,l}\left(\frac{u^{j+1} + u^{j}}{2}\right)_{k,l}\right]. \tag{6.42}$$

In the first of the above equations we use a *backward* difference, while a *forward* difference is used in the second one. Similarly for (6.41)

$$\frac{v_{k,l}^{j+1} - v_{k,l}^{j}}{\tau} = \frac{a_{k,l}}{h}\left[\left(\frac{\phi^{j+1} + \phi^{j+1/2}}{2}\right)_{k,l} - \left(\frac{\phi^{j+1} + \phi^{j+1/2}}{2}\right)_{k,l-1}\right],$$

$$\frac{\phi_{k,l}^{j+1} - \phi_{k,l}^{j+1/2}}{\tau} = \frac{1}{h}\left[a_{k,l+1}\left(\frac{v^{j+1} + v^{j}}{2}\right)_{k,l+1} - a_{k,l}\left(\frac{v^{j+1} + v^{j}}{2}\right)_{k,l}\right]. \tag{6.43}$$

The indices k and l in (6.42) and (6.43) run through positive integers up to $N - 1$.

Consider the case where, for instance, we have the following condition on the boundary of the region D:

$$\phi = 0 \quad \text{on} \quad \partial D \times D_t. \tag{6.44}$$

Let us project (6.44) on the $\partial D_h \times D_\tau$; we have

$$\phi_{0,l}^{j} = \phi_{N,l}^{j} = 0 \quad \text{and} \quad \phi_{k,0}^{j} = \phi_{k,N}^{j} = 0. \tag{6.45}$$

both for integer and fractional indices j.

Note that (6.42), (6.43) together with (6.45) can be used to compute $u_{k,l}^{j+1}$ and $v_{k,l}^{j+1}$ at all boundary net points, provided $\phi_{k,l}^{j}$, $\phi_{k,l}^{j+1/2}$, and $\phi_{k,l}^{j+1}$ are known.

Next, let us eliminate the variables $u_{k,l}^{j+1}$ and $v_{k,l}^{j+1}$ from (6.42) and (6.43) to obtain the difference equations for $\phi_{k,l}$. To this end, let us introduce the auxiliary variables

$$\phi_{k,l}^{j+1/4} = \tfrac{1}{2}(\phi_{k,l}^{j+1/2} + \phi_{k,l}^{j}), \qquad \phi_{k,l}^{j+3/4} = \tfrac{1}{2}(\phi_{k,l}^{j+1} + \phi_{k,l}^{j+1/2}). \tag{6.46}$$

Equations (4.42) and (6.43) can now be written in the form

$$\mu_{k+1,l}^{2}(\phi_{k+1,l}^{j+1/4} - \phi_{k,l}^{j+1/4}) - \mu_{k,l}^{2}(\phi_{k,l}^{j+1/4} - \phi_{k-1,l}^{j+1/4}) - \phi_{k,l}^{j+1/4} = -f_{k,l}^{j+1/4},$$

$$\mu_{k,l+1}^{2}(\phi_{k,l+1}^{j+3/4} - \phi_{k,l}^{j+3/4}) - \mu_{k,l}^{2}(\phi_{k,l}^{j+3/4} - \phi_{k,l-1}^{j+3/4}) - \phi_{k,l}^{j+3/4} = -f_{k,l}^{j+3/4},$$

$$\tag{6.47}$$

where

$$\mu_{k,l} = \frac{\tau a_{k,l}}{2h}$$

$$f_{k,l}^{j+1/4} = 2\phi_{k,l}^{j} + 2(\mu_{k+1,l}u_{k+1,l}^{j} - \mu_{k,l}u_{k,l}^{j}), \tag{6.48}$$

$$f_{k,l}^{j+3/4} = 2\phi_{k,l}^{j+1/2} + 2(\mu_{k,l+1}v_{k,l+1}^{k} - \mu_{k,l}v_{k,l}^{j}).$$

We have thus obtained the following algorithm for the numerical solution of (6.30).

First, determine the initial fields of functions $u_{k,l}^{0}$, $v_{k,l}^{0}$, and $\phi_{k,l}^{0}$, with $u_{k,l}^{0}$, $v_{k,l}^{0}$ satisfying the discrete analog of condition (6.29). Then using the first of the formulas of (6.48), find $f_{k,l}^{j+1/4}$ and solve the first of the difference equations (6.47) under condition (6.45) on the boundary of D^{h}. Next compute

$$\phi_{k,l}^{j+1/2} = 2\phi_{k,l}^{j+1/4} - \phi_{k,l}^{j},$$

where $\phi_{k,l}^{j+1/4}$ are found with the help of relations (6.46). Solve the second equation of equations (6.47) using condition (6.45). After this exploit (6.46) to obtain

$$\phi_{k,l}^{j+1} = 2\phi_{k,l}^{j+3/4} - \phi_{k,l}^{j+1/2}.$$

The quantities $\phi_{k,l}^{j+1}$ are further used for finding $u_{k,l}^{j+1}$ and $v_{k,l}^{j+1}$, with the help of the first relations from (6.42) and (6.43). Thus the algorithm is complete.

Finally, it is to be noted that this method is quite easily generalized to more complicated equations of hyperbolic type; usually it leads to absolutely stable, second-order approximation schemes with minimal requirements on the operators A_{α}.

CHAPTER 6

Richardson's Method for Increasing the Accuracy of Approximate Solutions

In the preceding chapters we have, for the most part, considered convergent difference schemes. In principle, these can be used to find the solution of a differential equation to any desired degree of precision, by choosing a sufficiently small mesh size. But then the size of the approximate problem and the cost of solving it both increase, especially for multi-dimensional problems. In many cases the essential limiting factors governing the accuracy of the approximate solution are the memory capacity and speed of the electronic computer.

The construction of high-accuracy approximations is therefore an urgent task for numerical mathematics. Several approaches are known; the most widely applied are the difference and variational-difference schemes of high order of accuracy, and the method of increasing the precision of the approximate solution via sequences of nets.

In this chapter we shall pay primary attention to the latter approach, which derives conceptually from Richardson [4], and was called by him *extrapolation to the limit*. The method consists in using sequences of nets and corresponding approximations (all of the same type), to construct solutions of a given order of accuracy. This approach allows us to limit ourselves to standard difference schemes of first- and second-order accuracy in our computations. Through the efforts of both Soviet and foreign mathematicians, extrapolatation to the limit may be validly applied to a wide variety of problems, including the nonlinear (cf. Volkov [4], Krylov, Bobkov, and Monastyrnyĭ [4]; Kuznetsov and Shaĭdurov [4]; Marchuk and Shaĭdurov [4]; Shaĭdurov [4]; Brezinski [4]; Joyce [4]; and others). We have chosen the simplest from among the variety of these problems for our exposition.

6.1 Ordinary First-Order Differential Equations

We shall make a detailed examination of the effect of Richardson's extrapolation with respect to the mesh size of the difference net, using a very simple
linear equation.

Let $u(t)$ be the solution of the differential equation

$$u' + a(t)u = f(t), \qquad t \in (0, 1), \tag{1.1}$$

with the initial condition

$$u(0) = u_0. \tag{1.2}$$

We assume that

$$a(t) \geq 0, \qquad t \in (0, 1), \tag{1.3}$$

and that all functions are sufficiently smooth to validate the later computations.

We subdivide the interval $[0, 1]$ uniformly by the "integer" vertices

$$t_j = j\tau, \qquad j = 0, 1, \ldots, M \tag{1.4}$$

(M is an integer) with mesh size $\tau = 1/M$ and with the intermediate net
points

$$t_{j+1/2} = (j + \tfrac{1}{2})\tau, \qquad j = 0, 1, \ldots, M - 1. \tag{1.5}$$

Following the Crank–Nicolson scheme (cf. Section 5.2) we replace the
original differential equation (1.1) at the intermediate net points by the
approximate system of algebraic equations

$$\frac{u^{j+1} - u^j}{\tau} + \alpha^{j+1/2} \frac{u^{j+1} + u^j}{2} = f^{j+1/2}, \qquad j = 0, 1, \ldots, M - 1. \tag{1.6}$$

If we adjoin the initial condition

$$u^0 = u_0, \tag{1.7}$$

all the u^j may be recursively defined:

$$u^j = \left(1 + \frac{\tau}{2} a^{j+1/2}\right)^{-1} \left[\left(1 - \frac{\tau}{2} a^{j-1/2}\right) u^{j-1} + \frac{\tau}{2} f^{j-1/2}\right],$$

$$j = 1, 2, \ldots, M. \tag{1.8}$$

As we have often remarked, the difference problem (1.6), (1.7) has a second-
order approximation. If we replace the positive definite matrix Λ^j by the
positive number $a^{j+1/2}$ in the discussion given in Section 5.2, we find the
stability estimate

$$|u^j| \leq |u_0| + \max_{0 \leq j \leq M-1} |f^{j+1/2}|. \tag{1.9}$$

Therefore, according to (1.4), the solution of the difference problem is a second-order approximation of the differential equation:

$$\max_{0 \le j \le M} |u^j - (u)^j| \le C_1 \tau^2, \tag{1.10}$$

where C_1 is a constant independent of τ.

Let us exemplify these theoretical estimates by some practical calculations on an electronic computer for the equation

$$u' + tu = t, \qquad t \in (0, 1), \tag{1.11}$$

with the initial condition

$$u(0) = 2. \tag{1.12}$$

It is easily verified that the exact solution of (1.1), (1.12) is the function

$$u(t) = e^{-t^2/2} + 1. \tag{1.13}$$

Our numerical experiment consisted of the construction of an approximate solution of the problem (1.4)–(1.8) for values of M equal successively to 10, 20, 50, 100, and 200, and in the determination of the quantities

$$\xi = \max_{0 \le j \le M} |u^j - (u)^j|. \tag{1.14}$$

A graphical representation of the dependence of ξ on M, as determined by the computations, is shown in Figure 6.1, in logarithmic coordinates.

As is easily seen from Figure 6.1, the numerical experiments support the estimate (1.10). Further comparison of the computed results with the slope of the theoretical line shows that (1.10) cannot be improved, i.e., that

$$\xi \ge C_2 \tau^2 \tag{1.15}$$

for some positive constant C_2 independent of τ. Thus the maximum approximation error (1.14) is of order τ^γ, with the exponent $\gamma = 2$.

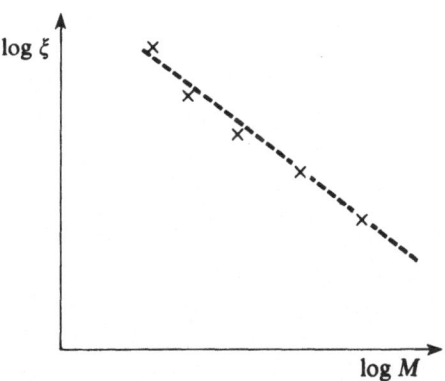

Figure 6.1 Approximation accuracy as a function of mesh size (logarithmic coordinates).

In 1910, Richardson made detailed observations of the approximation error at each point of the difference net, rather than the maximum error, and derived the following hypothesis on the behavior of the approximation error as $\tau \to 0$:

$$u^j - (u)^j = \tau^2(v)^j + \eta^j, \qquad j = 0, 1, \ldots, M. \tag{1.16}$$

The principal part of the error $\tau^2(v)^j$ is the product of τ^2, and the value on the net points of a function $v(t)$ independent of τ; the remainder terms η^j are for all $j = 0, 1, \ldots, M$ of the order $O(\tau^4)$.

In this section we shall establish this fact for the problem (1.1)–(1.3).

The justification is normally carried out in three stages. First, some necessary conditions for the satisfaction of (1.16) are established; next a boundary problem with the solution $v(t)$ is constructed from these; and finally the boundedness of the net functions η^j/τ^4 is seen to be proved.

Thus, if (1.16) is satisfied, we may write

$$u^j = (u)^j + \tau^2(v)^j + \cdots, \qquad j = 0, 1, \ldots, M, \tag{1.17}$$

where it is assumed that the remainder term is of order $O(\tau^4)$. We fix j and putting the value of u^j into (1.6) we obtain the equation

$$\frac{(u)^{j+1} - (u)^j}{\tau} + a^{j+1/2} \frac{(u)^{j+1} + (u)^j}{2}$$

$$+ \tau^2 \left(\frac{(v)^{j+1} - (v)^j}{\tau} + a^{j+1/2} \frac{(v)^{j+1} + (v)^j}{2} \right) + \cdots = f^{j+1/2}.$$

Using the Taylor expansion for the functions u and v,

$$(u)^{j+1/2 \pm 1/2} = \left(u \pm \frac{\tau}{2} u' + \frac{\tau^2}{8} u'' \pm \frac{\tau^3}{48} u''' \right)^{j+1/2} + \cdots,$$

$$(v)^{j+1/2 \pm 1/2} = \left(v \pm \frac{\tau}{2} v' \right)^{j+1/2} + \cdots, \tag{1.18}$$

we reduce each term to its value at the net point $j + \tfrac{1}{2}$:

$$(u' + au + \tau^2(\tfrac{1}{24} u''' + \tfrac{1}{8} au'') + \tau^2(v' + av))^{j+1/2} + \cdots = f^{j+1/2}.$$

Quantities of the order $O(\tau^0)$ are excluded because (1.1) is satisfied at the net point $j + \tfrac{1}{2}$. Therefore, if the above equation is to hold as $\tau \to 0$, we must have

$$(v' + av)^{j+1/2} = (-\tfrac{1}{24} u''' - \tfrac{1}{8} au'')^{j+1/2}, \qquad j = 0, 1, \ldots, M - 1. \tag{1.19}$$

We examine (1.17) separately for $j = 0$:

$$u^0 = u_0 + \tau^2(v)^0 + \cdots . \tag{1.20}$$

The initial conditions (1.2) and (1.7) imply that $u^0 = u_0$; hence as $\tau \to 0$ we arrive at the condition for v^0:

$$(v)^0 = 0. \tag{1.21}$$

This completes the construction of the necessary conditions arising from (1.17). We now require more in determining the function $v(t)$: let (1.19) be satisfied not only on the net points, but also on the whole interval $(0, 1)$, i.e., we require that

$$v'(t) + a(t)v(t) = -\tfrac{1}{24}u'''(t) - \tfrac{1}{8}au''(t), \qquad t \in (0, 1). \tag{1.22}$$

If we adjoin to this equation for $v(t)$ the initial condition

$$v(0) = 0, \tag{1.23}$$

which arises from (1.21), we shall have a differential problem, which has a unique sufficiently smooth solution $v(t)$. It is obvious that v is independent of τ.

Let us now show that the net function η, defined by the $M + 1$ relations

$$\eta^j = u^j - (u)^j - \tau^2(v)^j, \tag{1.24}$$

has values of the order $O(\tau^4)$.

For this purpose, we fix j and substitute the function η in the difference operator of the problem (1.6):

$$\frac{\eta^{j+1} - \eta^j}{\tau} + a^{j+1/2}\frac{\eta^{j+1} + \eta^j}{2} = \frac{u^{j+1} - u^j}{\tau} + a^{j+1/2}\frac{u^{j+1} + u^j}{2}$$
$$- \left(\frac{(u)^{j+1} - (u)^j}{\tau} + a^{j+1/2}\frac{(u)^{j+1} + (u)^j}{2}\right)$$
$$- \tau^2\left(\frac{(v)^{j+1} - (v)^j}{\tau} + a^{j+1/2}\frac{(v)^{j+1} + (v)^j}{2}\right). \tag{1.25}$$

We transform the right-hand side as follows: we find from (1.6) that the first two terms are equal to $f^{j+1/2}$. We reduce all the succeeding terms to their values at the net point $j + \tfrac{1}{2}$, using the expansion (1.18) with remainder terms in Lagrange's form:

$$\frac{\eta^{j+1} - \eta^j}{\tau} + a^{j+1/2}\frac{\eta^{j+1} + \eta^j}{2}$$
$$= (f - u' - au - \tau^2(\tfrac{1}{24}u''' + \tfrac{1}{8}au'') - \tau^2(v' + av))^{j+1/2}$$
$$- \tau^4\left(\frac{1}{16 \cdot 5!}u^V(\xi_j^1) + \frac{1}{16 \cdot 4!}a^{j+1/2}u^{IV}(\xi_j^2)\right)$$
$$- \tau^4(\tfrac{1}{24}v'''(\xi_j^3) + \tfrac{1}{8}a^{j+1/2}v''(\xi_j^1)),$$

where the ξ_j^i are points in the interval (t_j, t_{j+1}). This equation may be simplified by noting that u is a solution of (1.1) and v is a solution of (1.22):

$$\frac{\eta^{j+1} - \eta^j}{\tau} + a^{j+1/2}\frac{\eta^{j+1} + \eta^j}{2} = -\tau^4\left(\frac{1}{16 \cdot 5!}u^V(\xi_j^1) + \frac{1}{16 \cdot 4!}a^{j+1/2}u^{IV}(\xi_j^2)\right.$$
$$\left. + \tfrac{1}{24}v'''(\xi_j^3) + \tfrac{1}{8}a^{j+1/2}v''(\xi_j^4)\right). \tag{1.26}$$

If on the interval $[0, 1]$ the functions u and v have continuous derivatives appearing in (1.26), the right-hand side of (1.26) is bounded in absolute value by $C_3 \tau^4$, where C_3 is a constant independent of τ. This holds for all $j = 0, 1, \ldots, M$; using it and the estimate (1.9) we obtain the estimate

$$|\eta^j| \leq |\eta^0| + C_3 \tau^4.$$

It follows from (1.21), (1.7), and (1.23) that the definition of η^j implies $\eta^0 = 0$. Therefore,

$$|\eta^j| \leq C_3 \tau^4, \qquad j = 0, 1, \ldots, M, \tag{1.27}$$

and the hypothesis (1.16) is verified.

We may explain the method for increasing the approximation accuracy on the basis of this expansion.

Let u_1 and u_2 be solutions of the two problems (1.6), (1.7) with mesh sizes τ and $\tau/2$, respectively. Notwithstanding the fact that the precision of both solutions is of order $O(\tau^2)$ we may quite easily construct from them a solution with fourth-order precision on a net with mesh size τ. Let $t_j = \tau_i$ be an arbitrary point of the difference net with mesh size τ; then the linear combination

$$\bar{u}^j = \tfrac{4}{3}u_\tau^j - \tfrac{1}{3}u_{\tau/2}^{2j} \tag{1.28}$$

approximates the exact solution with fourth-order precision with respect to τ:

$$|\bar{u}^j - u(j\tau)| \leq C_4 \tau^4, \qquad j = 0, 1, \ldots, M, \tag{1.29}$$

with a constant C_4 independent of τ.

Let us prove this assertion. At the point j, the expansions (1.16) are valid for both approximate solutions

$$u_\tau^j = u(j\tau) + \tau^2 v(j\tau) + \eta_\tau^j,$$

$$u_{\tau/2}^{2j} = u(j\tau) + \frac{\tau^2}{4} v(j\tau) + \eta_{\tau/2}^{2j}.$$

Therefore, we have from (1.28)

$$\bar{u}^j = (\tfrac{4}{3} - \tfrac{1}{3})u(j\tau) + \tau^2(\tfrac{4}{3} \cdot \tfrac{1}{4} - \tfrac{1}{3})v(j\tau) + (\tfrac{4}{3}\eta_{\tau/2}^{2j} - \tfrac{1}{3}\eta_\tau^j).$$

Hence

$$\bar{u}^j = u(j\tau) + \tfrac{4}{3}\eta_{\tau/2}^{2j} - \tfrac{1}{3}\eta_\tau^j.$$

Taking account of the estimate (1.27) for the function η^j we obtain the inequality

$$\left|\tfrac{4}{3}\eta_{\tau/2}^{2j} - \tfrac{1}{3}\eta_\tau^j\right| \leq \tfrac{4}{3} \cdot C_3 \frac{\tau^4}{16} + C_3 \frac{\tau^4}{3} = \tfrac{5}{12}C_3 \tau^4,$$

whence (1.29) follows with $C_4 = \tfrac{5}{12}C_3$.

Note that our approach may be extended to the case of nonuniform nets, at the cost of a more complicated analysis.

We may generalize further by increasing the number of terms in (1.16), obtaining approximate solutions of precision greater than the fourth order. This assertion follows from the general theorems proved in the next section.

6.2 General Results

In the preceding section we constructed an example of the increase in approximation accuracy for a specific problem. We now present an abstract version of the sufficient conditions for an increase in the accuracy of a difference equation for a wide variety of problems.

6.2.1 The Decomposition Theorem

Let D be a bounded domain in the n-dimensional Euclidean space E_n, let $\bar{D}$ be its closure, and let ∂D be a portion of its boundary or the entire boundary. We consider the problem in mathematical physics

$$A\phi = f \quad \text{in} \quad D,$$
$$a\phi = g \quad \text{on} \quad \partial D,$$
(2.1)

where A and a are linear operators, and f, g, and ϕ are functions defined on D, ∂D, and $\bar{D}$, respectively. We make the natural assumption that this problem has a unique solution in the class of sufficiently smooth functions. In many problems of mathematical physics the conditions imposed on the right-hand sides imply the existence, uniqueness, and a certain smoothness of the solution. They are normally formulated in the following way.

Let F^k, G^k, and Φ^k be classes of functions defined, respectively, on D, ∂D, and $\bar{D}$, where k is some integer index. Then, in general, the following condition refers to the domain of definition of the problem and to the coefficients in equation (2.1), and guarantees the existence of a solution if the right-hand sides are smooth.

Condition A. For an arbitrary integer $k = 0, \ldots, m$, and an arbitrary pair of functions $f \in F^k$, $g \in G^k$, there exists a unique solution $\phi \in \Phi^k$ of the problem (2.1).

For instance, in the problem (1.1), (1.2) this notation takes the following form:

$$D = (0, 1), \quad \partial D = 0, \quad \bar{D} = [0, 1],$$
$$F^k = C^{2k+2}[0, 1], \quad \Phi^k = C^{2k+3}[0, 1],$$

and G^k coincides for arbitrary k with the set of real numbers $(-\infty, \infty)$. Then, for sufficiently smooth $a(t)$—for example, $a \in C^{2m+2}[0, 1]$—Condition A is obvious.

For a numerical solution of (2.1) we introduce a difference net $\bar{D}_h \in \bar{D}$ with a variable parameter h which may, in principle, be as small as we please. We replace the differential problem by a finite-difference system of (algebraic) equations defined on the net points of finite subsets $D_h \in D$ and $\partial D_h \in \partial D$. We seek an approximate solution in the space of net functions defined on the net points of the net $\bar{D}_h \in \bar{D}_h$. We have

$$A^h \phi^h = f \quad \text{on} \quad D_h,$$
$$a^h \phi^h = g \quad \text{on} \quad \partial D_h. \tag{2.2}$$

Here A^h and a^h are linear algebraic operators, and ϕ^h is a net function approximating the solution ϕ of the original differential equation at the net points $\bar{D}_h$. In the linear spaces Φ_h, F_h, and G_h of net functions defined on $\bar{D}_h$, D_h, and ∂D_h, we introduce the norms $\|\cdot\|_{\Phi_h}$, $\|\cdot\|_{F_h}$, and $\|\cdot\|_{G_h}$, respectively, and we require that the problem (2.2) must be stable.

Condition B. If the net function $u^h \in \Phi_h$ is a solution of the problem

$$A^h u^h = f^h \quad \text{on} \quad D_h,$$
$$a^h u^h = g^h \quad \text{on} \quad D_h, \tag{2.3}$$

where $f^h \in F_h$, $g^h \in G_h$, then

$$\|u^h\|_{\Phi_h} \leq C_1 \|f^h\|_{F_h} + C_2 \|g^h\|_{G_h}, \tag{2.4}$$

with constants not depending on h, f^h, and g^h.

Let us illustrate this condition for the sample problem giving in Section 6.1. We introduce the norms

$$\|u\|_{\Phi_h} = \max_{0 \leq j \leq M} |u^j|,$$
$$\|f\|_{F_h} = \max_{0 \leq j \leq M-1} |f^{j+1/2}|,$$
$$\|g\|_{G_h} = |g^0|.$$

Then the inequality (1.9) transcribes as (2.4), with $C_1 = C_2 = 1$. Then, Condition B is satisfied, i.e., the problem (1.6), (1.7) is stable.

One more condition bears on the approximation of differential operators by difference relations.

Condition C. The expansion

$$A^h u = Au + \sum_{j=1}^{k} h^j B_j + \sigma^h \quad \text{on} \quad D_h,$$
$$a^h u = au + \sum_{j=1}^{k} h^j b_j + \rho^h \quad \text{on} \quad \partial D_h, \tag{2.5}$$

holds for all $u \in \Phi^k$, $0 \le k \le m$, where B_j, b_j do not depend on h; $B_j \in F^{k-j}$, $b_j \in G^{k-j}$; and the following estimate holds for the remainder terms

$$\|\sigma^h\|_{F_h} \le C_3 h^{k+\beta}, \qquad \|\rho^h\|_{G_h} \le C_4 h^{k+\beta}, \tag{2.6}$$

where the constants C_3, C_4 do not depend on h, and $\beta > 0$ is independent of h, k, and u.

In the case $k = 0$ the upper bound of the sums in (2.5) is less than the lower bound, and both may be taken equal to zero; therefore, (2.5) becomes the condition that the operators A, a should be approximated by the difference operators A^h. a^h to the order β.

We note that Condition C is, as a rule, satisfied for the standard difference schemes. This is easily proved by the use of Taylor's theorem. Suppose, for instance, that $u \in C^{k+2}[0, 1]$. Then we have for the first difference the expansions

$$\frac{u(x + h) - u(x)}{h} = u'(x) + \sum_{j=1}^{k} h^j \frac{u^{(j+1)}(x)}{(j + 1)!} + \sigma_1^h(x),$$

$$\frac{u(x) - u(x - h)}{h} = u'(x) + \sum_{j=1}^{k} h^j \frac{(-1)^j u^{(j+1)}(x)}{(j + 1)!} + \sigma_2^h(x), \tag{2.7}$$

where

$$|\sigma_i(x)| \le \frac{h^{k+1}}{(k + 2)!} \max_{x \in [0, 1]} |u^{(k+2)}(x)|. \tag{2.8}$$

For the central difference the corresponding expansion contains only even powers of h:

$$\frac{u(x + h/2) - u(x - h/2)}{h} = u'(x) + \sum_{j=1}^{[k/2]} h^{2j} \frac{u^{(2j+1)}(x)}{(2j + 1)!4^j} + \sigma_3^h(x), \tag{2.9}$$

where

$$|\sigma_3^h(x)| \le \frac{h^{k+1}}{(k + 2)!2^{k+1}} \max_{x \in [0, 1]} |u^{k+2}(x)|.$$

Let us prove the following theorem.

Suppose that Conditions A, B, and C are satisfied for the problem (2.1), (2.2) and that $f \in F^m$, $g \in G^m$. Then for the difference solution ϕ^h we have the expansion

$$\phi^h = \phi + \sum_{k=1}^{m} h^k v_k + \eta^h \quad \text{on} \quad \bar{D}_h. \tag{2.10}$$

Here the functions v_k do not depend on h; $v_k \in \Phi^{m-k}$, and the estimate

$$\|\eta^h\|_{\Phi_h} \le C_5 h^{m+\beta}, \tag{2.11}$$

holds for the remainder term η^h, where the constant C_5 does not depend on h.

Let us now consider an arbitrary set of h independent functions $v_j \in \Phi^{m-j}$ $(j = 1, \ldots, m)$. Corresponding to these functions and two solutions ϕ and ϕ^h we define the net function

$$\eta^h = \phi^h - \phi - \sum_{j=1}^{m} h^j v_j \quad \text{on} \quad \bar{D}_h. \tag{2.12}$$

We solve for ϕ^h and substitute in (2.2), obtaining

$$A^h \phi + \sum_{j=1}^{m} h^j A^h v_j + A^h \eta^h = f \quad \text{on} \quad D_h,$$

$$a^h \phi + \sum_{j=1}^{n} h^j a^h v_j + a^h \eta^h = g \quad \text{on} \quad \partial D_h. \tag{2.13}$$

In accordance with Condition C we may write the expanded forms:

$$A^h \phi = f + \sum_{i=1}^{m} h^i B_{0,i} + \sigma_0^h \quad \text{on} \quad D_h,$$

$$a^h \phi = g + \sum_{i=1}^{m} h^i b_{0,i} + \rho_0^h \quad \text{on} \quad \partial D_h \tag{2.14}$$

and

$$A^h v_j = A v_j + \sum_{i=1}^{m-j} h^i B_{j,i} + \sigma_j^h \quad \text{on} \quad D_h,$$

$$a^h v_j = a v_j + \sum_{i=1}^{m-j} h^i b_{j,i} + \rho_j^h \quad \text{on} \quad \partial D_h \tag{2.15}$$

Here

$$B_{j,i} \in F^{m-j-i} \qquad b_{j,i} \in G^{m-j-i} \tag{2.16}$$

$B_{j,i}$ and $b_{j,i}$ are independent of h, and the remainder terms satisfy the inequalities

$$\|\sigma_j^h\|_{F_h} \le C_{j,1} h^{m-j+\beta}, \qquad \|\rho_j^h\|_{G_h} \le C_{j,2} h^{m-j+\beta}, \tag{2.17}$$

with constants $C_{j,1}$ and $C_{j,2}$ independent of h. Using the expansions (2.14), (2.15) we transform the equations (2.13) to the following form:

$$f + \sum_{j=1}^{m} h^j A v_j + \sum_{j=0}^{m} h^j \sum_{i=1}^{m-j} h^i B_{j,i} + \sum_{j=0}^{m} h^j \sigma_j^h + A^h \eta^h = f \quad \text{on} \quad D_h,$$

$$g + \sum_{j=1}^{m} h^j a v_j + \sum_{j=0}^{m} h^j \sum_{i=1}^{m-j} h^i b_{j,i} + \sum_{j=0}^{m} h^j \rho_j^h + a^h \eta^h = g \quad \text{on} \quad \partial D_h. \tag{2.18}$$

Writing

$$\xi^h = \sum_{j=0}^{m} h^j \sigma_j^h \qquad \zeta^h = \sum_{j=0}^{m} h^j \rho_j^h$$

and using the estimates (2.17) we find:

$$\|\xi^h\|_{F_h} \leq h^{m+\beta} \cdot \tilde{C}_1, \qquad \|\zeta\|_{G_h} \leq h^{m+\beta} \cdot \tilde{C}_2, \qquad (2.19)$$

where

$$\tilde{C}_1 = \sum_{j=0}^{m} C_{j,1}, \qquad \tilde{C}_2 = \sum_{j=0}^{m} C_{j,2}.$$

By a few simple transformations, and using the notation just introduced, we may reduce the relations (2.18) to the following form:

$$\sum_{j=1}^{m} h_j \left(Av_j + \sum_{i=1}^{j} B_{j-i,i} \right) + \xi^h + A^h \eta^h = 0 \quad \text{on} \quad D_h,$$

$$\sum_{j=1}^{m} h_j \left(av_j + \sum_{i=1}^{j} b_{j-i,i} \right) + \zeta^h + a^h \eta^h = 0 \quad \text{and} \quad \partial D_h. \qquad (2.20)$$

Thus, for an arbitrary choice of the functions $v_j \in \Phi^{m-j}$ and for the functions η^h defined by (2.12), we have derived equations (2.20) with remainder terms ξ^h and ζ^h satisfying the bounds (2.19).

We now choose the functions v_j $(j = 1, \ldots, m)$ to be solutions of the differential problem

$$Av_j = - \sum_{i=1}^{j} B_{j-i,i} \quad \text{on} \quad D,$$

$$av_j = - \sum_{i=1}^{j} b_{j-i,i} \quad \text{on} \quad \partial D. \qquad (2.21)$$

For instance, the function v_1 is found by solving the problem

$$Av_1 = -B_{0,1} \quad \text{on} \quad D,$$

$$av_1 = -b_{0,1} \quad \text{on} \quad \partial D.$$

By applying Condition C to the expansion (2.14) we find that $B_{0,1} \in F^{m-1}$ and $b_{0,1} \in G^{m-1}$. Therefore, v_1 is uniquely defined and $v_1 \in \Phi^{m-1}$ (cf. Condition A). We shall suppose that for $j = 1, \ldots, k$, where $1 \leq k \leq m$, we have already determined the functions $v_j \in \Phi^{m-j}$; then, by Condition C, the k expansions (2.15) are valid for $j = 1, \ldots, k$ and satisfy the conditions (2.16). Now write the problem (2.21) for $j = k + 1$:

$$Av_{k+1} = - \sum_{i=1}^{k+1} B_{k-i+1,i} \quad \text{on} \quad D,$$

$$av_{k+1} = - \sum_{i=1}^{k+1} b_{k-i+1,i} \quad \text{on} \quad \partial D. \qquad (2.22)$$

By (2.16), the right-hand sides belong to F^{m-k-1} and G^{m-k-1}, respectively. Therefore, Condition A implies that (2.22) has a unique solution $v_{k+1} \in \Phi^{m-k-1}$; obviously v_{k+1} is independent of h.

Thus, we have shown how to construct the h independent functions $v_j \in \Phi^{m-j}$ ($j = 1, \ldots, m$). For our specific choice of the v_j the identity (2.20) holds, with the bounds (2.19); by (2.21) the relations (2.20) take the form

$$A^h \eta^h = -\xi^h \quad \text{on} \quad D_h,$$

$$a^h \eta^h = -\zeta^h \quad \text{on} \quad D_h.$$

Condition B implies the inequality

$$\|\eta^h\|_{\Phi_h} \leq C(\|\xi^h\|_{F_h} + \|\zeta^h\|_{G_h})$$

Using (2.19), we obtain (2.11), where $C_5 = C(\tilde{C}_1 + \tilde{C}_2)$. Expressing ϕ^h via (2.12), we obtain the expansion (2.10) with the properties we require. This proves our theorem.

If we use only central differences in the difference approximations, then as a rule the coefficients B_j, b_j in the expansions (2.5) vanish for odd j. This occurred, for instance, in the preceding section, and also in the expansion (2.9). Consequently, the expansion (2.10) contains no odd powers. It is, therefore, worthwhile to formulate the analog of Condition C for this important case.

Condition D. For all $u \in \Phi^k$ ($0 \leq k \leq m$), the following expansions hold:

$$A^h u = Au + \sum_{j=1}^{k} h^{2j} B_j + \sigma^h \quad \text{on} \quad D_h,$$

$$a^h u = au + \sum_{j=1}^{h} h^{2j} b_j + \rho^h \quad \text{on} \quad \partial D_h$$
(2.23)

With bounds for the remainder terms as follows:

$$\|\sigma^h\|_{F_h} \leq C_6 h^{2k+\beta}, \qquad \|\rho^h\|_{G_h} \leq C_7 h^{2k+\beta}.$$
(2.24)

The corresponding version of the theorem on expansions is as follows:
Let Conditions A, B, and D be satisfied for the problem (2.1), (2.2). Then

$$\phi^h = \phi + \sum_{k=1}^{m} h^{2k} v_k + \eta^h \quad \text{on} \quad \bar{D}_h$$
(2.25)

and

$$\|\eta^h\|_{\Phi_h} \leq C_8 h^{2m+\beta}.$$
(2.26)

Then, as before, $B_j \in F^{k-j}$, $b_j \in G^{k-j}$, and $v_k \in \Phi^{m-k}$; B_j, b_j, v_k, C_6, C_7, and C_8 are independent of h, and $\beta > 0$ is independent of h, k, and u. The proof is like that for the case when Conditions A, B, and C are postulated.

Let us verify that the hypotheses are satisfied for the problems (1.1), (1.2). We have already verified Conditions A and B; Condition D remains to be verified. Applying Taylor's theorem to the function $u \in C^{2k+3}$ we obtain the expansion

$$\frac{u(x + h/2) - u(x - h/2)}{h} + a(x)\frac{u(x + h/2) + u(x - h/2)}{h}$$

$$= u'(x) + a(x)u(x) + \sum_{j=1}^{k} h^{2j}\left(\frac{u^{(2j+1)}(x)}{(2j+1)!2^{2j+1}} + \frac{a(x)u^{(2j)}(x)}{(2j)!2^{2j}}\right) + \sigma^h(x),$$

where

$$|\sigma^h(x)| \le h^{2k+2}\left(\frac{1}{(2k+3)!2^{2k+3}} \max_{x \in [0,1]} |u^{(2k+3)}(x)|\right.$$

$$\left. + \frac{1}{2^{2k+2}(2k+2)!} \max_{x \in [0,1]} |a(x)| \cdot \max_{x \in [0,1]} |u^{(2k+2)}(x)|\right).$$

Thus, if we put $F^k = C^{2k+2}[0, 1]$, and $\Phi^k = C^{2k+3}[0, 1]$, the first of the expansions (2.23) will be justified with the constant $\beta = 2$ and the functions

$$B_j(x) = \frac{u^{(2j+1)}(x)}{(2j+1)!2^{2j+1}} + \frac{a(x)u^{(2j)}(x)}{(2j)!2^{2j}}, \qquad j = 1, \ldots, k.$$

Since the initial condition (1.2) is approximated exactly, i.e., $a^h u(0) = au(0) = u(0)$, the second expansion in (2.23) will also hold if we set $b_j = 0$ ($j = 1, \ldots, k$) and $\rho^h = 0$. Thus Condition D is satisfied and, therefore, the expansion (2.25) and the estimate (2.26) are valid.

6.2.2 Acceleration of Convergence

Let us look at some applications of the expansions obtained in the preceding subsection. We shall suppose that the space Φ_h is endowed with the uniform norm, i.e.,

$$\|v\|_{\Phi_h} = \max_{x \in \bar{D}_h} |v(x)|. \tag{2.27}$$

Let the Condition A be satisfied for the problem (2.1), with integer $m \ge 1$, and suppose that x is a common point of the $m + 1$ difference nets $\bar{D}_{h_i}$ with mesh sizes $h_1, h_2, \ldots, h_{m+1}$. On each of these nets we construct the approximate problem

$$\begin{aligned}
A^{h_i}\phi^{h_i} &= f \quad \text{on} \quad D_{h_i} \\
a^{h_i}\phi^{h_i} &= g \quad \text{on} \quad \partial D_{h_i}, \qquad i = 1, \ldots, m+1.
\end{aligned} \tag{2.28}$$

If Condition B is satisfied the solution ϕ^{h_i} exists, and is unique. Thus, at the point x there exist $m + 1$ approximate solutions ϕ^{h_i}. Further, if Condition C is satisfied, each of these solutions may be expanded in the form

$$\phi^{h_i} = \phi + \sum_{k=1}^{m} h_i^k v_k + \eta^{h_i}, \qquad x \in \prod_{i=1}^{m+1} \overline{D}_{h_i} \tag{2.29}$$

with

$$|\eta^{h_i}(x)| \le C_5 h_i^{m+\beta}. \tag{2.30}$$

We consider the system

$$\sum_{i=1}^{m+1} \gamma_i = 1,$$

$$\sum_{i=1}^{m+1} \gamma_i h_i^k = 0, \qquad k = 1, \ldots, m. \tag{2.31}$$

The Vandermonde determinant

$$V(h_1, h_2, \ldots, h_{m+1}) = \begin{Vmatrix} 1 & 1 & \cdots & 1 \\ h_1 & h_2 & \cdots & h_{m+1} \\ \cdots\cdots\cdots\cdots\cdots\cdots \\ h_1^m & h_2^m & \cdots & h_{m+1}^m \end{Vmatrix}$$

divides the matrix of this system. But

$$V(h_1, h_2, \ldots, h_{m+1}) = \prod_{1 \le i < j \le m+1} (h_j - h_i).$$

It follows that the system (2.31) is nonsingular if the h_i are pairwise different. Let us now assume that the h_i are ordered by increasing magnitude, and that

$$\frac{h_{i+1}}{h_i} \ge 1 + C_9 \tag{2.32}$$

with a constant $C_9 > 0$. Then we apply Cramer's method to solve the system (2.31):

$$\gamma_i = \frac{V(h_1, \ldots, h_{i-1}, 0, h_{i+1}, \ldots, h_{m+1})}{V(h_1, \ldots, h_{i-1}, h_i, h_{i+1}, \ldots, h_{m+1})}$$

whence

$$\gamma_i = \prod_{1 \le k < i} \frac{-h_k}{h_i - h_k} \cdot \prod_{i < k \le m+1} \frac{h_k}{h_k - h_i}. \tag{2.33}$$

It follows from (2.32) that

$$\frac{h_k}{h_i - h_k} = \frac{1}{\dfrac{h_i}{h_k} - 1} \le \frac{1}{C_9}, \qquad\qquad\qquad \text{if} \quad k < i,$$

$$\frac{h_k}{h_k - h_i} = \frac{1}{1 - \dfrac{h_i}{h_k}} \le \frac{1}{1 - \dfrac{1}{1 + C_9}} \le 1 + \frac{1}{C_9}, \quad \text{if} \quad i < k.$$

Using this inequality, we find from (2.33) the estimate

$$|\gamma_i| \le \left(1 + \frac{1}{C_9}\right)^m \tag{2.34}$$

which proves the boundedness of the $|\gamma_i|$.

We form the linear combination

$$\bar{\phi}(x) = \sum_{i=1}^{m+1} \gamma_i \phi^{h_i}(x) \tag{2.35}$$

with the weights γ_i. We shall show that the solution $\bar{\phi}(x)$ is exact to the order $h_{m+1}^{m+\beta}$. To this end, we substitute the expansion (2.29) in (2.35):

$$\bar{\phi}(x) = \phi(x) \sum_{i=1}^{m+1} \gamma_i + \sum_{k=1}^{m} \left(v_k(x) \sum_{i=1}^{m+1} h_i^k \gamma_i\right) + \sum_{i=1}^{m+1} \gamma_i \eta^{h_i}(x).$$

Since the γ_i are solutions of the system (2.31) we may simplify this equation:

$$\bar{\phi}(x) = \phi(x) + \sum_{i=1}^{m+1} \gamma_i \eta^{h_i}(x).$$

Bringing in the estimates (2.30) for η^{h_i} and (2.34) for γ_i we find:

$$|\bar{\phi}(x) - \phi(x)| \le \sum_{i=1}^{m+1} C_5 h_i^{m+\beta} \left(1 + \frac{1}{C_9}\right)^m \le C_5(m+1)\left(1 + \frac{1}{C_9}\right)^m h_{m+1}^{m+\beta}. \tag{2.36}$$

Thus, if Conditions A, B, and C and the inequality (2.32) are satisfied, the linear combination $\bar{\phi}(x)$ is a solution exact to the order $h_{m+1}^{m+\beta}$.

Let us dwell for a moment on the choice of the sequence of parameters h_i. There are two predominant methods: one consists in setting $h_k = h/k$ ($h > 0$; $k = 1, \ldots, m+1$) and using these values of h to set up the sequence of nets $\bar{D}_{h_i}$. In this case, the condition (2.32) is satisfied with $C_9 = 1/m$, for arbitrary $h > 0$, and it follows from (2.33) that

$$\gamma_k = \frac{(-1)^{m-k+1} k^{m+1}}{k!(m-k+1)!}, \qquad k = 1, \ldots, m+1. \tag{2.37}$$

The second method sets $h_k = h/2^{k-1}$ ($h > 0$; $k = 1, \ldots, m+1$). Then (2.32) is satisfied with $C_9 = 1$ and arbitrary $h > 0$. Formula (2.33) provides an explicit expression for the weights γ_k, but the corresponding calculations are burdensome for large m. Therefore, we employ Romberg's rule to compute the sum

$$\sum_{i=1}^{m+1} \gamma_i \phi^{h_i}(x),$$

appearing on the right-hand side of (2.35). This rule consists in the successive computation of the quantities

$$T_i^{(1)} = 2\phi^{h_{i+1}}(x) - \phi^{h_i}(x), \qquad i = 1, \ldots, m.$$

We have a recursive process for accomplishing this:

$$T_i^{(k)} = \frac{2^k T_{i+1}^{(k-1)} - T_i^{(k-1)}}{2^k - 1}, \qquad i = 1, \ldots, m - k + 1; k = 2, \ldots, m.$$

As a result, we arrive at the expression

$$T_1^{(m)} = \sum_{i=1}^{m+1} \gamma_i \phi^{h_i}(x).$$

Let us now formulate the analogous result for the case in which the regular part of the expansion contains only even powers of h, i.e., when Condition D holds rather than Condition C. Then the solution ϕ^{h_i} of (2.28) has the expansion

$$\phi^{h_i} = \phi + \sum_{k=1}^{m} h_i^{2k} v_k + \eta^{h_i}, \qquad x \in \bigcap_{i=1}^{m+1} \overline{D}_{h_i} \tag{2.38}$$

and for the remainder term we have

$$|\eta^{h_i}(x)| \leq C_8 h_i^{2m+\beta}. \tag{2.39}$$

We determine the weights μ_i from the system

$$\sum_{i=1}^{m+1} \mu_i = 1,$$
$$\sum_{i=1}^{m+1} \mu_i h_i^{2k} = 0, \qquad k = 1, \ldots, m. \tag{2.40}$$

The determinant of this system is connected with the Vandermonde determinant by the relation

$$V(h_1^2, h_2^2, \ldots, h_{m+1}^2) = \prod_{1 \leq i < j \leq m+1} (h_j^2 - h_i^2).$$

Thus the system (2.40) is nonsingular if the h_i are pairwise distinct. The solution is found by Cramer's rule

$$\mu_i = \frac{V(h_1^2, \ldots, h_{i-1}^2, 0, h_{i+1}^2, \ldots, h_{m+1}^2)}{V(h_1^2, \ldots, h_{i-1}^2, h_i^2, h_{i+1}^2, \ldots, h_{m+1}^2)},$$

whence it follows that

$$\mu_i = \prod_{1 \leq k < i} \frac{-h_k^2}{h_i^2 - h_k^2} \cdot \prod_{1 < k \leq m+1} \frac{h_k^2}{h_k^2 - h_i^2}. \tag{2.41}$$

If, moreover, the inequality (2.32) is satisfied, the μ_i are estimated by the inequality

$$|\mu_i| \leq \left(1 + \frac{1}{2C_9}\right)^m, \tag{2.42}$$

which means that the $|\mu_i|$ are bounded.

With these weights we write the linear combination

$$\bar{\phi}(x) = \sum_{i=1}^{m+1} \mu_i \phi^{h_i}(x). \tag{2.43}$$

Let us show that the solution $\bar{\phi}(x)$ is exact to the order $h_{m+1}^{2m+\beta}$. We substitute the expansion (2.38) in (2.43) and use equations (2.40). In the end, we arrive at the expression

$$\bar{\phi}(x) = \phi(x) + \sum_{i=1}^{m+1} \mu_i \eta^{h_i}(x).$$

Bringing in the estimates (2.39), (2.42) we find the inequality

$$|\bar{\phi}(x) - \phi(x)| \leq C_8(m + 1)\left(1 + \frac{1}{2C_9}\right)^m h_{m+1}^{2m+\beta}. \tag{2.44}$$

Therefore, if Conditions A, B, and D, and the inequalities (2.32) are satisfied, the linear combination $\bar{\phi}(x)$ approximates the exact solution $\phi(x)$, to the order $h_{m+1}^{2m+\beta}$.

Let us consider a convenient rule for computing the weights μ_i for two special choices of the parameters h_i. When $h_i = h/i$ ($h > 0; i = 1, \ldots, m + 1$), formula (2.41) simplifies and becomes

$$\mu_i = 2\frac{(-1)^{m-i+1}i^{2m+2}}{(m + i + 1)!(m - i + 1)!}, \qquad i = 1, \ldots, m + 1. \tag{2.45}$$

In particular, we chose the weights $\mu_1 = -\frac{1}{3}$, $\mu_2 = \frac{4}{3}$ for the problem in Section 6.1, since m was equal to unity.

Our second choice of the parameters sets $h_i = h/2^{i-1}$ ($h > 0; i = 1, \ldots, m + 1$). In this case we may use both (2.30) and Romberg's rule. To compute the sum

$$\sum_{i=1}^{m+1} \mu_i \phi^{h_i}(x)$$

the rule is applied as follows: we first compute the quantities

$$K_i^{(1)} = \frac{4}{3}\phi^{h_{i+1}}(x) - \frac{1}{3}\phi^{h_i}(x), \qquad i = 1, \ldots, m.$$

Then we compute

$$K_1^{(m)} = \frac{4^j K_{i+1}^{(j-1)} - K_i^{(j-1)}}{4^j - 1}, \qquad i = 1, \ldots, m - j + 1; \quad j = 2, \ldots, m$$

and, in the end, we find

$$K_1^{(m)} = \sum_{i=1}^{m+1} \mu_i \phi^{h_i}(x).$$

Let us return to the problem (1.1)–(1.3) and the corresponding difference problem (1.6), (1.7). We have already shown that the condition

$a, f \in C^{2m+2}[0, 1]$ guarantees the existence of an expansion of the form (2.38), (2.39). For a solution u^{τ_i} of (1.6), (1.7) on a net with mesh size τ_i the expansion is as follows:

$$u^{\tau_i}(j\tau) = u(j\tau) + \sum_{k=1}^{m} \tau_i^{2k} v_k(j\tau) + \eta^{\tau_i}(j\tau), \qquad j = 1, \ldots, M_i, \quad M_i = 1/\tau_i, \quad (2.46)$$

where

$$|\eta^{\tau_i}(j\tau)| \leq C_8 \tau_i^{2m+2}. \tag{2.47}$$

Accordingly, a rule analogous to (2.40), (2.43) for increasing the precision of the approximation applies to this case.

Let us look at the specific case in which the parameters of the difference net are chosen as $\tau, \tau/2, \ldots, \tau/m + 1$. Then the sharpening rule is as follows:

$$\bar{u}(j\tau) = \sum_{i=1}^{m+1} \mu_i u^{\tau/i}(j\tau), \qquad j = 0, 1, \ldots, M, \tag{2.48}$$

where μ_i is defined by (2.45). Then $\bar{u}(j\tau)$ approximates the value of $u(j\tau)$ to the order $O(\tau^{2m+2})$.

It is easily seen that the refined solution is obtained on a relatively sparse net and only a small number of the values of $u^{\tau/i}$ needs to be used; for a different choice of the sequence τ_i of common points of several nets, the number may be even smaller. Therefore, in computing approximate values to a high order of precision at arbitrary points of the domain of definition of the solution of the original problem, it is advantageous to use interpolation. We shall consider a simple situation, applying Newton's interpolation polynomials.

Let t be an arbitrary point of the interval $[0, 1]$. We fix the arbitrary number i and consider a net with mesh size τ_i. On this net, at the point t, we choose $2m + 2$ nearby net points. Given the values of the approximate solution u^{τ_i} at these net points, we construct the Newton interpolation polynomial of degree $2m + 1$, which we shall denote by $P_i(x; u^{\tau_i})$. It follows from (2.46) that

$$P_i(x; u^{\tau_i}) = P_i(x; u) + \sum_{k=1}^{m} \tau_i^{2k} P_i(x; v_k) + P_i(x; \eta^{\tau_i}).$$

Since the Newton polynomial interpolates smooth functions to the order $O(\tau_i^{2m+2})$, and its coefficients are bounded, we have

$$P_i(t; u^{\tau_i}) = u(t) + \sum_{k=1}^{m} \tau_i^{2k} v_k(t) + O(\tau_i^{2m+2}).$$

This expansion is valid for $i = 1, \ldots, m + 1$; therefore, we may sharpen the difference solutions

$$\bar{u}(t) = \sum_{i=1}^{m+1} \mu_i P_i(t; u^{\tau_i}) \tag{2.49}$$

using the same weights μ_i as in (2.40). Then

$$|\bar{u}(t) - u(t)| \le \sum_{i=1}^{m+1} C_{10} |\gamma_i| \tau_i^{2m+2}$$

with a constant C_{10} containing the constants of the interpolation formula and the errors of the solution at the net points.

6.3 Simple Integral Equations

The integral equations we are about to present provide very simple illustrations of the general theorems given in Section 6.2; they are simple because there are no boundary conditions. Therefore, the set ∂D, on which boundary conditions are prescribed, and, therefore, also the set ∂D_h may be taken as empty, and all the conditions to be verified are much simpler than in the general case.

6.3.1 The Fredholm Equation of the Second Kind

Consider the equation

$$\phi(x) = \int_0^1 K(x, t)\phi(t) \, dt + f(x), \qquad x \in [0, 1]. \tag{3.1}$$

We constrain the function f and the kernel K by imposing the following conditions:

$$f \in C^{2m+2}[0, 1], \qquad K \in C^{2m+2}([0, 1] \times [0, 1]), \tag{3.2}$$

$$\kappa = \max_{\substack{x \in [0, 1] \\ t \in [0, 1]}} |K(x, t)| < 1, \tag{3.3}$$

where $m \ge 1$ is an integer. Then the solution ϕ of this problem exists and is unique in the class $C^{2m+2}[0, 1]$. Moreover, Condition A of Section 6.2.1. holds with respect to the classes $\Phi^k = F^k = C^{2k+2}[0, 1]$.

For an approximate solution of (3.1) we introduce the difference net

$$x_{i+1/2} = (i + \tfrac{1}{2})h, \qquad i = 0, 1, \ldots, N - 1 \tag{3.4}$$

with mesh size $h = 1/N$, and we replace the integral in (3.1) by a quadrature using mean rectangles. Then, at the net points we find the equations

$$\phi^h(x_{i+1/2}) = \sum_{j=0}^{N-1} hK(x_{i+1/2}, x_{j+1/2})\phi^h(x_{j+1/2}) + f(x_{i+1/2})$$

$$i = 0, 1, \ldots, N - 1. \tag{3.5}$$

Supposing that there exists at least one solution of this system, we derive an *a priori* estimate. Let the component of ϕ^h with maximum absolute value have the index k. Then it follows from (3.5) and (3.3) that

$$|\phi^h(x_{i+1/2})| \le \sum_{j=0}^{N-1} h|K(x_{k+1/2}, x_{j+1/2})||\phi^h(x_{j+1/2})| + |f(x_{k+1/2})|$$

$$\le \kappa|\phi^h(x_{k+1/2})| + \max_{0 \le j \le N-1} |f(x_{j+1/2})|.$$

Then

$$\max_{0 \le i \le N-1} |\phi^h(x_{i+1/2})| \le \frac{1}{1 - \kappa} \max_{0 \le i \le N-1} |f(x_{i+1/2})|. \qquad (3.6)$$

This estimate guarantees the solubility of (3.5) and the stability and uniqueness of the solution. This in turn verifies Condition B in Section 6.2.1, where

$$\|u\|_{\Phi_h} = \|u\|_{F_h} = \max_{0 \le i \le N-1} |u(x_{i+1/2})|.$$

The following property of mean rectangular quadratures is well known: if $u \in C^{2k}[0, 1]$,

$$\sum_{i=0}^{N-1} hu(x_{i+1/2}) = \int_0^1 u(x)\, dx$$

$$- \sum_{j=1}^{k} h^{2j} \frac{1 - 2^{-2j+1}}{(2j)!} B_{2j} u^{(2j-1)}(x)\Big|_{x=0}^{x=1} + h^{2k} \frac{B_{2k}}{(2k)!} u^{(2k)}(\xi), \qquad (3.7)$$

where $\xi \in [0, 1]$, $f(x)|_{x=0}^{x=1}$ is the difference $f(1) - f(0)$, and the B_j are the Bernoulli numbers $B_0 = 1, B_2 = \frac{1}{6}, B_4 = -\frac{1}{30}, B_6 = \frac{1}{42}, \ldots$. This expansion may be obtained, for example, as the sum of the analogous expansion with respect to the intervals $[ih, (i + 1)h]$, in which Taylor's formula is applied in order to obtain the values of u and its derivatives as sums of quantities evaluated at the point $x_{i+1/2}$.

We denote by A and A^h the following operators:

$$Au(x) = u(x) - \int_0^1 K(x, t)u(t)\, dt, \qquad x \in [0, 1],$$

$$A^h u(x_{i+1/2}) = u(x_{i+1/2}) - \sum_{j=0}^{N-1} K(x_{i+1/2}, x_{j+1/2})u(x_{j+1/2})h$$

$$i = 0, 1, \ldots, N - 1.$$

Then (3.7) implies for an arbitrary $u \in C^{2k+2}[0, 1]$ an expansion of the form (2.12):

$$A^h u(x_{i+1/2}) = Au(x_{i+1/2}) + \sum_{j=1}^{K} h^{2j} g_j(x_{i+1/2}) + \sigma^h(x_{i+1/2}),$$

$$i = 0, 1, \ldots, N - 1, \quad (3.8)$$

where

$$g_j(x) = \frac{1 - 2^{-2j+1}}{(2j)!} B_{2j} \frac{\partial^{2j-1}}{\partial x^{2j-1}} (K(x, t)u(t)) \Big|_{t=0}^{t=1}, \qquad x \in [0, 1],$$

$$|\sigma^h(x_{i+1/2})| \le C_1 h^{2m+2}.$$

These conditions are sufficient for the validity of the expansion

$$\phi^h(x_{i+1/2}) = \phi(x_{i+1/2}) + \sum_{k=1}^{m} h^{2k} v_k(x_{i+1/2}) + \eta^h(x_{i+1/2}),$$

$$i = 0, 1, \ldots, N - 1. \quad (3.9)$$

On this basis we may apply the sharpening method given in Section 6.2.2 to obtain an expansion in even powers of the parameter. To do this, we construct difference nets with mesh sizes $h_i = h \, [h/3, \ldots, h/(2m + 1)]$; on each of these we solve the system of linear equations (3.5). All the solutions $\phi^{h_i} \, (i = 1, \ldots, m + 1)$ are defined on the net (3.4) with mesh size h. We choose the μ_i as solutions of the system

$$\sum_{i=1}^{m+1} \mu_i = 1,$$

$$\sum_{i=1}^{m+1} \mu_i h_i^{2k} = 0, \qquad k = 1, 2, \ldots, m + 1.$$

With these we construct the linear combination

$$\bar{\phi}(x_{i+1/2}) = \sum_{k=1}^{m+1} \mu_k \phi^{h_k}(x_{i+1/2}), \qquad i = 0, 1, \ldots, N - 1.$$

Then, on the basis of Section 6.2.2, we have the estimate

$$|\phi(x_{i+1/2}) - \bar{\phi}(x_{i+1/2})| \le C_2 h^{2m+2}, \qquad i = 0, 1, \ldots, N - 1,$$

where C_2 is independent of h.

Note that the customary subdivision methods $h, h/2, h/3, \ldots$ and $h, h/2, h/4, \ldots$ yield no point in common with the sequence of nets (3.4).

6.3.2 The Volterra Equation of the First Kind

Consider the very simple integral equation (Volterra equation of the first kind)

$$\int_0^x K(x, t)\phi(t) \, dt = f(x), \qquad x \in [0, 1]. \tag{3.10}$$

We assume with respect to f and K that

$$f \in C^{2m+3}[0, 1], \qquad \frac{\partial K}{\partial x} \in C^{2m+2}(\bar{Q}), \tag{3.11}$$

where m is an integer, $m \geq 1$, and $\bar{Q}$ is the triangle $0 \leq t \leq x \leq 1$. To secure the uniqueness of the solution we assume also that

$$f(0) = 0, \qquad \min_{x \in [0, 1]} |K(x, x)| = k_1 \neq 0. \tag{3.12}$$

Then there exists a unique solution ϕ of (3.10) in the class $C^{2m+2}[0, 1]$.

For an approximate solution of (3.10) we introduce difference nets with integral and intermediate net points

$$x_i = ih, \qquad i = 0, 1, \ldots, N,$$

$$x_{i+1/2} = (i + \tfrac{1}{2})h, \qquad i = 0, 1, \ldots, N - 1$$

and use a mean rectangular quadrature. Then we arrive at the equations

$$\sum_{j=0}^{i} hK(x_{i+1}, x_{j+1/2})\phi^h(x_{j+1/2}) = f(x_{i+1}), \qquad i = 0, 1, \ldots, N - 1. \tag{3.13}$$

These form a system of linear algebraic equations in ϕ^h, and have a triangular matrix. The condition (3.12) implies, that for sufficiently small h, the diagonal elements exceed in absolute value some positive number, for example

$$|hK(x_{i+1}, x_{i+1/2})| \geq \frac{hk_1}{2} \tag{3.14}$$

Then the system (3.13) has a unique solution. Let us derive an *a priori* estimate for it: we find the difference of two equations in (3.13) with indices $i, i - 1$ and transform it. We have

$$K(x_{i+1}, x_{i+1/2})\phi^h(x_{i+1/2}) = - \sum_{j=0}^{i} \{K(x_{i+1}, x_{j+1/2})$$

$$- K(x_i, x_{j+1/2})\}\phi^h(x_{j+1/2})$$

$$+ \frac{f(x_{i+1}) - f(x_i)}{h}.$$

Taking account of the conditions (3.12), (3.14) we arrive at the inequality

$$|\phi^h(x_{i+1/2})| \leq \frac{2}{K_1}\left(hK_2 \sum_{j=0}^{i-1} |\phi^h(x_{j+1/2})| + f_1\right), \tag{3.15}$$

where

$$K_2 = \max_{(x, t) \in Q}\left|\frac{\partial K}{\partial x}(x, t)\right|, \qquad f_1 = \max_{0 \leq i \leq N-1} \frac{|f(x_{i+1}) - f(x_i)|}{h}.$$

It follows by mathematical induction on (3.15) that

$$|\phi^h(x_{i+1/2})| \leq \frac{2f_1}{K_1}\left(1 + \frac{2hk_2}{k_1}\right)^i.$$

Using the inequalities

$$1 + \frac{2hk_2}{K_1} \leq \exp\left(\frac{2hk_2}{k_1}\right), \qquad ih \leq 1,$$

we find the estimate

$$|\phi^h(x_{i+1/2})| \leq \frac{2f_1}{k_1} \exp\left(\frac{2k_2}{k_1}\right), \qquad i = 0, 1, \ldots, N - 1. \tag{3.16}$$

If we put

$$\|\phi\|_{\Phi_h} = \max_{0 \leq i \leq N-1} |\phi(x_{i+1/2})|,$$

$$\|f\|_{F_h} = \max_{0 \leq i \leq N-1} \frac{|f(x_{i+1}) - f(x_i)|}{h},$$

the estimate (3.16) may be written as

$$\|\phi\|_{\Phi_h} \leq C_3 \|f\|_{F_h}.$$

Condition C in Section 6.2 remains to be verified. For this, we bring in the smoothness classes

$$\Phi^k = C^{2k+2}[0, 1], \qquad F^k = C^{2k+3}[0, 1]$$

and write

$$Au(x) = \int_0^x k(x, t)u(t)\, dt, \qquad x \in [0, 1],$$

$$A^h u(x_i) = \sum_{j=0}^i k(x_i, x_{j+1/2})u(x_{j+1/2})h, \qquad i = 1, \ldots, N.$$

Using an expansion of the form (3.7) in the interval $[0, ih]$ we expand the function $u \in C^{2k+2}[0, 1]$:

$$A^h u(x_i) = Au(x_i) + \sum_{j=1}^k h^{2j} g_j(x_i) + \sigma^h(x_i), \qquad i = 1, 2, \ldots, N$$

with $g_j \in C^{2k-2j+3}[0, 1]$, and

$$|\sigma^h(x_i)| \leq C_4 h^{2k+2}.$$

Putting $\sigma^h(0) = 0$, we have

$$\|\sigma^h\|_{F_h} \leq 2C_4 h^{2k+1}.$$

Therefore, the theorem of Section 6.2.1 holds on the expansion of the solution of (3.13) in even powers of h with $\beta = 1$. We employ this expansion just as in the preceding section.

We choose h so that (3.14) is satisfied, and we construct difference nets with mesh sizes $h_i = h, h/3, \ldots, h/(2m + 1)$. On each net we obtain a solution ϕ^{h_i} of (3.13); at the net points of the net with mesh size h we construct the linear combination

$$\bar{\phi}(x_{i+1/2}) = \sum_{k=1}^{m+1} \mu_k \phi^{h_k}(x_{i+1/2}), \qquad i = 0, 1, \ldots, N - 1, \qquad (3.17)$$

with weights defined by the system

$$\sum_{i=1}^{m+1} \mu_i = 1,$$

$$\sum_{i=1}^{m+1} \mu_i h_i^{2k} = 0, \qquad k = 1, 2, \ldots, m + 1.$$

Then by the theorem given in Section 6.2.2 we have the estimate

$$|\phi(x_{i+1/2}) - \bar{\phi}(x_{i+1/2})| \le C_5 h^{2m+1}, \qquad i = 0, 1, \ldots, N - 1.$$

The precision attained here is less than that for the Fredholm equation of the second kind, although our conditions on the smoothness of the right-hand side of the equation are the same.

6.4 The One-Dimensional Diffusion Equation

In this section we shall study the Dirichlet problem for the very simple diffusion equation

$$-\frac{d^2u}{dx^2} + q(x)u = f(x), \qquad x \in (0, 1), \qquad (4.1)$$

$$u(0) = 0, \qquad u(1) = 0, \qquad (4.2)$$

under the assumption that

$$q(x) \ge 0, \qquad (4.3)$$

and that both q and f (and therefore u) are smooth enough to permit the subsequent calculations.

First, using the results of Section 6.2 concerning difference schemes on sequences of nets, we construct approximate solutions of order $O(h^{2k})$ for $k \ge 2$ and $h \to 0$. Then, using the example (4.1), (4.2), we consider a modification of the extrapolation to the limit, worked through in an application to the Galerkin variational-difference method.

6.4.1 The Difference Method

We construct the difference

$$x_k = kh, \qquad k = 0, 1, \ldots, N \tag{4.4}$$

with the uniform mesh size $h = 1/N$, and we pass from (4.1), (4.2) to the approximate problem

$$\frac{-u^h_{k-1} + 2u^h_k - u^h_{k+1}}{h^2} + q_k u^h_k = f_k, \qquad k = 1, 2, \ldots, N-1, \tag{4.5}$$

$$u^h_0 = u^h_N = 0$$

with second-order approximation.

Let us verify Conditions A, B, and D of Section 6.2 for the problems (4.1), (4.2), and (4.5). Condition A, which characterizes the existence and uniqueness of the solution of (4.1), (4.2) is related to the smoothness of the coefficient q. We require that $q \in C^{2m+2}[0, 1]$, where m is an integer, $m \geq 1$. Then, for an arbitrary right-hand side $f \in C^{2m+2}[0, 1]$, there exists a unique solution $u \in C^{2m+4}[0, 1]$, satisfying the condition (4.2). Thus Condition A will be satisfied if $F^k = C^{2k+2}[0, 1]$ and Φ^k consists of functions in $C^{2k+4}[0, 1]$ vanishing at the endpoints of the interval $[0, 1]$.

To test Condition B we derive an estimate. For each value of k we multiply the corresponding equation of the system (4.5) by $h u^h_k$, and add. This yields:

$$\frac{1}{h} \sum_{k=1}^{N} \{(u^h_k - u^h_{k-1})^2 + q_k(u^h_k)^2\} = \sum_{k=1}^{N-1} f_k u^h_k h.$$

Removing the positive terms $q_k(u^h_k)^2$ from the left-hand side and replacing the f_k, u^h_k by their maximum values, we find

$$\frac{1}{h} \sum_{k=1}^{N} (u^h_k - u^h_{k-1})^2 \leq \max_{1 \leq k \leq N-1} |f_k| \cdot \max_{1 \leq k \leq N-1} |u^h_k|. \tag{4.6}$$

For a lower bound, we employ a relation derived from the Cauchy-Bunyakovskiĭ inequality for a net function y vanishing at $x = 0$:

$$|y_k| = \left| \sum_{j=1}^{k} (y_j - y_{j-1}) \right| \leq k^{1/2} \left(\sum_{j=1}^{k} (y_j - y_{j-1})^2 \right)^{1/2}$$

$$\leq N^{1/2} \left(\sum_{j=1}^{N} (y_j - y_{j-1})^2 \right)^{1/2}.$$

Since $u^h_0 = 0$, we put $y_k = u^h_k$ and obtain the inequality

$$\max_{1 \leq k \leq N-1} |u^h_k|^2 \leq \frac{1}{h} \sum_{k=1}^{N} (u^h_k - u^h_{k-1})^2.$$

Combining this with (4.6) we have

$$\max_{1 \leq k \leq N-1} |u^h_k|^2 \leq \max_{1 \leq k \leq N-1} |f_k| \cdot \max_{1 \leq k \leq N-1} |u^h_k|.$$

This implies the estimate

$$\|u^h\|_{\Phi^h} \leq \|f\|_{F^h}, \tag{4.7}$$

where

$$\|u^h\|_{\Phi^h} = \max_{0 \leq k \leq N} |u^h_k|, \qquad \|f\|_{F^h} = \max_{1 \leq k \leq N-1} |f_k|.$$

The estimate (4.7) implies that Condition B is satisfied. Moreover, Condition D is satisfied: in fact, suppose $u \in C^{2k+4}[0, 1]$. We write

$$Au(x) = -u''(x) + q(x)u(x), \qquad x \in (0, 1),$$

$$A^h u(x_k) = \frac{-u(x_{k-1}) + 2u(x_k) - u(x_{k+1})}{h^2} + q(x_k)u(x_k),$$

$$k = 1, 2, \ldots, N - 1.$$

Using Taylor's formula,

$$u(x_{k\pm 1}) = \sum_{r=0}^{2k+1} \frac{(\pm h)^r}{r!} u^{(r)}(x_k) + O(h^{2k+2}),$$

we arrive at the expansion

$$A^h u(x_k) = Au(x_k) + \sum_{j=0}^{k} h^{2j}g_j(x_k) + \sigma^h(x_k), \qquad k = 1, 2, \ldots, N - 1,$$

where

$$\|\sigma^h\|_{F^h} \leq C_1 h^{2k+2}.$$

Thus, we have verified the theorem of Section 6.2.1 on the expansion of the network solution u^h in even powers of h with $\beta = 2$:

$$u^h_k = u(x_k) + \sum_{j=0}^{m} h^{2j}v_j(x_k) + \eta^h_k, \tag{4.8}$$

where

$$\|\eta^h\|_{\Phi_h} \leq C_2 h^{2m+2}.$$

Let us now look at the method for increasing the accuracy of the difference solutions of the problem (4.5). Let x be a common point of the $m + 1$ difference nets with mesh sizes $h_1 > h_2 > \cdots > h_{m+1}$, with

$$h_j/h_{j+1} \geq 1 + C_3, \qquad C_3 > 0.$$

On each of these nets we construct an approximate problem (4.5) and find its solution u^{h_j}, for instance by the sweep method. Thus, at the point x we have $m + 1$ approximate values $u^{h_j}(x)$. Let the coefficients μ_j be solutions of the system

$$\sum_{j=1}^{m+1} \mu_j = 1, \qquad \sum_{j=1}^{m+1} h_j^{2k}\mu_j = 0, \qquad k = 1, 2, \ldots, m + 1.$$

We construct the linear combination

$$\bar{u}(x) = \sum_{j=1}^{m+1} \mu_j u^{h_j}(x).$$

By the theorem in Section 6.2.2 we have the estimate

$$|u(x) - \bar{u}(x)| \leq C_4 h^{2m+2}.$$

for this solution.

The procedures we have just described are easily generalized to the third boundary problem, and to equations with a variable coefficient

$$-(ku')' + qu = f,$$

including the case in which k is discontinuous. In the latter situation we must use a piecewise-linear mapping of the independent variable such that the points of discontinuity become integer vertices of the difference nets with mesh sizes $h_1, h_2, \ldots, h_{m+1}$.

6.4.2 The Galërkin Method

In contrast to the high-precision variational-difference methods, based on the use of an abnormally large number of very smooth basis functions (an example of such basis functions was given in Section 2.4), we now present a method for increasing the precision of approximate solutions of variational-difference problems with differing mesh sizes, using only piecewise-linear basis functions.

To construct the approximate problems we divide the interval $[0, 1]$ into N equal subintervals, of length $1/N$, by the points

$$x_k = kh, \qquad k = 0, 1, \ldots, N, \tag{4.9}$$

and we construct the basis functions ω_k defined in Section 2.3.2:

$$\omega_k(x) = \begin{cases} 0, & \text{if } x \in (-\infty, x_{k-1}], \\[2mm] 1 + \dfrac{x - x_k}{h}, & \text{if } x \in (x_{k-1}, x_k], \\[2mm] 1 - \dfrac{x - x_k}{h}, & \text{if } x \in (x_k, x_{k+1}), \\[2mm] 0, & \text{if } x \in [x_{k+1}, +\infty), \qquad k = 1, 2, \ldots, N-1. \end{cases}$$

We multiply (4.1) by each of these basis functions and integrate the resulting equations. If the function $u(x)$ is a solution of the problem (4.1), (4.2), we obtain the identities:

$$-\int_{x_{k-1}}^{x_{k+1}} \left(\frac{d^2 u}{dx^2} \omega_k + qu\omega_k \right) dx = \int_{x_{k-1}}^{x_{k+1}} f\omega_k\, dx, \qquad k = 1, \ldots, N-1. \tag{4.10}$$

We transform these, integrating by parts and using the conditions $\omega_k(x_{k\pm1}) = 0$:

$$\int_{x_{k-1}}^{x_{k+1}} \left(\frac{du}{dx}\frac{d\omega_k}{dx} + qu\omega_k \right) dx = \int_{x_{k-1}}^{x_{k+1}} f\omega_k \, dx.$$

We introduce some notation that will simplify our later calculations:

$$(v, w) = \int_0^1 v(x)w(x) \, dx, \qquad [v, w] = \int_0^1 \left(\frac{dv}{dx}\frac{dw}{dx} + qvw \right) dx.$$

In this notation the preceding equations become

$$[u, \omega_k] = (f, \omega_k), \qquad i = 1, 2, \ldots, N - 1. \tag{4.11}$$

Using these identities, we shall seek the approximate solution $u^h(x)$ in the form

$$u^h(x) = \sum_{l=1}^{N-1} \alpha_l^h \omega_l(x) \tag{4.12}$$

with a set of constants α_i^h defined by the equations that result when we substitute $u^h(x)$ in (4.11) in place of $u(x)$:

$$\sum_{l=1}^{N-1} \alpha_l^h [\omega_l, \omega_k] = (f, \omega_k), \qquad k = 1, 2, \ldots, N - 1. \tag{4.13}$$

The system (4.13) may be written, for better comprehensibility, in the matrix form

$$A\alpha \equiv \begin{Vmatrix} b_1 & c_1 & 0 & \cdots & 0 & 0 \\ a_2 & b_2 & c_2 & \cdots & 0 & 0 \\ 0 & a_3 & b_3 & \cdots & 0 & 0 \\ \multicolumn{6}{c}{\dotfill} \\ 0 & 0 & 0 & \cdots & b_{N-2} & c_{N-2} \\ 0 & 0 & 0 & \cdots & a_{N-1} & b_{N-1} \end{Vmatrix} \begin{Vmatrix} \alpha_1^h \\ \alpha_2^h \\ \vdots \\ \\ \alpha_{N-2}^h \\ \alpha_{N-1}^h \end{Vmatrix} = \begin{Vmatrix} g_1 \\ g_2 \\ \vdots \\ \\ g_{N-2} \\ g_{N-1} \end{Vmatrix} \tag{4.14}$$

with the elements

$$a_k = -\frac{1}{h} + \int_{x_{k-1}}^{x_k} \left(1 + \frac{x - x_k}{h} \right) \frac{x_k - x}{h} q \, dx = [\omega_{k-1}, \omega_k],$$

$$b_k = \frac{2}{h} + \int_{x_{k-1}}^{x_k} \left(1 + \frac{x - x_k}{h} \right)^2 q \, dx + \int_{x_k}^{x_{k+1}} \left(1 - \frac{x - x_k}{h} \right)^2 q \, dx = [\omega_k, \omega_k],$$

$$c_k = a_{k+1} = [\omega_{k+1}, \omega_k],$$

$$g_k = \int_{x_{k-1}}^{x_k} \left(1 + \frac{x - x_k}{h} \right) f \, dx + \int_{x_k}^{x_{k+1}} \left(1 - \frac{x - x_k}{h} \right) f \, dx = (f, \omega_k).$$

It is clear that a function which is not a polynomial of degree less than 2 cannot be uniformly approximated on the interval $[0, 1]$ with accuracy of the order $O(h^2)$ by piecewise-linear functions with a basis drawn from the $\omega_i^h(x)$. Let us therefore transform the Richardson postulate into the following: the coefficients α_k^h may be expanded, as $h \to 0$, in the form

$$\alpha_k^h = u(kh) + h^2 v(kh) + \eta_k^h, \qquad k = 0, 1, \ldots, N, \qquad (4.15)$$

where $u(x)$ is a solution of (4.1), (4.2); $v(x)$ is some smooth function independent of h; and the η_k^h are net functions whose absolute values are uniformly bounded, with respect to k, by a quantity of order $O(h^4)$:

$$|\eta_k^h| \leq C_1 h^4 \qquad (4.16)$$

with a constant C_1 independent of h and k.

The use of the general methods given in Section 6.2, without taking account of the specifics of the Galërkin method, leads to a significant increase in the constraints that must be imposed on the functions g and f. We therefore develop further the foundation for the expansion (4.15), with special attention to the method of finite elements, attaining fourth-order accuracy with much weaker constraints on the smoothness of the coefficients in (4.1) than in the difference method.

We develop the expansion (4.15) in three stages, as in Section 6.1. We begin with a discussion that explains the choice of the function $v(x)$, then we determine the function, and end with a proof of the inequality (4.16) for the remainder term.

Thus, given (4.15), we single out only the two leading terms, throughout our calculations, and obtain the expression

$$\alpha_k^h = u(x_k) + h^2 v(x_k) + O(h^4). \qquad (4.17)$$

We substitute this in the system (4.14)

$$a_k u(x_{k-1}) + b_k u(x_k) + c_k u(x_{k+1}) + h^2(a_k v(x_{k-1})$$
$$+ b_k v(x_k) + c_k v(x_{k+1})) + O(h^4) = g_k, \qquad k = 1, 2, \ldots, N - 1, \quad (4.18)$$

where, for generality, we have set $a_1 = c_{n-1} = 0$. The magnitude of the remainder term is not immediately evident, but we shall later show that it is even smaller than $O(h^4)$.

Using the values of $u(x)$ and $v(x)$ at the net points of the difference net, we construct the continuous functions

$$\tilde{u}(x) = \sum_{l=1}^{N-1} u(x_l)\omega_l(x), \qquad \tilde{v}(x) = \sum_{l=1}^{N-1} v(x_l)\omega_l(x),$$

which are called the piecewise-linear completions of $u(x)$ and $v(x)$, respectively. Recalling the form of the coefficients a_k, b_k, c_k, and g_k we replace (4.18) by the equations

$$[\tilde{u}, \omega_k] + h^2[\tilde{v}, \omega_k] + O(h^4) = (f, \omega_k), \qquad k = 1, 2, \ldots, N - 1. \quad (4.19)$$

Now we integrate by parts, and easily arrive at the equations

$$-\frac{1}{h}u(x_{k-1}) + \frac{2}{h}u(x_k) - \frac{1}{h}u(x_{k+1}) = \int_0^1 \frac{du}{dx}\frac{d\omega_k}{dx}\,dx = \int_0^1 \frac{d\tilde{u}}{dx}\frac{d\omega_k}{dx}\,dx,$$

$$(4.20)$$

$$k = 1, 2, \ldots, N - 1.$$

We now make use of the Taylor expansions of the functions $u(x)$ and $v(x)$:

$$u(x) = u(x_k) + (x - x_k)\frac{du}{dx}(x_k) + \frac{(x - x_k)^2}{2}\frac{d^2u}{dx^2}(x_k) + O(h^3),$$

$$(4.21)$$

$$q(x) = q(x_k) + (x - x_k)\frac{dq}{dx}(x_k) + \frac{(x - x_k)^2}{2}\frac{d^2q}{dx^2}(x_k) + O(h^3)$$

on each of the intervals (x_{k-1}, x_k) in order to prove that

$$(q\tilde{u}, \omega_k) = (qu, \omega_k) + h^2(w, \omega_k) + O(h^3),$$

$$(4.22)$$

where

$$w(x) = -\tfrac{1}{12}q(x)\frac{d^2u}{dx^2}(x).$$

Since

$$\tilde{u}(x) = \begin{cases} \dfrac{1}{h}(u(x_k)(x - x_{k-1}) + u(x_{k-1})(x_k - x)), & x \in (x_{k-1}, x_k), \\[2mm] \dfrac{1}{h}(u(x_k)(x_{k+1} - x) + u(x_{k+1})(x - x_k)), & x \in (x_k, x_{k+1}) \end{cases}$$

we have

$$\tilde{u}(x) = \begin{cases} u(x_k) + (x_k - x)u'(x_k) + (x_k - x)\dfrac{h}{2}u''(x_k) + O(h^3), \\[2mm] \qquad\qquad\qquad\qquad\qquad\qquad\qquad \text{for} \quad x \in (x_{k-1}, x_k), \\[2mm] u(x_k) + (x - x_k)u'(x_k) + (x - x_k)\dfrac{h}{2}u''(x_k) + O(h^3), \\[2mm] \qquad\qquad\qquad\qquad\qquad\qquad\qquad \text{for} \quad x \in (x_k, x_{k+1}). \end{cases}$$

Therefore, using the latter equation and the expansions (4.21), and integrating the polynomials under the integral sign, we find that

$$(q\tilde{u}, \omega_k) = hu(x_k)q(x_k)$$
$$- h^3(\tfrac{1}{12}u(x_k)q''(x_k) + \tfrac{1}{12}u'(x_k)q(x_k) + \tfrac{1}{12}u''(x_k)q(x_k)) + O(h^3),$$

$$(qu, \omega_k) = hu(x_k)q(x_k) - h^3\tfrac{1}{12}(u(x_k)q(x_k))'' + O(h^4),$$

$$h^2(w, \omega_k) = h^3w(x_k) + O(h^4).$$

By comparing the coefficients for terms of orders h and h^3 we prove the equation (4.22). Combining (4.20) and (4.22) we find that

$$[\tilde{u}, \omega_k] = [u, \omega_k] + h^2(w, \omega_k) + O(h^4). \qquad (4.23)$$

If we carry these calculations through for $v(x)$, with a smaller number of terms, we obtain the relation

$$h^2(\tilde{v}, \omega_k) = h^2(v, \omega_k) + O(h^4).$$

We use the identity (4.20), which holds for $v(x)$, to transform this equation into the equation

$$h^2[\tilde{v}, \omega_k] = h^2[v, \omega_k] + O(h^4). \qquad (4.24)$$

Finally, using (4.23) and (4.24) in the left-hand side of (4.19), we obtain the equations

$$[u, \omega_k] + h^2(w, \omega_k) + h^2[v, \omega_k] + O(h^4) = (f, \omega_k),$$
$$k = 1, 2, \ldots, N - 1. \quad (4.25)$$

Since we are seeking an identity for all h, we require that the coefficients of the leading powers of h must be equal. A comparison of the coefficients of h^0 and h^2 yields the equations

$$[u, \omega_k] = (f, \omega_k), \qquad (4.26)$$

$$[v, \omega_k] = -(w, \omega_k), \qquad k = 1, 2, \ldots, N - 1. \qquad (4.27)$$

These equations complete the construction of the necessary conditions. The first group coincides with (4.11), and are consequences of (4.1). The second group will be satisfied if we choose $v(x)$ to be a solution of the boundary problem

$$-\frac{d^2v}{dx^2} + qv = -w(x), \qquad (4.28)$$

$$v(0) = v(1) = 0. \qquad (4.29)$$

We have now completed the construction of the equations needed for the definition of $v(x)$. We have still to prove (4.16), i.e., the boundedness of the net functions defined by

$$\eta_k^h = \alpha_k^h - u(x_k) - h^2 v(x_k), \qquad k = 0, 1, \ldots, N. \qquad (4.30)$$

For the proof we construct the piecewise-linear completion of the values of η_l^h at the net points of the difference net:

$$\tilde{\eta}^h(x) = \sum_{l=1}^{N-1} \eta_l^h \omega_l(x).$$

Then (4.30) may be rewritten as

$$\tilde{\eta}^h(x) = u^h(x) - \tilde{u}(x) - h^2 \tilde{v}(x),$$

which implies the equality of the integrals

$$[\tilde{\eta}(x), \omega_k] = [u^h, \omega_k] - [\tilde{u}, \omega_k] - h^2[\tilde{v}, \omega_k]. \tag{4.31}$$

Supposing that u and q have continuous derivatives on the interval $[0, 1]$ through the fourth order, we obtain the Taylor expansions:

$$u(x) = \sum_{l=0}^{3} \frac{d^l u}{dx^l}(x_k) \frac{(x - x_k)^l}{l!} + \frac{(x - x_k)^4}{4!} \frac{d^4 u}{dx^4}(\zeta_k^x), \qquad x \in (x_{k-1}, x_{k+1}),$$

for u, and a similar expansion for q; for v we have an expansion two terms shorter:

$$v(x) = v(x_k) + (x - x_k)\frac{dv}{dx}(x_k) + \frac{(x - x_k)^2}{2}\frac{d^2 v}{dx^2}(\rho_k^x), \qquad x \in (x_{k-1}, x_{k+1}),$$

where the points ζ_k^x and ρ_k^x lie in the interval (x_{k-1}, x_{k+1}). Using these expansions, we repeat the calculations carried out above in deriving the necessary conditions. Then (4.31) becomes

$$[\tilde{\eta}, \omega_k] = \sigma_k, \tag{4.32}$$

where $|\sigma_k| \le C_2 h^5$ (stipulated in order to preserve all terms of order h and h^3, in view of our choice of $v(x)$). We recall that $v(x)$ is the solution of (4.28), (4.29).

We multiply the equations in (4.32) by the corresponding η_l^h and sum over l:

$$[\tilde{\eta}^h, \tilde{\eta}^h] = \sum_{l=1}^{N-1} \sigma_l \eta_l^h.$$

Further, we replace the values of σ_l and η_l^h by their maxima. Then we have

$$[\tilde{\eta}^h, \tilde{\eta}^h] \le NC_2 h^5 \max_{1 \le j \le N-1} |\eta_l^h| = C_2 h^4 \max_{0 \le x \le 1} |\tilde{\eta}^h(x)|. \tag{4.33}$$

Let us now obtain a lower bound. We apply an inequality stemming from the Cauchy-Bunyakovskiĭ inequality for the function $y(x)$, which vanishes at the point $x = 0$:

$$|y(x)| = \left| \int_0^x y'(t)\, dt \right| \le \left(\int_0^x 1\, dt \right)^{1/2} \left(\int_0^x (y'(t))^2\, dt \right)^{1/2}$$

$$\le \sqrt{x} \left(\int_0^1 (y'(t))^2\, dt \right)^{1/2} \le \left(\int_0^1 \{(y'(t))^2 + q(y(t))^2\}\, dt \right)^{1/2}.$$

Since $\tilde{\eta}^h(x)$ vanishes at $x = 0$,

$$|\tilde{\eta}^h(x)|^2 \le [\tilde{\eta}^h, \tilde{\eta}^h], \qquad x \in [0, 1],$$

and we find from the estimate (4.33) that

$$\max_{0 \le x \le 1} |\tilde{\eta}^h(x)|^2 \le [\tilde{\eta}^h, \tilde{\eta}^h] \le C_2 h^4 \max_{0 \le x \le 1} |\tilde{\eta}^h(x)|.$$

Thus

$$\max_{0 \le x \le 1} |\tilde{\eta}^h(x)| = \max_{1 \le l \le N} |\tilde{\eta}_l^h(x)| \le C_2 h^4, \qquad (4.34)$$

which completes the derivation of the expansion (4.15).

The remaining arguments are analogous to those presented in Section 6.1, namely, we construct two uniform nets with mesh sizes h and $h/2$, and find the corresponding solutions of the two problems (4.13). If $x_k = kh$ is a net point, the fourth-order solution is constructed as a linear combination

$$\bar{u}_k = \tfrac{4}{3}\alpha_{2k}^{h/2} - \tfrac{1}{3}\alpha_k^h.$$

The proof is practically the same as that for ordinary first-order differential equations, and yields the estimate

$$|\bar{u}_k - u(x_k)| \le \tfrac{5}{12}C_2 h^4.$$

The method we have presented can be generalized to the case of other boundary conditions. Moreover, it can be applied to the solution of boundary problems for quasi-linear equations, and equations with variable coefficients.

6.5 Nonstationary Problems

In this section we shall first consider an increase in the precision of the solution for the heat equation with one spatial variable. With this as an example, we shall study the compatibility equations arising in nonstationary problems for agreement between the initial and boundary conditions. After that, we shall apply our general results, on increasing the precision of difference solutions on a sequence of nets, to the splitting-up method for systems of differential equations, and in particular, those obtained by replacing spatial derivatives by difference operators.

6.5.1 The Heat Equation

Consider the problem of the cooling of a heated rod

$$\frac{\partial u}{\partial t} = a^2 \frac{\partial^2 u}{\partial x^2}, \qquad x \in [0, 1]; t \in [0, T], \qquad (5.1)$$

having a known initial temperature distribution

$$u(x, 0) = u_0(x), \qquad x \in [0, 1]; \qquad (5.2)$$

the ends of the rod are maintained at temperature zero:

$$u(0, t) = u(1, t) = 0, \qquad t \in (0, T]. \qquad (5.3)$$

Suppose

$$u_0 \in C^\infty[0, 1]. \tag{5.4}$$

Then, at each point of the domain $Q = (0, 1) \times (0, T)$, there exist continuous partial derivatives of the solution u of all orders. However, the continuity of a function on an open set does not ensure its boundedness. Also, supplementary conditions are required to ensure continuity on the closed set $\bar{Q}$. In fact, if the solution $u(x, t)$ is continuous on $\bar{Q}$, comparison of equations (5.2) and (5.3) at two corners of the rectangle $\bar{Q}$ implies that

$$u_0(0) = 0, \qquad u_0(1) = 0.$$

We shall call these the compatibility equations of order zero. If the derivatives $\partial u/\partial t$, $\partial^2 u/\partial x^2$ are continuous on $\bar{Q}$, the evaluation of (5.1)–(5.3) at the corners $(0, 0)$ and $(0, 1)$ yields the first-order compatibility equations

$$\frac{\partial^2 u_0}{\partial x^2}(0) = 0, \qquad \frac{\partial^2 u_0}{\partial x^2}(1) = 0.$$

Differentiating (5.1) under the assumption that all the derivatives so obtained are continuous, we find the equations

$$\frac{\partial^k u}{\partial t^k} = a^{2k} \frac{\partial^{2k} u}{\partial x^{2k}} \quad \text{on} \quad \bar{Q}, k = 1, 2, \dots.$$

Using these, and evaluating (5.2), (5.3) at two corners of the rectangle $\bar{Q}$, we obtain the equations

$$\frac{\partial^{2k} u_0}{\partial x^{2k}}(0) = 0, \qquad \frac{\partial^{2k} u_0}{\partial x^{2k}}(1) = 0, \tag{5.5}$$

which we refer to as the compatibility equations of order $k = 0, 1, 2, \dots$. But these, together with (5.4), ensure the continuity (and boundedness) on $\bar{Q}$ of all derivatives of the form

$$\frac{\partial^{k+1} u}{\partial t^k \partial x^1}, \quad \text{where} \quad 2k + 1 \leq 6. \tag{5.6}$$

For an approximate solution of (5.1)–(5.3) we construct a difference net consisting of the intersections of the lines

$$x_k = kh, \qquad k = 0, 1, \dots, N,$$

$$t_j = j\tau, \qquad j = 0, 1, \dots, M.$$

Here $h = 1/N$, $\tau = T/M$ are the mesh sizes of the difference net with respect to x and t. We replace (5.1) by the implicit difference scheme

$$\frac{v_k^j - v_k^{j-1}}{\tau} = a^2 \frac{v_{k-1}^j - 2v_k^j + v_{k+1}^j}{h^2}, \quad k = 1, \dots, N-1; j = 1, \dots, M.$$

$$\tag{5.7}$$

The initial and boundary conditions are replaced by the system of equations

$$v_k^0 = u_0(x_k), \qquad k = 0, 1, \ldots, N, \tag{5.8}$$

$$v_0^j = v_N^j = 0, \qquad j = 1, \ldots, M. \tag{5.9}$$

We need an *a priori* estimate for this problem with an inhomogeneous right-hand side and zero boundary values. Let η_k^j be a solution of the problem

$$\frac{\eta_k^j - \eta_k^{j-1}}{\tau} = a^2 \frac{\eta_{k-1}^j - 2\eta_k^j + \eta_{k+1}^j}{h^2} + \sigma_k^j,$$

$$k = 1, \ldots, N - 1; j = 1, \ldots, M, \quad (5.10)$$

$$\eta_k^0 = 0, \qquad k = 0, 1, \ldots, N, \tag{5.11}$$

$$\eta_0^j = \eta_N^j = 0, \qquad j = 1, \ldots, M. \tag{5.12}$$

We prove by mathematical induction that

$$\max_{0 \le k \le N} |\eta_k^j| \le \tau_j \max_{\substack{1 \le k \le N-1 \\ 1 \le j' \le j}} |\sigma_k^{j'}|. \tag{5.13}$$

In fact, for $j = 0$ the assertion (5.13) follows from the initial condition (5.11). Suppose it is satisfied on the jth strip. We consider the component of maximum absolute value η_k^{j+1} on the strip $j + 1$. Let it have index number k_0. Clearly, $k_0 \ne 0$ and $k_0 \ne N$. Then it follows from (5.10) that

$$\left(\frac{1}{\tau} + \frac{2a^2}{h^2}\right)|\eta_{k_0}^{j+1}| = \left|\frac{1}{\tau} \eta_{k_0}^j + \frac{a^2}{h^2} \eta_{k_0-1}^{j+1} + \frac{a^2}{h^2} \eta_{k_0+1}^{j+1} + \sigma_{k_0}^{j+1}\right|$$

$$\le \frac{2a^2}{h^2} |\eta_{k_0}^{j+1}| + (j + 1) \max_{\substack{1 \le k \le N-1 \\ 1 \le j' \le j+1}} |\sigma_k^{j'}|.$$

Hence

$$\max_{0 \le k \le N} |\eta_k^{j+1}| = |\eta_{k_0}^{j+1}| \le (j + 1)\tau \max_{\substack{1 \le k \le N-1 \\ 1 \le j' \le j+1}} |\sigma_k^{j'}|$$

and therefore the estimate (5.13) is valid.

It follows that the problem (5.7)–(5.9) is uniquely soluble since the homogeneous problem $[u_0(x_k) = 0 \ (k = 0, 1, \ldots, N)]$ can have no solutions other than the trivial ones.

We shall prove that the solution of the difference problem (5.7)–(5.9) may be expanded as follows:

$$v_k^j = u(x_k, t_j) + h^2 w(x_k, t_j) + \tau z(x_k, t_j) + \eta_k^j,$$

$$k = 0, 1, \ldots, N; j = 0, \ldots, M, \tag{5.14}$$

where the functions w and z are independent of τ and h, while the net function η_k^j is small:

$$\max_{\substack{0 \le k \le N \\ 0 \le j \le N}} |\eta_k^j| \le C_1(h^4 + \tau^2). \tag{5.15}$$

We shall omit some nonrigorous preliminary arguments, leading to the formulation of problems for w and z; and we shall immediately seek these functions as solutions of the problems

$$\frac{\partial w}{\partial t} = a^2 \frac{\partial^2 w}{\partial x^2} + \frac{a^2}{12} \frac{\partial^4 u}{\partial x^4} \quad \text{on} \quad Q,$$

$$w(x, 0) = 0, \qquad x \in [0, 1], \tag{5.16}$$

$$w(1, t) = w(0, t) = 0, \qquad t \in (0, T],$$

$$\frac{\partial z}{\partial t} = a^2 \frac{\partial^2 z}{\partial x^2} - \frac{1}{2} \frac{\partial^2 u}{\partial t^2} \quad \text{on} \quad Q,$$

$$z(x, 0) = 0, \qquad x \in [0, 1], \tag{5.17}$$

$$z(1, t) = z(0, t) = 0, \qquad t \in (0, T].$$

Since the function u has continuous derivatives of the form (5.6) on $\bar{Q}$, the solution of (5.16) exists and is unique. Moreover, it satisfies the compatibility conditions of orders 0, 1, and 2. In fact, those arising from the continuity of w on $\bar{Q}$ are automatically satisfied. Those of order 1, which are necessary and sufficient for the continuity of the derivatives $\partial w/\partial t$, $\partial^2 w/\partial x^2$ on $\bar{Q}$, have the form

$$\frac{\partial^4 u}{\partial x^4}(0, 0) = 0, \qquad \frac{\partial^4 u}{\partial x^4}(1, 0) = 0.$$

These follow from (5.5) with index $k = 2$. To derive the compatibility conditions of order 2 we differentiate the equation in (5.16). We have

$$\frac{\partial^2 w}{\partial t^2} = a^2 \frac{\partial^2}{\partial x^2}\left(\frac{\partial w}{\partial t}\right) + \frac{a^2}{12} \frac{\partial^4}{\partial x^4}\left(\frac{\partial u}{\partial t}\right)$$

$$= a^2 \frac{\partial^2}{\partial x^2}\left(a^2 \frac{\partial^2 w}{\partial x^2} + \frac{a^2}{12} \frac{\partial^4 u}{\partial x^4}\right) + \frac{a^2}{12} \frac{\partial^4}{\partial x^4} a^2 \frac{\partial^2 u}{\partial x^2}$$

$$= a^4 \frac{\partial^4 w}{\partial x^4} + \frac{a^4}{6} \frac{\partial^6 u}{\partial x^6}.$$

Taking account of the zero initial and boundary conditions, we arrive at the second-order compatibility conditions:

$$\frac{a^4}{6} \frac{\partial^6 u}{\partial x^6}(0, 0) = 0, \qquad \frac{a^4}{6} \frac{\partial^6 u}{\partial x^6}(1, 0) = 0.$$

These are evidently satisfied because of (5.5) with index $k = 3$. Thus, all the compatibility conditions of orders 0, 1, and 2 are satisfied for (5.16). They suffice for the continuity on $\bar{Q}$ of the derivatives

$$\frac{\partial^{k+1}w}{\partial t^k \partial x^1}, \quad \text{where} \quad 2k + 1 \le 4. \tag{5.18}$$

A similar assertion holds for the solution of (5.17).

Thus, the functions v_k^j, u, w, and z are uniquely defined at the vertices of the net (x_k, t_j). We introduce the net function

$$\eta_k^j = v_k^j - u_k^j - h^2 w_k^j - \tau z_k^j, \quad k = 0, 1, \ldots, N; j = 0, 1, \ldots, M \tag{5.19}$$

and prove that its values are of the order $\tau^2 + h^4$. To this end we solve (5.19) for the v_k^j and substitute in the difference equation (5.7). We have

$$\frac{(u + h^2 w + \tau z)_k^j - (u + h^2 w + \tau z)_k^{j-1}}{\tau} + \frac{\eta_k^j - \eta_k^{j-1}}{\tau}$$

$$= a^2 \frac{(u + h^2 w + \tau z)_{k-1}^j - 2(u + h^2 w + \tau z)_k^j + (u + h^2 w + \tau z)_{k+1}^j}{h^2}$$

$$+ a^2 \frac{\eta_{k-1}^j + 2\eta_k^j + \eta_{k+1}^j}{h^2}.$$

We expand the functions u, w, and z in a Taylor series, and arrive at the equation

$$\left(\frac{\partial u}{\partial t} + h^2 \frac{\partial w}{\partial t} + \tau \frac{\partial z}{\partial t} + \frac{\tau}{2} \frac{\partial^2 u}{\partial t^2}\right)_k^j + \rho_k^j + \frac{\eta_k^j - \eta_k^{j-1}}{\tau}$$

$$= a^2 \left(\frac{\partial^2 u}{\partial x^2} + h^2 \frac{\partial^2 w}{\partial x^2} + \tau \frac{\partial^2 z}{\partial x^2} + \frac{1}{12} \frac{\partial^4 u}{\partial x^4}\right)_k^j + E_k^j$$

$$+ a^2 \frac{\eta_{k-1}^j + 2\eta_k^j + \eta_{k+1}^j}{h^2}. \tag{5.20}$$

Here ρ_k^j and E_k^j are the remainder terms in Taylor's formula; on account of the boundedness of the corresponding derivatives of u, w, and z we have

$$|\rho_k^j| \le C_2 \tau^2 + C_3 \tau h^2 \le \left(C_2 + \frac{C_3}{2}\right) \tau^2 + \frac{C_3}{2} h^4,$$

$$|E_k^j| \le C_4 h^4 + C_5 \tau h^2 \le \frac{C_5}{2} \tau^2 + \left(C_4 + \frac{C_5}{2}\right) h^4. \tag{5.21}$$

Taking account of (5.1), (5.16), and (5.17) we reduce the parenthesized expressions in (5.20). We obtain the equations

$$\frac{\eta_k^j - \eta_k^{j-1}}{\tau} = a^2 \frac{\eta_{k-1}^j - 2\eta_k^j + \eta_{k+1}^j}{h^2} + \sigma_k^j,$$

$$k = 1, \ldots, N - 1; j = 1, \ldots, M. \tag{5.22}$$

We obtain from (5.21) the estimate

$$|\sigma_k^j| \le C_6(\tau^2 + h^4), \qquad k = 1, \dots, N; j = 1, \dots, M \qquad (5.23)$$

for $\sigma_k^j = E_k^j - \rho_k^j$. Furthermore, it follows from (5.19) and the boundary conditions for u, w, z, and v_k^j that

$$\eta_k^0 = 0, \qquad k = 0, 1, \dots, N,$$

$$\eta_0^j = \eta_N^j = 0, \qquad j = 1, 2, \dots, M.$$

Therefore, the *a priori* estimate (5.13) is valid; this implies the inequality

$$|\eta_k^j| \le TC_6(\tau^2 + h^4), \qquad k = 0, 1, \dots, N; j = 0, 1, \dots, M. \qquad (5.24)$$

Therefore, the estimate (5.15) and the expansion (5.14) have the properties we need.

The solution u has different smoothnesses with respect to x and t. It is therefore quite natural that the expansion (5.14) contains τ^2 and h^4.

We now present a method for increasing the precision of the solution, based on (5.14). Choose integers $M \ge 2$ and $N \ge 2$. Construct the difference net with time mesh $\tau = T/M$ and spatial mesh $h = 1/N$. Then find the solution of (5.7)–(5.9) and denote it by u_h^τ. Next construct a difference net with temporal mesh $\tau/4$ and spatial mesh $h/2$, and again solve (5.7)–(5.9); denote the new solution by $u_{h/2}^{\tau/4}$. We now have at the net points of the net (x_k, t_j) with mesh sizes τ, h two approximate solutions u_h and $u_{h/2}^{\tau/4}$; we form the linear combination

$$\bar{u}(x_k, t_j) = \tfrac{4}{3} u_{h/2}^{\tau/4}(x_k, t_j) - \tfrac{1}{3} u_h^\tau(x_k, t_j),$$

$$k = 0, 1, \dots, N; j = 0, 1, \dots, M. \qquad (5.25)$$

We shall show that the sharpened difference solution $\bar{u}$ approximates the exact solution to the order $\tau^2 + h^4$. In fact, at each net point (x_k, t_j) we have

$$u_h^\tau(x_k, t_j) = u(x_k, t_j) + h^2 w(x_k, t_j) + \tau z(x_k, t_j) + O(\tau^2 + h^4),$$

$$u_{h/2}^{\tau/4}(x_k, t_j) = u(x_k, t_j) + \frac{h^2}{4} w(x_k, t_j) + \frac{\tau}{4} z(x_k, t_j) + O(\tau^2 + h^4).$$

Since the functions u, w, and z do not depend on h and τ, the combination with the weights $-\tfrac{1}{3}, \tfrac{4}{3}$ yields the equation

$$\bar{u}(x_k, t_j) = u(x_k, t_j) + O(\tau^2 + h^4).$$

Thus the difference solution (5.25) obtained by a linear combination of the approximate solutions with order of approximation $\tau + h^2$, approximates the exact solution to order $\tau^2 + h^4$.

6.5.2 The Splitting-Up Method for the Evolutionary Equation

In this section we shall study the differential equation

$$\frac{du}{dt} + A(t)u = f(t), \qquad t \in (0, 1),$$
$$u(0) = u_0, \tag{5.26}$$

where $A(t)$ is a matrix with $n \times n$ elements (functions) and $u(t)$ and $f(t)$ are vector functions with n components. We shall assume that A is representable as a sum

$$A(t) = A_1(t) + A_2(t),$$

where $A_i(t)$ is a positive semidefinite matrix for all $t \in [0, 1]$.

We subdivide the interval $[0, 1]$ into M equal parts by the points

$$t_j = j\tau, \qquad j = 0, 1, \dots, M, \tag{5.27}$$

with mesh size $\tau = 1/M$. To solve (5.26) numerically we consider an implicit splitting-up scheme:

$$(E + \tau A_1(t_j))u^{j-1/2} = u^{j-1} + \tau f^j,$$
$$(E + \tau A_2(t_j))u^j = u^{j-1/2}, \qquad j = 1, 2, \dots, M; u^0 = u_0, \tag{5.28}$$

where E is the unit matrix.

We introduce the norm

$$\|v\| = \left(\sum_{i=1}^{n} v_i^2 \right)^{1/2}$$

for n-dimensional vectors. We derive from Section 4.4.3 an estimate characterizing the stability of our scheme:

$$\max_{0 \le j \le M} \|u^j\| \le \|u_0\| + \max_{0 \le j \le M} \|f^j\|. \tag{5.29}$$

Our aim is to prove that the expansion

$$u^j = \sum_{k=0}^{l-1} \tau^k v_k(t_j) + \eta_\tau^j, \qquad j = 0, 1, \dots, M, \tag{5.30}$$

is valid, where the vector functions v_k are smooth and independent of τ, while the vector functions η_τ satisfy the inequality

$$\|\eta_\tau^j\| \le C_1 \tau^l \tag{5.31}$$

with a constant C_1 independent of τ and k.

To this end we rewrite the system (5.28), excluding the intermediate values $u^{j-1/2}$:

$$\frac{1}{\tau}(E + \tau A_1(t_j))(E + \tau A_2(t_j))u^j - \frac{1}{\tau}u^{j-1} = f^j, \qquad j = 1, 2, \dots, M. \tag{5.32}$$

Let us test Conditions A, B, and C of Section 6.2. Let F^k be a set of n-dimensional vector functions with components in $C^{k+1}[0, 1]$, Φ^k a set of n-dimensional vector functions with components in $C^{k+2}[0, 1]$, and G^k a set of n-dimensional vectors with real coefficients, for all k. Then the assumption that the elements of A belong to $C^{m+1}[0, 1]$ immediately implies Condition A. Condition B is equivalent to the inequality (5.29) with the following notation:

$$\|u\|_{\Phi_h} = \max_{0 \le j \le M} \|u^j\|, \qquad \|f\|_{F_h} = \max_{1 \le j \le M} \|f^j\|, \qquad \|u\|_{G_h} = \|u^0\|.$$

Condition C remains to be tested. As before, we use a Taylor series expansion. Then some elementary calculations lead to the expansions (2.5) of Section 6.2, with $h = \tau$

$$B_1 = A_1 A_2 u + \frac{1}{2} \frac{d^2 u}{dt^2},$$

$$B_l = \frac{(-1)^l}{(l+1)!} \frac{d^{l+1} u}{dt^{l+1}}, \qquad 1 = 2, \dots, k,$$

$$b_1 = 0, \qquad l = 1, \dots, k$$

Thus, all three of Conditions A, B, and C are satisfied, and the theorem on the expansion of the difference solution in powers of h (here τ) is validated, with $\beta = 1$. The expansion so obtained is used in the cutomary manner, on the difference nets with mesh sizes $\tau_i = \tau/i$ $(i = 1, \dots, m+1)$, $= 1/M$. We solve the system (5.28) and denote the solutions by $u^{\tau}i$. These are all defined on the net with mesh size τ. We choose the γ_i to be the solution of the system

$$\sum_{i=1}^{m+1} \gamma_i = 1, \qquad \sum_{i=1}^{m+1} \gamma_i h_i^k = 0, \qquad k = 1, 2, \dots, m+1.$$

With these, we construct the linear combination

$$\ddot{u}(t_j) = \sum_{k=1}^{m+1} \gamma_k u^{\tau_k}(t_j), \qquad j = 0, 1, \dots, M.$$

Then on the basis of the theorem given in Section 6.2.2 we have the estimate

$$\|u(t_j) - \bar{u}(t_j)\| \le C_2 \tau^{m+1}, \qquad j = 0, 1, \dots, M.$$

6.6 Richardson's Extrapolation for Multi-Dimensional Problems

In multi-dimensional problems, the application of extrapolation methods to even the simplest difference schemes is hindered by a number of difficult circumstances. We shall review the principal difficulties and indicate several approaches that circumvent them.

We shall use the two-dimensional Poisson equation

$$\Delta u = f \quad \text{in} \quad D \tag{6.1}$$

as our model, with the boundary condition

$$u = 0 \quad \text{on} \quad \partial D. \tag{6.2}$$

When the region D has a curvilinear boundary ∂D, the elementary difference analogs do not, near the boundary, allow us to expand the approximation error in powers of the mesh size—as we do in the interior of the region. For, in this case the representation

$$u^h(x) = u(x) + h^2 v_1(x) + h^4 v_2(x) + \cdots, \tag{6.3}$$

which is quite natural for the one-dimensional case, has no meaning for a curvilinear boundary ∂D. The v_i in (6.3) are discontinuous and do not lead to smooth solutions that would permit a recursive definition of the necessary v_k $(k < i)$.

Therefore, in the neighborhood of the boundary we must reject the elementary difference analogs, and adopt more complex ones, which usually lead to a large number of nonzero coefficients in the equations near the boundary. If we do this, we shall succeed in obtaining expressions for the approximation error which agree in the interior, and in the strip near the boundary. The first such construction is due to Volkov [4]; it specified the necessary form. As to the method of finite elements, the requirement that the coefficients in the expansion of the approximation error should be continuous up to the boundary leads to the choice of basis functions that are highly complicated in the strip near the boundary, and this in turn leads to the presence of a large number of unknowns in the equations in the boundary strip. Examples of such finite elements are given by Strang and Fix [3].

Even when the boundary consists of straight line segments, and it is possible to construct simple regular difference nets and finite elements up to the boundary, it is difficult to apply the extrapolation method because of the unbounded growth of the derivatives of the solution near the corners of the region. The same state of affairs prevails with respect to the smoothness of the solutions near the intersections of the boundary, with lines of discontinuities of the first kind in the coefficients of the equation. Figure 6.2 displays four types of singular points, of which the first three may fail to allow the customary precision of order h in the solution.

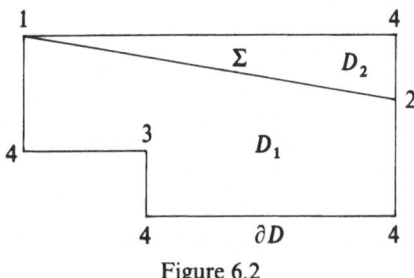

Figure 6.2

The difficulties that we have indicated may be avoided in the following way. In constructing difference schemes by the Galërkin method, we replace the customary scalar product by a weighted product, with weights tending to zero near the singular points. For instance, suppose a singularity exists at the origin of coordinates. Then, instead of using the natural norm for (6.1), (6.2):

$$\|u\|_{\overset{\circ}{W}\frac{1}{2}} = \left[\int_{\Omega} \left[\left(\frac{\partial u}{\partial x} \right)^2 + \left(\frac{\partial u}{\partial y} \right)^2 \right] d\Omega \right]^{1/2}, \tag{6.4}$$

and functions vanishing on the boundary of the region, we employ the following:

$$\|u\|^0_{\alpha, \overset{\circ}{W}\frac{1}{2}} = \left[\int_{\Omega} \left(\frac{\partial u}{\partial x} \right)^2 + \left(\frac{\partial u}{\partial y} \right)^2 \right] (x^2 + y^2)^{\alpha} d\Omega \right]^{1/2}. \tag{6.5}$$

Here α is the order of the singularity, supposedly known *a priori*. We carry through the customary Galërkin construction and arrive at an approximate solution u^h which satisfies the inequality

$$\|u - u^h\|_{\alpha, \overset{\circ}{W}\frac{1}{2}} \le C\|u - \tilde{u}\|_{\alpha, \overset{\circ}{W}\frac{1}{2}}, \tag{6.6}$$

where $\tilde{u}$ is an interpolant of the function u in the space of basis functions.

Note that the admissibility of this method is limited by the fact that α may not be chosen greater than 1.

Another possibility consists in concentrating the difference net or the finite elements so that the approximation error becomes acceptable. This error is made up of integrals over elementary regions and is estimated by quantities of the form

$$Ch_i^{\beta}\|u\|_{k, i}, \tag{6.7}$$

where h_i is the diameter of the ith elementary region or network cell; $\|u\|_{k, i}$ is the norm of the function in this region, and contains the kth order derivative; and C is a constant independent of these factors. We may then demand, for instance, that the quantities (6.7) should be identical throughout D. To this end, we may make the mesh size h_i inversely proportional to the $\|u\|_{k, i}$ as we approach a singularity. We must, however, avoid excessive condensation of the nets since the conditioning number of the algebraic system in the Ritz and Galërkin methods depends in an essential manner on the ratio

$$h_{\max}/h_{\min} \left(h_{\min} = \min_i h_i \right),$$

and the more strongly we condense the network, the larger this ratio becomes. An exception exists in the simple piecewise-linear basis functions on triangles; for these we may normalize the algebraic system by a diagonal matrix in such a way that the ratio in question has no significant influence on the conditioning number.

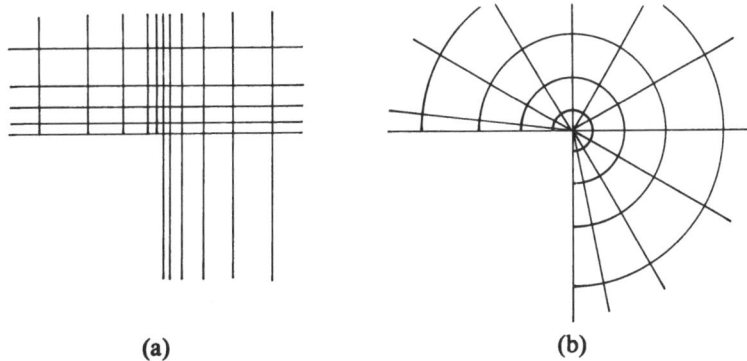

(a)   (b)

Figure 6.3

To condense the nets near singularities we may use an approach outlined schematically in Figure 6.3. Process (a) provides a simpler algorithm, since it involves an uncomplicated computer program; process (b), on the other hand, is more complex, since it involves the pasting together of a polar and a rectangular net, or of two or more polar nets in the case of several singularities (Volkov [4]). Process (b) is more economical as to the number of net points, since the condensation occurs only in the immediate neighborhood of the singularity, while in process (a) the condensation occurs along entire lines, of which a portion is distant from the singularity.

There exists yet another method for increasing the precision of the approximation (Strang and Fix [3]; Oganesyan, Rivkind, and Rukhovets [5]), which consists in adding to the set of standard basis functions described in Chapter 2, another set which describes in detail the behavior of the derivatives of the solution near the singularities. Such an approach is dictated when the right-hand side of the problem (6.1), (6.2) belongs to $W_2^k(D)$ and the solution u is representable as a sum of smooth functions $v \in W_2^{k+2}(D)$ and a finite number of terms which can be represented in polar coordinates centred on a singularity, in the form

$$w_i = \mu_i(\phi)r^{\gamma_i}\ln^{p_i}r,$$

where $\mu(\phi)$ is an analytic function of the angular coordinate, ϕ, γ_i is a real positive constant, and p_i is a nonnegative integer. This form of the singular functions was developed by Nikol'skii [1]. The proof, in each specific case, usually determines these functions only to within a multiplicative constant. Thus, in the Galërkin method, one seeks not only the weights to be used in the expansion of the smooth functions in terms of basis functions, but also the weights to be used for the singular functions. The fundamental difficulty in applying this method consists in the fact that the functions w_i complicate the structure of the expansion of the nonzero elements in the algebraic system, since the supports of these functions embrace a large number of elementary regions.

The use of singular basis functions, to describe in detail the behavior of the solution near the singularities, allows us to use simple elements in the interior of the region, with successive extrapolations with respect to the mesh size, to attain high precision. For the problem (6.1), (6.2) we shall present only the general scheme of the algorithm and its foundation, since the two-dimensional case permits many distinct modifications that cannot be considered here.

We must note that the study of the behavior of the precision of the approximations has generated extensive information about the smoothness of the solutions of elliptic equations in regions of different forms; such information may be found in Volkov [4], Lebedev [4], and others.

Suppose for simplicity that our region is a rectangle with sides of lengths a and b. We use two families of parallel lines, containing the sides of the rectangle, to construct a uniform rectangular net with mesh sizes $h_x = a/N$ and $h_y = b/N$, respectively. (We assume $N > 2$.) To complete the triangulation, we draw a diagonal in each cell, for instance, from the left upward. For convenience we introduce the parameter $h = 1/N$ to characterize the dimensions of the net, and write $h_x = ah$, $h_y = bh$.

At each inner net point X_i we introduce a basis frunction ϕ_i, see Chapter 2, which is linear on every triangle. Apart from these $(N - 1)^2$ functions we introduce the supplementary basis functions ψ_k, equal to the w_i in some neighborhood of the singularities (the corners of the rectangle, and the intersections of the boundaries with possible lines of discontinuity of the coefficients, which are included as lines of the difference net) and smoothly extended on the remainder of the region so that they satisfy the boundary condition (6.2). The number of such supplementary functions depends, in each specific case, on the degree of precision we wish to attain.

Using the Ritz or Galërkin method we obtain the solution

$$u^h(X) = \sum_{i=1}^{(N-1)^2} \alpha_i^h \phi_i(X) + \sum_{k=1}^{M} \beta_k^h \psi_k(X), \tag{6.8}$$

which may be expanded in powers of h at the net points of the difference net (the vertices of the triangles):

$$u^h(X) = u(X) + \sum_{l=1}^{m-1} h^{2l} v_l(X) + h^{2m} \xi_h(X) \tag{6.9}$$

where the functions $v_l(X)$ do not depend on h and the net functions $\xi_h(X)$ are bounded. When this expansion is validated, it is necessary to discard terms of the orders $h^2, \ldots, h^{2m-2}$, obtaining a solution $u(x)$ of precision h^{2m}.

This expansion rests on the possibility of expanding the weights α_i^h and β_k^h occurring in (6.8) in powers of h:

$$\alpha_i^h = v(X_i) + \sum_{l=1}^{m-1} h^{2l} Z_1(X_i) + h^{2m} \eta_h(X_i),$$

$$\beta_k^h = \gamma_k + \sum_{l=1}^{m-1} h^{2l} \eta_{k,l} + h^{2m} \xi_{k,h}, \tag{6.10}$$

where $v(X)$ is the smooth component of $u(X)$; the functions $Z_1(X)$ do not depend on h; the net functions η_h are bounded as $h \to 0$; and γ_k, $\eta_{k,l}$, and $\xi_{k,h}$ are numbers; γ_k and $\eta_{k,l}$ do not depend on h, and $\xi_{k,h}$ is bounded as $h \to 0$.

If we carry through the calculations implicit in the expansion (6.10) we may formulate the conditions (supplementary differential problems) defining the functions $Z_1(X)$ and the constants $\eta_{k,l}$. These appear as coefficients of the $\phi_k(X)$ in the expanded solutions of the supplementary problems for the smooth components $v_l(X)$ and the singular functions $\phi_k(X)$. This schematic implementation is, of course, only indicative of the potential approach; in every specific case it must be carefully justified.

CHAPTER 7

Numerical Methods for Some Inverse Problems

In the current literature, the term *inverse problems* designates various kinds of problems of mathematical physics. Broad classes of inverse problems have been studied by Levitan [16], Lavrentiev, Romanov and Vasilyev [16], and others.

We will consider two types of inverse problems. The first type involves determining past states of a process. For example, the problem of finding the initial distribution of temperature on an object, given the current field of temperatures. In the second type of problem we try to identify the coefficients of the operator with a known structure in terms of information provided by some functionals of the solution. An instance of this is the inverse problem for the Sturm–Liouville equation, in which it is necessary to determine the coefficients of a second-order differential equation using the properties of the spectral function corresponding to a certain boundary value problem.

Inverse problems of mathematical physics are often *ill-posed* (in the classical sense): small perturbations in the observed functionals may result into large changes in the corresponding solutions. The notion of a *well-posed* problem as well as an example of an ill-posed problem were given by Hadamard at the beginning of the century.

For a long time ill-posed problems of mathematical physics were considered uninteresting and drew little attention. Intensive research into these problems has been triggered by a need to interpret geophysical data. Those who have significantly contributed to the theory of ill-posed problems of mathematical physics (in the classical sense of Hadamard) include Tikhonov [16], Lavrentiev [2, 16], Ivanov [16], Turchin [16], and others.

It has been shown that for ill-posed problems to possess stable solutions relative to data perturbations one has to impose certain additional restrictions on the admissible solutions. These problems have become known as *conditionally well-posed*.

A need for approximate solutions of conditionally well-posed problems with inaccurate data has led Tikhonov to the notion of a *regularization family*: along with the conditionally well-posed problem in question we construct a parametrized family of well-posed problems (the regularization family), which has the property that for a parameter approaching its limit the corresponding sequence of solutions of these problems approaches the solution of the conditionally well-posed problem. It has been shown that, depending on the accuracy of the data, the parameter (regularization parameter) can be chosen in such a way that the approximate solution of the corresponding problem from the regularization family will turn out to solve approximately the original conditionally well-posed problem.

Broad classes of conditionally well-posed problems have been studied by the authors just cited. We shall consider only some of them. Section 7.1 introduces the fundamental concepts in the general theory of conditionally well-posed problems. In Sections 7.2 and 7.3 we consider regularization of the problem of determining the initial data in evolution equations. In the rest of the chapter we use perturbation theory to study the problem of establishing the structure of linear and nonlinear operators.

7.1 Fundamental Definitions and Examples

Consider the problem of finding a solution of the equation

$$A\phi = f, \tag{1.1}$$

where A is a linear operator in a Banach space F with an unbounded inverse. The problem (1.1) is ill-posed, since on the one hand the solution for an arbitrary $f \in F$ may fail to exist, and on the other hand small perturbations of f may induce arbitrarily large changes in the solution ϕ when it does exist. We can restore the stability of (1.1) by restricting the class $\{\phi\}$ of solutions. Let M be a set contained in the space F. We shall say that (1.1) is conditionally well-posed (well-posed on M) if the restriction A_M of the operator A on M has a bounded inverse, that is, if we have for all $\phi \in M$ the *a priori* bound

$$\|\phi\| \le \omega(\|A\phi\|), \qquad \phi \in M, \tag{1.2}$$

where $\omega(\varepsilon)$ is a continuous function and $\omega(0) = 0$. The set M is called the *correctness set*.

The choice of M is influenced by physical considerations connected with the formulation of the problem, by the capabilities of the electronic computer to be used, and by the accuracy required of the results to be obtained.

This definition suggests a method for obtaining a stable solution of (1.1)—minimization of the functional $\|A\phi - f\|$ on the correctness set M. An element ϕ_0 that yields a minimum of this functional is called a quasi-solution.

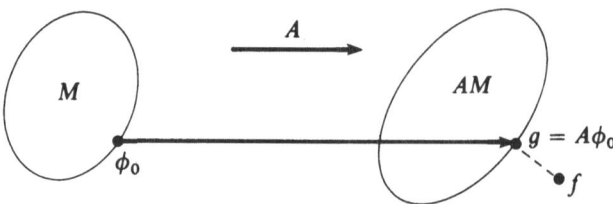

Figure 7.1

It can be shown (Ivanov [16]) that under some additional postulates (M is convex and compact, F is strongly convex, and, in particular, a Hilbert space) the quasi-solution exists, is unique, and depends continuously on the right-hand side $f \in F$. Thus the notion of the quasi-solution makes the problem (1.1) well-posed. For example, Figure 7.1 represents a situation in which the right-hand side $f \notin AM$. Then the quasi-solution ϕ_0 is found as the solution of the equation $A\phi_0 = g$, where g is the projection of f on the set AM.

The correctness set M can often be defined with the aid of some nonnegative homogeneous functional l:

$$M = \{\phi \in F \mid l(\phi) \leq m\}, \qquad l(\phi) \geq 0; \, l(\lambda\phi) = |\lambda| l(\phi), \qquad (1.3)$$

(it is possible that for some elements $\phi \in F$ we have $l(\phi) = \infty$).

In this case there is a natural alternative to the quasi-solution method (Tikhonov [16]): minimization of the functional $l(\phi)$ on the set $\|A\phi - f_\varepsilon\| \leq \varepsilon$, where ε is a number characterizing the accuracy of the initial data f_ε, i.e., $\|f - f_\varepsilon\| \leq \varepsilon$ represents either errors in the right-hand side f, or in the operator A. Usually, the magnitude of the functional l^2 (called the stabilizing functional) characterizes the smoothness of the solution ϕ. As a typical example, we take $F = L_2$, $l(\phi) = \|\phi\|_{W_2^k}$. Then the method selects the "smoothest" ("up to order k") solution of the inequality $\|A\phi - f_\varepsilon\| \leq \varepsilon$. It can be shown that this method is equivalent to the minimization of the functional

$$\psi_\alpha[\phi] = \|A\phi - f_\varepsilon\|^2 + \alpha l^2(\phi)$$

over the whole space; the positive parameter α is defined by the residual of the equation

$$\|A\phi_\alpha - f_\varepsilon\| = \varepsilon,$$

where ϕ_α is the extremal of the functional $\psi_\alpha[\phi]$. In this method there is no need to know the value of the number m; it suffices that the set defined by the inequality $l(\phi) \leq 1$ should be compact, and that on the exact solution of (1.1) the functional l should be finite.

If the set M is chosen in accord with (1.3) it may be shown (Buchheĭm [16]) that the estimate (1.2) is equivalent to the linear estimate

$$\|\phi\| \leq \alpha l(\phi) + c(\alpha)\|A\phi\|, \qquad \alpha \in (0, \alpha_0); \, \phi \in F, \qquad (1.4)$$

which holds for all $\phi \in F$ and all α in some interval $(0, \alpha_0)$; $c(\alpha)$ is a positive continuous function of $\alpha \in (0, \alpha_0)$, related to the function $\omega(\varepsilon)$ by the formula

$$\omega(\varepsilon) = \inf_{\alpha \in (0, \alpha_0)} (\alpha \cdot m + c(\alpha)\varepsilon).$$

Normally, $\lim_{\alpha \to 0} c(\alpha) = \infty$ in ill-posed problems.

The singular case, when M has the form (1.3) with $m = 0$, is also of great interest. Here M is a surface Φ in the space F, given by $l(\phi) = 0$. For $\phi \in \Phi$, the estimate (1.4) becomes

$$\|\phi\| \leq c\|A\phi\|, \qquad \phi \in \Phi.$$

First, the surface Φ is some (finite dimensional) subspace of F, i.e., the functional l is defined by a linear projection P: $l(\phi) = \|P\phi\|$. For this situation, let us analyze a simple instance of the organization of an iterative numerical algorithm that does not derive a sequence of approximate solutions in the given subspace Φ.

Let F be a Hilbert space, of which each element f is representable as a Fourier series in some complete biorthogonal system of functions $\{u_n\}$, $\{u_n^*\}$. Thus

$$f = \sum_{n=1}^{\infty} f_n u_n, \tag{1.5}$$

where

$$f_n = (f, u_n^*).$$

We prescribe the subspace Φ in such a way that it includes only those elements f of the Hilbert space for which the expansion (1.5) contains no more than N nonzero components, corresponding to the perturbations of greatest amplitude:

$$f = \sum_{n=1}^{N} f_n u_n.$$

We shall solve (1.1) by the iterative process

$$\phi^{j+1} = \phi^j - \tau(A\phi^j - f), \qquad \phi^0 = 0, \tag{1.6}$$

or, more compactly,

$$\phi^{j+1} = T\phi^j + \tau f, \qquad \phi^0 = 0,$$

where $T = E - \tau A$ is the transition operator.

We suppose that the solution of (1.1) exists, is unique, and belongs to Φ. We suppose, further, that the norm over the whole Hilbert space satisfies

$$\|T\|_F^2 = \sup_{\phi \in F} \frac{(T\phi, T\phi)}{(\phi, \phi)} = \beta(T^*T) > 1,$$

and that on the subspace Φ

$$\|T\|_\Phi^2 = \sup_{\phi \in \Phi} \frac{(T\phi, T\phi)}{(\phi, \phi)} < 1. \tag{1.7}$$

Under these assumptions the iterative process (1.6), as applied to the functions ϕ^j in Φ, will converge, but will diverge on functions over the whole space F. Therefore, if we wish to obtain a convergent process we must take care that, at every step of the iteration, the approximate solution ϕ^j belongs to Φ. It is very easy to do this constructively.

Suppose that at some step $\phi^{j-1} \in \Phi$. Then by the recursive relation (1.6) we find a new element ϕ^j. The function $T\phi^{j-1}$ may fail to belong to the subspace Φ. To ensure that $\phi^j \in \Phi$, we must expand the ϕ^j in a Fourier series

$$\phi^j = \sum_{n=1}^{\infty} \phi_n^j u_n$$

and throw out all terms with index $n > N$. This aim is easily achieved by defining only the first N Fourier coefficients

$$\phi_k^j = (\phi^j, u_k^*), \qquad k = 1, 2, \ldots, N$$

and constructing the finite sum

$$\phi^j = \sum_{k=1}^{N} \phi_k^j u_k. \tag{1.8}$$

If we extend this process by iteration, all the functions ϕ^j appearing as approximate solutions of our problem will belong to the subspace Φ. With some additional assumptions (for instance, that the basis $\{u_n\}$ is orthogonal) the sequence $\{\phi^j\}$ will converge to some function ϕ^∞, which may be taken as the approximate solution of (1.4).

Let us now turn our attention to the other side of the question of conditionally well-posed problems, namely, the accuracy of the input data. Usually, in a problem of mathematical physics, we must at least deal with errors arising from the approximation of the problem (1.1) by a difference problem, or from the imprecision of our information about the operator A and the function f.

Suppose $\bar{A}$ is the exact operator, $\bar{f}$ the exact vector, and $\bar{\phi}$ the exact solution of the problem

$$\bar{A}\bar{\phi} = \bar{f},$$

and that

$$A = \bar{A} + \delta A, \qquad f = \bar{f} + \delta f;$$

suppose also that $A\phi = f$ is well-posed on the subspace Φ of the Hilbert space F, $A\Phi \subseteq \Phi$, and that A is a positive symmetric operator. Suppose that on the subspace Φ we know the *a priori* magnitudes of δA and δf :

$$\|\delta A\|_\Phi \leq \varepsilon_1, \qquad \|\delta f\|_\Phi \leq \varepsilon_2. \tag{1.9}$$

Then the solution of the equation $A\phi = f$ reduces to the development of an iterative process that will produce new approximations belonging to Φ. If A is a positive definite matrix, then as we showed in Chapter 4, there exists a complete kit of iterative processes converging to the solution of the problem $A\phi = f$. The subspace Φ of approximate solutions coincides with the whole (Euclidean) Hilbert space F. This, incidentally, is one of the pleasant properties of problems with positive matrices. With an appropriate choice of the parameter τ the iterative process (1.6) will converge and the process may be optimized, for instance by using the *a priori* information (1.9) in the choice of the number of steps j_0 in the iterative process (cf. Section 4.6).

Let us now suppose that in the equation $A\phi = f$ the operator A has both positive and negative parts in its spectrum. Analysis shows that (1.6) will then diverge. In fact, let

$$\phi = \sum_n \phi_n u_n, \qquad f = \sum_n f_n u_n \tag{1.10}$$

and let $\{u_n\}$ be the complete orthonormal system of eigenfunctions of A. Putting (1.10) in (1.6), and carrying out a scalar multiplication by u_n, we arrive at a recursive relation for the Fourier coefficients

$$\phi_n^{j+1} = \phi_n^j - \tau(\lambda_n \phi_n^j - f_n), \qquad \phi_n^0 = 0$$

or for the residuals $\xi^j = A\phi^j - f$

$$\xi_n^{j+1} = (1 - \tau\lambda_n)\xi_n^j, \qquad \xi_n^0 = -f_n. \tag{1.11}$$

Solving (1.11) we find

$$\xi_n^j = -(1 - \tau\lambda_n)^j f_n.$$

Accordingly,

$$\xi^j = -\sum_n (1 - \tau\lambda_n)^j f_n u_n.$$

The process (1.6) will converge only if

$$\lim_{j \to \infty} \xi^j = 0.$$

If the operator A had only positive eigenvalues, in the interval

$$\alpha(A) \leq \lambda_n(A) \leq \beta(A),$$

then choosing $\tau > 0$ to lie in the interval

$$0 < \tau < 2/\beta \tag{1.12}$$

would cause (1.6) to converge. In the case we are considering, however, the matrix A has both positive and negative eigenvalues. Suppose τ is chosen to lie in the interval (1.12); then all the components of the residual corresponding to positive values of λ will be damped from one iteration to the next with speed T_n^j, where $T_n^j = (1 - \tau\lambda_n)^j < 1$ and j is an exponent. For the components corresponding to negative eigenvalues we shall have

$$T_n^j = (1 - \tau\lambda_n)^j > 1$$

and such components of the residual will grow, leading to divergence of the iterative process.

Thus, the process (1.6) with the sequence of trial functions ϕ^j belonging to the whole Hilbert space will diverge.

An example of a process that will not generate a sequence of approximate solutions in Φ is provided by the two-step method of minimal residuals (cf. Section 4.3.2):

$$\phi^{j+1} = \phi^j - \tau_j(A\phi^j - f) - \gamma_j A^*(A\phi^j - f).$$

Let us consider the practical approach to the numerical solution of conditionally well-posed problems. These problems generally reduce to systems of linear equations with ill-conditioned matrices of a general form. As a rule, they are solved by a multi-step method of minimal residuals (cf. Section 4.3.2), which ensures rapid convergence of the iterative process. The conjugate gradient method may also be applied. after the equations have been symmetrized by a Gauss transformation. We shall take up this latter method later, in connection with the solution of inverse evolution problems (cf. Section 7.3). An iterative solution process must be broken off when the norm of the residuals is approximately equal to the *a priori* error in the initial data, i.e., when $\|\xi^j\| = \varepsilon_1 + \varepsilon_2$. Then, as we noted in Section 4.6, we arrive at the greatest possible accuracy in the solution, for a given *a priori* error in the input data.

The iterative-variational method introduced above for the stable solution of equation (1.1) is a particular instance of the so-called regularization algorithms. The general definition is as follows.

A family of linear operators R_α in the space F, depending on a numerical parameter α, $0 \leq \alpha \leq \alpha_0$, is called a regularizing family (algorithm) for equation (1.1) on the set M_R, if the following condition

$$\|R_\alpha\| < \infty \tag{1.13}$$

holds for all $\alpha \in (0, \alpha_0]$ and

$$\|R_\alpha A\phi - \phi\| \to 0 \tag{1.14}$$

as $\alpha \to 0$ and for all $\phi \in M_R$. The set M_R, on which (1.14) holds, may be the total correctness set M of the problem (1.1). This is normally the case when the approach to zero in (1.14) is uniform with respect to $\phi \in M_R$. We shall show how the use of the regularization family R_α enables us to find a stable process for obtaining an approximate solution of (1.1) under the condition that the exact solution $\phi \in M_R$ and that the right-hand side contains, instead of f, a known ε-approximation f_ε:

$$\|f - f_\varepsilon\| \leq \varepsilon.$$

We write $\phi_{\alpha\varepsilon} = R_\alpha f_\varepsilon$, and we estimate the norm $\|\phi - \phi_{\alpha\varepsilon}\|$. The triangle inequality implies that

$$\|\phi - \phi_{\alpha\varepsilon}\| = \|\phi - R_\alpha f + R_\alpha(f - f_\varepsilon)\| \leq \|\phi - R_\alpha A\phi\|$$
$$+ \|R_\alpha\| \cdot \|f - f_\varepsilon\| \leq \|\phi - R_\alpha A\phi\| + \|R_\alpha\| \cdot \varepsilon. \quad (1.15)$$

We choose $\alpha(\varepsilon)$ so that $\|R_\alpha\| \cdot \varepsilon \to 0$ and $\alpha(\varepsilon) \to 0$ as $\varepsilon \to 0$.

Then the first term in (1.15) tends to zero as $\varepsilon \to 0$ by virtue of (1.14), and the second term tends to zero by the construction of $\alpha(\varepsilon)$. Thus, $\phi_{\alpha(\varepsilon)\varepsilon} \to \phi$ as $\varepsilon \to 0$.

Therefore, the regularizing family R_α yields, in principle, the possibility of a stable solution of the conditionally well-posed problem (1.1) with an approximately determined right-hand side. The parameter α is called the regularization parameter. Its role in (1.6) is played by the number of iterations j.

For a given problem (1.1) we can usually find an infinite number of regularizing algorithms. In practical computations, we naturally look for such characteristics as simplicity in the computational process, attention to the properties of the input data, optimality in one sense or another, etc.

Let us apply these notions to a simple integral equation—Volterra's equation of the first kind

$$A\phi(t) \equiv \int_0^t A(t, \tau)\phi(\tau)\, d\tau = f(t), \qquad t \in [0, T]. \quad (1.16)$$

The interpretation of meter readings in many physical problems reduces to the solution of (1.16). The operator A in (1.16) is assumed to act on the space $C[0, T]$ of continuous functions ϕ with norm

$$\|\phi\| \equiv \|\phi\|_T = \max_{t \in [0, T]} |\phi(t)|.$$

The kernel $A(t, \tau)$ is assumed to be continuously differentiable with respect to t, continuous in τ, and different from zero on the diagonal $t = \tau$. For simplicity we shall assume that

$$A(t, t) \equiv 1. \quad (1.17)$$

Under these conditions equation (1.16), differentiated with respect to t, yields a Volterra equation of the second kind:

$$\phi(t) + \int_0^t D_t A(t, \tau)\phi(\tau)\, d\tau = f'(t), \quad (1.18)$$

which is known to be soluble by the method of successive approximations. However, if instead of f we know only its ε-approximation f_ε in the norm of $C[0, T]$:

$$\|f - f_\varepsilon\| \leq \varepsilon,$$

the differentiation problem becomes improperly-posed. We shall show that the family R_α of operators which map the functions $f(t)$ into the solutions $\phi(t)$ of the equation

$$\phi_\alpha(t) + \int_0^t \Delta_\alpha A(t, \tau)\phi_\alpha(\tau)\, d\tau = \Delta_\alpha f(t), \qquad t \in [0, T_0], \qquad (1.19)$$

$$T_0 = T - \alpha_0, \qquad \alpha \in (0, \alpha_0],$$

where

$$\Delta_\alpha f(t) = \frac{f(t + \alpha) - f(t)}{\alpha}, \qquad \alpha_0 < T$$

is a regularizing family on the interval $[0, T_0]$ for all continuously differentiable solutions ϕ of (1.16). In fact, the solution ϕ_α of the Volterra equation of the second kind (1.19) exists, is unique, and satisfies the inequality

$$\|\phi_\alpha\|_{T_0} \leq \exp(K T_0)\|\Delta_\alpha f\|_{T_0},$$

where

$$K = \max_{0 \leq \tau \leq t \leq T_0} |\Delta_\alpha A(t, \tau)|,$$

and since

$$\|\Delta_\alpha f\|_{T_0} \leq \frac{2}{\alpha}\|f\|_T,$$

we have

$$\|\phi_\alpha\|_{T_0} \equiv \|R_\alpha f\|_{T_0} \leq \frac{2\exp(K T_0)}{\alpha}\|f\|_T < \infty \qquad (1.20)$$

for 0, and therefore, condition (1.13) is satisfied. To verify (1.14) it suffices to show that

$$\|\phi_\alpha - \phi\|_{T_0} \to 0 \quad \text{as} \quad \alpha \to 0.$$

We apply the operator Δ_α to (1.19), and, taking account of (1.17) we easily obtain the equations

$$\phi(t) + \int_0^t \Delta_\alpha A(t, \tau)\phi(\tau)\, d\tau = \Delta_\alpha f(t) - g_\alpha(t), \qquad (1.21)$$

$$g_\alpha(t) = \frac{1}{\alpha}\int_t^{t+\alpha} (A(t + \alpha, \tau) - A(\tau, \tau))\phi(\tau)\, dt + \frac{1}{\alpha}\int_t^{t+\alpha} (\phi(\tau) - \phi(t))\, d\tau.$$

$$(1.22)$$

Subtracting (1.21) from (1.19) to obtain the residual $u_\alpha = \phi_\alpha - \phi$, we arrive at the equation

$$u_\alpha + \int_0^t \Delta_\alpha A(t, \tau) u_\alpha(\tau) \, d\tau = g_\alpha(t), \qquad t \in [0, T_0]. \tag{1.23}$$

It follows from (1.23) and (1.22) that

$$\|\phi - \phi_\alpha\|_{T_0} \equiv \|u_\alpha\|_{T_0} \le \exp(K T_0) \|g_\alpha\|_{T_0}, \tag{1.24}$$

$$\|g_\alpha\|_{T_0} \le \frac{1}{\alpha} K\alpha \|\phi\|_T \alpha + \frac{1}{\alpha} \|\phi'\|_T \alpha^2 = \alpha(K\|\phi\|_T + \|\phi'\|_T) \to 0, \tag{1.25}$$

as $\alpha \to 0$. Thus, we have shown that the operators R_α form a regularizing family for equation (1.16) on the set of continuously differentiable functions ϕ. An important property of this regularizing algorithm is that—as opposed to the standard variational algorithms—it conserves the Volterran character of the initial equation.

We define a functional l by the formula

$$l(\phi) \equiv (K\|\phi\|_T + \|\phi'\|_T) \exp(K T_0).$$

Since $\|\phi\|_{T_0} \le \|\phi_\alpha\|_{T_0} + \|\phi_\alpha - \phi\|_{T_0}$, it follows from the inequalities (1.20), (1.24), and (1.25) that

$$\|\phi\|_{T_0} \le \alpha l(\phi) + \frac{2 \exp(KT)}{\alpha} \|f\|_T, \qquad 0 < \alpha \le \alpha_0.$$

Thus, by (1.4) the problem is conditionally well-posed on the set $M = \{\phi : l(\phi) \le m\}$.

7.2 Solution of the Inverse Evolution Problem with a Constant Operator

In this section we shall consider two stable methods for solving the inverse evolution equation:

$$\frac{d\phi}{dt} - A\phi = 0, \qquad \phi(0) = g; A \ge 0.$$

The first is based on the use of Fourier series and reduces in practice to the solution of a spectral problem; in the second, an initially ill-posed problem is reduced to the solution of a sequence of well-posed (direct) evolution equations.

7.2.1 The Fourier Method

Let A be a time-invariant positive matrix with a real spectrum in the interval $\alpha(A) \leq \lambda \leq \beta(A)$, and let the vector function ϕ solve the following Cauchy problem:

$$\frac{d\phi}{dt} - A\phi = 0, \quad 0 \leq t \leq t_0,$$

$$\phi = g \quad \text{for} \quad t = 0, \tag{2.1}$$

where g is a given vector. Consider the spectral problems

$$Au = \lambda u; \quad A^*u^* = \lambda u^*. \tag{2.2}$$

and suppose they define biorthogonal bases $\{u_n\}$, $\{u_n^*\}$. Writing

$$\phi = \sum_n \phi_n u_n, \quad g = \sum_n g_n u_n, \tag{2.3}$$

substituting into (2.1), and multiplying the result by u_n^* (in the sense of inner product), we obtain the following system of ordinary differential equations for the Fourier coefficients:

$$\frac{d\phi_n}{dt} - \lambda_n \phi_n = 0,$$

$$\phi_n = g_n \quad \text{for} \quad t = 0 \tag{2.4}$$

$$n = 1, 2, \ldots, N.$$

Solving each of the above equations separately yields

$$\phi_n = g_n e^{\lambda_n t}, \quad n = 1, 2, \ldots, N, \tag{2.5}$$

and hence the solution of (2.1) can be represented in the form

$$\phi(t) = \sum_{n=1}^{N} g_n e^{\lambda_n t} u_n. \tag{2.6}$$

Thus, our solution is given in terms of a Fourier sum, each term of which grows exponentially in time, the rate of growth being dependent on the corresponding eigenvalue λ_n.

Suppose we are interested in a solution which is physically meaningful. Thus consider a similar problem as (2.1), but now well-posed:

$$\frac{d\phi}{dt} - A\phi = 0, \quad 0 \leq t \leq t_0,$$

$$\phi = h \quad \text{for} \quad t = t_0. \tag{2.7}$$

Similarly as above, we obtain

$$\phi = \sum_{n=1}^{N} h_n e^{-\lambda_n(t_0 - t)} u_n. \tag{2.8}$$

Let us require, that for $t = 0$ the solution of (2.8) coincides with the vector g from (2.1). From this we obtain the relation between the Fourier coefficients of h and g:

$$g_n = h_n e^{-\lambda_n t_0}. \tag{2.9}$$

In this manner we have derived a rather simple formula for g in terms of h:

$$g = \sum_{n=1}^{N} h_n e^{-\lambda_n t_0} u_n. \tag{2.10}$$

Moreover, small errors in h (or h_n) do not cause large errors in g. Our problem, however, is just the opposite to the above: what we have at our disposal is the information regarding g, while h is to be computed in terms of g by the formula

$$h = \sum_{n=1}^{N} g_n e^{\lambda_n t_0} u_n. \tag{2.11}$$

If we had accurate information about the function g and could compute without error, the function h could be reconstructed via equation (2.11) without difficulty. In our situation, however, the function g is known only approximately, with the corresponding error bound given a priori. At the same time the numerical procedures are subject to round-off errors in the computer. In view of these two limitations the computation of h by equation (2.11) becomes more difficult.

Suppose first of all that the system of eigenvectors u_n is known and that the Fourier expansion of the initial data may be used to draw useful information including sufficiently accurate error estimates for the Fourier components g_n.

If the problem is of a statistical nature and can be repeated many times, well-developed correlation techniques allow a significant increase in the accuracy of the data entering g_n, even if the error of a single measurement drastically excedes the relevant information. In any case, the preliminary processing of the observation data may serve to draw conclusions regarding the systematic (or random, if only one single measurement is available) error in g_n. Thus we have that for any n

$$g_n = \bar{g}_n(1 + \delta_n),$$

where g_n denotes the exact value (not known a priori), and δ_n is a relative inaccuracy which we consider as known.

The error estimate δ_n is usually small for the low-frequency components, but grows quickly with the frequency as a rule. Therefore, starting with some index, the coefficients g_n describe in fact the errors in the input data. Returning now back to equation (2.11), it follows that the highest-frequency components grow at the highest exponential rate. Hence, if we do not take notice and leave these parasitic frequencies in, we may obtain a wrong result: these components contain no practical useful information, and being multiplied by large factors $\exp(\lambda_n t_0)$, they may sometimes cause irreparable damage to the solution of the problem. The primary task is therefore to determine the information content of the coefficients g_n.

Suppose that using the *a priori* information it has been concluded that the relative inaccuracies in the first n_0 coefficients g_n is smaller than η, i.e., $\delta_n < \eta$, where η is the largest admissible error. Algorithm (2.11) then becomes similar to the one we have already considered when constructing the sub-space Φ for problem (2.1). The only thing we need is to drop the parasitic components in (2.11):

$$h = \sum_{n=1}^{n_0} g_n e^{\lambda_n t_0} u_n. \tag{2.12}$$

The algorithm for solving a particular spectral problem is based on the iteration method described in Section 1.1. If we must construct the set of leading (most significant) eigenfunctions u_n or u_n^* and the corresponding eigenvalues, we may use the orthogonalization algorithm given in Section 1.1.

7.2.2 Reduction to the Solution of a Direct Equation†

We again consider the Cauchy problem

$$\frac{d\phi}{dt} - A\phi = 0, \qquad \phi(0) = g, \qquad 0 \le t \le t, \tag{2.13}$$

where A is a self-adjoint unbounded positive operator on the Hilbert space F. In other words, as opposed to what we have just discussed, we do not develop any preliminary finite-difference approximation with respect to the spatial variable. In this case problem (2.13) is ill-posed in the classical sense. We shall suppose that the solution ϕ of (2.13) exists and belongs to the set M but, instead of an exact initial condition g we are given the approximation g_ε

$$\|g - g_\varepsilon\| \le \varepsilon. \tag{2.14}$$

It is known (Kreĭn [16]) that

$$\|\phi(t)\| \le \|\phi(0)\|^{1-(t/t_0)} \|\phi(t_0)\|^{t/t_0}, \tag{2.15}$$

and therefore, for arbitrary fixed $t \in (0, t_0)$ the problem of determining $\phi(t)$ is conditionally well-posed, since by (2.15)

$$\|\phi(t)\| \le m^{t/t_0} \|g\|^{1-(t/t_0)}, \qquad \phi \in M. \tag{2.16}$$

We show that the operators R_α defined by the formula

$$R_\alpha = (e^{-At_0} + \alpha E)^{-t/t_0}, \qquad \alpha > 0 \tag{2.17}$$

form a regularizing family on the correctness set M.

† See Shishatskii [16].

In fact, since the operator R_α is a function of the self-adjoint positive operator A, by (2.17), we use a spectral resolution and find that

$$\|R_\alpha\| \leq \max_{\lambda \geq 0} (e^{-\lambda t_0} + \alpha)^{-t/t_0} = \alpha^{-t/t_0} < \infty. \tag{2.18}$$

To verify condition (1.14) of Section 7.1 (the operator e^{-At} plays the role of A, and $f \equiv g$, so that $\phi(t) = e^{At}g$, $e^{-At}\phi(t) = g$) we have only to show that

$$\|R_\alpha e^{-At}\phi(t) - \phi(t)\| \to 0 \quad \text{as} \quad \alpha \to 0, \qquad \phi \in M. \tag{2.19}$$

Again, using the spectral resolution of the operator A, we have

$$\|\phi(t) - R_\alpha e^{-At}\phi(t)\| = \|e^{At}g - R_\alpha g\|$$

$$\leq \max_{\lambda \geq 0} e^{\lambda(t-t_0)}[1 - (1 + \alpha e^{\lambda t_0})^{-t/t_0}] \cdot \|e^{At_0}g\|.$$

Further, to estimate these quantities we use the inequality

$$(1 + x)^\tau - 1 \leq \tau x (1 + x)^\tau (1 + \tau x)^{-1},$$

which is easily proved for $x \geq 0$ and $\tau \in [0, 1]$ by a differential computation. Noting also that $\phi \in M$, i.e., that $\|\phi(t_0)\| = \|e^{At_0}g\| \leq m$, we find that:

$$\max_{\lambda \geq 0} e^{\lambda(t-t_0)}[1 - (1 + \alpha e^{\lambda t_0})^{-t/t_0}]\|e^{At_0}g\|$$

$$\leq \alpha^{1-(t/t_0)} \max_{\alpha \leq x < \infty} x^{(t/t_0)-1}[1 - (1 + x)^{-t/t_0}] \cdot m$$

$$\leq \alpha^{1-(t/t_0)} \max_{\alpha \leq x < \infty} \frac{t}{t_0} x^{t/t_0}\left(1 + \frac{t}{t_0}\right)^{-1} \cdot m = \frac{t}{t_0}\left(1 - \frac{t}{t_0}\right)^{1-(t/t_0)} \alpha^{1-(t/t_0)} \cdot m.$$

Thus

$$\|\phi(t) - R_\alpha g\| \leq \frac{t}{t_0}\left(1 - \frac{t}{t_0}\right)^{1-(t/t_0)} \alpha^{1-(t/t_0)} \cdot m \to 0 \tag{2.20}$$

as $\alpha \to 0$, with $t < t_0$. This proves that $\{R_\alpha\}$ is a regularizing family. Using (2.16), (2.20), and (2.22) and the triangle inequality, we have

$$\|\phi(t) - R_\alpha g_\varepsilon\| \leq \|\phi(t) - R_\alpha g\| + \|R_\alpha\| \cdot \|g - g_\varepsilon\|$$

$$\leq \frac{t}{t_0}\left(1 - \frac{t}{t_0}\right)^{1-(t/t_0)} \alpha^{1-(t/t_0)} \cdot m + \alpha^{-t/t_0} \cdot \varepsilon.$$

Some elementary computations will show that the right-hand side of this inequality is minimized when

$$\alpha_0 = \left(1 - \frac{t}{t_0}\right)^{-2+(t/t_0)} \cdot \frac{\varepsilon}{m}.$$

Since for $t \in (0, t_0)$

$$\left(1 - \frac{t}{t_0}\right)^{-[1 - (t/t_0)]^2} \leq e^{1/2e},$$

we have

$$\|\phi(t) - R_{\alpha_0}g\| \leq e^{1/2e}m^{t/t_0} \cdot \varepsilon^{1 - (t/t_0)}. \tag{2.21}$$

The inequality (2.21) is an estimate of the distance between the approximate solution of the problem (2.15), as constructed with the aid of the operator R_{α_0}, and the exact solution $\phi(t)$. It differs from (2.18), the *a priori* estimate of the stability on M, only by the insignificant multiplier $e^{1/2e} \approx 1.21$. In this sense our current method for the solution of (2.15) is optimal.

For practical computation of the element

$$R_{\alpha_0}g_\varepsilon \equiv (e^{-At_0} + \alpha_0)^{-t/t_0}g_\varepsilon$$

it is appropriate to write

$$R_{\alpha_0}g_\varepsilon \approx Q_n(e^{-At_0})g_\varepsilon,$$

where $Q_n(x)$ is the nth degree polynomial of best approximation to the function $(x + \alpha_0)^{-t/t_0}$ on the interval $0 \leq x \leq 1$. The operator $Q_n(e^{-At_0})$ is a polynomial in e^{-At_0}, i.e., to compute the element

$$Q_n(e^{-At_0})g_\varepsilon$$

it suffices to know how to compute the elements $e^{-kAt_0}g_\varepsilon$ for $k = 1, 2, \ldots, n$. But $e^{kAt_0}g_\varepsilon$ is the solution of the well-posed Cauchy problem

$$\frac{d\psi}{dt} + A\psi = 0, \qquad \psi(0) = g_\varepsilon, \qquad t \geq 0 \tag{2.22}$$

for $t = kt_0$. We have, therefore, reduced the ill-posed problem (2.15) to a sequence of well-posed problems (2.22); effective methods for dealing with these were given in Chapter 5. It can be shown that the final error in the method, namely $\|\phi(t) - Q_n(e^{-At_0}g_\varepsilon)\|$ is bounded from above by the quantity

$$\left(1 - \frac{t}{t_0}\right)^{-[1 - (t/t_0)]^2} \varepsilon^{1 - (t/t_0)} \cdot m^{t/t_0} + \frac{2^{2(t + t_0) + 1}}{\Gamma(t/t_0)}(n + 1)^{(t/t_0) - 1}$$

$$\times \frac{\beta^{n + 1 + (t/t_0)}}{(1 - \beta^2)^{1 + (t/t_0)}} \|g_\varepsilon\|, \tag{2.23}$$

where $\beta = 1 - 2\alpha_0 - 2\sqrt{\alpha_0 + \alpha_0^2}$, and Γ is the gamma function. We must emphasize the fact that in cases of practical interest the degree of the polynomials Q_n is small. As (2.23) implies, for

$$\frac{\varepsilon}{\|g_\varepsilon\|} = 0.1, \qquad \frac{\|g_\varepsilon\|}{m} = 0.1$$

the choice of Q_2 as our polynomial will ensure an accuracy of

$$2\varepsilon^{1-(t/t_0)} \cdot m^{t/t_0};$$

for

$$\frac{\varepsilon}{\|g_\varepsilon\|} = 0.05, \qquad \frac{\|g_\varepsilon\|}{m} = 0.1$$

Q_4 suffices. We have explicit formulas for computing the coefficients of the polynomials $Q_n(x)$; they depend only on the ratios ε/m and t/t_0, and are independent of the operator A and the initial datum g_ε.

7.3 Inverse Evolution Problems with Time-Varying Operators

Consider the evolution problem

$$\frac{d\phi}{dt} - A(t)\phi = 0, \qquad 0 \le t \le t_0,$$

$$\phi = g \quad \text{for} \quad t = 0,$$

(3.1)

with $A > 0$ depending on time. We assume, as before, that (3.1) resulted by reducing a problem of mathematical physics to a system of ordinary differential equations. Since the Fourier method is no longer applicable in this case, (3.1) must be solved by numerical methods.

Let us consider one of the possible algorithms. In correspondence to (3.1) let us define the following *model* [a problem in a sense close to (3.1)]:

$$\frac{d\bar{\phi}}{dt} - \bar{A}\bar{\phi} = 0, \qquad 0 \le t \le t_0,$$

$$\bar{\phi} = g \quad \text{for} \quad t = 0$$

(3.2)

where the operator $\bar{A} > 0$ is now time invariant, has a positive spectrum $\alpha(\bar{A}) \le \lambda(\bar{A}) \le \beta(\bar{A})$, and in some sense is close to the operator $A(t)$. To be specific, let

$$A(t) = \bar{A} + \delta A(t),$$

(3.3)

where

$$\|\delta A(t)\| \ll \|\bar{A}\|$$

(3.4)

for any t from the interval $0 \le t \le t_0$. Equations (3.2) will be used for obtaining a necessary *a priori* information to be used later in designing the numerical algorithm for solving the original problem (3.1).

Using the methods for time-invariant problems from Section 3.2, we first determine the eigenvectors u_n and u_n^* $(n = 1, \ldots, m)$ with the major information content (from the viewpoint of the input data errors) and the corresponding eigenvalues λ_n. The remaining eigenvectors are not needed, since the Fourier components g_n $(n = m + 1, \ldots, N)$ are to be dropped as the numerical error involved exceeds (sometimes quite considerably) the useful information they may carry. Thus

$$g = \sum_{n=1}^{m} g_n u_n, \qquad (3.5)$$

where

$$g_n = (g, u_n^*).$$

As a result, the solution $\bar{\phi}$ corresponding to the model of (3.2) can be written in the form

$$\bar{\phi}(t) = \sum_{n=1}^{m} g_n e^{\lambda_n t} u_n, \qquad 0 \le t \le t_0. \qquad (3.6)$$

Let us try to solve the model problem (3.2) numerically. For this consider, for instance, the following difference scheme of second-order accuracy in $\Delta t = \tau$:

$$\frac{\bar{\phi}^{j+1} - \phi^j}{\tau} - \bar{A} \frac{\bar{\phi}^{j+1} + \bar{\phi}^j}{2} = 0, \qquad \bar{\phi}^0 = g, \qquad j = 1, 2, \ldots, j_0. \qquad (3.7)$$

Assume we know the set of eigenelements u_n and u_n^*. (This assumption is made only for theoretical purposes, in order to obtain an a priori information regarding the solution). Using the Fourier method, we obtain

$$\bar{\phi}^j = \sum_{n=1}^{N} \bar{\phi}_n^j u_n, \qquad (3.8)$$

and combining further with (3.7) we have

$$\bar{\phi}_n^{j+1} = \frac{1 + \dfrac{\tau \lambda_n}{2}}{1 - \dfrac{\tau \lambda_n}{2}} \bar{\phi}_n^j, \qquad \bar{\phi}_n^0 = g_n, \qquad j = 1, 2, \ldots, j_0. \qquad (3.9)$$

Hence

$$\bar{\phi}_n^j = \left(\frac{1 + \dfrac{\tau \lambda_n}{2}}{1 - \dfrac{\tau \lambda_n}{2}} \right)^j g_n. \qquad (3.10)$$

Thus

$$\bar{\phi}^j = \sum_{n=1}^{N} T_n^j g_n u_n, \tag{3.11}$$

where

$$T_n = \frac{1 + \dfrac{\tau \lambda_n}{2}}{1 - \dfrac{\tau \lambda_n}{2}}.$$

Let τ be such that the denominator in (3.11) does not become zero for any n. Then

$$\tau < \frac{2}{\beta(\bar{A})} \tag{3.12}$$

Note that this condition is compatible with the emphasis we place on the components with the highest information content.

A formal analysis of (3.11) shows that $T_n > 1$ for all n, and that the high-frequency components which correspond to the large indices n grow quickly in amplitude as n increases. For these indices we have, therefore, $T_n \gg 1$, and in particular, $T_n^j \gg 1$. In the course of processing the input data g we have dropped all the Fourier components in (3.5) beginning with the index $n = m + 1$; hence it may seem at first that we are guaranteed that the Fourier sum

$$\bar{g} = \sum_{n=1}^{m} g_n u_n$$

generates a solution with the same number of components, i.e.,

$$\bar{\phi}^j = \sum_{n=1}^{m} T_n^j g_n u_n. \tag{3.13}$$

This would indeed be the case if the computations were made with the infinite accuracy. Since this is not the case, we will immediately observe the components g_n with $n > m$. Although small, they have large weights (proportional to $T_n^j \gg 1$), and eventually they may corrupt significantly the desired solution. In order to avoid the catastrophic growth of errors in the high-frequency Fourier components, we need to find a construction which would automatically take the elements from F into a certain subspace Φ.

Let us define Φ as follows: an element will be considered to belong to Φ if its last $(N - m)$ components in the Fourier sum do not grow faster in amplitude than several amplitudes of the last informative component indexed by m. With Φ constructed in this way, the round-off errors will not grow on its elements faster than the amplitudes of the mth component. This guarantees that the computational scheme is well-posed.

Lavrentiev [2, 16] and Lions and Lattes [16] have suggested a modification of (3.2) by taking $\bar{A} = \bar{A} - \varepsilon\bar{A}^2$ instead of $\bar{A}$. In this case we have

$$\frac{d\bar{\phi}_\varepsilon}{dt} - \bar{A}\bar{\phi}_\varepsilon = -\varepsilon\bar{\phi}^2\bar{\phi}_\varepsilon, \qquad 0 \le t \le t_0,$$

$$\bar{\phi}_\varepsilon = g \qquad t = 0,$$

(3.14)

where ε is arbitrary at the moment. This parameter will be chosen from the requirement that the solution stay in Φ. For simplicity, assume that $\bar{A} = \bar{A}^*$ and consider the difference scheme

$$\frac{\bar{\phi}_\varepsilon^{j+1} - \bar{\phi}_\varepsilon^j}{\tau} - (\bar{A} - \varepsilon\bar{A}^2)\frac{\bar{\phi}_\varepsilon^{j+1} + \bar{\phi}_\varepsilon^j}{2} = 0,$$

$$\bar{\phi}_\varepsilon^0 = g.$$

(3.15)

The above equation will be solved using the Fourier series expansion along the eigenfunctions of the operator $\bar{A}$. We obtain

$$\bar{\phi}_\varepsilon^j = \sum_{n=1}^N \left[\frac{1 + \dfrac{\tau\lambda_n}{2} - \varepsilon\dfrac{\tau\lambda_n^2}{2}}{1 - \dfrac{\tau\lambda_n}{2} + \varepsilon\dfrac{\tau\lambda_n^2}{2}} \right]^j g_n u_n.$$

(3.16)

In order to choose ε, let us require that the relative error in the mth component, due to the operator εA^2, does not exceed η (usually $\eta < 1$ depending on the ratio between the useful information and the inaccuracies which can not be accounted for (the noise) in the mth component). The above requirement implies

$$\frac{\tau\lambda_m}{2} = \frac{\eta\varepsilon\tau\lambda_m^2}{2}.$$

(3.17)

Hence

$$\varepsilon = \frac{1}{\eta\lambda_m}.$$

(3.18)

In this fashion we have determined an important *a priori* quantity which will be needed later. It is easily seen that for ε given by (3.18) the amplitudes of all the components with indices $n > m$ will increase with time at a rate not higher than T_m^j.

We will have occasion to use yet another *a priori* quantity. Consider

$$\bar{\phi}^j = \sum_{n=1}^m g_n e^{\tau\lambda_n j} u_n$$

(3.19)

and

$$\bar{\phi}_\varepsilon^j = \sum_{n=1}^N g_n T_n^j(\varepsilon) u_n,$$

(3.20)

where

$$T_n(\varepsilon) = \frac{1 + \dfrac{\tau\lambda_n}{2} - \varepsilon\dfrac{\tau\lambda_n^2}{2}}{1 - \dfrac{\tau\lambda_n}{2} + \varepsilon\dfrac{\tau\lambda_n^2}{2}}.$$

Since $\bar{\phi}_\varepsilon^j$ belongs to Φ, we will not introduce a large error by replacing it with

$$\bar{\phi}_\varepsilon^j = \sum_{n=1}^m g_n T_n^j(\varepsilon) u_n, \tag{3.21}$$

where we limit ourselves only to the first m members in the series. Using the above form, the solution can be found in a constructive fashion (the system of functions u_n and u_n^* ($n = 1, \dots, m$), has already been found). Next, we compute $\bar{\phi}^j$ and $\bar{\phi}_\varepsilon^j$ ($j = 1, \dots, j_0$), using (3.19) and (3.21). Let us introduce the new vectors

$$\bar{\phi} = \begin{vmatrix} \bar{\phi}^1 \\ \bar{\phi}^2 \\ \cdots \\ \bar{\phi}^{j_0} \end{vmatrix}, \qquad \bar{\phi}_\varepsilon = \begin{vmatrix} \bar{\phi}_\varepsilon^1 \\ \bar{\phi}_\varepsilon^2 \\ \cdots \\ \bar{\phi}_\varepsilon^{j_0} \end{vmatrix}$$

and compute the norm

$$\|\bar{\phi} - \bar{\phi}_\varepsilon\| = \delta. \tag{3.22}$$

The above norm is the last *a priori* quantity we need. The other two, τ and ε, are given by (3.12) and (3.18).

Let us formulate the numerical algorithm for solving the original problem (3.1). Taking into account the above analysis, let us construct the following approximation problem:

$$\frac{\phi^{j+1} - \phi^j}{\tau} - (A_j - \varepsilon A_j^2)\frac{\phi^{j+1} + \phi^j}{2} = 0, \qquad \phi^0 = g, \tag{3.23}$$

$$\tau = \frac{2}{\beta(\bar{A})}; \qquad \varepsilon = \frac{1}{\eta\lambda_m(\bar{A})}. \tag{3.24}$$

Let

$$\phi = \begin{vmatrix} \phi^1 \\ \phi^2 \\ \cdots \\ \phi^{j_0} \end{vmatrix}, \qquad f = \begin{vmatrix} -R_0 g \\ 0 \\ \cdots \cdots \\ 0 \end{vmatrix},$$

$$\Lambda = \begin{Vmatrix} -S_0 & 0 & 0 & 0 & \cdots & 0 & 0 \\ R_1 & -S_1 & 0 & 0 & \cdots & 0 & 0 \\ 0 & R_2 & -S_2 & 0 & \cdots & 0 & 0 \\ 0 & 0 & R_3 & -S_3 & \cdots & 0 & 0 \\ \cdots\cdots\cdots\cdots\cdots\cdots\cdots\cdots\cdots\cdots \\ 0 & 0 & 0 & 0 & \cdots & R_{j_0-1} & -S_{j_0-1} \end{Vmatrix},$$

where

$$S_j = E - \frac{\tau}{2}(A_j - \varepsilon A_j^2); \qquad R_j = E + \frac{\tau}{2}(A_j - \varepsilon A_j^2);$$

$$A_j = A(t_{j+1/2}).$$

The problem can then be written as

$$\Lambda\phi = f. \tag{3.25}$$

Next let us symmetrize by multiplying with Λ^*, i.e.,

$$\Lambda^*\Lambda\phi = \Lambda^*f, \tag{3.26}$$

and formulate an iterative process. In particular, we can use conveniently the conjugate gradients method, which doesn't require computation of the spectral boundaries of $\Lambda^*\Lambda$.

Having specified the successive approximations, we still have to determine the optimal number of iterations k_0, so that we reach the highest attainable accuracy under the given *a priori* conditions. Since such a number cannot be computed with a high precision, we will assume that the *a priori* estimate of (3.22), obtained for the model, is also acceptable for (3.1). Thus suppose

$$\|\phi - \phi_\varepsilon\| = \delta, \tag{3.27}$$

where ϕ is the exact solution of (3.1) at the net points and ϕ_ε solves the difference problem with the regularization operator. Under these conditions it is natural to continue the iterative process (3.23) as long as the iteration error stays larger than the approximation error of (3.27). This can be easily implemented. Introduce the residual ζ^k according to the formula

$$\zeta^k = \Lambda^*(\Lambda\phi^k - f) = \Lambda^*\Lambda(\phi^k - \phi). \tag{3.28}$$

We have then

$$\|\zeta^k\| \le \|\Lambda^*\Lambda\| \|\phi^k - \phi\|. \tag{3.29}$$

Clearly, the approximation error $\|\phi - \phi_\varepsilon\| = \delta$ is equivalent to the residual

$$\|\zeta^k\| \le \delta\|\Lambda^*\Lambda\|. \tag{3.30}$$

This means that the numerical process is to be continued until the residual no longer comes close to the right-hand side of (3.30) in norm. Thus we have the following parametric estimate for k_0:

$$\|\zeta^{k_0}\| \le \beta(\Lambda^*\Lambda)\delta. \tag{3.31}$$

As we have seen, inverse evolution problems require a formidable preliminary analysis of various simple models, which eventually allow one to obtain a necessary *a priori* information needed for the qualitative construction of numerical algorithms. In particular cases we may face even more complex situations. Nevertheless, the above development is sufficient to give an idea of how simple models and error analysis can be used in order to formulate numerical methods. Although the regularization process has been looked at from only one point of view, it is enough to see various possible numerical approaches to the inverse problems. Deeper considerations can be found in the monograph by Lavrentiev [2].

In conclusion, let us note that the methods and ideas described above can also be applied to Cauchy problems for elliptic equations. These problems are ill-posed in the classical sense and require methods from the theory of the conditionally well-posed problems. A sizable research in this direction has been carried out by Tikhonov [16], Lavrentiev [16], Ivanov [16], and others.

7.4 Methods of Perturbation Theory for Inverse Problems

Some inverse problems can be formulated in the framework of the theory of conjugate functions and perturbation theory. This approach is beginning to play an ever-increasing role in forming numerical algorithms, especially when dealing with complicated problems of mathematical physics in which is difficult to estimate *a priori* the effects of various factors on the solution of the problem. The approach has acquired a special significance in the problem of planning an experiment, where the objective is to obtain the functionals with the highest information content.

Important results in this direction have been obtained in the theory of radiation by Fuchs [16], Usachev [16], Kadomtsev [16] and Marchuk and Orlov [16]. The inverse problems are also actively studied in the areas of pattern recognition, identification, and optimization theory. These problems are studied in detail in Pontryagin [3], Balakrishnan [3], Lions [2], and others.

7.4.1 Some Problems of the Linear Theory of Measurements

The theory of measurements has become greatly important for the purposes of organizing information systems. The measurement techniques allow one to obtain data (functionals) regarding the process and to analyze and control the process. We will not discuss the particular elementary measurements such as voltage or current measurements on various segments of an electrical network. Instead we will be interested in complex physical phenomena and processes which are to be understood and quantitatively evaluated within a required accuracy. Such problems are emerging all the time, especially in new branches of technology. It is impossible to design methods for measuring the coefficient of neutron multiplication in the reactor if the physical process of the chain reaction and the diffusion of neutrons are not clearly understood in detail, or if we do not know the equations describing the behavior of the nuclear reactor under various changing conditions.

There is no doubt that the measurement methods, as well as the measuring instruments themselves, improve considerably, as the theory of the physical process develops. As a rule, advances in the theory and experiment are accompanied by new or improved measurement methods.

A question arises whether one could not formulate at this time more or less general approaches to measurement methods as applied to various processes with the possibility of a formal mathematical description of the algorithm. It turns out that such an approach can be indeed formulated at least for problems with linear operators. This particular class of problems will be, in fact, the topic of our discussion in what follows.

It may be conjectured that the theory of variation measurements of physical quantities can be based on perturbation theory. The essence of the matter is as follows: suppose we are studying a complicated physical process using an instrument with known physical characteristics. Its readings are in relation with the field of physical quantities under investigation, and are functionals on this field. In most cases, however, the experimenter is interested not in the field itself, but rather in its deviations caused by some (usually small) perturbations. This requires that the measurements be made with enough accuracy as to permit the observations of the field deviations from some "standard" state. Thus, assume that this first requirement on the instrument is met, so that we have at our disposal sufficiently accurate measurements of the deviations from the standard. We now ask whether this information will suffice for a satisfactory interpretation of the experiment, and whether we can reconstruct the information about the perturbed state of the system with a sufficient accuracy. Unfortunately, this question is usually very difficult to answer. The reason is that the problem of reconstructing the information about the field of a physical quantity using measurements is, as a rule, an ill-posed problem of mathematical physics.

In order to sidestep this essential difficulty of data processing, it is necessary that the deviations of the instrument reading are related from the very beginning to the deviations in the physical parameters of the process under consideration. In this case, the error in the characteristic considered becomes proportional to the error in the deviation of the instrument reading (the variation of the functional), and hence, we can use the maximal information provided by the instrument for interpretation purposes. Taking the above point of view, we now develop the theory as based on the results of Marchuk and Orlov [16].

7.4.2 Conjugate Functions and the Notion of Value

Consider a function $\phi(x)$ satisfying the equation

$$L\phi(x) = q(x), \qquad (4.1)$$

where L is some linear operator and $q(x)$ is a source distribution in the medium. By x we represent all the variables of the problem (the time and space coordinates, energy, direction, and velocity. We also assume that the operator L and the function ϕ are real, and that $\phi \in \Phi$.

In order to be specific, we will assume that the process is related to a diffusion or to the propagation of a substance; the results, however, reach far beyond the scope of these types of problems.

Let us introduce a Hilbert space of functions endowed with the inner product

$$(g, h) = \int g(x)h(x)\, dx, \qquad (4.2)$$

where the integration is performed over the region D of definition of g and h.

Various physical problems are usually solved with the objective of evaluating eventually certain functionals of the flow $\phi(x)$. Any quantity which is linearly related to $\phi(x)$ can be represented in the form of this inner product. For instance, while observing some process, we may account for the characteristic $\Sigma(x)$ of the measuring instrument by taking

$$J_\Sigma[\phi] = \int \phi(x)\Sigma(x)\, dx = (\phi, \Sigma). \qquad (4.3)$$

Hence, we will consider physical quantities which can be expressed as linear functionals of $\phi(x)$:

$$J_p[\phi] = (\phi, p),$$

where p designates the physical process we are interested in. Along with L consider its adjoint L^*, defined by

$$(g, Lh) = (h, L^*g), \qquad (4.4)$$

for any g and h, and introduce (formally at this moment) the nonhomogeneous adjoint equation

$$L^*\phi_p^* = p(x), \qquad (4.5)$$

where the function $p(x)$ is yet to be determined, and $\phi_p^* \in \Phi^*$. The original equation (4.1) will be called fundamental. Take for g and h in (4.4) the solutions ϕ and ϕ_p^* of (4.1) and (4.5). We obtain

$$(\phi_p^*, L\phi) = (\phi, L^*\phi_p^*) \qquad (4.6)$$

or, using (4.1) and (4.5) again,

$$(\phi_p^*, q) = (\phi, p) \tag{4.7}$$

In other words, $J_q[\phi_p^*] = J_p[\phi]$. Hence, in order to find $J_p[\phi]$, we can proceed in two ways: either we solve (4.1) and use the formula

$$J_p[\phi] = (\phi, p), \tag{4.8}$$

or we solve (4.5) and take

$$J_p[\phi] = J_q[\phi_p^*] = (\phi^*, q). \tag{4.9}$$

Thus each linear functional can be put into a correspondence with a function $\phi_p^*(x)$ satisfying (4.5), where the free element in this equation is chosen to be $p(x)$—the function characterizing the process we are interested in.

Suppose that at the point x_0 of the medium there is a "source of unit power":

$$q(x) = \delta(x - x_0). \tag{4.10}$$

Since

$$(\phi(x), \delta(x - x_0)) = \phi(x_0), \tag{4.11}$$

we have

$$J_p[\phi] = J_{q=\delta(x-x_0)}[\phi_p^*] = \phi_p^*(x_0). \tag{4.12}$$

Consequently, the conjugate function $\phi_p^*(x)$ describes the dependence of the functional $J_p[\phi] = (\phi, p)$ on the location of the source.

Imagine a physical system (or instrument) which involves measurements of a certain linear functional $J_p[\phi]$ of the solution, related, for instance, to the density of the particles in the medium. Assume there occur emissions of a certain number of particles at some point (or, conversely, absorption of these particles). Then the measured value of $J_p[\phi]$ will correspondingly increase or decrease, and this change will depend on the location of the emission (or absorption) point. As can be seen from the foregoing, this dependence is described by the conjugate function $\phi_p^*(x)$, which in turn satisfies (4.5). Consequently, the conjugate function $\phi_p^*(x)$ describes the effect of depositing the particles at a given point. The function $\phi_p^*(x)$ can be called a *value* of the substance at the point x, relative to the functional $J_p[\phi] = (\phi, p)$. (The term value fits well in problems of the theory of radiation. A more suitable term may well be found for another application.)

The interpretation of the conjugate function $\phi_p^*(x)$ as the value of the substances also helps to clarify the exposition of perturbation theory for arbitrary functionals $J_p[\phi]$. Indeed, suppose we change the number of particles in an element of volume Δx, surrounding the point x, by an increment of ΔN. The corresponding change in J_p can then be expressed by the formula

$$\delta J_p = \delta N \phi_p^*(x). \tag{4.13}$$

If the system under consideration is subject to some small changes in the parameters, so that the operator L becomes $L + \delta L$, then the number of particles in each element Δx changes correspondingly by the amount of $\Delta N = -\Delta x\, \delta L\phi$. The overall change of J_p is thus given by

$$\delta J_p = -\int \phi_p^*(x)\, \delta L\phi(x)\, dx. \tag{4.14}$$

A rigorous derivation of this result will be given later.

Relation (4.13) facilitates the measurements of the value distribution function of the system. It can be accomplished by changing in a certain manner the number of particles at various points x, while measuring the corresponding changes in the quantity L_p. The notion of a value can also be useful in the theory of measuring instruments. An instrument is usually designed to measure a single variable J_p. Hence for each instrument we can introduce a well-defined value function $\phi_p^*(x)$, which need be calculated only once. If the distributions of the substance and of its value are known, relation (4.14) can be used in the following two ways.

First, it can be used for determining the quantities δL, i.e., various characteristics of the mutual interaction between the particles and the matter, by taking measurements of δJ_p for various changes in the parameters of δL. For instance, we can measure sections of the interacting neutrons for various figures (shapes) by accommodating these figures into an instrument and evaluating $\delta\Sigma = \delta L$ using the increments of J_p.

Second, (4.14) allows one to correct the effect of various perturbing factors of the instrument on the functional J_p.

Finally, the notion of value makes it possible to derive the equations for ϕ_p^* directly from its physical meaning; the procedure is the same as when deriving the equation for the flow of neutrons using the law of preserving neutrons.

The above formulas can also be used to prove the reciprocity theorem for the Green functions $G(x, x_0)$ and $G^*(x, x_1)$ corresponding to the fundamental and adjoint equations, respectively. Indeed, the function $G(x, x_0)$ satisfies (4.1) for $q(x) = \delta(x - x_0)$, and $G^*(x, x_1)$ satisfies (4.5) for $p(x) = \delta(x - x_1)$. Substituting $\phi(x) = G(x, x_0)$, $\phi_p^*(x) = G^*(x, x_1)$ and the above equations for q and p in formula (4.7), we obtain

$$G(x_1, x_0) = G^*(x_0, x_1), \tag{4.15}$$

which is the reciprocity theorem.

7.4.3 Perturbation Theory for Linear Functionals

If the properties of the medium interacting with the field are subject to changes, i.e., of the operator in (4.1) become

$$L' = L + \delta L,$$

then both the field $\phi(x)$ itself, and the functional $J_p[\phi]$ are also changing:

$$\phi(x) \to \phi'(x), \qquad J_p[\phi] \to J'_p = J_p + \delta J_p.$$

Let us find the relation between the increments δL and δJ_p. The perturbed system is described by the equation

$$L'\phi' = (L + \delta L)\phi' = q. \tag{4.16}$$

The conjugate function of the unperturbed system corresponding to J_p is given by the equation

$$L^*\phi_p^* = p. \tag{4.17}$$

Take the inner products of (4.16) and ϕ^*, and (4.17) and ϕ'. Then subtracting, we obtain

$$(\phi_p^*, L'\phi') - (\phi', L^*\phi_p^*) = (\phi_p^*, \delta L\phi') \tag{4.18}$$

on one hand, and

$$(\phi_p^*, q) - (\phi', p) = J_p[\phi] - J_p[\phi'] = -\delta J_p \tag{4.19}$$

on the other hand [in agreement with (4.7)].

A comparison of (4.18) and (4.19) gives the following relation for the increment of the functional:

$$\delta J_p = -(\phi_p^*, \delta L\phi'). \tag{4.20}$$

If instead of (4.16) and (4.17) we consider the perturbed adjoint equation

$$(L^* + \delta L^*)\phi_p^{*'} = p \tag{4.21}$$

and the unperturbed fundamental equation (4.1), respectively, a similar procedure yields

$$\delta J_p = -(\phi, \delta L^*\phi_p^{*'}), \tag{4.22}$$

which is, of course, equivalent to (4.20).

Note an important aspect of applying the formulas of perturbation theory. These formulas are written in a form applicable to the variation of the functional, the admissible inaccuracy of the variation usually being in the limits of several percent. Therefore the calculation of the variations indicated does not require precise knowledge of the fundamental and adjoint problems; it is enough to use their approximate solutions.

If the perturbation of L (and L^* for the matter) is small enough, so that the functions ϕ and ϕ_p^* are not seriously distorted, then in (4.20) and (4.22) one can put $\phi' \approx \phi$, $\phi^{*'} \approx \phi^*$, and obtain the following mutually equivalent formulas:

$$\delta J_p = -(\phi_p^*, \delta L\phi), \tag{4.23}$$

$$\delta J_p = -(\phi, \delta L^*\phi_p^*). \tag{4.24}$$

Besides their direct application in estimating various effects and their use in measurement analysis, the obtained formulas of perturbation theory have yet another quite important application.

In theoretical considerations, as well as practical computations, the original complicated problem is often approximated by a simplified model. For this it is clearly necessary that the replacement leave unchanged certain characteristics of the system which are fundamental for the process under consideration. As an example consider the approximation of a time-varying differential equation by a time invariant equation. An instance of this approach is the method of efficient boundary conditions: essentially, we replace the true boundary conditions by simpler ones, but such that they result in correct values of a certain functional.

The formulas of perturbation theory obtained above facilitate formulations of quite general approaches to various problems. Consider a system characterized by the operator L, and assume that the most significant quantity for our purposes is the functional $J_p[\phi]$. If the sought simplified model can be characterized by the operator $L' = L + \delta L$, then for J_p not to be affected by the replacement of the true system with the model it is enough that

$$\delta J_p = -(\phi_p^*, [L' - L]\phi') = 0, \quad \text{i.e.,} \; (\phi_p^*, L'\phi') = (\phi_p^*, L\phi'). \quad (4.25)$$

For more quantities J_{p_1}, J_{p_2}, etc., we similarly obtain conditions of the type of (4.25) with the solutions $\phi_{p_1}^*$, $\phi_{p_2}^*$, etc.

Condition (4.25) does not determine the desired model uniquely. But, being the necessary condition, it may still help in finding the model in conjunction with other conditions. In particular, if the model operator L' involves one or more parameters, (4.25) can serve to find their values. (The form of L' can be determined from physical considerations.)

7.4.4 Numerical Methods for Inverse Problems and Design of Experiment

Suppose that we have at our disposal a set of functionals (measurements) J_{p_i} $(i = 1, \ldots, n)$. Suppose that the measurements are essentially different (for instance, the measurements are taken at different points of the domain of the solution using the same single instrument; or, several instruments are used to register different characteristics of the considered phenomenon). For the sake of simplicity, the random errors are assumed to be already removed from the measurements by preliminary data processing.

Let us put each J_{p_i} in correspondence with the respective value function for the unperturbed problem, i.e., for the model in which the operator L and its domain are considered to be known. Solving now a total of n different problems

$$L^*\phi_{p_i}^* = p_i, \qquad i = 1, 2, \ldots, n, \quad (4.26)$$

we find the value functions $\phi_{p_i}^*$. Having done this, let us solve the single fundamental problem with the model ("unperturbed") operator L, adjoint to L^*:

$$L\phi = f. \tag{4.27}$$

We will assume that $\phi \in \Phi$ and $\phi^* \in \Phi^*$, where Φ and Φ^* are the domains of L and L^*, respectively. Next consider the total of n formulas of the theory of small perturbations

$$(\phi_{p_i}^*, \delta L\phi) = -\delta J_{p_i}, \qquad i = 1, 2, \ldots, n, \tag{4.28}$$

where δL designates the difference between L' and L. Suppose that L is known:

$$L = \sum_{k=1}^{m} [\alpha_k A_k + B_k(\beta_k C_k)], \tag{4.29}$$

where A_k, B_k, and C_k are elementary linear operators (for example, differentiation, or integration, or their various combinations); the coefficients $\alpha_k(x)$ and $\beta_k(x)$ are to be determined (their rough approximations in the unperturbed problem are usually known).

Our problem now is to reconstruct the coefficients α_k' and β_k' appearing in the expression

$$L' = \sum_{k=1}^{m} [\alpha_k' A_k + B_k(\beta_k' C_k)]. \tag{4.30}$$

Using (4.29) and (4.30) we get

$$\delta L' = \sum_{k=1}^{m} [\delta\alpha_k A_k + B_k(\delta\beta_k C_k)], \tag{4.31}$$

where

$$\delta\alpha_k = \alpha_k' - \alpha_k, \qquad \delta\beta_k = \beta_k' - \beta_k.$$

We substitute (4.31) into (4.28), and under the corresponding assumptions obtain the following system of equations:

$$\sum_{k=1}^{m} [(\phi_{p_i}^*, \delta\alpha_k A_k \phi) + (B_k^* \phi_{p_i}^*, \delta\beta_k C_k \phi)] = -\delta J_{p_i}, \qquad i = 1, 2, \ldots, n. \tag{4.32}$$

The next procedure is to parametrize the variations $\delta\alpha_k$ and $\delta\beta_k$. To begin with, consider the simplest case where $\delta\beta_k = 0$ and $\delta\alpha_k$ are constant. Under these conditions system (4.32) becomes a problem of linear algebra

$$\sum_{k=1}^{m} \delta\alpha_k(\phi_{p_i}^*, A_k \phi) = -\delta J_{p_i}, \qquad i = 1, 2, \ldots, n. \tag{4.33}$$

Here $(\phi_{p_i}^*, A_k \phi)$ are elements of a matrix which can be calculated.

Let y be a vector with the components $\delta\alpha_k$, let F be another vector with the components $-\delta J_{p_i}$, and finally let $a_{ik} = (\phi_{p_i}^*, A_k \phi)$ be the entries of a matrix Λ. Then

$$\Lambda y = F. \tag{4.34}$$

If the number of functionals n equals the number of variations of the coefficients α_k to be determined, then $\delta\alpha_k$ can be found, in principle, from (4.34). If $n > m$, (4.34) is overdefined, and its solution (if it exists) can usually be found with the help of the least-squares method under the assumption that the quadratic functional below achieves its minimum on y:

$$\|\Lambda y - F\|^2 = \min. \tag{4.35}$$

The minimizing vector y is sometimes referred to as a *quasi-solution* of (4.34). If $n = m$, system (4.34) can be solved by methods we have discussed in Chapter 4 in connection with the analysis of numerical methods with inaccurate data.

In the case where $\delta\alpha_k(x)$ and $\delta\beta_k(x)$ are functions, the inverse problem can be solved by various parametric methods, the essence of which consists of the following: it is assumed that based on an *a priori* analysis of the behavior of physical parameters (usually as a result of statistical and correlation analysis) we have found a certain complete orthogonal system of functions $u_{k,l}(x)$ and $v_{k,l}(x)$ such that for a small number $n(k)$ they give a reasonable approximation of α_k and β_k:

$$\delta\alpha_k(x) = \sum_{l=1}^{n(k)} \alpha_{k,l} u_{k,l}(x);$$

$$\delta\beta_k(x) = \sum_{l=1}^{n(k)} b_{k,l} v_{k,l}(x), \tag{4.36}$$

where $a_{k,l}$ and $b_{k,l}$ are yet to be determined.

Let us substitute the expressions (4.36) into (4.32). There results

$$\sum_{k=1}^{m} \sum_{l=1}^{n(k)} [\alpha_{k,l}(\phi_{p_i}^*, u_{k,l} A_k \phi) + b_{k,l}(B_k^* \phi_{p_i}^*, v_{k,l} C_k \phi)] = -\delta J_{p_i}, \tag{4.37}$$

$$i = 1, 2, \ldots, n.$$

Using a suitable ordering, let us relabel the coefficients $a_{k,l}$ and $b_{k,l}$ as y_j $(j = 1, 2, \ldots)$ and introduce further a matrix Λ so that the equation

$$\Lambda y = F$$

is equivalent to (4.37). In this fashion we again obtain a problem in linear algebra, from which we eventually find $a_{k,l}$ and $b_{k,l}$, and consequently $\delta\alpha_k$ and $\delta\beta_k$.

So far, we have only considered the case where the solution of the model is close to reality; thus one could interchange ϕ' and ϕ and use the theory of small perturbations. If the unperturbed (model) state of the process differs from that of the true process, the above algorithm can be taken only as a first approximation of the solution to the inverse problem. Once the variations $\delta\alpha_k$ and $\delta\beta_k$ are found, one can modify the coefficients α_k and β_k and find

$$\alpha_k' = \alpha_k + \delta\alpha_k,$$

$$\beta_k' = \beta_k + \delta\beta_k.$$

Next one has to solve the "perturbed" problem

$$L'\phi' = f \tag{4.38}$$

with the operator

$$L' = \sum_{k=1}^{m} [\alpha_k' A_k + B_k(\beta_k' C_k)],$$

and then switch to a new approximation in the solution of the inverse problem, replacing (4.32) by a more general perturbation formula

$$\sum_{k=1}^{m} [(\phi_{p_i}^*, \delta\alpha_k A_k \phi') + (B_k^* \phi_{p_i}^*, \delta\beta_k C_k \phi')] = -\delta J_{p_i}, \quad i = 1, 2, \ldots, n \tag{4.39}$$

One must repeat the computational cycle in order to improve the accuracy of the variations $\delta\alpha_k$ and $\delta\beta_k$. This is called a *second approximation* in solving the inverse problem. It is understood that the above process can be continued. The successive approximations may be shown to converge, depending on specific information about the elementary operators A_k and the domain of the operators L, L^*.

Let us illustrate our algorithm with a simple example, namely,

$$-\frac{d}{dx} \beta(x) \frac{d\phi'}{dx} + \alpha(x)\phi' = f(x), \qquad \phi'(0) = \phi'(1) = 0, \tag{4.40}$$

where the unknown coefficients $\alpha(x)$ and $\beta(x)$ are *a priori* assumed to be, for instance, continuous on $0 \le x \le 1$, and approximately equal to $\bar\alpha$ and $\bar\beta$, i.e.,

$$\alpha(x) = \bar\alpha + \delta\alpha, \qquad \beta(x) = \bar\beta + \delta\beta(x). \tag{4.41}$$

Of course, if the values of $\bar\alpha(x)$ and $\bar\beta(x)$ can be specified with more precision using an *a priori* information, then the approximation by the constants $\bar\alpha$ and $\bar\beta$ becomes void.

By a preliminary analysis we eventually conclude that $\delta\alpha(x)$ and $\delta\beta(x)$ may be represented in the form of the finite sums

$$\delta\alpha(x) = \sum_{l=1}^{n} a_l u_l(x), \qquad \delta\beta(x) = \sum_{l=1}^{n} b_l v_l(x), \tag{4.42}$$

where $\{u_k(x)\}$ and $\{v_k(x)\}$ form complete systems of orthonormal functions (for example, trigonometric polynomials, or Legendre polynomials).

Let $p_1(x)$, $p_2(x), \ldots, p_n(x)$ be the measurement characteristics, so that each of the instruments registers the functional

$$J'_{p_i}[\phi'] = \int_0^1 p_i(x)\phi'(x)\,dx, \qquad i = 1, 2, \ldots, n. \qquad (4.43)$$

The functions p_i can be viewed as the characteristics of the instrument.

Consider next the unperturbed problem (model) corresponding to problem (4.40):

$$-\frac{d}{dx}\beta\frac{d\phi}{dx} + \bar{\alpha}\phi = f, \qquad \phi(0) = \phi(1) = 0. \qquad (4.44)$$

Along with (4.44) consider also the adjoint problems which correspond to the model chosen:

$$-\frac{d}{dx}\beta\frac{d\phi_{p_i}^*}{dx} + \bar{\alpha}\phi_{p_i}^* = p_i(x), \qquad \phi_{p_i}^*(0) = \phi_{p_i}^*(1) = 0, \qquad i = 1, 2, \ldots, n. \qquad (4.45)$$

According to the general theory, we have

$$J_{p_i}[\phi] = \int_0^1 p_i(x)\phi(x)\,dx = \int_0^1 f(x)\phi_{p_i}^*(x)\,dx. \qquad (4.46)$$

Assume now that the model problems (4.44) and (4.45) have been solved, and let us find the variations δJ_{p_i} from the formula

$$\delta J_{p_i} = J'_{p_i} - J_{p_i}, \qquad i = 1, 2, \ldots, n, \qquad (4.47)$$

where J'_{p_i} is the measurement with the characteristic p_i (see (4.43), where ϕ' is unknown); J_{p_i} can be expressed by means of any of the formulas in (4.46). The measurement accuracy must be such as to guarantee that the variation δJ_{p_i} can be evaluated.

Consider now the formulas of the theory of small perturbations (4.33):

$$A = E; \qquad B = -\frac{d}{dx}; \qquad C = \frac{d}{dx}.$$

Taking into account the boundary conditions for $\phi_{p_i}^*$ and ϕ, we obtain

$$\int_0^1 \left(\delta\alpha\phi\phi_{p_i}^* + \delta\beta\frac{d\phi}{dx}\frac{d\phi_{p_i}^*}{dx} \right) dx = -\delta J_{p_i}. \qquad (4.48)$$

Substituting (4.42) into (4.48), we obtain

$$\sum_{l=1}^{n\,(1)} \left(a_l \int_0^1 u_l\phi\phi_{p_i}^*\,dx + b_l \int_0^1 v_l\frac{d\phi}{dx}\frac{d\phi_{p_i}^*}{dx}\,dx \right) = -\delta J_{p_i}, \qquad i = 1, 2, \ldots, n. \qquad (4.49)$$

If $n = 2n(1)$ the system is completely determined.

Solving this system, we find the coefficients a_l and b_l, and based on the representation (4.42), we obtain the first approximation for α' and β'. These quantities can be made more precise by the successive approximation method described earlier. In the same manner we can formulate and solve more complicated inverse problems, including the problem of determining the source perturbations δf.

Here we have to design a rather complicated experiment. The problem can be formulated as follows: consider a family of measurements (which is practically realizable). We are required to choose a measurement from this family with the highest information content, having in mind the specific inverse problem of reconstructing the required characteristics of the medium (the coefficients of the equation). In the general framework of optimization this problem turns out to be very difficult. Nevertheless, we can consider certain particular approaches to its solution.

Thus, let us suppose that prior to the execution of the experiment we have constructed a model of the unperturbed problem, which is subsequently used to describe the linear functionals of the solution; using the *a priori* information about the measurement accuracy we then make conclusions regarding the necessary measurement accuracy of the functionals. Suppose that the necessary measurement accuracy requirements on δJ_{p_i} are met. Next, we consider various families of measurements and choose those which give the best conditioning of the matrix Λ. The system of linear equations obtained thereby can be easily inverted; thus we have produced a sort of optimal design of experiment (of course, we have omitted economy considerations which sometimes may play a decisive role). One may fail to achieve the high requirements regarding the accuracy of the measurements of the functionals J_{p_i} (chosen so as to maximize the conditioning of the matrix B). In this case we face a more complicated problem of experiment planning: we are constrained by a given accuracy compatible with the instrument technology. This belongs to another class of problems, namely optimization problems with constraints.

We have not considered here the problems of statistical processing of the empirical data. These problems are dealt with in sufficient detail in the relevant literature, and in any case, they do not add any major complication to the theory of inverse problems (see Marchuk and Drobyshev [16]).

7.5 Perturbation Theory for Complex Nonlinear Models

It is normally not easy to construct mathematical models reflecting complex processes and phenomena. Such models must, as a rule, take into account many different effects, some of them not described with the desired accuracy. This means that from moment to moment we may use one simplified model after another, neglecting through our abstractions many details, sometimes very important ones. Nevertheless, as a rule, and in general, it is possible to describe physical processes with sufficient accuracy. As was shown in the preceding section, an estimate of the influence of effects not comprised in the model can be obtained by using perturbation theory in a special way.

In this section we shall be concerned with a fairly general approach to the construction and analysis of mathematical models. In contrast to our attention in the preceding section to linear processes, we shall here consider processes characterized by nonlinear equations. We shall show that the associated problems may be solved and analyzed by the use of approaches characteristic of linear problems; naturally, of course, these approaches are based on various linearization methods.

7.5.1 Fundamental and Adjoint Equations

Consider a stationary process described by the following operator equation:

$$A\phi = f, \tag{5.1}$$

where A is a matrix differential operator depending on the solution (the vector function ϕ) and on the initial data $\alpha_1, \alpha_2, \ldots, \alpha_n$ which are functions of the coordinates; f is a given vector of sources, a function of the coordinates, and of given parameters $\beta_1, \beta_2, \ldots, \beta_n$. Then we have

$$A = A(\phi, \alpha_1, \alpha_2, \ldots, \alpha_n)$$

and

$$f = f(\beta_1, \beta_2, \ldots, \beta_n).$$

Let $\bar{\alpha}_1, \bar{\alpha}_2, \ldots, \bar{\alpha}_n$ and $\bar{\beta}_1, \bar{\beta}_2, \ldots, \bar{\beta}_n$ be the data corresponding to some standard state of the system, which we shall call the unperturbed state. This state will obviously be described by the equation

$$A(\bar{\phi}, \bar{\alpha}_1, \bar{\alpha}_2, \ldots, \bar{\alpha}_n)\bar{\phi} = f(\bar{\beta}_1, \bar{\beta}_2, \ldots, \bar{\beta}_n). \tag{5.2}$$

If we write

$$\bar{A} = A(\bar{\phi}, \bar{\alpha}_1, \bar{\alpha}_2, \ldots, \bar{\alpha}_n), \quad \bar{f} = f(\bar{\beta}_1, \bar{\beta}_2, \ldots, \bar{\beta}_n),$$

equation (5.2) may be rewritten as

$$\bar{A}\bar{\phi} = \bar{f}. \tag{5.3}$$

We assume that (5.3) is known.

We also assume that the actual state—or as we shall call it, the perturbed state—of the system is described by (5.1), with parameters differing little from those of the standard state:

$$a_i = \bar{\alpha}_i + \delta\alpha_1, \quad \beta_j = \bar{\beta}_j + \delta\beta_j,$$

i.e., the deviations $\delta\alpha_i$ and $\delta\beta_j$ are small compared to $\bar{\alpha}_i$ and $\bar{\beta}_j$, respectively.

Then in place of (5.2) we have

$$A(\bar{\phi} + \delta\phi, \bar{\alpha}_i + \delta\alpha_i)(\bar{\phi} + \delta\phi) = f(\bar{\beta}_j + \delta\beta_j), \qquad (5.4)$$

where we have used the notation

$$\phi = \bar{\phi} + \delta\phi.$$

We assume *a priori* that $\delta\phi$ is much smaller than $\bar{\phi}$. If A, ϕ, and the initial data are all sufficiently smooth, we may consider the expansion

$$A(\bar{\phi} + \delta\phi, \bar{\alpha}_i + \delta\alpha_i) = \bar{A} + \frac{\overline{\partial A}}{\partial\phi}\delta\phi + \frac{\overline{\partial A}}{\partial\alpha_i}\delta\alpha_i + \cdots,$$

$$f(\bar{\beta}_i + \delta\beta_j) = \bar{f} + \frac{\overline{\partial f}}{\partial\beta_j}\delta\beta_j. \qquad (5.5)$$

Substituting (5.5) in (5.4) and including only first-order terms we obtain the equation

$$\bar{A}\bar{\phi} + \left(\bar{A} + \frac{\overline{\partial A}}{\partial\phi}\bar{\phi}\right)\delta\phi + \frac{\overline{\partial A}}{\partial\alpha_i}\bar{\phi}\delta\alpha_i = \bar{f} + \frac{\overline{\partial f}}{\partial\beta_j}\delta\beta_j. \qquad (5.6)$$

Making use of (5.3) we have

$$\left(\bar{A} + \frac{\overline{\partial A}}{\partial\phi}\bar{\phi}\right)\delta\phi = \frac{\overline{\partial f}}{\partial\beta_j}\delta\beta_j - \frac{\overline{\partial A}}{\partial\alpha_i}\bar{\phi}\delta\alpha_i. \qquad (5.7)$$

This is the fundamental equation for defining small deviations of the solution ϕ from its unperturbed state.

We now suppose that the operator $\bar{A}$ and the source function $\bar{f}$ in the unperturbed state are known with a certain accuracy,

$$\begin{aligned} \bar{A} &= \bar{\Lambda} + \varepsilon, \\ \bar{f} &= \bar{F} + \xi. \end{aligned} \qquad (5.8)$$

Here ε is the error operator of the model:

$$\varepsilon = \bar{A} - \Lambda$$

and ξ is the error (vector function) of the source:

$$\xi = \bar{f} - \bar{F}.$$

Suppose that

$$\begin{aligned} \|\bar{A}\| &\gg \|\varepsilon\|, \\ \|\bar{f}\| &\gg \|\xi\|. \end{aligned} \qquad (5.9)$$

(The norms of the operator and the vector function are defined with respect to their corresponding metric spaces.)

Substituting (5.8) in (5.7) we have

$$\left(\bar{\Lambda} + \frac{\overline{\partial \Lambda}}{\partial \phi}\bar{\phi}\right)\delta\phi = \frac{\overline{\partial F}}{\partial \beta_j}\delta\beta_j - \frac{\overline{\partial \Lambda}}{\partial \alpha_i}\bar{\phi}\delta\alpha_i + \eta, \qquad (5.10)$$

where

$$\eta = \frac{\overline{\partial \xi}}{\partial \beta_j}\delta\beta_j - \frac{\overline{\partial \varepsilon}}{\partial \alpha_i}\bar{\phi}\delta\alpha_i - \left(\varepsilon + \frac{\overline{\partial \varepsilon}}{\partial \phi}\bar{\phi}\right)\delta\phi. \qquad (5.11)$$

Given (5.9) we may rewrite (5.10) in the form:

$$\left(\bar{\Lambda} + \frac{\overline{\partial \Lambda}}{\partial \phi}\bar{\phi}\right)\delta\phi = \frac{\overline{\partial F}}{\partial \beta_j}\delta\beta_j - \frac{\overline{\partial \Lambda}}{\partial \alpha_i}\bar{\phi}\delta\alpha_i + O(\|\varepsilon\| + \|\xi\|).$$

Accordingly, to within small quantities

$$\left(\bar{\Lambda} + \frac{\overline{\partial \Lambda}}{\partial \phi}\bar{\phi}\right)\delta\phi = \frac{\overline{\partial F}}{\partial \beta_j}\delta\beta_j - \frac{\overline{\partial \Lambda}}{\partial \alpha_i}\bar{\phi}\delta\alpha_i. \qquad (5.12)$$

This is our model for calculating the unperturbed state when the initial conditions are perturbed by the amounts $\delta\alpha_i$ and $\delta\beta_j$.

We introduce the notation

$$L = \bar{\Lambda} + \frac{\overline{\partial \Lambda}}{\partial \phi}\bar{\phi},$$

$$\delta F = \frac{\overline{\partial F}}{\delta \beta_j}\delta\beta_j - \frac{\overline{\partial \Lambda}}{\partial \alpha_i}\bar{\phi}\delta\alpha_i. \qquad (5.13)$$

Then, finally, we have

$$L\delta\phi = \delta F, \qquad (5.14)$$

and the formal solution of this problem is

$$\delta\phi = L^{-1}\delta F. \qquad (5.15)$$

7.5.2 The Adjoint Equation in Perturbation Theory

Formula (5.15) is most suitable in perturbation theory for the case in which we must determine the deviation of the solution for a single set of initial conditions only. In planning the direction of motion of the model, it is extremely important that we have a series of trial computations. We must remember that to assess the stability of a model against changes in various parameters or to obtain an optimal relation among the parameters we need a large number of solutions. We must, therefore, attempt to construct a more nearly-universal theory of perturbations for functionals that will permit us to make a many-sided study of a mathematical model, in the presence of variations, in the initial data.

As we did in the linear case, we replace the fundamental equation

$$L\,\delta\phi = \delta F \tag{5.16}$$

by the adjoint equation

$$L^*\phi^* = p, \tag{5.17}$$

where L and L^* are conjugate in the Lagrangian sense:

$$(Lg, h) = (g, L^*h) \tag{5.18}$$

Here g and h are elements of a Hilbert space in the domain of definition of the operators L and L^*, respectively. The function p we leave for the time being undefined.

We take the scalar products of (5.16) and (5.17) by ϕ^* and $\delta\phi$, respectively, and subtract the second equation from the first. Then we have

$$(\phi^*, L\delta\phi) - (\delta\phi, L^*\phi^*) = (\phi^*, \delta F) - (p, \delta\phi) \tag{5.19}$$

By (5.18), the left-hand side of this equation vanishes, and we have

$$(p, \delta\phi) = (\phi^*, \delta F) \tag{5.20}$$

Now consider the collection of linear functionals

$$\delta J_n = (p_n, \delta\phi). \tag{5.21}$$

If, in particular, $p_n = \delta(x - x_n)$, we have as in the linear case

$$\delta J_n = \delta\phi(x_n). \tag{5.22}$$

The function ϕ^* corresponding to p_n will be denoted by ϕ_n^*.

Then, on the basis of (5.20) we have the set of functionals

$$\delta J_n = (\phi_n^*, \delta F) \tag{5.23}$$

We shall suppose that we have preselected N of the functionals $J_1, J_2, \ldots, J_N$, and have solved N corresponding adjoint problems

$$L^*\phi_n^* = p_n, \qquad n = 1, 2, \ldots, N. \tag{5.24}$$

It follows from (5.23) that we need not compute the variations $\delta\phi$ corresponding to the several sets of parameters $\delta\alpha_i$ and $\delta\beta_j$, since if the ϕ_n^* $(n = 1, 2, \ldots, N)$ are given, we may compute the values of the functionals δJ_n directly for arbitrary perturbations in the initial data.

7.5.3 Perturbation Theory for Nonstationary Problems

Now consider a nonstationary problem in the abstract form

$$\frac{\partial\phi}{\partial t} + A\phi = f, \qquad \phi = g \quad \text{for} \quad t = 0, \tag{5.25}$$

and set up the corresponding adjoint problem

$$-\frac{\partial \phi^*}{\partial t} + A^*\phi^* = f^*, \qquad \phi^* = g^*, \quad \text{for} \quad t = T. \tag{5.26}$$

Here f^* and g^* are vector functions, currently undefined, and to be specified later. Multiply (5.25) and (5.26) by ϕ^* and ϕ, respectively; subtract one result from the other and integrate with respect to t over the interval $0 \le t \le T$. The result is as follows:

$$\int_0^T \frac{\partial}{\partial t}(\phi^*, \phi)\, dt + \int_0^T [(\phi^*, A\phi) - (\phi, A^*\phi^*)]\, dt$$

$$= \int_0^T [(f, \phi^*) - (f^*, \phi)]\, dt. \tag{5.27}$$

Since the operators A and A^* are conjugate, i.e.,

$$(\phi^*, A\phi) = (\phi, A^*\phi^*) = 0,$$

the expression (5.27), if we consider the initial conditions, may be written as

$$(g^*, \phi_T) - (g, \phi_0^*) = \int_0^T [(f, \phi^*) - (f^*, \phi)]\, dt, \tag{5.28}$$

where

$$\phi_T = \phi|_{t=T}, \qquad \phi_0^* = \phi^*|_{t=0}.$$

We now suppose that we are to find a linear functional of the solution of (5.25), in the form

$$J = (g^*, \phi_T) + \int_0^T (f^*, \phi)\, dt. \tag{5.29}$$

Using the identity (5.28), we may rewrite it as

$$J = (g, \phi_0^*) + \int_0^T (f, \phi^*)\, dt. \tag{5.30}$$

We suppose further that the initial data in (5.25) are perturbed, that is, we consider $g' = g + \delta g, f' = f + \delta f$ in place of f and g. Then we find, as an expression for the variation of the functional

$$\delta J = (\delta g, \phi_0^*) + \int_0^T (\delta f, \phi^*)\, dt. \tag{5.31}$$

Therefore, in order to compute the deviation of the functional J corresponding to various perturbations of the initial data, we do not need to solve a large number of equations of the form

$$\frac{\partial \phi'}{\partial t} + A\phi' = f', \qquad \phi' = g' \quad \text{for} \quad t = 0 \tag{5.32}$$

for various values of f' and g'. Instead, we need solve only one adjoint equation (5.26) and make use of (5.31), which enables us to construct the inverse problem of finding δg and δf by means of the set of functionals δJ.

In many mathematical models of nonlinear problems, we are not interested in functionals of the solutions, but in their variations. Then we may use the method given above in Section 5.5.2 to arrive at the fundamental equation

$$\frac{\partial}{\partial t}\delta\phi + L\,\delta\phi = \delta F, \qquad \delta\phi = \delta\phi_0 \quad \text{for} \quad t = 0 \tag{5.33}$$

and the adjoint

$$-\frac{\partial\phi^*}{\partial t} + L^*\phi^* = p, \qquad \phi^* = \phi_T^* \quad \text{for} \quad t = T. \tag{5.34}$$

The procedure for solving the inverse problems is the same as that considered above in connection with nonstationary problems.

7.5.4 Spectral Methods in Perturbation Theory

In this section we consider another approach to perturbation theory, which is preferable when we must determine a deviation from the solution of the unperturbed problem, rather than the variation of the functionals J.

Then we begin from (5.14):

$$L\,\delta\phi = \delta F \tag{5.35}$$

and the spectral problem

$$Lw = \lambda w. \tag{5.36}$$

Since L is not self-adjoint, we must consider also the conjugate problem

$$L^*w^* = \lambda w^*. \tag{5.37}$$

Let (5.36) determine the complete set of eigenfunctions $\{w_n\}$ corresponding to the eigenvalues $\{\lambda_n\}$, and let (5.37) determine $\{w_n^*\}$ and $\{\lambda_n\}$. Then, our two sets of vector functions are orthogonal:

$$(w_n, w_m^*) = 0 \quad \text{for} \quad m \neq n, \tag{5.38}$$

and may be normalized

$$(w_n, w_m^*) = \begin{cases} 1, & n = m, \\ 0, & n \neq m. \end{cases} \tag{5.39}$$

The solution of (5.35) is representable as a Fourier series in the eigenfunctions (5.36):

$$\delta\phi = \sum_n \delta\phi_n w_n, \tag{5.40}$$

and similarly

$$\delta F = \sum_n \delta F_n w_n, \tag{5.41}$$

where

$$\delta \phi_n = (\delta \phi, w_n^*), \qquad \delta F_n = (\delta F, w_n^*).$$

Substituting (5.40) and (5.41) into (5.35) and taking the scalar product with w_m^* we obtain the relationship

$$\lambda_m \delta \phi_m = \delta F_m, \qquad m = 1, 2, \dots . \tag{5.42}$$

The solution of (5.35) may be written as

$$\delta \phi = \sum_n \frac{\delta F_n}{\lambda_n} w_n, \tag{5.43}$$

or as

$$\delta \phi = \sum_n \frac{(\delta F, w_n^*)}{\lambda_n} w_n. \tag{5.44}$$

Thus, we have obtained the deviation of the solution from that of the unperturbed state, as a Fourier series. If we are interested in a linear functional rather than in the solution itself, we have

$$\delta J = (p, \delta \phi). \tag{5.45}$$

Then

$$\delta J = \sum_n \frac{(\delta F, w_n^*)}{\lambda_n} (p, w_n) = (\delta F, \phi^*), \tag{5.46}$$

where

$$\phi^* = \sum \frac{(p, w_n)}{\lambda_n} w_n^*.$$

Here ϕ^* is the solution of the problem $L^* \phi^* = p$ (cf. (5.17)).

We have considered the stationary case. The nonstationary case may be handled in the same way, yielding the corresponding perturbation formulas.

CHAPTER 8

Methods of Optimization

The rapid developments of methods for mathematical modelling of problems in science and technology has stimulated a wide-ranging search for effective algorithms to optimize the parameters of mathematical models with respect to various functional criteria. Some general approaches to the problem of optimization have arisen in the theory of variational computations; these were partially described in Section 2.2. The set of optimizing algorithms is, however, now so large that one may regard it as a specialized field in numerical mathematics. It includes the methods of linear and quadratic programming, convex and dynamic programming, and Pontrjagin's maximum principle. The fundamental results in this field are due to Kantorovich [21], Kuhn and Tucker [21], Frank and Wolfe [2], Courant [21], Fiacco and MacCormick [21], Bellman [22], Pontrjagin [22]. All these methods have their own peculiar formulations and implementations for the problem of optimization. They will be discussed in this chapter.

8.1 Convex Programming

We begin by recalling a definition. A function f on a convex set X is said to be convex if, for all $\alpha \in (0, 1)$ and $x', x'' \in X$, it satisfies the inequality

$$f(\alpha x' + (1 - \alpha)x'') \leq \alpha f(x') + (1 - \alpha)f(x'').$$

If a convex function f is continuous and differentiable, it satisfies the inequality

$$f(x) \geq f(x_*) + (f'(x_*), x - x_*) \tag{1.1}$$

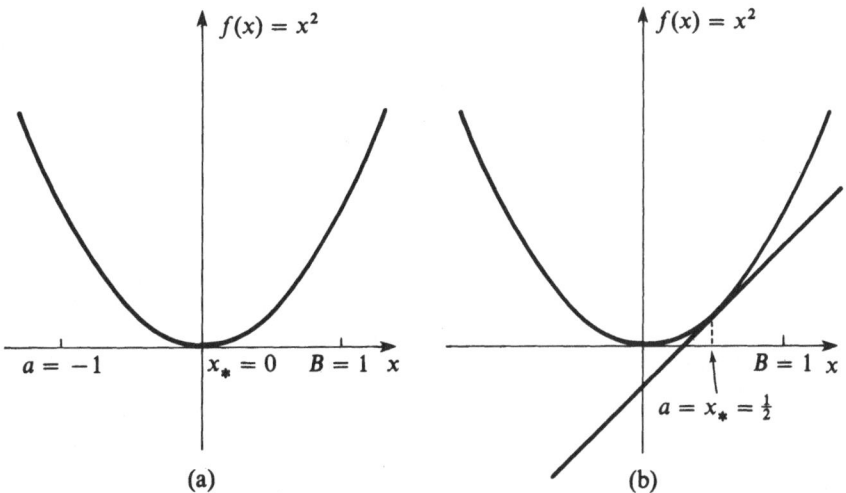

Figure 8.1

for all $x, x_* \in X$. Here the quantity on the right-hand side contains the scalar product of the vectors $f'(x_*)$ and $(x - x_*)$. On the other hand, if a differentiable function f satisfies (1.1) for all $x, x_* \in X$, it is convex.

The geometric interpretation of (1.1) is as follows: construct the graph of f and draw the tangent hyperplane at the point $(x_*, f(x_*))$. Then for every abscissa $x \in X$ the right-hand side of (1.1) coincides with the corresponding ordinate of the tangent hyperplane; that is, the graph of the convex function lies above (or, more correctly, not below) its tangent hyperplane. In the one-dimensional case two typical situations are shown in Figure 8.1. In Figure 8.1(a) the point x_* lies at an absolute minimum of f, the derivative $f'(x_*)$ vanishes, and the tangent line is horizontal. The general case is shown by Figure 8.1(b).

In the one-dimensional case the inequality (1.1) assumes the form

$$f(x) \geq f(x_*) + f'(x_*) \cdot (x - x_*). \tag{1.2}$$

The right-hand sides of (1.1) and (1.2) contain the first two terms of the Taylor series expansion of f near the point $x = x_*$. We note that a twice-differentiable function may be represented as

$$f(x) = f(x_*) + f'(x_*)(x - x_*) + \tfrac{1}{2}f''(\xi)(x - x_*)^2, \tag{1.3}$$

where ξ is some point in the interval $[x, x_*]$ (or $[x_*, x]$ if $x_* < x$). Thus, for a twice-differentiable function, convexity is expressed by the condition

$$f''(x) \geq 0. \tag{1.4}$$

A similar argument applies in the multi-dimensional case. The condition (1.4) is replaced by the requirement that the matrix of the second derivatives $f''(x)$ should be positive semidefinite.

In defining an admissible set on which we seek the minimum of f, we generally constrain the domain of definition X in some way. For instance, in the cases depicted in Figure 8.1, we might use the inequalities $x \geq a$ and $x \leq b$ to restrict the whole real axis $-\infty < x < +\infty$. These inequalities may be written as

$$a - x \leq 0,$$
$$x - b \leq 0. \tag{1.5}$$

From now on we shall write the constraints, of which there may be more than one, say m of them, in the universal form

$$g_i(x) \leq 0, \qquad i = 1, 2, \ldots, m. \tag{1.6}$$

For a broad class of problems with a given set X and constraints (1.6), it is convenient to write

$$D = \{x \in X : g_i(x) \leq 0, i = 1, 2, \ldots, m\}.$$

Then D is the admissible set. We remark that the convexity of D can be guaranteed if all the g_i are convex. The search for the infimum,

$$\inf\{f(x) : x \in D\} \tag{1.7}$$

is called the convex programming problem. We may take as X either the whole space or, at times, its positive orthant, the set of vectors with nonnegative components.

Let us now consider an important concept—the Lagrangian, defined as follows:

$$L(x, \lambda) = f(x) + \sum_{i=1}^{m} \lambda_i g_i(x), \qquad x \in X, \lambda_i \geq 0, \qquad i = 1, \ldots, m. \tag{1.8}$$

Here $\lambda = (\lambda_1, \lambda_2, \ldots, \lambda_m)$. The condition that all the components of such a vector shall be nonnegative will be expressed by the vector inequality $\lambda \geq 0$. To avoid various singular cases that may arise when the left-hand side of (1.6) is nonlinear, we shall assume from now on that the following condition is satisfied.

The Regularity Condition. There exists a point $\xi \in D$ at which all the constraints (1.6) reduce to the strict inequality

$$g_i(\xi) < 0, \qquad i = 1, 2, \ldots, m.$$

We have the following important theorem (the Kuhn–Tucker theorem; for a proof see Karmanov [21]).

Theorem. *In order that the point $x_* \in X$ yield a minimum of the function f under the constraints (1.6) it is sufficient—and under the regularity condition neces-*

sary—that there exists a vector $\lambda^* = (\lambda_1^*, \lambda_2^*, \ldots, \lambda_m^*)$ *such that for all* $x \in X$
and $\lambda \geq 0$ *the inequality*

$$L(x_*, \lambda) \leq L(x_*, \lambda^*) \leq L(x, \lambda^*) \tag{1.9}$$

is satisfied.

The point (x_*, λ^*) is called a *saddle point* of the Lagrange function (on the set $X \times R_+^m$). It is important to note that the set on which we seek it has a simple form, i.e., the constraints (1.6) do not appear in its description and are automatically satisfied for its x-components. We note also that when the g_i in (1.6) are linear, the regularity condition may be dropped from the Kuhn–Tucker theorem.

If the pair (x_*, λ^*) is a saddle point of the Lagrange function (and, consequently, x_* solves the convex programming problem) the following complementarity condition is fulfilled:

$$\lambda_i^* \cdot g_i(x_*) = 0, \qquad i = 1, 2, \ldots, m. \tag{1.10}$$

In fact, if we had $\lambda_k^* > 0$ and $g_k(x_*) < 0$ for some k, then by writing $\lambda_i = \lambda_i^*$ for $i \neq k$ and $\lambda_k = \lambda_k^*/2$, we would violate the left-hand inequality in (1.9). The admissibility of x_* and the complementarity condition are obviously sufficient for the resolution of the left-hand inequality in (1.19). The condition (1.10) implies that

$$L(x_*, \lambda^*) = f(x_*). \tag{1.11}$$

We now introduce the important notion of duality in convex programming problems. We shall suppose that the Lagrange function has the saddle point (x_*, λ^*). Write

$$\phi(\lambda) = \inf\{L(x, \lambda): x \in X\}, \qquad \lambda \geq 0. \tag{1.12}$$

For some λ, the function ϕ may assume the singular value $-\infty$. By (1.9) and (1.11), however, $\phi(\lambda^*) = f(x_*)$. Now, if we write

$$\tilde{f}(x) = \begin{cases} +\infty, & x \notin D, \\ f(x), & x \in D, \end{cases}$$

it turns out that

$$\tilde{f}(x) = \sup\{L(x, \lambda): \lambda \geq 0\}, \qquad x \in X.$$

The problems $\inf\{\tilde{f}(x): x \in X\}$ and (1.7) are equivalent. We see from (1.9) and (1.11) that we have

$$f(x_*) = \min\{\tilde{f}(x): x \in X\} = \max\{\phi(\lambda): \lambda \geq 0\}. \tag{1.13}$$

Thus, if we are to find $f(x_*) = \min\{f(x): x \in D\}$, we may instead solve the dual problem, i.e., find $\max\{\phi(\lambda): \lambda \geq 0\}$.

Let us illustrate the above material by a simple example, taking $X = R$ and $f(x) = x^2$. In Figure 8.1(a), $g_1(x) = -1 - x$, $g_2(x) = x - 1$, and the Lagrange function has the form

$$L(x, \lambda) = x^2 - \lambda_1(1 + x) + \lambda_2(x - 1). \tag{1.14}$$

Since at the point $x_* = 0$ the strict inequality is satisfied in both constraints, (1.10) implies that $\lambda_1^* = 0$ and $\lambda_2^* = 0$. Let us test the necessary and sufficient condition (1.9): in our case

$$L(x_*, \lambda) = L(0, \lambda) = -\lambda_1 - \lambda_2, \qquad L(x_*, \lambda^*) = L(0, 0) = 0,$$

$$L(x, \lambda^*) = L(x, 0) = x^2,$$

and the conditions (1.9) become the inequalities $-(\lambda_1 + \lambda_2) \leq 0 \leq x^2$, which are trivial for $\lambda_1, \lambda_2 \geq 0$. In Figure 8.1(b), $g_1(x) = \frac{1}{2} - x, g_2(x) = x - 1$, and the Lagrange function is

$$L(x, \lambda) = x^2 + \lambda_1(\tfrac{1}{2} - x) + \lambda_2(x - 1). \tag{1.15}$$

Now $x_* = \frac{1}{2}$ and only the second constraint becomes a strict inequality. Therefore, we must put $\lambda_2^* = 0$ and define λ_1^* by the condition that $L(x, \lambda^*)$ attains a free maximum at $x = x_* = \frac{1}{2}$. Because all the relevant functions are differentiable, we are led to the system

$$\lambda_2^* = 0,$$

$$\frac{\partial}{\partial x} [x^2 + \lambda_1^*(\tfrac{1}{2} - x) + \lambda_2^*(x - 1)]|_{x = 1/2} = 0,$$

whence we find that $\lambda_1^* = 1$. Then the conditions (1.9) become: $\frac{1}{4} - \frac{1}{2} \cdot \lambda_2 \leq \frac{1}{4} \leq x^2 + \frac{1}{2} - x$. The left-hand inequality is satisfied because λ_2 is nonnegative, and the right-hand inequality is satisfied because $x^2 - x + \frac{1}{4} = (x - \frac{1}{2})^2$ is nonnegative for all x.

Let us now use the duality concept to solve the same problems. The functions (1.14) and (1.15) attain their minimum with respect to x at $2x = \lambda_1 - \lambda_2$; therefore, a minimum can be reached for all values of λ_1 and λ_2. To compute the solution ϕ in the first problem we must find a free minimum (since $X = R$) of (1.14) with respect to x, for fixed values of λ_1 and λ_2. This minimum is attained at $x = (\lambda_1 - \lambda_2)/2$, and

$$\phi(\lambda) = -\left[\frac{(\lambda_2 - \lambda_1)^2}{4} + \lambda_1 + \lambda_2\right]. \tag{1.16}$$

We see easily that for nonnegative λ the maximum of (1.16) is attained at $\lambda_1^* = \lambda_2^* = 0$. This is the solution of the dual problem, with $\phi(0) = 0 = f(0)$, i.e., we have arrived at the relation (1.13). In exactly the same way we find for (1.15) that

$$\phi(\lambda) = -\left[\frac{(\lambda_2 - \lambda_1)^2}{4} + \lambda_2 - \tfrac{1}{2}\lambda_1\right]. \tag{1.17}$$

Here the dual problem consists in maximizing (1.17) under the constraints $\lambda_1 \geq 0, \lambda_2 \geq 0$. We can show that the maximum is attained at $\lambda_1 = 1, \lambda_2 = 0$, and we again arrive at (1.13).

8.2 Linear Programming

The linear programming problem consists in finding the maximum or minimum of a linear function under a finite number of linear constraints. It arises in many different circumstances, but usually as a component part of optimization methods in nonlinear cases when we are making a stage-by-stage linearization. Certain canonical forms have been adopted for the notation in linear programming problems. In many practical cases the conditions $x_i \geq 0$ $(i = 1, 2, \ldots, n)$, are naturally imposed, and therefore, the following format is widely used:

$$\sum_{j=1}^{n} \overline{\overline{C}}_j x_j \rightarrow \min,$$

$$\sum_{j=1}^{n} a_{ij} x_j \geq b_i, \qquad i = 1, 2, \ldots, l,$$

$$x_i \geq 0, \qquad i = 1, 2, \ldots, n.$$

For our subsequent work, however, the following format is more convenient:

$$\sum_{j=1}^{n} \overline{\overline{C}}_j x_j \rightarrow \min,$$

$$\sum_{j=1}^{n} a_{ij} x_j \geq b_i, \qquad i = 1, 2, \ldots, m, \qquad m \geq n, \tag{2.1}$$

including the nonnegativity coefficients (if they are present) in the total list of indices. By introducing the vectors $b = (b_1, \ldots, b_n)$, $\overline{\overline{C}} = (\overline{\overline{C}}_1, \ldots, \overline{\overline{C}}_n)^T$, $x = (x_1, \ldots, x_n)^T$ for the matrix

$$A = \{a_{ij}, i = 1, 2, \ldots, m; j = 1, 2, \ldots, n\}$$

we may rewrite the problem (2.1) in the form

$$\min\{(\overline{\overline{C}}, x): Ax \geq b\}. \tag{2.2}$$

Let us apply the results of the preceding section to establish the dual problem. We note that here

$$f(x) = (c, x), \qquad g_i(x) = b_i - (Ax)_i, \qquad i = 1, 2, \ldots, m.$$

Therefore, the Lagrange function for the linear programming problem as formulated is:

$$L(x, \lambda) = \sum_{j=1}^{n} c_j x_j + \sum_{i=1}^{m} \lambda_i \left(b_i - \sum_{j=1}^{n} a_{ij} x_j \right), \tag{2.3}$$

or, if we introduce the vector of Lagrange multipliers $\lambda = (\lambda_1, \lambda_2, \ldots, \lambda_m)^T$ the Lagrange function becomes

$$L(x, \lambda) = (c, x) + (\lambda, b - Ax) = (\lambda, b) + (x, c - A^T \lambda). \qquad (2.4)$$

The function $\phi(\lambda)$ in (1.12) becomes

$$\phi(\lambda) = \inf\{(\lambda, b) + (c - A^T \lambda, x) \colon \in R^n\}, \qquad \lambda \geq 0.$$

Therefore, since x is arbitrary, the function ϕ will take on finite values only under the condition that λ satisfies the linear equations

$$c - A^T \lambda = 0$$

or

$$\sum_{i=1}^{m} a_{ij} \lambda_i = c_j, \qquad j = 1, 2, \ldots, n. \qquad (2.5)$$

Thus the dual of the problem (2.1) is the following linear programming problem:

$$\sum_{i=1}^{m} b_i \lambda_i \to \max,$$

$$\sum_{i=1}^{m} a_{ij} \lambda_i = \overline{\overline{C}}_j, \qquad j = 1, 2, \ldots, n, \qquad (2.6)$$

$$\lambda_i \geq 0, \qquad i = 1, 2, \ldots, m,$$

or

$$\max\{(\lambda, b) \colon A^T \lambda = \overline{\overline{C}}, \lambda \geq 0\}. \qquad (2.7)$$

Let us consider various questions concerning the numerical solutions of linear programming problems; from now on, we assume that these problems are nonsingular. First of all, this means that the equations in the system (2.5) are linearly independent (rank $A = n$), and the system of constraints in the problem (2.1) is compatible. We assume, that is, that the multiply-bounded set described by the system of constraints in (2.1) has corner points. Every such point satisfies all the constraints, and in some n of them with linearly independent left-hand sides the inequality is a strict equality, i.e., at the corner point $\bar{x}$ we can find a subset of the indices $I \equiv \{i_1, i_2, \ldots, i_n\}$ such that

$$\sum_{j=1}^{n} a_{ij} \bar{x}_j = b_i, \qquad i \in I,$$

$$\sum_{j=1}^{n} a_{ij} \bar{x}_j \leq b_i, \qquad i \notin I, \qquad (2.8)$$

where the square submatrix $\{a_{ij}, i \in I; j = \overline{1, n}\}$ is nonsingular. If, for some $i \notin I$, an equality holds in (2.8), the corresponding corner point is said to be degenerate. In a nondegenerate linear programming problem all the vertices are assumed nondegenerate, i.e., for $i \notin I$, only the strict inequality holds. The simplex method, formally described below, can be applied to degenerate problems, but can be guaranteed to be finite only in nondegenerate cases.

The basis of the simplex method is the fact that a linear function attains its minimum or maximum at a corner of a multiply bounded set. We may, therefore, organize an enumeration of these corners (they are finite in number) and choose the one that interests us. Since in real-life problems the number of corners is very large, a complete enumerative search is unrealizable. The simplex method consists in a directed search of the corners, and results in the practical necessity of inspecting only a small number of them.

Thus, suppose we have chosen the corner $\bar{x}$ and the corresponding index set $I \equiv \{i_1, i_2, \ldots, i_n\}$. Since the complementarity condition (1.10) must hold at a saddle point, and since, with the above assumptions on the nondegeneracy of all corners, this condition is equivalent to the condition $\lambda_i = 0$, $i \notin I$, we seek a vector $\bar{\lambda}$ satisfying the system

$$\bar{\lambda}_i = 0, \qquad i \notin I.$$

By (2.5), the remaining components $\bar{\lambda}_i$ ($i \in I$), must satisfy the system

$$\sum_{i \in I} a_{ij} \lambda_i = \bar{\bar{C}}_j, \qquad j = 1, 2, \ldots, n. \tag{2.9}$$

But this system has a matrix that is the transpose of the matrix in (2.8) and, therefore, together with the latter, is square and nonsingular. This leads to the following rule for testing $\bar{x}$ for optimality: solve (2.9) and test its unique solution for nonnegativity. If $\bar{\lambda}_i \geq 0$, for all $i \in I$, the results of the preceding subsection imply that the pair $(\bar{x}, \bar{\lambda})$ is a saddle point of the Lagrange function and the problem is solved: $x_* = \bar{x}$ is the solution of (2.1) and $\lambda^* = \bar{\lambda}$ is the solution of the dual problem (2.6). If, however, $\bar{\lambda}_{i_k} < 0$ for some $i_k \in I$ we must go to another corner and therefore to a different set of indices for the system (2.8)

The transition is to be accomplished as follows: the number i_k for which $\bar{\lambda}_{i_k} < 0$ is excluded from the set I. Then in place of (2.8) we have the system

$$\sum_{j=1}^{n} a_{ij} x_j = b_i, \qquad i \in \{i_1, \ldots, i_{k-1}, i_{k+1}, \ldots, i_n\}$$

The solution is a line passing through the point $\bar{x}$. The direction vector z satisfies the corresponding homogeneous system, and, when normalized, will satisfy the square nonsingular system

$$\sum_{j=1}^{n} a_{ij} z_j = 0, \qquad i \in \{i_1, i_2, \ldots, i_{k-1}, i_{k+1}, \ldots, i_n\},$$

$$\sum a_{i_k j} z_j = 1. \tag{2.10}$$

We have set up the normalization so that the system (2.10) differs from (2.8) only in its right-hand side. Since

$$(c, z) = \sum_{j=1}^{n} c_j z_j = \sum_{j=1}^{n} \left(\sum_{i=1}^{m} a_{ij} \bar{\lambda}_i \right) z_j = \sum_{i=1}^{m} \left(\sum_{j=1}^{n} a_{ij} z_j \right) \bar{\lambda}_i = \bar{\lambda}_{i_k} < 0,$$

we find, that as the parameter ε increases from 0 to $+\infty$, the value of the minimand function at the points $\bar{x} + \varepsilon z$ decreases strictly:

$$(c, \bar{x} + \varepsilon z) = (c, \bar{x}) + \varepsilon \lambda_{i_k}.$$

Two cases arise. If

$$\sum_{j=1}^{n} a_{ij} z_j \geq 0, \qquad i = 1, 2, \ldots, m,$$

then, together with $\bar{x}$, the points of the ray

$$\{\bar{x} + \varepsilon z : \varepsilon \geq 0\} \tag{2.11}$$

satisfy the constraints of (2.1). Since the minimand function decreases to $-\infty$ along the ray, the linear programming problem has no solution, since there is no lower bound to the value of the minimand function. The dual problem has an inconsistent set of constraints.

In the second case the set

$$S = \left\{ i : \sum_{j=1}^{n} a_{ij} z_j < 0 \right\}$$

is not empty, and we may not proceed along the ray (2.11), without leaving the admissible set further than the point corresponding to

$$\bar{\varepsilon} = \min\{\Delta_i / g_i : i \in S\}, \tag{2.12}$$

where we use the abbreviations

$$\Delta_i = \sum_{j=1}^{n} a_{ij} x_j - b_i,$$

$$g_i = -\sum_{j=1}^{n} a_{ij} z_j.$$

We note that $S \cap I = \varnothing$, and for $i \in S$ we always have $g_i > 0$ and $\Delta_i > 0$. If the minimum in (2.12) is attained for $i' \in S$, the new corner point is

$$\bar{\bar{x}} = \bar{x} + \bar{\varepsilon} z$$

and the new set $\bar{I}$ corresponding to it is

$$\bar{I} = \{i_1, \ldots, i_{k-1}, i', i_{k+1}, \ldots, i_n\},$$

which is derived from I by substituting i' for i_k. It may now be taken as the initial point for the next step. The value of the minimand function is decreased in the transition from i_k to i', by the amount $(-\bar{\varepsilon}\lambda_{i_k})$, which is strictly positive if $\bar{\varepsilon} > 0$, and this is guaranteed in the nondegenerate case. In this case, therefore, the value of the minimand function strictly decreases at each step, no corner can be revisited, and since there are only a finite number of corners the method must terminate.

The following problem is an illustration:

$$x_1 + x_2 \rightarrow \min, \qquad -x_1 \geq -1,$$
$$x_1 \geq 0,$$
$$-x_2 \geq -1,$$
$$x_2 \geq 0.$$

The admissible set D is a square with the corners $(0, 0)$, $(0, 1)$, $(1, 0)$, and $(1, 1)$. The solution is the origin of coordinates. Suppose we have chosen the vertex $(1, 1)$. This corresponds to the set $I = \{1, 3\}$ since the first and third constraints reduce to equalities there. The system (2.9) for determining $\bar{\lambda}_1$ and $\bar{\lambda}_3$ becomes

$$-\lambda_1 = 1,$$
$$-\lambda_3 = 1,$$

so that $\bar{\lambda}_1 = \bar{\lambda}_3 = -1$, and we may choose either 1 or 3 as the value of the index i_k. Setting $i_k = 1$, we arrive at the system

$$-z_1 = 1,$$
$$-z_2 = 0,$$

into which (2.10) is converted. Thus $z = (-1, 0)$, $g_2 = 1$, $g_4 = 0$, $S = \{2\}$ and by (2.12)

$$\bar{\varepsilon} = \Delta_2/g_2 = 1/1 = 1.$$

Since the set S contains only one element, we have not selected the minimum of (2.12), and we must adopt the only possible selection: $i' = 2$. Then we find

$$\bar{\bar{x}} = \bar{x} + \bar{\varepsilon}z = (1, 1) + 1 \cdot (-1, 0) = (0, 1),$$

$$\bar{I} = \{2, 3\}.$$

For the new set $\bar{I}$ we obtain the system (2.9)

$$\lambda_2 = 1,$$
$$-\lambda_3 = 1,$$

so that $i_k = 3$. The vector z is now defined by the system

$$z_1 = 0,$$
$$-z_2 = 1,$$

and we have: $z = (0, -1)$, $g_1 = 0$, and $g_4 = 1$. Since $\Delta_4 = 1$, we find $\bar{\varepsilon} = 1$, $i' = 4$,

$$x_* = \bar{x} + \bar{\varepsilon}z = (0, 1) + 1 \cdot (0, -1) = (0, 0),$$

$$I^* = \{2, 4\}.$$

Finally, from the system

$$\lambda_2 = 1,$$
$$\lambda_4 = 1,$$

we obtain $\lambda_2^* = \lambda_4^* = 1 \geq 0$ for the set I^*, and we conclude that $x_* = (0, 0)$ is the solution of our problem. The set $\lambda = (0, 1, 0, 1)$ solves the corresponding problem.

$$-\lambda_1 - \lambda_3 \to \max,$$

$$-\lambda_1 + \lambda_2 = 1,$$

$$-\lambda_3 + \lambda_4 = 1,$$

$$\lambda_1, \lambda_2, \lambda_3, \lambda_4 \geq 0.$$

We conclude by noting that special methods have been developed for choosing the initial vertex when it is not determined by supplementary considerations. Also, the simplex method can be applied in another form to the solution of linear programming problems, and various numerical schemes may be used to implement it. Detailed information on these points may be found in monographs by Bulavskii, Zvyagina, and Yakovleva [21], Yudin and Goldstein [21].

8.3 Quadratic Programming

The convex quadratic programming problem consists in the minimization of a second-order polynomial, in the presence of a finite number of linear constraints on the unknowns. It is usually assumed that the quadratic form appearing in the minimand function is positive semidefinite, so that the convexity of the function is guaranteed. To simplify the following exposition we shall assume more, namely, that the quadratic form is positive definite. Although the quadratic problem is more complex than the linear, we may yet develop a finite method for its solution, using approaches like those employed in the preceding subsection.

It will be convenient to consider the quadratic programming problem in the following form:

$$f(x) = \frac{1}{2} \sum_{j=1}^{n} \sum_{k=1}^{n} \overline{\overline{C}}_{jk} x_j x_k + \sum_{j=1}^{n} d_j x_j \to \min,$$

$$\sum_{j=1}^{n} a_{ij} x_j \geq b_i, \qquad i = 1, 2, \ldots, m,$$

or in matrix notation

$$f(x) = \frac{1}{2}(x, Cx) + (d, x) \to \min,$$

$$Ax \geq b. \tag{3.1}$$

Here $C = \{c_{j,k}\}$ is a symmetric positive definite matrix, $A = \{a_{i,j}\}$ is an $m \times n$ matrix, d is a vector of dimension n, and b a vector of dimension m. In our current case the Lagrange function is

$$L(x, \lambda) = \tfrac{1}{2}(x, Cx) + (d, x) + (\lambda, b - Ax), \qquad (3.2)$$

and the saddle point is to be sought for $\lambda \geq 0$ and all x (as before, $X = R^n$). Therefore, the right-hand inequality in the optimality criterion (1.9) reduces to the following condition:

$$\left. \frac{\partial}{\partial x} L(x, \lambda^*) \right|_{x = x_*} = 0,$$

or

$$Cx_* + d - A^T \lambda^* = 0. \qquad (3.3)$$

The left-hand inequality in the optimality criterion (1.9) reduces, as before, to the admissibility of x_* and the validation of the completeness condition (1.10). In our case this means that

$$Ax_* \geq b \qquad (3.4)$$

$$\lambda_i^*(b_i - (Ax_*)_i) = 0, \qquad i = 1, \ldots, m. \qquad (3.5)$$

The Lagrange multipliers must of course also be nonnegative:

$$\lambda_i^* \geq 0, \qquad i = 1, 2, \ldots, m. \qquad (3.6)$$

Thus, if we choose the pair (x_*, λ^*) to satisfy the conditions (3.3)–(3.6), x_* will be the solution of (3.1) and λ^* will be the solution of the dual problem.

We now describe an analog of the simplex method. The general features of the method are based on the following considerations: the multiply bounded set described by the constraint system in problem (3.1) is to be put up into faces of different dimensionalities, including the interior as the face of highest dimensionality. Each face consists of admissible points satisfying some system of equations of the form

$$A_I x = b_I. \qquad (3.7)$$

Here A_I denotes a matrix consisting of the rows of A with indices $i \in I \equiv \{i_1, i_2, \ldots, i_l\}$, i.e.,

$$A_I = \{(A)_i, i \in I\}.$$

The column vector b_1 is made up in the same way. We may suppose that the rows of A_1 are linearly independent, by decreasing I if necessary. However, the presence of faces described by a linear system (3.7) with dependent equations will be considered a degeneracy, and then we cannot guarantee the finiteness of our method (just as in the linear programming case).

Since the desired point x_* belongs to the (relative) interior of some face— which, in particular, may be a zero-dimensional face, i.e., a vertex—it should yield a minimum of our quadratic function on the whole set of solutions of

(3.7). We may therefore proceed as follows: we inspect various subsystems (3.7) and obtain a minimum of our quadratic function on their solutions; then we test to see whether our minimum satisfies the constraints and the optimality criterion. As in the linear case, the inspection of the sets I must be controlled so that we avoid a complete count of all faces which would render the method unrealizable. Moreover, the scheme we have in mind is valid only for positive definite matrices C, since, otherwise, the minimum may fail to be reached for some sets I.

The formal description of the method is as follows: suppose we have an admissible point $\bar{x}$ and we have selected a set $I \equiv \{i_1, i_2, \ldots, i_1\}$ for which the rows of the matrix A_I are linearly independent. Suppose also that $\bar{x}$ satisfies (3.7).

Two cases are possible:

(a) The point $\bar{x}$ yields a minimum of f under the constraints (3.7). Then, by the classic Langrangian criterion, there exist multipliers $\bar{\lambda}_I = \{\bar{\lambda}_i, i \in I\}$ for which

$$\frac{\partial}{\partial x}\left[f(x) + \bar{\lambda}_I(b_I - A_I x)\right]\Bigg|_{x=\bar{x}} = 0,$$

or, in our case,

$$C\bar{x} + d - A_I^T \bar{\lambda}_I. \tag{3.8}$$

We note that the vector $\bar{\lambda}_I$ is uniquely defined by (3.8), since the rows of A_I are linearly independent. If the $\bar{\lambda}_I$ are nonnegative, then by setting $\lambda_i = 0$ for $i \notin I$, we find that $x_* = \bar{x}$ and $\lambda^* = \bar{\lambda}$ satisfy all the conditions (3.3)–(3.6), and we have found a solution. If, however, for some $i_k \in I$ we have $\bar{\lambda}_{i_k} < 0$, we remove that index from I and for the same point $\bar{x}$ we have a new set of indices $I = \{i_1, \ldots, i_{k-1}, i_{k+1}, \ldots, i_1\}$, that is, we have moved to a new face, whose dimension is higher by one than the dimension of the original face.

(b) The function f attains a minimum on the solutions of (3.7) at some point $x_0 \neq \bar{x}$. Then we attempt to move to the point x_0, or, if we are prevented from reaching it by the constraints in (3.1), then to come as close as possible. Algorithmically, this means the following: set $z = x_0 - \bar{x}$, and

$$g_i = -(Az)_i = -\sum_{j=1}^{n} a_{ij} z_j, \qquad i \notin I, \tag{3.9}$$

$$\Delta_i = (A\bar{x})_i - b_i = \sum_{i=1}^{n} a_{ij} \cdot \bar{x}_j - b_i, \qquad i \notin I \tag{3.10}$$

and, defining ε_0 by the formula

$$\varepsilon_0 = \min\{\Delta_i/g_i : g_i > 0\}, \tag{3.11}$$

we set $\bar{\varepsilon} = \min\{\varepsilon_0, 1\}$. Then if $\bar{\varepsilon} = 1$ we have reached the point x_0 and may preserve the set I. If $\bar{\varepsilon} < 1$, the index i' at which the minimum is realized in (3.11) is to be included in I, that is, we pass to a new set of indices $I = \{i_1, i_2, \ldots, i_l, i'\}$. In both cases the point $\bar{x}$ is replaced by $\bar{x} + \bar{\varepsilon}z$.

Note that to find the point x_0 we must solve the system

$$Cx + d - A_I^T \lambda_I = 0,$$
$$A_I x = b_I, \tag{3.12}$$

which has a nonsingular matrix since C is positive definite and the rows of A_I are linearly independent.

The method of solution, as a whole, consists of a succession of computational steps, of two kinds. Since the number of elements in the set I increases in step (b), only a finite number of steps of type (b) may intervene between two successive steps of type (a). On the other hand, at every step of type (a) the value of f at the point $\bar{x}$ is uniquely determined by the set I, and if it turns out that f is strictly decreasing from one step to another, a given set I cannot be encountered twice in steps of type (a), and the method terminates after a finite number of steps. With this end in view, we require that the following nondegeneracy condition should be satisfied: no bounding hyperplane (among the constraints) with an index $i \notin I$ shall pass through a minimum point of the function f on the set of solutions of any system (3.7). In this case each step of type (a) is followed by a step of type (b) with $\bar{\varepsilon} > 0$ (since all the $\Delta_i > 0$ for $i \notin I$) and the value of f decreases strictly (we have in fact moved away from the point $\bar{x}$). Since the value of f does not increase on the remaining steps, the finiteness of the method is guaranteed.

The following example illustrates the method. We are to find

$$\min\{\tfrac{1}{2}(x_1 - \tfrac{3}{4})^2 + \tfrac{1}{2}(x_2 - \tfrac{3}{4})^2 : x_1 \geq 0, x_2 \geq 0, -x_1 - x_2 \geq -1\}. \tag{3.13}$$

Beginning at the point $(0, 0)$, which corresponds to $I = \{1, 2\}$ and to the system $x_1 = 0$, $x_2 = 0$, we execute a step of type (a), since our face consists of a single point and a trivial minimum of f is encountered at $(0, 0)$. Since in our example

$$C = \begin{pmatrix} 1 & 0 \\ 0 & 1 \end{pmatrix}, \qquad d = (-\tfrac{3}{4}, -\tfrac{3}{4})^T,$$

$$A_I = \begin{pmatrix} 1 & 0 \\ 0 & 1 \end{pmatrix}, \qquad b_I = \begin{pmatrix} 0 \\ 0 \end{pmatrix},$$

we find from (3.8) that

$$\lambda_1 = \lambda_2 = -\tfrac{3}{4},$$

since either index in I may be chosen as i_k. We set $i_k = 1$ and pass to the next step, with $\bar{x} = (0, 0)$ and $I = \{2\}$. Now the system (3.7) consists of the equation $x_2 = 0$ and the set of its solutions consists of the axis of abscissas. The minimum of f on this set is attained at the point $(\tfrac{3}{4}, 0)$, which satisfies the

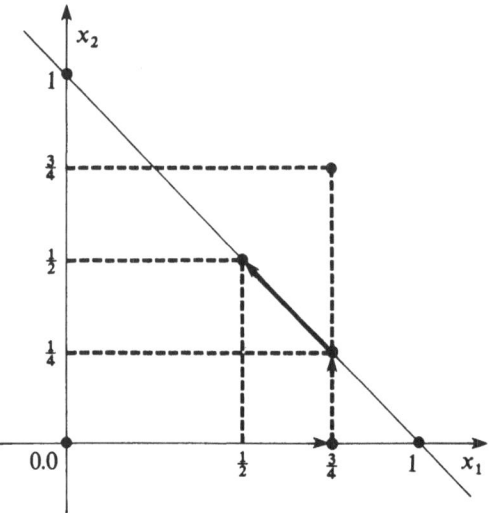

Figure 8.2

system of constraints. Therefore, as a result of a type (b) step we move to this point, conserving the set $I = \{2\}$. Next we execute a step of type (a), for which the system (3.8) becomes

$$\frac{3}{4} - \frac{3}{4} = 0,$$
$$0 - \frac{3}{4} - \bar{\lambda}_2 = 0,$$

and we conclude that $\bar{\lambda}_2 = -\frac{3}{4}$, so that we set $i_k = 2$. We pass to the next step with $\bar{x} = (\frac{3}{4}, 0)$ and $I = \varnothing$ (the empty set); accordingly, the minimum of f is to be sought over the entire plane, and we find $x_0 = (\frac{3}{4}, \frac{3}{4})$. However, this point does not satisfy the third constraint. We move toward it, setting $z = (0, \frac{3}{4})$, and reach the point $\bar{x} = (\frac{3}{4}, \frac{1}{4})$ and the set $I = \{3\}$ for the value $\bar{\varepsilon} = \frac{1}{3}$. We again execute a step of type (b), since the minimum point $x_0 = (\frac{1}{2}, \frac{1}{2})$ on the set of solutions of the equation $x_1 + x_2 = 1$ does not coincide with $\bar{x}$. However, since it is an admissible point, it turns out that $\varepsilon = 1$, and we have obtained the solution $(\frac{1}{2}, \frac{1}{2})$. The sequence of points we have obtained is shown in Figure 8.2. The arrows indicate the passages from one point to another.

8.4 Numerical Methods in Convex Programming Problems

We now consider infinite iterative methods, used as a rule in a case that is more general than the ones considered in the preceding sections. We shall merely outline the basic schemes of these methods and illustrate them via the example (3.13) of the preceding section. The minimand function is in each case supposed differentiable.

(a) *The Conditional Gradient Method*

Given a point $\bar{x}$ in the admissible set D we linearize the minimand function and consider the problem

$$\min\{f(\bar{x}) + (f'(\bar{x}), x - \bar{x}): x \in D\}. \tag{4.1}$$

If x^0 is the solution of this problem, we search for a minimum of f on the segment joining the points $\bar{x}$ and x^0; that is, we solve the following one-dimensional minimization problem:

$$\min\{f(\bar{x} + \alpha(x^0 - \bar{x})): \alpha \in [0, 1]\}. \tag{4.2}$$

If this minimum is attained with $\alpha = \bar{\alpha}$, we choose the point

$$\bar{x} + \bar{\alpha}(x^0 - \bar{x}). \tag{4.3}$$

as the next approximation. Since the set D is convex the point (4.3) is admissible along with $\bar{x}$ and x^0. Note that if D has a complicated representation (i.e., the functions g_i are nonlinear) the problem (4.1) is a little less difficult than the original problem. In practice, therefore, this method is applied only when the constraints are linear, and the nonlinearity is concentrated in the function f. Then (4.1) becomes a linear programming problem and can be effectively solved. In the notation of Section 8.2 we have $c = f'(\bar{x})$; the constant term $f(\bar{x}) + (f'(\bar{x}), \bar{x})$, of course, plays no role in the optimization.

The character of the work to be performed in this method can be illustrated by the example (3.13). If, at the point $\bar{x} = (\bar{x}_1, \bar{x}_2)$, it turns out that $\bar{x}_1 > \bar{x}_2$, i.e., if the point lies below the diagonal of the first square, we take the point $x_1^0 = 0$, $x_2^0 = 1$ as the solution of (4.1), which here takes the form

$$\min\{(\bar{x}_1 - \tfrac{3}{4})x_1 + (\bar{x}_2 - \tfrac{3}{4})x_2 : x_1 \geq 0, x_2 \geq 0, x_1 + x_2 \leq 1\},$$

after we discard the constant term. If, however, $\bar{x}_1 < \bar{x}_2$ we have $x_1^0 = 1$ and $x_2^0 = 0$. The iteration proceeds as shown in Figure 8.3, where we display four successive points and the paths from one to another.

(b) *The Gradient Projection Method*

We know that if $f'(x) \neq 0$ the function f increases as we move along the gradient. Therefore, to find a free minimum we often use a method consisting in displacements from the successive points $\bar{x}$ in the "antigradient" direction $z = -f'(\bar{x})$. In the presence of constraints such a displacement may take us out of the admissible region. We must, therefore, provide some sort of method for returning, for example, by projection onto the set D. The numerical procedure is as follows: choosing a step length α (generally speaking, separately for each step) we compute the point $x^0 = \bar{x} - \alpha f'(\bar{x})$ and then find the value of x that solves the problem

$$\min\{\tfrac{1}{2}(x - x^0, x - x^0): x \in D\}. \tag{4.4}$$

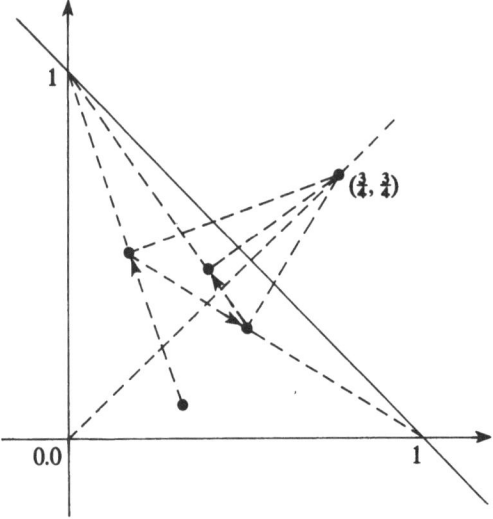

Figure 8.3

The problem (4.4) can be quite complex if D is arbitrary. However, if D has a single boundary, the problem is one of quadratic programming, which can be solved by a finite number of operations. Note that (4.4) has a specific characteristic: the matrix C of the quadratic form is unitary. This fact may be taken into account in the solution of a quadratic programming problem. The nature of the work to be done in this method is illustrated in Figure 8.4. Moving toward the point $(\frac{3}{4}, \frac{3}{4})$—in our example this is the direction of the antigradient—and projecting onto the admissible set, we successively

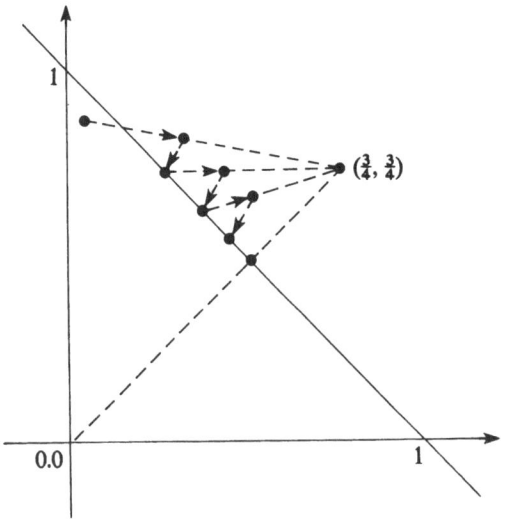

Figure 8.4

approximate the point $(\frac{1}{2}, \frac{1}{2})$. It is worth noting that here, as in general for many iterative methods, the tactics for choosing a step-length play an essential role both in attaining convergence and in determining the rate of convergence.

(c) *The Method of Admissible Directions*

Both the preceding methods are fundamentally oriented toward the problem of linear constraints. We now look at a method applicable to general non-linear constraints. Suppose we have an admissible point $\bar{x}$. We write

$$S(\bar{x}) = \{i: g_i(\bar{x}) = 0\}. \tag{4.5}$$

Constraints with indices in the set (4.5) will be referred to as active at the point $\bar{x}$. We seek a direction z from the point $\bar{x}$ which shall make an obtuse angle with the gradient of the function f and with the outward normals to the active constraints, i.e., with the vectors $g_i'(\bar{x})$ ($i \in S$). This choice of the direction z guarantees on the one hand that the minimand function will decrease along z and, on the other hand, that motion in this direction is possible without transgressing the (curvilinear) boundary of D. Technically this idea is implemented by imposing the linear constraints

$$(f'(\bar{x}), z\sigma | f'(\bar{x}) | \leq 0,$$
$$(g_i'(\bar{x}), z) + \sigma | g_i'(\bar{x}) | \leq 0, \qquad i \in S, \tag{4.6}$$

where $| f'(\bar{x}) |$ and $| g_i'(\bar{x}) |$ are the Euclidean lengths of the corresponding vectors. The value of σ is to be maximized. However, in view of the homogeneity of the constraints (4.6), we must still add some sort of normalization condition. This is usually a constraint on either the Euclidean norm of the vector z

$$(z, z) \leq 1$$

or the cubic norm

$$-1 \leq z_j \leq 1, \qquad j = 1, \ldots, n.$$

In the latter case a linear programming problem is used to determine z and σ. The direction z is found by solving the one-dimensional minimization

$$\min_\alpha \{ f(\bar{x} + \alpha z): g_i(\bar{x} + \alpha z) \leq 0, i = 1, 2, \ldots, m \}.$$

We point out that in such a simple case the method of possible directions may fail to converge. To guarantee convergence we must replace the set (4.5) in the system (4.6) by the set

$$S = \{i: g_i(\bar{x}) \geq -\varepsilon\},$$

where ε is some small positive number, which may be decreased as we approach the solution. Figure 8.5 shows, qualitatively, the nature of the work done in this method, for the problem (3.13).

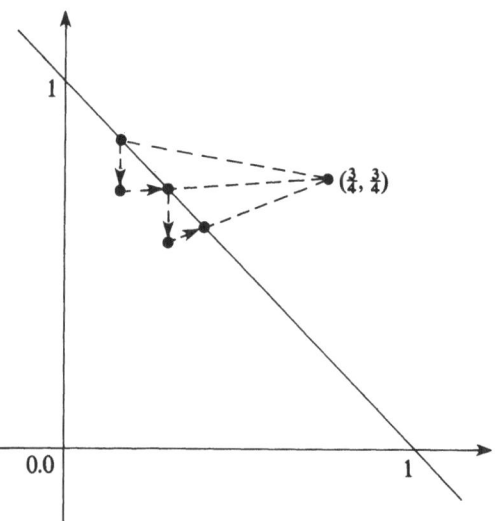

Figure 8.5

(d) *The Method of Cost Functions*

This is also a method intended, in general, for use with nonlinear constraints. The idea consists in not applying the constraints explicitly, but rather in adding to the minimand function a term that imposes a cost for violating the constraints. If the magnitude of the cost is great enough we may expect that the point representing the free minimum of the modified function will not greatly violate the constraints. This scheme is often implemented in the following way: write

$$\Phi(x) = \sum_{i=1}^{m} (\max\{g_i(x), 0\})^2,$$

and seek the free minimum of the function

$$f(x) + r\Phi(x), \tag{4.7}$$

where r is some sufficiently large positive number.

For (3.13) the function Φ takes the form

$$\Phi(x) = (\max\{-x_1, 0\})^2 + (\max\{-x_2, 0\})^2 + (\max\{x_1 + x_2 - 1, 0\})^2.$$
$$\tag{4.8}$$

Since, however, the condition that the variables take on nonnegative values does not conflict with the minimization of f, these constraints will be satisfied at the free minimum of (4.7). Therefore, in fact, the first two terms of (4.8) will vanish.

The third term, on the other hand, will not vanish, since the third constraint prevents us from reaching the absolute minimum of the function f. These qualitative (not completely rigorous) arguments show that the minimum of (4.7) will in fact coincide with the minimum of the function

$$\tfrac{1}{2}(x_1 - \tfrac{3}{4})^2 + \tfrac{1}{2}(x_2 - \tfrac{3}{4})^2 + r(x_1 + x_2 - 1)^2.$$

Equating to zero the partial derivatives of this function with respect to x_1 and x_2 we obtain the system

$$x_1 - \tfrac{3}{4} + 2r(x_1 + x_2 - 1) = 0,$$
$$x_2 - \tfrac{3}{4} + 2r(x_1 + x_2 - 1) = 0.$$

From this we find that

$$x_1 = x_2 = \frac{\tfrac{3}{4} + 2r}{1 + 4r}.$$

For large r we have:

$$x_1 = x_2 \approx \tfrac{1}{2} + \frac{1}{16 \cdot r},$$

which approximates the solution of the problem.

8.5 Dynamic Programming

In the two preceding sections we have dealt with the problems of optimal controls. An arbitrary problem of this type, in connection with a given mathematical model, is provided by the equations of state and the boundary conditions describing the behavior of some object. We may always separate the variables into two groups—a set of dependent variables describing the state of the object, and a set of control functions which are immediately accessible from the outside, and may be changed, and have values belonging to a given set of control values. The optimal control problem is as follows: we are required to choose control values from the admissible set that will minimize a given functional (depending in the general case on the controls and on the solution of the equations).

We shall look at the optimal control problem in the light of an example in which the behavior of the controlled object is described by a system of ordinary differential equations:

$$\frac{d\phi}{dt} = f(\phi, u), \qquad 0 \leq t \leq T \tag{5.1}$$

$$\phi(0) = \phi_0,$$

where

$$\phi = \{\phi_1, \ldots, \phi_n\}, \qquad f = \{f_1, \ldots, f_n\}, \qquad u = \{u_1, \ldots, u_m\}.$$

The admissible controls will be taken as arbitrary piecewise-continuous measurable functions $u = u(t)$, taking on values in the closed region $U \subset E^m$.

We are required to find a control function $u(t)$ belonging to the class of admissible controls and the corresponding solution $\phi(t)$ of (5.1), such that some given functional $J[u]$ is minimized:

$$J[u] = \int_0^T f_0(\phi, u)\, dt = \min_{u \in U}. \tag{5.2}$$

It is assumed that every admissible control uniquely defines a solution of (5.1).

The dynamic programming method is based on Bellman's optimality principle, which may be formulated as follows:

An optimal control is at each instant independent of the prior history of the system and is defined only by the state of the system at that instant and by the goal of the control.

If we introduce the notation

$$Q(\phi, t) = \min_{u \in U} \int_t^T f_0(\phi, u)\, d\tau, \tag{5.3}$$

we have from the optimality principle

$$Q(\phi(t), t) = \min_u \left\{ \int_t^{t+\Delta t} f_0(\phi, u)\, d\tau + \min_u \int_{t+\Delta t}^T f_0(\phi, u)\, d\tau \right\} \tag{5.4}$$

The second term in the braces is by definition $Q(\xi + \Delta\xi, t + \Delta t)$, where

$$\Delta\xi = \int_t^{t+\Delta t} f(\phi, u)\, d\tau \tag{5.5}$$

We expand both terms in a Taylor series and let Δt tend to zero; then we obtain from (5.4) a partial differential equation (Bellman's equation):

$$-\frac{\partial Q}{\partial t} = \min \left[f_0(\phi, u) + \left(f(\phi, u), \frac{\partial Q}{\partial \phi} \right) \right], \tag{5.6}$$

$$Q(\phi, T) = 0.$$

Suppose that the minimum on the right-hand side of (5.6) is attained only on the unique element $u^* \in U$; then u^* is a function of ϕ and $\partial Q/\partial \phi$,

$$u^* = u^*\left(\phi, \frac{\partial Q}{\partial \phi} \right). \tag{5.7}$$

Substituting this function in (5.6) we have a nonlinear system of equations

$$-\frac{\partial Q}{\partial t} = f_0\left(\phi, u^*\left(\phi, \frac{\partial Q}{\partial \phi} \right) \right) + \left(f\left(\phi, u^*\left(\phi, \frac{\partial Q}{\partial \phi} \right) \right), \frac{\partial Q}{\partial \phi} \right). \tag{5.8}$$

If we suppose that $u^*(\ )$ is a function of ϕ and t then (5.8) will be a hyperbolic system of equations, with characteristics directed from $t = 0$ to $t = T$. A rigorous foundation for the dynamic programming method, applicable to continuous problems of optimal control, has been given by Boltyanskii [21], yielding necessary and sufficient conditions for optimality in terms of $Q(\phi, t)$.

The idea underlying the dynamic programming method is that of embedding a given concrete problem in a family of simpler problems. This is most clearly seen in the derivation of the equations of dynamic programming for processes described by a system of difference equations

$$\phi_{i+1} = g(\phi_i, u_i), \qquad i = 0, 1, \ldots, N - 1, \tag{5.9}$$

where $\phi_i \in E_n$ is an n-dimensional state vector and $u_i \in E_m$ is an m-dimensional control vector.

The difference equations (5.9) may arise either from a physical description of the process or from the discretization of the system (5.1). We require that the solutions of (5.9) minimize a functional of the form

$$J(u) = \sum_{i=0}^{N-1} \theta(\phi_i, u_i) \to \min_{\{u_0, \ldots, u_{N-1}\}}. \tag{5.10}$$

It is clear, from the formulation of the problem, that if the solution of (5.9), (5.10) exists the optimum of the functional depends on the initial state ϕ_0 and the number of steps N. Denoting this optimal value by $Q_N(\phi_0)$, we write the minimization problem in the following form:

$$Q_N(\phi_0) = \min_{u_0} \min_{\{u_1, \ldots, u_{N-1}\}} \left[\theta(\phi_0, u_0) + \sum_{i=1}^{N-1} \theta(\phi_i, u_i) \right] \tag{5.11}$$

By the structure of the system (5.9) a change in $(u_1, \ldots, u_{N-1})$ can have no effect on ϕ_0 and on the choice of u_0; therefore (5.11) may be rewritten as

$$Q_N(\phi_0) = \min_{u_0} \left[\theta(\phi_0, u_0) + \min_{\{u_1, \ldots, u_{N-1}\}} \sum_{i=1}^{N-1} \theta(\phi_i, u_i) \right]. \tag{5.12}$$

By definition, the second term in the braces is $Q_{N-1}(\phi_1)$ and we have

$$Q_N(\phi_0) = \min_{u_0} [\theta(\phi_0, u_0) + Q_{N-1}(\phi_1)]. \tag{5.13}$$

Continuing our argument in similar fashion we obtain the following recurrence relations:

$$\phi_0 \text{ is given,}$$

$$Q_{N-j}(\phi_j) = \min_{u_j \in U} [\theta(\phi_j, u_j) + Q_{N-j-1}(\phi_{j+1})], \tag{5.14}$$

$$j = 0, \ldots, N - 2; \qquad \phi_{j+1} = g(\phi_j, u_j),$$

$$Q_1(\phi_{N-1}) = \min_{u_{N-1} \in U} [\theta(\phi_{N-1}, u_{N-1})], \tag{5.15}$$

$$\phi_{N-1} = g(\phi_{N-2}, u_{N-2}).$$

It follows from (5.14), (5.15) that, knowing ϕ_{N-1} and solving a relatively simple problem of minimizing a function of m variables, we may successively solve problems of the type of (5.14) and determine $u_{N-2}, \ldots, u_0$ and $Q_N(\phi_0)$. But since the system (5.9) successively defines the $\phi_1, \phi_2, \ldots, \phi_{N-1}$ we have in fact obtained a two-point boundary problem typical of optimal control problems. The equations (5.14), (5.15) yield the necessary and sufficient conditions for optimality of the control sequence $(u_1, \ldots, u_{N-1})$; they are consequences of the structure of the system (5.9)—for known ϕ_j, u_j, the solution of (5.9) does not depend on $\phi_{j-1}, u_{j-1}, \ldots$ (i.e., it is a Markov system) —and of the additivity of the functional (5.10).

Let us apply the dynamic programming method to a simple example. The process is defined by a single equation

$$\phi = u$$
$$\phi(0) = \phi_0, \qquad |u| \le 1 \tag{5.16}$$

and we are to minimize the functional

$$J[u] = \int_0^T \phi^2 \, dt \to \min_u.$$

Bellman's equation (5.6) is written as follows:

$$-\frac{\partial Q}{\partial t} = \min_{|u| \le 1} \left[\phi^2 + u \frac{\partial Q}{\partial \phi} \right]. \tag{5.17}$$

Since a linear function attains its minimum on the boundary of an interval,

$$u^* = -\text{sign}\left(\frac{\partial Q}{\partial \phi}\right). \tag{5.18}$$

We discretize the problem (5.16) as follows: $(T = 5, N = 5, \tau = T/N = 1)$:

$$\phi_{i+1} = \phi_i + \tau u_i \qquad i = 0, 1, \ldots, N-1, \qquad |u_i| \le 1 \tag{5.19}$$

$$J = \sum_{i=0}^{N-1} \tau \phi_i^2 \to \min_{\{u_0, \ldots, u_{N-1}\}}.$$

The expression for $Q_1(\phi_{N-1})$ has the form

$$Q_1(\phi_4) = \min_{|u_4| \le 1} \tau \phi_4^2 = \tau \phi_4^2, \tag{5.20}$$

u_4^* is arbitrary, with $|u_4^*| \le 1$. The equation for $Q_2(\phi_3)$ has the form

$$Q_2(\phi_3) = \min_{|u_3| \le 1} [\phi_3^2 + Q_1(\phi_3 + u_3)]. \tag{5.21}$$

We shall vary ϕ_i ($i = 3, 2, 1, 0$) from $\phi = +5$ to $\phi = -5$ with step -1. For each ϕ_3 we find u_3^* from (5.21) and compute the corresponding value of $Q_2(\phi_3)$. In a similar fashion we determine the table of values of u_i^* and $Q_{5-i}(\phi_i)$ for $i = 2, 1, 0$. The values so obtained are displayed in Table 8.1.

Table 8.1

ϕ	$Q_5(\phi_0)$	u_0^*	$Q_4(\phi_1)$	u_1^*	$Q_3(\phi_2)$	u_2^*	$Q_2(\phi_{N-2})$	u_{N-2}^*	$Q_1(\phi_{N-1})$	u_{N-1}^*
5	55	-1	54	-1	50	-1	41	-1	25	
4	30	-1	30	-1	29	-1	25	-1	16	
3	14	-1	14	-1	14	-1	13	-1	9	
2	5	-1	5	-1	5	-1	5	-1	4	
1	1	-1	1	-1	1	-1	1	-1	1	
0	0	0	0	0	0	0	0	0	0	
-1	1	1	1	$+1$	1	$+1$	1	1	1	
-2	5	1	5	$+1$	5	$+1$	3	1	4	
-3	14	1	14	$+1$	14	$+1$	13	1	0	
-4	30	1	30	$+1$	29	$+1$	25	1	16	
-5	55	1	54	$+1$	50	$+1$	41	1	25	

Let us find the solution of (5.19) for $\phi_0 = 3$. We obtain from Table 8.1 the values $Q_5(3) = 14$, $u_0^*(\phi_0 = 3) = -1$. We find from the equation that $\phi_1 = \phi_0 + u_0^* = 3 - 1 = 2$, and from the table that $u_{1(2)}^* = -1$. Continuing the operation, we find

$$\phi_2 = 1, \qquad u_2^*(1) = -1, \qquad \phi_3 = 0, \qquad u_3^*(0) = 0, \qquad \phi_4 = 0, \qquad u_4^* = 0.$$

Table 8.1 yields, simultaneously, the solutions of (5.19) for $\phi_0 = +5, \ldots$, -5. As in other areas of numerical analysis, the solution of a dynamic programming problem brings up questions of approximation, stability, and convergence of the algorithms. The complexity of the equations that arise in dynamic programming hinders the practical application of the method. We may apply to problems of the type of (5.1) and (5.2) a simpler maximum principle.

8.6 Pontrjagin's Maximum Principle

Suppose that the motion of some object is defined by the system of differential equations

$$\frac{d\phi}{dt} = f(\phi, u), \qquad 0 \le t \le T \tag{6.1}$$

with the boundary conditions

$$\phi(0) \in S_0, \qquad \phi(T) \in S_1, \tag{6.2}$$

where

$$\phi = (\phi_1, \ldots, \phi_n), \qquad f = (f_1, \ldots, f_n), \qquad u = (u_1, \ldots, u_m),$$

and S_0, S_1 are given manifolds which, in particular, may independently degenerate into a point or may coincide with the whole Euclidean n-dimensional space E_n. Let $U \subset E_m$ be a given closed set and let there be defined an epoch T and a piecewise-continuous control function $u = u(t) \in U$ such that the corresponding trajectory $\phi = \phi(t, u)$ satisfies the conditions (6.1), (6.2) and the functional

$$J[u] = \int_0^T f_0(\phi, u) \, dt = \min_{u \in U}. \tag{6.3}$$

is minimized.

We shall suppose that the functions $f_i(\phi, u)$ are defined and continuous on the ensemble of variables (ϕ, u) together with their partial derivatives

$$\partial f_i/\partial \phi_j, \qquad i = 0, 1, 2, \ldots, n; j = 1, 2, \ldots, n,$$

and that the manifolds S_0 and S_1 are defined by the relations

$$S_0 = \{\phi : \phi_i(0) = \phi_i^0, i = 1, 2, \ldots, n\}, \tag{6.4}$$

$$S_1 = \{\phi : h_k(\phi(t)) = 0; k = 1, 2, \ldots, l, l \le n\}, \tag{6.5}$$

where the $h_k(x)$ are functions having continuous partial derivatives and the system of vectors

$$\frac{\partial h_k(x)}{\partial x} \equiv \text{grad } h_k(x), \qquad k = 1, 2, \ldots, l$$

is linearly independent for arbitrary $x \in S_1$. In particular, if $l = n$, we can, in general, find from (6.5) a set of isolated points $\phi = (\phi_1, \ldots, \phi_n)$ which may serve as coordinates of the right-hand end of the trajectory. We, therefore, naturally say that when $l = n$ the problem is one of optimal control (6.1)–(6.3) with fixed right-hand end. The conditions (6.4) provide the simplest example of fixed left-hand end. Further, if $S_1 \equiv E_n$, we speak of an optimal problem (6.1)–(6.3) with free right-hand end. For $0 < l < n$, of course, we speak of a problem with moving right-hand end. Under the assumptions that we have made, the manifold S_1 has dimension $n - l$ independently of whether we have a problem with fixed, free, or moving ends.

Theorem 1 (The Maximum Principle). *Suppose that for the controlled object*

$$\frac{d\phi}{dt} = f(\phi, u), \quad u \in U, \quad S_0 = \{\phi(0) = \phi^0\},$$

$$S_1 = \{h_k(\phi(T)) = 0, \quad k = 1, 2, \ldots, l\} \tag{A}$$

all the above postulates are satisfied. Let $\{\phi(t), u(t)\}, 0 \le t \le T$ be an optimal process carrying the object from a given state ϕ^0 into the state $\phi^1 \in S_1$, and suppose we have introduced the auxiliary function

$$H(\phi, \psi, u) = \sum_{i=0}^n \psi_i f_i(\phi, u). \tag{B}$$

Then there exists a nontrivial *vector function*

$$\psi(t) = \{\psi_0, \psi_1(t), \ldots, \psi_n(t)\}, \qquad \psi_0 = \text{const} \leq 0,$$

satisfying the system of equations

$$\frac{\partial \psi_i}{\partial t} = -\frac{\partial H(\phi(t), \psi, u(t))}{\partial \phi_i}, \qquad i = 1, 2, \ldots, n \tag{C}$$

with the boundary conditions

$$\psi_i(T) = \sum_{k=1}^{1} \gamma_k \frac{\partial h_k(\phi(T))}{\partial \phi_i}, \qquad i = 1, 2, \ldots, n, \tag{D}$$

where $\gamma_1, \gamma_2, \ldots, \gamma_l$ *are numbers such that at any epoch* $t, (0 \leq t \leq T)$, *the maximum condition*

$$H(\phi(t), \psi(t), u(t)) = \max_{u \in U} H(\phi(t), \psi(t), u), \tag{E}$$

is satisfied and, if the finite epoch T *is not fixed, then the auxiliary equation*

$$H_T = H(\phi(T), \psi(T), u(T)) = 0. \tag{F}$$

is satisfied.

A proof may be found in Pontrjagin *et al.*, *Matematiceskaya Teoriya Optimal'nyh Processov*, Moscow, Nauka, 1976. (*Mathematical Theory of Optimal Processes*).

This theorem is central to the theory of optimal controls. It yields several variants of the maximum principle for different ways of stating the boundary conditions and the functionals. We note further that if the left-hand end is moving, there exist for it relations similar to (D) for the right-hand end.

The case $f_0(\phi, u) \equiv 1$ is often met, and is important in technical applications to the problem of high-speed operations. Theorem 1 may be very simply stated for this case: the Hamiltonian function takes the form

$$\mathcal{H} = \sum_{i=1}^{n} \psi_i f_i(\phi, u) + \psi_0 \cdot 1 = H + \psi_0. \tag{6.6}$$

The adjoint system becomes

$$\frac{d\psi}{dt} = \frac{\partial H}{\partial \phi_i}, \quad i = 1, 2, \ldots, n, \qquad H = \sum_{i=1}^{n} \psi_i f_i(\phi, u). \tag{6.7}$$

since $\partial \mathcal{H}/\partial \phi_i = \partial H/\partial \phi_i$. The transversality conditions (D) and (F) are unchanged:

$$\psi_i(T) = \sum_{k=1}^{1} \gamma_k \frac{\partial h_k(\phi(T))}{\partial \phi_i}, \qquad i = 1, 2, \ldots, n, \tag{6.8}$$

$$\mathcal{H}_T = H_T + \Psi_0 = 0.$$

Since $\Psi_0 \le 0$, the latter condition may be written as an inequality

$$H(\phi(T), \psi(t), u(T)) \ge 0. \tag{6.9}$$

And, finally, the maximum condition (E) may be written for the function $\mathscr{H} = H + \psi_0$ as

$$H(\phi(t), \psi(t), u(t)) = \max_{u \in U} H(\psi(t), \psi(t), u). \tag{6.10}$$

Consequently we are led to the following:

Theorem 2. *If* $\{u(t), \phi(t)\}$, $0 \le t \le T$, *are optimal for the rapid solution of* (6.1)–(6.3), *assuming* $f_0(\phi, u) \equiv 1$, *there exists a nontrivial vector function* $\psi = (\psi_1, \ldots, \psi_n)$ *satisfying the system of equations* (6.7) *and the conditions* (6.8), (6.9) *and such that the maximum condition* (6.10) *is satisfied at all times.*

We can prove by an uncomplicated argument that the vector $(\psi_1, \ldots, \psi_n)$ is nontrivial. In fact, if it were not we could derive from the condition $\mathscr{H}_T = H_T + \psi_0 = 0$ the consequence $\phi_0 = 0$, and this contradicts the nontriviality of the vector $(\psi_0, \psi_1, \ldots, \psi_n)$.

The maximum principle yields a necessary condition for optimality and, in combination with various numerical methods that allow us to compute trajectories and controls that approximately satisfy it, is one of the most important tools for the practical solution of many problems in the field of optimal control. We shall limit ourselves to the analysis of a single example.

Consider the controlled object

$$\frac{d^2\phi}{dt^2} = u \tag{6.11}$$

with the constraint

$$-1 \le u \le 1. \tag{6.12}$$

on the control. If we introduce new phase variables

$$\phi_1 = \phi, \qquad \phi_2 = \frac{d\phi}{dt},$$

we obtain a system of equations equivalent to (6.11):

$$\frac{d\phi_1}{dt} = \phi_2, \qquad \frac{d\phi_2}{dt} = u. \tag{6.13}$$

For this system we pose the problem of fastest arrival from an arbitrary point (ϕ_1, ϕ_2) of the phase space to the origin of coordinates $\phi_1 = 0$, $\phi_2 = 0$. That is, our object (6.11) is to make the least time transit to the origin, and remain there.

We use Theorem 2. We write out the Hamiltonian

$$H = \psi_1\phi_2 + \psi_2 u, \tag{6.14}$$

and the adjoint system of equations:

$$\frac{d\psi_1}{dt} = -\frac{\partial H}{\partial \phi_1} = 0, \qquad \frac{d\psi_2}{dt} = -\frac{\partial H}{\partial \phi_2} = -\psi_1. \tag{6.15}$$

Hence

$$\psi_1 = d_1, \qquad \psi_2 = -d_1 t + d_2,$$

where d_1, d_2 are constants of integration. From the maximum condition (D), it follows immediately that

$$u = \begin{cases} +1, & \text{if } \psi_2 > 0, \\ -1, & \text{if } \psi_2 < 0 \end{cases} \tag{6.16}$$

and since ψ_2 is linear, i.e. can change sign only once, the desired control function $u(t)$ has only one switching point.

Let us consider the case in which $u = +1$. Integrating (6.13) we have

$$\phi_2 = C_2 + t, \qquad \phi_1 = \tfrac{1}{2}(C_2 + t)^2 + C_1,$$

or

$$\phi_1 = \tfrac{1}{2}\phi_2^2 + C_1. \tag{6.17}$$

Under the control $u = +1$ the trajectories are the parabolas (6.17), depicted in Figure 8.6. The direction of motion is upward, since $d\phi_2/dt = +1$, i.e., ϕ_2 increases. A similar argument shows that when $u = -1$ the trajectories are parabolas of the form

$$\phi_2 = -\tfrac{1}{2}\phi_2^2 + C_1^1, \tag{6.18}$$

as represented in Figure 8.7. Further, since our object must at some epoch pass through the origin of coordinates, i.e., move on one of the parabolas

$$\phi_1 = \tfrac{1}{2}\phi_2^2, \quad \text{or} \quad \phi_2 = -\tfrac{1}{2}\phi_1,$$

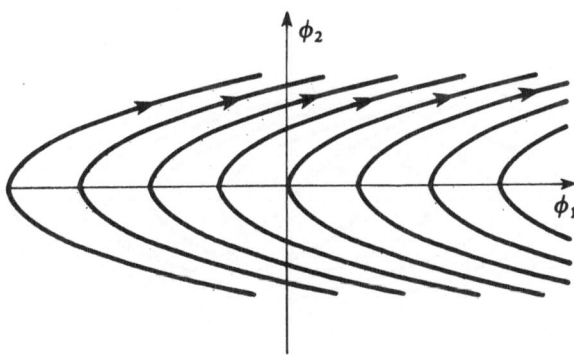

Figure 8.6

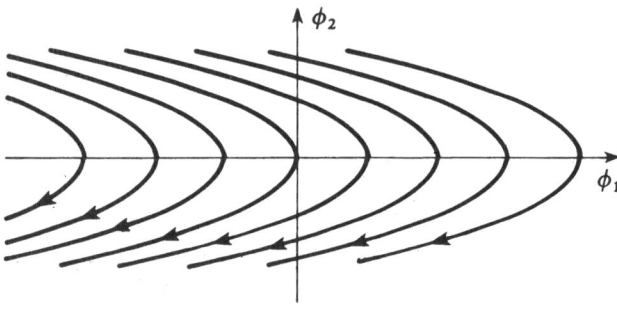

Figure 8.7

the final map of the possible trajectories has the form shown in Figure 8.8. Thus, if the initial point (ϕ_1, ϕ_2) lies above the curve AOB, the object will move under the action of the control $u = -1$ until it reaches the parabola $\phi_1 = \frac{1}{2}\phi_2^2$; then the control switches to $u = +1$ and the point moves on that parabola until it reaches the origin of coordinates. Its behavior when the initial point lies on AOB is easily found.

Thus the optimal control may have only the following forms:

$$u = \begin{cases} +1, & \text{under } AOB \text{ and on the parabola } \phi_1 = \frac{1}{2}\phi_2^2, \\ -1, & \text{above } AOB \text{ and on the parabola } \phi_1 = -\frac{1}{2}\phi_2^2. \end{cases}$$

If we know the initial point, we may easily find the trajectory, the time of motion, and the epoch at which the control switches from one boundary value to the other.

We have shown that if the optimal solution exists, it must have the form shown in Figure 8.8, since the maximum principle is a necessary condition for optimality. It is possible to show that these trajectories are in fact optimal.

In conclusion, we note that the classical problem of variational computation, namely to minimize the functional

$$J = \int_0^T f_0\left(\phi, \frac{d\phi}{dt}, t\right) dt \tag{6.19}$$

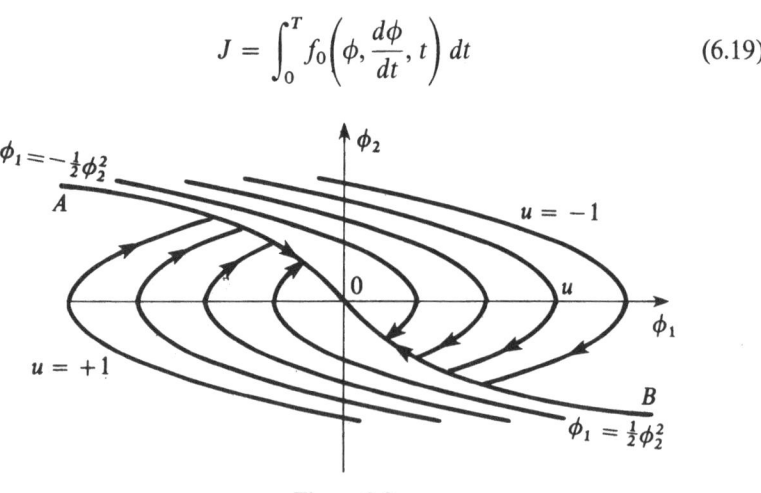

Figure 8.8

over the class of piecewise-smooth functions and under some boundary constraints

$$\phi(t_0) \in S_0, \qquad \phi(T) \in S_1,$$

is a simple particular case of the problem (6.1)–(6.3). In fact, we are to find a minimum of the functional

$$J = \int_{t_0}^{T} f_0(\phi, u, t)\, dt \qquad (6.20)$$

under the conditions

$$\frac{d\phi}{dt} = u, \qquad u \in U \equiv E_n. \qquad (6.21)$$

Given this equation we may invoke the maximum principle to derive all the classically known results in the field of variational analysis: the Euler equations, the Weierstrass–Erdman conditions at breakpoints in the extremals, the Legendre condition, and the Weierstrass condition.

Together with the problems formulated above, the theory of optimal controls also considers problems with delays, with integral boundary conditions, with parameters, with discrete time, and optimal control equations with partial derivatives.

8.7 Extremal Problems with Constraints and Variational Inequalities

In Chapter 2 we studied certain variational problems of mathematical physics and the methods for approximate solution of the corresponding extremal problems. A distinctive feature of this work was the contemplation of extremal problems in an entire Hilbert space, with no constraints on the solutions. However, a broad class of problems arises in connection with physics, mechanics, geophysics, optimal control, and economics, which lead to extremal problems whose solutions are to be sought in a class of functions that is narrower than that allowed by the domain of definition of the corresponding functionals. Examples of such problems, with descriptions of various approaches to their theory and approximate solution, may be found in studies by Lions [22], Duvaur and Lions [21], Stampacchia [22], Bredis and Stampacchia [5], Chernouskko and Banichuk [4], Bensusan, Lions, and Teschama [4], Glovinski [21], Glovinski, Lions, and Tremolières [21], and others.

In this section we:

(a) present some results on the existence and uniqueness of solutions of extremal problems with constraints;
(b) characterize these solutions by means of variational inequalities;

(c) give a number of examples applying variational inequalities to analyze the properties of the solutions of variational problems with constraints; and

(d) evaluate the possible approaches to the approximate solution of such tasks.

8.7.1 Elements of the General Theory

Let U be a Hilbert space over the field of real numbers, with the scalar product $(\cdot)_U$ and the norm $\|\cdot\|_U$. We assume that we are given a continuous symmetric bilinear form $\pi(u, v)$ on U, a continuous linear form $L(v)$, and a convex closed set $U_d \subseteq U$. We consider the quadratic functional

$$J(v) = \pi(v, v) - 2L(v), \tag{7.1}$$

under the assumption that $\pi(v, v)$ is positive definite on U, i.e., that there exists a constant $\alpha > 0$ such that

$$\pi(v, v) \geq \alpha \|v\|_U^2 \tag{7.2}$$

for arbitrary $v \in U$.

A detailed proof of the following assertion can be found in the monograph by Lions [2].

Under the above assumptions, there exists a unique element $u \in U_d$ which satisfies the equation

$$J(u) = \inf_{v \in U_d} J(v). \tag{7.3}$$

One of the most widely used and well-developed approaches to an intuitive statement of the problem (7.3), aiming at a study of the properties of the solution, consists in the use of variational inequalities. Let us formulate and prove an assertion that underlies this approach.

Suppose the assumptions described above are satisfied. Then an element $u \in U_d$ is a solution of (7.3) if and only if the inequality

$$\pi(u, v - u) \geq L(v - u) \tag{7.4}$$

holds for all $v \in U_d$.

We first prove the necessity. Let u satisfy (7.3). Then for arbitrary $v \in U_d$ and $\Theta \in (0, 1)$ the inequality

$$J(u) \leq J((1 - \Theta)u + \Theta v) \tag{7.5}$$

is satisfied. (Here we use the convexity of U_d, i.e., the fact that $\phi = (1 - \theta)w + \Theta v$ belongs to U_d for all $w, v \in U_d$ and $\Theta \in (0, 1)$.) From this inequality we have

$$\frac{J(u + \Theta(v - u)) - J(u)}{\Theta} \geq 0. \tag{7.6}$$

Letting Θ tend to zero and using the specific form of the functional $J(v)$ we find that

$$\lim_{\Theta \to 0} \frac{J(u + \Theta(v - u)) - J(u)}{\Theta}$$

$$= \lim_{\Theta \to 0} \{2[\pi(u, v - u) - L(v - u)] + \Theta\pi(v - u, v - u)\}$$

$$= 2[\pi(u, v - u) - L(v - u)] \geq 0 \qquad (7.7)$$

for all $v \in U_d$. This proves that (7.4) is satisfied.

We now prove the sufficiency of the condition. Let u satisfy (7.4). The quadratic functional $J(v)$ is convex, i.e., for all $v, w \in U$ and $\Theta \in (0, 1)$ we have the inequality

$$J((1 - \Theta)w + \Theta v) \leq (1 - \Theta)J(w) + \Theta J(v)$$

or the equivalent inequality

$$J(v) - J(w) \geq \frac{J((1 - \Theta)w + \Theta v) - J(w)}{\Theta}. \qquad (7.8)$$

Substituting u for w in (7.8) we have

$$J(v) - J(u) \geq \frac{J((1 - \Theta)u + \Theta v) - J(u)}{\Theta}.$$

Now letting Θ tend to zero in this latter inequality, and taking account of (7.7) and (7.4) we have

$$J(v) - J(u) \geq 2[\pi(u, v - u) - L(v - u)] \geq 0.$$

Consequently,

$$J(u) \leq J(v)$$

for all $v \in U_d$, i.e., u is the solution of (7.3).

Let us look at two concrete applications of this theorem. Let U_d be a subspace of U, closed in the U-norm. Then, putting $v = u \pm \phi$ in (7.4), where ϕ is an arbitrary element of U_d, we arrive at the two inequalities

$$\pi(u, \phi) \geq L(\phi),$$
$$-\pi(u, \phi) \geq -L(\phi).$$

It follows that under our given assumptions the element u is a solution of (7.3) if and only if

$$\pi(u, \phi) = L(\phi) \qquad (7.9)$$

for all $\phi \in U_d$. But this is precisely the Euler equation for the variational problem

$$J(u) = \inf_{v \in U_d} J(v). \qquad (7.10)$$

Now suppose that U_d is a cone. (We recall that a cone is a closed convex set $K \subset U$ containing for every $\phi \in K$ the multiple $\gamma\phi$ for all $\gamma \geq 0$, and having the property that if it contains ϕ and $-\phi$, then ϕ is the null element of U.) If we substitute $v + u$ for v in (7.7) we arrive at the inequality

$$\pi(u, v) \geq L(v), \qquad (7.11)$$

which holds for all $v \in U_d$. Here we have used the fact that if v and w are contained in U_d, so is their sum. Further, setting $v = 0$ in (7.7) and $v = u$ in (7.11) we obtain the corresponding inequalities

$$-\pi(u, u) \geq -L(u),$$
$$\pi(u, u) \geq L(u),$$

whence

$$\pi(u, u) = L(u). \qquad (7.12)$$

We shall show that (7.11) and (7.12) are equivalent to (7.4); in fact, we need only show that (7.4) follows from them. Subtracting (7.12) from (7.11) we obtain the inequality

$$\pi(u, v) - \pi(u, u) \geq L(v) - L(u),$$

when, since π is bilinear and L is linear, we have the inequality

$$\pi(u, v - u) \geq L(v - u),$$

which is satisfied for all $v \in U_d$. Thus, if U_d is a cone (7.11) and (7.12) together are equivalent to (7.4).

8.7.2 Examples of Extremal Problems

In illustrating the above results we shall confine ourselves to the one-dimensional case, being easier to understand intuitively. Then the elements $u \in U$ are functions of one variable x, and we take U as a Sobolev space $W_2^1(0, 1)$. The scalar product and the norm in U are, respectively, defined by the formulas

$$(u, v)_U = \int_0^1 \left[\frac{du}{dx} \frac{dv}{dx} + uv \right] dx,$$

$$\|u\|_U = (u, u)_U^{1/2}. \qquad (7.13)$$

Consider the bilinear functional

$$\pi(u, v) = (u, v)_U \qquad (7.14)$$

and the linear functional

$$L(v) = \int_0^1 fv \, dx, \qquad (7.15)$$

where $f = f(x)$ is some fixed element of the space $L_2(0, 1)$. Clearly, $\pi(u, v)$ is symmetric, continuous, and positive definite, while L is continuous, as may be seen from the inequalities

$$\pi(u, v) \leq \|u\|_U \cdot \|v\|_U,$$
$$\pi(v, v) = \|v\|_U^2, \tag{7.16}$$

$$L(v) \leq \|f\|_{L_2(0, 1)} \cdot \|v\|_{L_2(0, 1)} \leq \|f\|_{L_2(0, 1)} \cdot \|v\|_U. \tag{7.17}$$

Thus the quadratic functional

$$J(v) = (v, v)_U - 2 \int_0^1 fv \, dx \tag{7.18}$$

will satisfy all the conditions of the theorem in the preceding section and, therefore, if $U_d \subseteq U$ is a closed convex set, there exists a unique element $u \in U_d$ which solves the extremal problem

$$J(u) = \inf_{v \in U_d} J(v). \tag{7.19}$$

It is uniquely defined by the inequality

$$\pi(u, v - u) \geq L(v - u), \tag{7.20}$$

satisfied for all $v \in U_d$. We should note that since an arbitrary function $v(x) \in W_2^1(0, 1)$ is also an element of the Banach space of continuous functions $C(0, 1)$, the solution $u(x)$ of (7.19) is always continuous (cf. Sobolev [1]). Moreover, there exists a constant α such that for all $v(x) \in U \subset W_2^1(0, 1)$

$$\|v\|_{C(0, 1)} \leq \alpha \|v\|_U, \tag{7.21}$$

i.e., the convergence of a sequence of functions $v_k(x)$ to a function $v^\infty(x)$ in the norm of $W_2^1(0, 1)$ implies convergence in the norm of $C(0, 1)$ to the same function $v^\infty(x)$.

Let us consider some concrete examples.

Let $U_d = U$. Then U_d is an improper subspace of U (it coincides with U) and, by the first example in Section 8.7.1, the solution $u(x)$ of (7.19) satisfies the equation

$$(u, v)_U = L(v)$$

for all $v(x) \in U$. Hence, by a standard argument, we conclude that $u(x)$ is a generalized solution of the von Neumann boundary problem

$$-\frac{d^2u}{dx^2} + u = f(x), \quad 0 < x < 1,$$

$$\frac{du}{dx} = 0 \quad \text{for} \quad x = 0,$$

$$\frac{du}{dx} = 0 \quad \text{for} \quad x = 1. \tag{7.22}$$

If, for example, $f(x)$ is continuous, $u(x)$ is the classical solution of this problem.

Now suppose $U_d = \mathring{W}^1_2(0, 1)$. This is a closed subspace of $W^1_2(0, 1)$, since it is complete in the metric of the latter. Therefore, in accordance with the first example in Section 8.7.1, the solution $u(x) \in \mathring{W}^1_2(0, 1)$ of (7.19) satisfies the equation

$$(u, v)_U = L(v)$$

for all $v(x) \in \mathring{W}^1_2(0, 1)$. Hence, as in the preceding case, we conclude that $u(x)$ is a generalized solution of the Dirichlet problem

$$-\frac{d^2 u}{dx^2} + u = f(x), \qquad 0 < x < 1,$$

$$u(0) = u(1) = 0. \tag{7.23}$$

Now suppose that U_d is not a subspace of U, i.e., that it has a more complicated structure. Define the set U_d by the relation

$$U_d = \{u(x): u(x) \in \mathring{W}^1_2(0, 1), u(x) \geq 0 \text{ for } x \in (0, 1)\}, \tag{7.24}$$

that is, U_d consists of those functions in $W^1_2(0, 1)$ that are nonnegative in $(0, 1)$ and vanish at the endpoints. We shall show that U_d is a cone in $W^1_2(0, 1)$. We must prove that:

(1) U_d is convex;
(2) it is closed;
(3) if $v(x) \in U_d$ then $\gamma v(x) \in U_d$ for all $\gamma \geq 0$; and
(4) if $v(x)$ and $-v(x)$ both belong to U_d, then $v(x) \equiv 0$.

The convexity condition means that if $u(x)$ and $v(x)$ belong to U_d then $w(x) = (1 - a)u(x) + av(x)$ belongs to U_d for all $a \in (0, 1)$. The truth of this statement follows from the nonnegativity of $w(x)$, as a linear combination of nonnegative elements with positive coefficients.

U_d is closed: let $v^k(x)$ ($k = 0, 1, \ldots$) be a sequence of functions belonging to U_d and converging in the norm of $W^1_2(0, 1)$ to some function $v^\infty(x) \in W^1_2(0, 1)$. Since $v^\infty(x) \in C(0, 1)$ we know, by (7.21), that the $\{v^k(x)\}$ converge in the norm of $C(0, 1)$ to the same function $v^\infty(x)$. Then $v^\infty(x)$ is nonnegative, vanishes at the endpoints of the interval $(0, 1)$, and therefore belongs to U_d.

The third and fourth requirements are trivially satisfied.

Thus, if the set U_d is as defined in (7.24), then the results in Section 8.7.1 imply that the solution $u(x)$ of (7.19) exists, is unique, and is characterized by the fact that for all $v(x) \in U_d$

$$\pi(u, v) \geq L(v) \tag{7.25}$$

and, moreover,

$$\pi(u, u) = L(u). \tag{7.26}$$

Let us investigate the properties of $u(x)$. Suppose $u(x)$ is positive at some point $\xi \in (0, 1)$. Then, since it is continuous, there exists an interval $(\xi^-, \xi^+) \subseteq$

$(0, 1)$ such that $u(x)$ is positive within it, and $u(\xi^-) = u(\xi^+) = 0$ (in particular, it may turn out that $\xi^- = 0$ or $\xi^+ = 1$, or both may occur). Denote by $D(\xi^-, \xi^+)$ the set of functions that are infinitely differentiable on $(0, 1)$, finite in (ξ^-, ξ^+) and identically zero outside (ξ^-, ξ^+). We know (cf. Sobolev [1]) that $D(\xi^-, \xi^+)$ is dense in $\mathring{W}_2^1(\xi^-, \xi^+)$ and an arbitrary function belonging to it has the property that for sufficiently small $\varepsilon > 0$ the functions

$$v_\varepsilon(x) = u(x) \pm \varepsilon\phi(x) \tag{7.27}$$

belong to U_d. If we substitute $v(x) = v_\varepsilon(x)$ in the inequality (7.25) we have

$$\pi(u, u) \pm \varepsilon\pi(u, \phi) \geq L(u) \pm \varepsilon L(\phi). \tag{7.28}$$

Hence, by (7.26) we obtain

$$\pi(u, \phi) \geq L(\phi), \tag{7.29}$$

$$-\pi(u, \phi) \geq -L(\phi), \tag{7.30}$$

so that

$$\pi(u, \phi) = L(\phi) \tag{7.31}$$

for all $\phi \in D(\xi^-, \xi^+)$.

Now rewrite (7.31), taking account of the fact that $\phi(x) \equiv 0$ for $x \notin D(\xi^-, \xi^+)$:

$$\int_{\xi^-}^{\xi^+} \left[\frac{du}{dx}\frac{d\phi}{dx} + u\phi \right] dx = \int_{\xi^-}^{\xi^+} f\phi \, dx, \tag{7.32}$$

where $\phi(x)$ is an arbitrary element of $D(\xi^-, \xi^+)$ and $u(x) \in \mathring{W}_2^1(\xi^-, \xi^+)$. Since $D(\xi^-, \xi^+)$ is dense in this latter space, equation (7.32) will be satisfied also for $\phi(x) \in \mathring{W}_2^1(\xi^-, \xi^+)$. Hence, by the second example in this subsection, it follows that on the segment $[\xi^-, \xi^+]$ the function $u(x)$ is a generalized solution of the Dirichlet problem

$$-\frac{d^2u}{dx^2} + u = f(x), \qquad \xi^- < x < \xi^+$$
$$u(\xi^-) = u(\xi^+) = 0. \tag{7.33}$$

Thus to an arbitrary point $\xi \in (0, 1)$, at which the solution $u(x)$ of (7.19) is positive, there corresponds an interval (ξ^-, ξ^+) on which $u(x)$ is a generalized solution of the problem (7.33).

To sum up, we conclude that we may represent the interval $\Omega = (0, 1)$ in the form

$$\Omega = \Omega^0 \cup \Omega^+, \tag{7.34}$$

where Ω^+ is an open set consisting of the union of some (possibly countable) number of intervals on which $u(x)$ is positive and is the generalized solution of problems of the form (7.33), while Ω^0 is the (possibly empty) complement of Ω^+ in Ω, on which $u(x)$ vanishes identically.

This problem is related to the free boundary problems (Lions [5]), which arise, for example, in the mechanics of continuous media. The unknowns are then: (a) the endpoints of the intervals making up Ω^+, and (b) the function $u(x)$ on these intervals.

In conclusion we consider an interesting example in which U_d is neither a subspace of U nor a cone in it. We assume that we are given on the interval $(0, 1)$ two functions $F_1(x)$ and $F_2(x)$ belonging to $W_2^1(0, 1)$ and satisfying the inequality

$$F_1(x) < F_2(x) \tag{7.35}$$

for all $x \in (0, 1)$, and also the inequalities

$$\begin{aligned} F_1(0) &\le 0 \le F_2(0), \\ F_1(1) &\le 0 \le F_2(1). \end{aligned} \tag{7.36}$$

We define the set U_d by the relation

$$U_d = \{u(x): u(x) \in \mathring{W}_2^1(0, 1), F_1(x) \le u(x) \le F_2(x) \text{ for all } x \in (0, 1)\}. \tag{7.37}$$

It consists of those functions in $W_2^1(0, 1)$ that vanish at the endpoints of the interval $(0, 1)$ and satisfy the inequalities

$$F_1(x) \le u(x) \le F_2(x) \tag{7.38}$$

within it. By an analysis similar to that carried out in the preceding section we may show that U_d is a closed convex set in $W_2^1(0, 1)$; hence, and from Section 8.7.1, it follows that for this choice of U_d equation (7.19) has a unique solution.

Now assume that at some point $\xi \in (0, 1)$

$$F_1(\xi) < u(\xi) < F_2(\xi). \tag{7.39}$$

Then, since $u(x)$ is continuous there exist points ξ^-, $\xi^+ \in [0, 1]$ such that $\xi \in (\xi^-, \xi^+)$ and $u(x)$ satisfy the inequalities

$$F_1(x) < u(x) < F_2(x) \tag{7.40}$$

in the interval (ξ^-, ξ^+), and at the endpoints satisfy the equations

$$\begin{aligned} u(\xi^-) &= \begin{cases} F_1(\xi^-), & \text{or} \\ F_2(\xi^-), \end{cases} \\ u(\xi^+) &= \begin{cases} F_1(\xi^+), & \text{or} \\ F_2(\xi^+). \end{cases} \end{aligned} \tag{7.41}$$

Thus the interval $\Omega = (0, 1)$ is representable in the form

$$\Omega = \Omega^0 \cup \Omega^+, \tag{7.42}$$

where Ω^+ is the union of intervals in which $u(x)$ satisfies (7.40) and Ω^0 is the complement of Ω^+ in Ω; on Ω^0, $u(x)$ is equal to either $F_1(x)$ or $F_2(x)$. Note that if $F_1(x)$ and $F_2(x)$ do not vanish at the endpoints of $(0, 1)$, Ω^0 is a closed set, that is, it is the union of a finite number of intervals and isolated points lying in $(0, 1)$. On each of these intervals $u(x)$ is equal to only one of the functions $F_1(x)$ or $F_2(x)$.

Let us now investigate the properties of $u(x)$ on an interval $(\xi^-, \xi^+) \subset \Omega^+$. If $u(x)$ belongs to U_d, so will $v_\varepsilon(x) = u(x) \pm \varepsilon\phi(x)$, if $\varepsilon > 0$ is sufficiently small and $\phi(x) \in D(\xi^-, \xi^+)$; therefore, substituting $v_\varepsilon(x)$ for $v(x)$ in (7.20) we find that

$$\pi(u, \phi) \geq L(\phi),$$
$$-\pi(u, \phi) \geq -L(\phi).$$

We conclude that

$$\pi(u, \phi) = L(\phi) \tag{7.43}$$

for all $\phi(x) \in D(\xi^-, \xi^+)$, or, equivalently,

$$\int_{\xi^-}^{\xi^+} \left[\frac{du}{dx}\frac{d\phi}{dx} + u\phi \right] dx = \int_{\xi^-}^{\xi^+} f\phi \, dx, \tag{7.44}$$

where $u(x)$ satisfies the conditions

$$u(\xi^-) = g^-,$$
$$u(\xi^+) = g^+, \tag{7.45}$$

and g^-, g^+ are defined by the relations

$$g^- = \begin{cases} F_1(\xi^-), & \text{or} \\ F_2(\xi^-), \end{cases}$$

$$g^+ = \begin{cases} F_1(\xi^+), & \text{or} \\ F_2(\xi^+). \end{cases} \tag{7.46}$$

On the basis of the examples considered earlier, we conclude that in the interval (ξ^-, ξ^+) the function $u(x)$ is a generalized solution of the Dirichlet problem

$$-\frac{d^2u}{dx^2} + u = f(x), \qquad \xi^- < x < \xi^+,$$

$$u(\xi^-) = g^-, \tag{7.47}$$

$$u(\xi^+) = g^+.$$

Thus, as opposed to the earlier problems, this problem requires for its solution not only finding the endpoints of the intervals in the set Ω^+ and computing the function $u(x)$ in these intervals, but also defining the values of $u(x)$ at the endpoints. It is also related to problems with free boundaries.

8.7.3 Numerical Methods in Extremal Problems

The choice of method for solving an extremal problem is closely bound up with the specific properties of the quadratic functional $J(v)$ and the set $U_d \subset U$ in which the solution is sought. In the first and second problems in the preceding section the search for an approximate solution of the extremal problem reduced to the search for a generalized or classical solution of a linear boundary problem for a differential equation. It was natural to solve the differential equation immediately. This approach is much more difficult, from a practical viewpoint in the third and fourth examples, since we are attempting to solve a boundary problem without knowing either the boundaries of the region in which the solution is sought, nor even the boundary conditions. Yet there are ways of solving such problems, for instance by regularization, by the penalty method, or by nonlinear programming (Lions [5]). In this section we shall study yet another approach, which consists in first reducing the problem (7.3) to a finite-dimensional problem, and then solving an extremal problem for a quadratic function of n variables over a convex bounded set in n-dimensional space.

Let U^h be a sequence of finite-dimensional subspaces of U, of dimension $n(h)$—we suppose $n(h) \to \infty$ as $h \to 0$—and let U^h approximate U in the following sense: for all $v(x) \in U$ and all $\Delta > 0$ there exists an h_Δ such that for all $h < h_\Delta$ an element $v^h \in U^h$ can be found which approximates $v(x)$ with precision Δ, i.e., for all $h < h_\Delta$ we have

$$\inf_{\phi \in U^h} \|v - \phi\| \le \Delta. \tag{7.48}$$

We replace (7.3) by the approximate problem

$$J(u^h) = \inf_{v \in U_d^h} J(v), \tag{7.49}$$

where $U_d^h = U_d \cap U^h$ is a closed convex set. It is clear that the problem (7.49) also has a unique solution (cf. Section 8.7.1) which, because U^h is by assumption complete in U, converges to a solution of the original problem (7.3) as $h \to 0$.

Now assume that the functions $\psi_k^h(x)$ $(k = 1, \ldots, n)$, form a basis in the space U^h and that for some $v(x)$ we have the expansion

$$v(x) = \sum_{k=1}^{n} \alpha_k^h \psi_k^h(x). \tag{7.50}$$

Then

$$J(v) \equiv \Phi(\alpha) = (A\alpha, \alpha) - 2(g, \alpha), \tag{7.51}$$

where A is a symmetric positive definite matrix of order n, with the elements

$$a_{ij} = \pi(\psi_i^h, \psi_j^h) \tag{7.52}$$

and g and α are n-dimensional vectors with the components

$$g_i = L(\psi_i^h),$$
$$\alpha_i = \alpha_i^h, \tag{7.53}$$

respectively; by $(\ ,\ )$ we denote the customary scalar product in the space of real n-dimensional vectors. Then (7.49) becomes

$$\Phi(\beta) = \inf_{\alpha \in V_d} \Phi(\alpha), \tag{7.54}$$

where V_d is defined as follows: the vector α belongs to V_d if and only if

$$v(x) = \sum_{k=1}^{n} \alpha_k \psi_k^h(x) \tag{7.55}$$

belongs to the set U_d^h. It is clear that V_d is a closed convex set of n-dimensional real vectors, and the problem (7.55) has the unique solution β, and that the function

$$u^h(x) = \sum_{k=1}^{n} \beta_k \psi_k^h(x) \tag{7.56}$$

is a solution of the problem (7.49).

There are currently a large number of algorithms for the solution of (7.54). The most widely used are different variants, the method of steepest descent, the gradient method, the method of possible directions, the method of local variations, and others. Detailed discussions will be found in the monographs by Pshenichnyi and Danilin [21], Chernous'ko and Banichuk [22], and Polak [3].

Let us consider first one of the variants of the relaxation method.

Let $\alpha^{(s)} = (\alpha_1^{(s)}, \ldots, \alpha_n^{(s)}) \in V_d$ be the sth approximation to the desired solution of the problem (7.54). Then the computation of the $(s + 1)$st approximation consists of n intermediate steps carried out according to the formula

$$\alpha^{s + (i/n)} = \alpha^{s + [(i-1)/n]} - q_i^s r_i^s e_i, \qquad i = 1, \ldots, n, \tag{7.57}$$

where the e_i are vectors with the components $e_{ij} = \delta_{ij}$ (δ_{ij} is the Kronecker symbol), the r_i^s are given by

$$r_i^s = \sum_{i=1}^{n} a_{ij} \alpha_i^{s + [(i-1)/n]} - g_i = \left.\frac{\partial \Phi(\alpha)}{\partial \alpha_i}\right|_{\alpha = \alpha^{s + [(i-1)/n]}}, \tag{7.57'}$$

and the q_i^s are solutions of the following one-dimensional extremal problem:

$$\Phi(\alpha^{s + (i/n)}) = \inf \Phi(\alpha^{s + [(i-1)/n]} - q r_i^s e_i),$$

$$\alpha^{s + (i/n)} \in V_d. \tag{7.58}$$

This relaxation method relates to the methods of coordinate-by-coordinate descent and, if V_d coincides with the whole space, it goes over into the well-known Gauss–Seidel solution of a system $A\alpha = g$ of linear equations with a symmetric positive definite matrix A.

We can clarify what we have just said, by an example in which we solve (7.19) under the assumption that the set U_d is defined by (7.37) with an additional constraint, that the function $F_1(x)$ is concave and $F_2(x)$ is convex, i.e., that for all $x, x' \in [0, 1]$ and $\Theta \in [0, 1]$ we have the inequalities

$$F_1((1 - \Theta)x' + \Theta x'') \geq (1 - \Theta)F_1(x') + \Theta F_1(x''),$$
$$F_2((1 - \Theta)x' + \Theta x'') \leq (1 - \Theta)F_2(x') + \Theta F_2(x''). \tag{7.59}$$

For a given $n \geq 1$ we consider the uniform net

$$\Omega_h = \left\{ x_k : x_k = k \times h, k = 0, 1, \ldots, N + 1, h = \frac{1}{N + 1} \right\} \tag{7.60}$$

on the segment $[0, 1]$ and we introduce a system of basis functions $\psi_k(x)$ ($k = 1, \ldots, n$), according to formula (3.14) of Chapter 2. The space U^h is defined as the linear hull of the system $\{\psi_k(x)\}_{k=1}^n$. In other words, U^h is the space of piecewise-linear completions of the net functions defined on the vertices of the nets Ω_h and vanishing at the points $x_0 = 0$ and $x_{N-1} = 1$.

With these assumptions concerning $F_1(x)$, $F_2(x)$, and the space U^h, the set $U_d^h = U_d \cap U^h$ is the subset of U^h consisting of the functions $v(x) \in U^h$ satisfying the inequalities

$$F_1(x_i) \leq v(x_i) \leq F_2(x_i), \qquad i = 1, \ldots, n.$$

Then the set V_d will be the n-dimensional rectangular parallelepiped whose coordinates $\alpha = (\alpha_1, \ldots, \alpha_n)$ satisfy the inequalities

$$F_1(x_i) \leq \alpha_i \leq F_2(x_i), \qquad 1, \ldots, n.$$

If we do not make the assumptions of concavity and convexity regarding $F_1(x)$ and $F_2(x)$ the statement of the requirement that $v(x)$ belong to U_d^h (the set of vectors α belonging to V_d) becomes much more complex.

To illustrate the method of local variation we consider a simple one-dimensional variational problem. Suppose we have the variational functional

$$J(\phi) = \int_0^1 \pi(x, \phi, \phi_x) \, dx. \tag{7.61}$$

We are to find a function $\phi(x)$ yielding the minimum of (7.61) and satisfying the constraints

$$F_1(x) \leq \phi(x) \leq F_2(x). \tag{7.62}$$

Note that at some points of the segment $[0, 1]$ the constraints may be missing; then we suppose formally that we have the natural bounds

$$-\infty < \phi$$

or

$$\phi < \infty.$$

With this treatment of the boundary conditions, it is natural to assume also that (x) is bounded at $x = 0$ and $x = 1$, and we shall do so.

The simplest variant among the implementations of the method is as follows: we divide the interval $[0, 1]$ into n equal parts by the points $x = i + x$, $x = 1/n$ $(i = 0, 1, \ldots, n)$. The variational functional (7.61) is represented as the sum of the partial functionals

$$J = \sum_{i=0}^{n-1} \int_{x_i}^{x_{i+1}} \pi(x, \phi, \phi_x) \, dx. \tag{7.63}$$

We approximate each of the integrals in (7.63) by the rectangular formula

$$\int_{x}^{x_{i+1}} \pi(x, \phi, \phi_x) \, dx \approx \Delta x \pi(x_{i+1/2}, \phi_{i+1/2}, \phi_{i+1/2}),$$

where

$$\phi_{i+1/2} = \frac{\phi_i + \phi_{i+1}}{2}, \qquad \phi'_{i+1/2} = \frac{\phi_{i+1} - \phi_i}{\Delta x}, \qquad x_{i+1/2} = x_i + \frac{\Delta x}{2}.$$

Thus the expressions for the partial functionals contain the values of the desired function only at the points ϕ_i and ϕ_{i+1}. We introduce the following notation:

$$V_i(\phi_i, \phi_{i+1}) = \Delta x \pi(x_{i+1/2}, \phi_{i+1/2}, \phi'_{i+1/2}).$$

We have the approximation

$$J(\phi) \approx V = \sum_{i=0}^{n-1} V_i(\phi_i, \phi_{i+1}). \tag{7.64}$$

This means we have arrived at the problem of finding a set of numbers $\{\phi_i\}$ which satisfy the given constraints

$$F_1(x_i) \le \phi_i \le F_2(x_i) \tag{7.65}$$

and yield a minimum of the functional V.

We use the method of successive approximations to solve this problem. Suppose we know the approximations up to the kth, say ϕ_i^k. The new approximation ϕ_i^{k+1} will be found in the following way: we introduce a number h and for every x_i we consider the triplet $\phi_i^k - h$, ϕ_i^k, $\phi_i^k + h$. By hypothesis ϕ_i^k satisfies the constraints. We must test the other two quantities. If one or other of the constraints is not satisfied, we reject the corresponding quantity, and are left with a pair of numbers, either $\phi_i^k - h$, ϕ_i^k or ϕ_i^k, $\phi_i^k + h$.

As in the Gauss–Seidel method, we suppose that we already know the values of the solution, at the $(k + 1)$st approximation, for all indices less than i, that is, we know $\phi_0^{k+1}, \phi_1^{k+1}, \ldots, \phi_{i-1}^{k+1}$. Naturally, we use this information to speed up the convergence of the process. Therefore, together with the partial functionals V_i it is convenient to introduce more functionals:

$$
\begin{aligned}
G_i &= V_{i-1}(\phi_{i-1}^{k+1}, \phi_i^k) + V_i(\phi_i^k, \phi_{i+1}^k), \\
U_i^+ &= V_{i-1}(\phi_{i-1}^{k+1}, \phi_i^k + h) + V_i(\phi_i^k + h, \phi_{i+1}^k), \qquad (7.66) \\
U_i^- &= V_{i-1}(\phi_{i-1}^{k+1}, \phi_i^k - h) + V_i(\phi_i^k - h, \phi_{i-1}^k).
\end{aligned}
$$

In the sum (7.64) these expressions take into account the two terms that depend on ϕ_i and correspond to the three possible variations $\phi_i^k - h$, ϕ_i^k, and $\phi_i^k + h$.

There remains to determine which of the three numbers satisfies the constraints and yields the minimum of one of the three functionals (7.66). That number is taken as the new approximate value ϕ_i^{k+1}. After we have determined the values of the ϕ_i^{k+1}, for all $i = 0, \dots, n$, we may pass to the next approximation.

After the approximate solution has been obtained, with the given precision, for given values of the parameters Δx and h, the process must be repeated with a new value of h, for instance, half the preceding value, and so on. Of course, in setting up the new process we take advantage of our knowledge regarding the approximate solution we have obtained, taking the latter as our initial approximation. Then we diminish Δx and obtain a better approximation to the solution of the original equation.

We have considered only the simplest kind of constraints, in which the given requirements are placed on the solution itself. In the more general case, where the constraints are given by a relation

$$
B(x, \phi, \phi_x) \leq 0,
$$

we must approximate the solution for arbitrary x_i, and at each step test for fulfillment of the approximate constant. The problems of convergence and precision in the method of local variation have been studied in detail in the monograph of Chernous'ko and Banichuk [22].

The method of local variation may be extended to multi-dimensional variational problems with constraints, conserving in principle all the stages of the above algorithm. For instance, in a two-dimensional region we are not dealing with a triplet $\phi_i - h, \phi_i, \phi_i + h$ whose members are to be compared with regard to a functional and to certain constraints; instead, we deal with quantities $\phi_{ij} - h, \phi_{ij}, \phi_{ij} + h$ and the partial functional U_{ij} will consist of four terms

$$
U_{ij} = V_{ij} + V_{i-1, j} + V_{i, j-1} + V_{i-1, j-1},
$$

which contain the quantity ϕ_{ij}^k. We define U_{ij}^+ and U_{ij}^- in a similar fashion, where ϕ_{ij} is replaced by $\phi_{ij} + h$ and $\phi_{ij} - h$, respectively. For problems with more variables the algorithm is constructed in the same way.

In conclusion, we should note that the method of local variation has nothing to do with the choice of a basis and is, therefore, applicable to regions of arbitrary form. However, we must not lose sight of the fact that it may converge to a local, rather than a global minimum. The convergence to a (local) minimum is guaranteed by the nature of the algorithm.

CHAPTER 9

Some Problems of Mathematical Physics

We will consider a number of simple and at the same time typical problems of mathematical physics, and use them to illustrate the fundamental methods of numerical mathematics.

9.1 The Poisson Equation

Let us attempt to illustrate the numerical approaches to some problems related to elliptic equations.

9.1.1 The Dirichlet Problem for the One-Dimensional Poisson Equation

To begin with, let us consider the Dirichlet problem for the one-dimensional Poisson equation

$$-\frac{d^2\phi}{dx^2} = f, \qquad 0 < x < 1,$$

$$\phi(0) = a, \qquad \phi(1) = b, \tag{1.1}$$

where $f(x)$ represents the sources, and a and b are some given constants.

Let us divide the interval $0 \leq x \leq 1$ into N equal subintervals $x_k \leq x \leq x_{k+1}$ of length $h = x_k - x_{k-1}$. Writing down the second-order approximation of (1.1) with respect to h, we obtain

$$\frac{-\phi_{k-1} + 2\phi_k - \phi_{k+1}}{h^2} = f_k \qquad k = 1, 2, \ldots, N-1, \tag{1.2}$$

$$\phi_0 = a, \qquad \phi_N = b.$$

In agreement with the general principles of constructing difference schemes, we need to eliminate first the boundary points from the difference equation (1.2). This can be done with the help of the boundary conditions. Of course, the operator of the problem changes in the process, and we obtain the following problem:

$$\frac{2\phi_1 - \phi_2}{h^2} = \frac{a}{h^2} + f_1,$$

$$\frac{-\phi_{k-1} + 2\phi_k - \phi_{k+1}}{h^2} = f_k, \qquad k = 2, 3, \ldots, N-2, \tag{1.3}$$

$$\frac{-\phi_{N-2} + 2\phi_{N-1}}{h^2} = \frac{b}{h^2} + f_{N-1}.$$

Thus we have a system of $N-1$ equations with the unknowns $\phi_1, \phi_2, \ldots, \phi_{N-1}$. It can be written in the matrix form

$$A\phi = g, \tag{1.4}$$

where

$$A = \frac{1}{h^2} \left\| \begin{matrix} 2 & -1 & 0 & 0 & \cdots & 0 & 0 \\ -1 & 2 & -1 & 0 & \cdots & 0 & 0 \\ 0 & -1 & 2 & -1 & \cdots & 0 & 0 \\ \multicolumn{7}{c}{\dotfill} \\ 0 & 0 & 0 & 0 & \cdots & -1 & 2 \end{matrix} \right\|,$$

$$\phi = \left\| \begin{matrix} \phi_1 \\ \phi_2 \\ \cdots \\ \phi_{N-1} \end{matrix} \right\|, \qquad g = \left\| \begin{matrix} g_1 \\ g_2 \\ \cdots \\ g_{N-1} \end{matrix} \right\|,$$

$$g_k = \begin{cases} f_1 + \dfrac{a}{h^2}, & k = 1, \\[2mm] f_k, & k = 2, 3, \ldots, N-2, \\[2mm] f_{N-1} + \dfrac{b}{h^2}, & k = N-1. \end{cases}$$

It is not difficult to verify that for $\phi \neq 0$

$$(A\phi, \phi) > 0.$$

This guarantees that the problem is uniquely soluble. In order to solve (1.3), we use the factorization method. The result is

$$\beta_{k+1} = \frac{1}{2 - \beta_k},$$

$$z_{k+1} = \beta_{k+1}(z_k + h^2 g_k), \qquad k = 1, 2, \ldots, N - 1, \tag{1.5}$$

$$\phi_k = \beta_{k+1}\phi_{k+1} + z_{k+1}, \qquad k = N - 1, N - 2, \ldots, 1.$$

In order to obtain the initial conditions β_2 and z_2, let us specialize the third of equations (1.5) to $k = 1$:

$$\phi_1 = \beta_2 \phi_2 + z_2,$$

and choose β_2 and z_2 so that the above equation becomes identical with the first equation in (1.3). Thus

$$\beta_2 = \tfrac{1}{2}, \qquad z_2 = (a + f_1 h^2)/2. \tag{1.6}$$

Initial conditions for ϕ_k are taken as

$$\phi_N = 0. \tag{1.7}$$

9.1.2 The One-Dimensional von Neumann Problem

Consider now the von Neumann problem

$$-\frac{d^2\phi}{dx^2} = f,$$

$$\frac{d\phi}{dx} = a \quad \text{for} \quad x = 0, \tag{1.8}$$

$$\frac{d\phi}{dx} = b \quad \text{for} \quad x = 1,$$

where a and b are given constants.

Let us integrate the above equation over the entire domain of definition of the solution, while making use of the boundary conditions. There results

$$a - b = \int_0^1 f(x)\, dx \tag{1.9}$$

This relation represents a necessary condition for solvability of (1.8). If $a = b$, it says that the total number of sources sums to zero, i.e., each emission source is counterbalanced by an absorption.

Thus, let (1.8) be solvable and let $_0(x)$ be its solution. We may convince ourselves that $(x) = _0(x) + C$ (C an arbitrary constant) is also a solution, and that this family includes all solutions.

In order to obtain the second-order accuracy difference analog for the problem, the solution of the problem (which must be sufficiently smooth) is extended to the additional intervals of length h adjoining its domain $0 \le x \le 1$ from the right and left. This means that we are considering a net region $x_k = kh$ $(k = -1, 0, \ldots, N, N + 1)$, $h = 1/N$. Let us define on this net the approximation of the problem as follows:

$$\frac{-\phi_{k-1} + 2\phi_k - \phi_{k+1}}{h^2} = f_k \qquad k = 0, 1, \ldots, N,$$

$$\frac{\phi_1 - \phi_{-1}}{2h} = a, \qquad \frac{\phi_{N+1} - \phi_{N-1}}{2h} = b. \tag{1.10}$$

As a preliminary work, let us remove the boundary conditions by eliminating ϕ_{-1} and ϕ_{N+1} from (1.10). Solving for ϕ_{-1}, ϕ_{N+1}, we obtain

$$\phi_{-1} = \phi_1 - 2ha; \qquad \phi_{N+1} = \phi_{N-1} + 2hb. \tag{1.11}$$

Substituting into (1.10), we obtain

$$\frac{\phi_0 - \phi_1}{h^2} = \frac{f_0}{2} - \frac{a}{h},$$

$$\frac{-\phi_{k-1} + 2\phi_k - \phi_{k+1}}{h^2} = f_k, \qquad k = 1, \ldots, N - 1, \tag{1.12}$$

$$\frac{-\phi_{N-1} + \phi_N}{h^2} = \frac{f_N}{2} + \frac{b}{h}.$$

Introduce the matrix

$$A = \frac{1}{h^2} \begin{Vmatrix} 1 & -1 & 0 & \cdots & 0 & 0 \\ -1 & 2 & -1 & \cdots & 0 & 0 \\ 0 & -1 & 2 & \cdots & 0 & 0 \\ \multicolumn{6}{c}{\cdots\cdots\cdots\cdots\cdots\cdots\cdots\cdots\cdots} \\ 0 & 0 & 0 & \cdots & 2 & -1 \\ 0 & 0 & 0 & \cdots & -1 & 1 \end{Vmatrix} \tag{1.13}$$

and the vectors

$$\phi = \begin{vmatrix} \phi_0 \\ \phi_1 \\ \phi_2 \\ \cdots \\ \phi_N \end{vmatrix}, \qquad g = \begin{vmatrix} g_0 \\ g_1 \\ g_2 \\ \cdots \\ g_N \end{vmatrix},$$

$$g_k = \begin{cases} \dfrac{f_0}{2} - \dfrac{a}{h}, & k = 0, \\[2mm] f_k, & k = 1, 2, \ldots, N - 1, \\[2mm] \dfrac{f_N}{2} + \dfrac{b}{h}, & k = N. \end{cases}$$

We can then write

$$A\phi = g. \tag{1.14}$$

The first question to be resolved when analyzing the matrix A is that regarding its definiteness. It is well known that in the von Neumann difference problem the matrix A is singular, since its smallest eigenvalue is zero. This can be seen rather easily by recognizing that the spectral problem

$$Au = \lambda u$$

has for its eigenvector the vector u_0 with identical components, the corresponding eigenvalue being $\lambda = 0$.

It is easy to show that $(A\phi, \phi) \geq 0$. We studied a problem of linear algebra like this, in detail, in Section 4.5. There we proved the necessary and sufficient conditions for solvability; in our current case they reduce to the orthogonality of g and the vector u_0, i.e.,

$$\sum_{k=0}^{N} g_k = 0. \tag{1.15}$$

The left-hand side of (1.15) may be written in the form

$$\sum_{k=0}^{N} g_k = \frac{1}{h}\left[\frac{h}{2}(f_0 + 2f_1 + \cdots + 2f_{N-1} + f_N) - a + b\right]. \tag{1.16}$$

But

$$\frac{h}{2}(f_0 + 2f_1 + \cdots + 2f_{N-1} + f_N)$$

is the well-known trapezoidal quadrature formula for the integral

$$\int_0^1 f(x)\,dx.$$

Accordingly, the square brackets in (1.16) either vanish or are small, depending on the error of the quadrature formula. To bring (1.14) into correspondence with the solvability condition (1.15) we must correct a potential discrepancy and take instead of g_k the quantity $\tilde{g}_k = g_k - \hat{g}$, where

$$\hat{g} = \frac{1}{N+1}\sum_{k=0}^{N} g_k.$$

This yields

$$\sum_{k=0}^{N} \tilde{g}_k = 0. \tag{1.17}$$

We set forth in Section 4.5 an iterative method for solving the consistent system

$$A\phi = \tilde{g}. \tag{1.18}$$

However, the system (1.18), with a tri-diagonal matrix, may be solved by h factorization if we introduce a slight change in the algorithm, as required by the singularity of the matrix A. In our exemplary problem we can easily carry the whole analysis of the factorization method through, and perceive the fundamental thrust of the algorithm.

Thus, let us turn to the factorization formulas. In our specific instance,

$$b_0 = c_0 = b_N = a_N = 1,$$

$$b_k = 2, \qquad a_k = c_k = 1, \qquad k = 1, 2, \ldots, N - 1,$$

and therefore

$$\beta_0 = 0, \qquad \beta_1 = \beta_2 = \cdots = \beta_N = 1,$$

$$z_0 = 0, \qquad z_{k+1} = z_k + g_k, \qquad k = 0, 1, \ldots, N - 1.$$

From the last equation we easily find that

$$z_{m+1} = \sum_{k=0}^{m} \tilde{g}_k$$

and in particular

$$z_N = \sum_{k=0}^{N-1} \tilde{g}_k.$$

Finally,

$$z_{N+1} = \frac{\tilde{g}_N + z_N}{b_N - a_N \beta_N} = \frac{\tilde{g}_N + \sum_{k=0}^{N-1} \tilde{g}_k}{b_N - a_N \beta_N} = \frac{\sum_{k=0}^{N} \tilde{g}_k}{1 - 1} = \frac{0}{0}.$$

Thus, we encounter an indeterminacy in the factorization algorithm. This is quite natural, as we see if we consider that $\phi_N = z_{N+1}$ may be any number, in view of the nonuniqueness of the solution. We may take an arbitrary number for ϕ_N and arrive at the solution we want. We must be mindful of peculiarities of this nature in the solutions of von Neumann's problem with operators more complicated than d^2/dx^2.

9.1.3 The Two-Dimensional Poisson Equation

We now turn to the multi-dimensional problems of mathematical physics, noting at the outset that their approximation is a far from trivial task. We can best explain the problem by considering the two-dimensional Poisson equation. If the domain of definition of the solution has a smooth boundary, and if the functions defined on the boundary are sufficiently smooth, then, given that the solution is stable, its second-order precision is guaranteed by a second-order approximation of the equation and of the boundary conditions. If, however, either the boundary or the functions prescribed on it are not smooth, very serious errors occur in the solution in the neighborhood of the singularities.

In fact, let the domain of definition of the solution be a triangle or an L-shaped region (cf. Figure 2.11). Then, as is well-known, logarithmic or fractional singularities normally occur in the solutions of problems with elliptic operators. Therefore, a uniform net and second-order approximation of the problem with respect to both variables within the region does not guarantee a solution having second-order accuracy. In such cases, one uses either a condensed net near the singularities of the solution, or a preliminary isolation of the singularities with a subsequent numerical solution of a regular problem, thus guaranteeing second-order accuracy. We must note, however, that for certain correspondences between the boundary conditions and the right-hand side of the equation, the solution may be smooth at the singular points of the boundary, and then the supplementary problem of approximation does not arise (see, for instance, Godunov [2]).

These questions were touched upon in Chapter 6 and should be kept in mind with regard to the numeric algorithms for problems of mathematical physics. Here we shall concentrate our attention on methods for the solution of difference equations arising in connection with multi-dimensional problems of mathematical physics.

Consider a Poisson equation in two dimensions

$$-\left(\frac{\partial^2 \phi}{\partial x^2} + \frac{\partial^2 \phi}{\partial y^2}\right) = f \quad \text{in} \quad D,$$

$$\phi = g \quad \text{on} \quad \partial D,$$

(1.19)

where the domain D is the square $0 \leq x \leq 1$, $0 \leq y \leq 1$, and $g(x, y)$ is a sufficiently smooth function defined on the boundary of D.

Let us choose a grid of net points in D, defined by coordinate lines $x = x_k$, $y = y_l$. Consider the corresponding difference equation

$$\frac{-\phi_{k-1,l} - \phi_{k+1,l} - \phi_{k,l-1} - \phi_{k,l+1} + 4\phi_{k,l}}{h^2} = f_{k,l}$$

$$\phi_{0,l} = a_l, \qquad \phi_{N,l} = b_l,$$

$$\phi_{k,0} = c_k, \qquad \phi_{k,N} = d_k$$

$$k, l = 1, 2, \ldots, N - 1,$$

(1.20)

where

$$a_l = g_{0,l}, \qquad b_l = g_{N,l}, \qquad c_k = g_{k,0}, \qquad d_k = g_{k,N}, \qquad x_k = kh,$$

$$y_l = lh.$$

Next, we eliminate the boundary conditions in (1.20) and write the result in matrix form. For this purpose let us introduce the following matrix A and the vectors $\{\phi_l; g_l\}_{l=1}^{N-1}$:

$$A = \frac{1}{h^2} \begin{Vmatrix} 2 & -1 & 0 & -1 & \cdots & 0 & 0 \\ -1 & 2 & -1 & -0 & \cdots & 0 & 0 \\ 0 & -1 & 2 & -1 & \cdots & 0 & 0 \\ \multicolumn{7}{c}{\cdots\cdots\cdots\cdots\cdots\cdots\cdots} \\ 0 & 0 & 0 & 0 & \cdots & -1 & 2 \end{Vmatrix}, \qquad \phi_i = \begin{Vmatrix} \phi_{1,l} \\ \phi_{2,l} \\ \phi_{3,l} \\ \cdots \\ \phi_{N-1,l} \end{Vmatrix},$$

$$g_l = \begin{Vmatrix} f_{1,l} + \dfrac{a_l}{h^2} \\[2mm] f_{2,l} \\[2mm] f_{3,l} \\[1mm] \cdots\cdots\cdots \\[1mm] f_{N-1,l} + \dfrac{b_l}{h^2} \end{Vmatrix}.$$

Define

$$B = h^2 A + 2E,$$

and rewrite (1.20) in the form

$$\frac{1}{h_2}(-\phi_{l-1} + B\phi_l - \phi_{l+1}) = g_l \qquad l = 1, 2, \ldots, N-1, \qquad (1.21)$$

assuming

$$\phi_0 = c, \qquad \phi_N = d, \qquad (1.22)$$

where

$$c = \begin{Vmatrix} c_1 \\ c_2 \\ \cdots \\ c_{N-1} \end{Vmatrix}, \qquad d = \begin{Vmatrix} d_1 \\ d_2 \\ \cdots \\ d_{N-1} \end{Vmatrix}.$$

Eliminating the boundary conditions (1.22) from (1.21), we obtain

$$\frac{1}{h^2}(B\phi_1 - \phi_2) = g_1 + \frac{1}{h^2} c,$$

$$\frac{1}{h^2}(-\phi_{l-1} + B\phi_l - \phi_{l+1}) = g_l, \qquad l = 2, 3, \ldots, N-2, \qquad (1.23)$$

$$\frac{1}{h^2}(-\phi_{N-2} + B\phi_{N-1}) = g_{N-1} + \frac{1}{h^2} d.$$

Let us rewrite the system (1.23) using the block matrix and block vector notation:

$$
\Lambda = \frac{1}{h^2}
\begin{Vmatrix}
B & -E & 0 & 0 & \cdots & 0 & 0 \\
-E & B & -E & 0 & \cdots & 0 & 0 \\
0 & -E & B & -E & \cdots & 0 & 0 \\
\hdotsfor{7} \\
0 & 0 & 0 & 0 & \cdots & -E & B
\end{Vmatrix},
\qquad
\phi =
\begin{Vmatrix}
\phi_1 \\
\phi_2 \\
\vdots \\
\phi_{N-1}
\end{Vmatrix}
$$

$$
F =
\begin{Vmatrix}
\dfrac{1}{h^2} c + g_1 \\
g_2 \\
\vdots \\
\dfrac{1}{h^2} d + g_{N-1}
\end{Vmatrix},
$$

where E is the unit matrix. System (1.23) then becomes

$$\Lambda \phi = F. \tag{1.24}$$

Write next the matrix Λ as a sum of two matrices,

$$\Lambda = \Lambda_1 + \Lambda_2$$

where

$$
\Lambda_1 =
\begin{vmatrix}
A & 0 & 0 & \cdots & 0 \\
0 & A & 0 & \cdots & 0 \\
0 & 0 & A & \cdots & 1 \\
\hdotsfor{5} \\
0 & 0 & 0 & \cdots & A
\end{vmatrix},
$$

$$
\Lambda_2 = \frac{1}{h^2}
\begin{vmatrix}
2E & -E & 0 & 0 & \cdots & 0 \\
-E & 2E & -E & 0 & \cdots & 0 \\
0 & -E & 2E & -E & \cdots & 0 \\
\hdotsfor{6} \\
0 & 0 & 0 & 0 & \cdots & 2E
\end{vmatrix}.
$$

It is not difficult to verify that

$$
\Lambda_1 \phi =
\begin{vmatrix}
A\phi_1 \\
A\phi_2 \\
\vdots \\
A\phi_{N-1}
\end{vmatrix},
\qquad
\Lambda_2 \phi = \frac{1}{h^2}
\begin{vmatrix}
2\phi_1 - \phi_2 \\
\vdots \\
-\phi_{l-1} + 2\phi_l - \phi_{l+1} \\
\vdots \\
-\phi_{N-2} + 2\phi_{N-1}
\end{vmatrix}.
\tag{1.25}
$$

Introduce new vectors as follows:

$$(\Lambda_1\phi)_l = A\phi_l, \qquad l = 1, 2, \ldots, N-1,$$

$$(\Lambda_2\phi)_k = \begin{cases} \dfrac{1}{h^2}(2\phi_1 - \phi_2)_k, & l = 1, \\[2mm] \dfrac{1}{h^2}(-\phi_{l-1} + 2\phi_l - \phi_{l+1})_k, & l = 2, 3, \ldots, N-2, \quad (1.26) \\[2mm] \dfrac{1}{h^2}(-\phi_{N-2} + 2\phi_{N-1})_k, & l = N-1. \end{cases}$$

We see that in this notation $(\Lambda_1\phi)_l$ and $(\Lambda_2\phi)_k$ are components of the vectors $\Lambda_1\phi$ and $\Lambda_2\phi$, while having the following convenient representations:

$$(\Lambda_1\phi)_l = \frac{1}{h^2}\left\|\begin{array}{c} 2\phi_{1,l} - \phi_{2,l} \\ \vdots \\ -\phi_{k-1,l} + 2\phi_{k,l} - \phi_{k+1,l} \\ \vdots \\ -\phi_{N-2,l} + 2\phi_{N-1,l} \end{array}\right\|,$$

$$((\Lambda_2\phi)_l)_k = \begin{cases} \dfrac{1}{h^2}(2\phi_{k,l} - \phi_{k,l+1}), & k = 1, \\[2mm] \dfrac{1}{h^2}(-\phi_{k,l-1} + 2\phi_{k,l} - \phi_{k,l+1}), & k = 2, \ldots, N-2, \\[2mm] \dfrac{1}{h^2}(-\phi_{k,l-1} + 2\phi_{k,l}), & k = N-1, \end{cases}$$

$$l = 1, 2, \ldots, N-1.$$

Similarly, let us represent F in the form

$$F = \left\|\begin{array}{c} F_1 \\ F_2 \\ \vdots \\ F_{N-1} \end{array}\right\|, \quad F_1 = \left\|\begin{array}{c} f_{1,1} + \dfrac{c_1}{h^2} + \dfrac{a_1}{h^2} \\[2mm] f_{2,1} + \dfrac{c_2}{h^2} \\[2mm] f_{3,1} + \dfrac{c_3}{h^2} \\ \vdots \\ f_{N-1,1} + \dfrac{c_{N-1}}{h} + \dfrac{b_1}{h^2} \end{array}\right\|, \quad F_l = \left\|\begin{array}{c} f_{1,l} + \dfrac{a_l}{h^2} \\[2mm] f_{2,l} \\[2mm] f_{3,l} \\ \vdots \\ f_{N-1,l} + \dfrac{b_l}{h^2} \end{array}\right\|,$$

$$(l = 2, \ldots, N-2),$$

$$F_{N-1} = \begin{vmatrix} f_{1,\,N-1} + \dfrac{d_1}{h^2} + \dfrac{a_{N-1}}{h^2} \\[2ex] f_{2,\,N-1} + \dfrac{d_2}{h^2} \\[2ex] f_{3,\,N-1} + \dfrac{d_3}{h^2} \\[2ex] \vdots \\[2ex] f_{N-1,\,N-1} + \dfrac{d_{N-1}}{h^2} + \dfrac{b_{N-1}}{h^2} \end{vmatrix}.$$

As a result we obtain the component version of the problem

$$(\Lambda_1 \phi)_l + (\Lambda_2 \phi)_l = F_l. \tag{1.27}$$

Finally, we obtain the vector matrix form of (1.16)

$$(\Lambda_1 + \Lambda_2)\phi = F, \tag{1.28}$$

by taking $l = 1, 2, \ldots, N - 1$ in (1.27).

One can easily verify that each component of (1.28) corresponds to a difference equation from (1.20) with the boundary conditions already accounted for.

In order to form the algorithm, let us now compute the upper and lower spectral bounds of the operator Λ.

Considering the form of the eigenfunctions of Λ (see Section 1.1.4), that is,

$$u_{kl}^{mp} = \sin m\pi kh \sin p\pi lh, \qquad k, l = 1, 2, \ldots, N - 1, \tag{1.29}$$

we find that

$$\lambda_{mp} = \frac{4}{h^2}\left(\sin^2 \frac{m\pi h}{2} + \sin^2 \frac{p\pi h}{2}\right). \tag{1.30}$$

Hence

$$\alpha = \frac{8}{h^2}\sin^2 \frac{\pi h}{2}, \qquad \beta = \frac{8}{h^2}\cos^2 \frac{\pi h}{2}, \tag{1.31}$$

where

$$\alpha = \alpha(\Lambda), \qquad \beta = \beta(\Lambda).$$

Let us use one of the iterative methods considered in Chapter 4:

$$\phi^{j+1} = \phi^j - \tau_j(\Lambda \phi^j - F), \tag{1.32}$$

or, equivalently,

$$\phi^{j+1} = \phi^j - \tau_j \xi^j,$$
$$\xi^j = \Lambda \phi^j - F.$$

The iterative process is to be continued until we have the inequality

$$\|\phi^j - \phi\| \le \varepsilon,$$

where ε is an *a priori* constant. This estimate holds if

$$\|\xi^j\| \le \alpha(\Lambda)\varepsilon. \tag{1.33}$$

Generally speaking, the rate of convergence can be improved by replacing iterations (1.28) by successive approximations with the Chebyshev acceleration:

$$\phi^{j+1} = \phi^j - \tau_j B^{-1}(\Lambda\phi^j - F), \tag{1.34}$$

where

$$B = (E + \sigma\Lambda_1)(E + \sigma\Lambda_2); \qquad \sigma = \frac{2}{\sqrt{\alpha\beta}}; \qquad \alpha = \alpha(\Lambda); \beta = \beta(\Lambda).$$

Let us find the spectral bounds of the operator $B^{-1}\Lambda$. Since the matrices Λ_1 and Λ_2 have a common basis, the eigenvalues λ_1 and λ_2 of the matrices Λ_1 and Λ_2 correspondingly are related to the eigenvalues of the problem

$$B^{-1}\Lambda u = \lambda(B^{-1}\Lambda)u$$

by means of the following formula:

$$\lambda(B^{-1}\Lambda) = \frac{\lambda_1 + \lambda_2}{\left(1 + \dfrac{2\lambda_1}{\sqrt{\alpha\beta}}\right)\left(1 + \dfrac{2\lambda_2}{\sqrt{\alpha\beta}}\right)}, \tag{1.35}$$

$$\frac{\alpha}{2} \le \lambda_1 \le \frac{\beta}{2}, \qquad \frac{\alpha}{2} \le \lambda_2 \le \frac{\beta}{2}.$$

Expression (1.35) can be written as

$$\lambda(B^{-1}\Lambda) = \frac{\sqrt{\alpha\beta}}{2} f(x, y), \tag{1.36}$$

where

$$x = \frac{2\lambda_1}{\sqrt{\alpha\beta}}, \qquad y = \frac{2\lambda_2}{\sqrt{\alpha\beta}}; \qquad f(x, y) = \frac{x + y}{(1 + x)(1 + y)}. \tag{1.37}$$

Thus, in order to define the spectral bounds of the matrix $B^{-1}\Lambda$, it is enough to find the maximum and minimum of the function $f(x, y)$ on the square $\sqrt{\alpha/\beta} \le x, y \le \sqrt{\beta/\alpha}$. Looking at the derivatives of this function, i.e.,

$$\frac{\partial f}{\partial x} = \frac{1 - y}{1 + y} \cdot \frac{1}{(1 + x)^2}, \qquad \frac{\partial f}{\partial y} = \frac{1 - x}{1 + x} \cdot \frac{1}{(1 + y)^2}, \tag{1.38}$$

is not difficult to show, that the maximum of $f(x, y)$ is attained at two of the corners:

$$\max_{\sqrt{\alpha/\beta} \le x, y \le \sqrt{\beta/\alpha}} f(x, y) = f\left(\sqrt{\frac{\alpha}{\beta}}, \sqrt{\frac{\beta}{\alpha}}\right) = f\left(\sqrt{\frac{\beta}{\alpha}}, \sqrt{\frac{\alpha}{\beta}}\right), \quad (1.39)$$

while the other two corners yield the minimum:

$$\min_{\sqrt{\alpha/\beta} \le x, y \le \sqrt{\beta/\alpha}} f(x, y) = f\left(\sqrt{\frac{\alpha}{\beta}}, \sqrt{\frac{\alpha}{\beta}}\right) = f\left(\sqrt{\frac{\beta}{\alpha}}, \sqrt{\frac{\beta}{\alpha}}\right). \quad (1.39a)$$

From this and from (1.36) it follows that

$$\alpha(B^{-1}\Lambda) = \frac{\alpha}{\left(1 + \sqrt{\frac{\alpha}{\beta}}\right)^2}, \quad \beta(B^{-1}\Lambda) = \frac{1}{2}\sqrt{\frac{\alpha}{\beta}} \cdot \frac{(\alpha + \beta)}{\left(1 + \sqrt{\frac{\alpha}{\beta}}\right)^2}. \quad (1.40)$$

Asymptotically, this means that for $\beta(\Lambda) \gg \alpha(\Lambda)$ we have

$$p(B^{-1}\Lambda) = \frac{\beta(B^{-1}\Lambda)}{\alpha(B^{-1}\Lambda)} = \frac{1}{2} \cdot \frac{\alpha + \beta}{\sqrt{\beta\alpha}} \cong \frac{1}{2}\sqrt{\frac{\beta}{\alpha}} = \frac{1}{2}[p(\Lambda)]^{1/2}, \quad (1.41)$$

and

$$s = \frac{2}{\sqrt{p(B^{-1}\Lambda)}} = 2^{3/2}p^{-1/4}(\Lambda). \quad (1.42)$$

Consider now the realization scheme of the iterative process (1.34):

$$\xi^j = \Lambda\phi^j - F, \quad (E + \sigma\Lambda_1)\xi^{j+1/2} = \xi^j,$$

$$(E + \sigma\Lambda_2)\xi^{j+1} = \xi^{j+1/2}, \quad \phi^{j+1} = \phi^j - \tau_j\xi^{j+1}, \quad (1.43)$$

where the choice of τ_j depends on the particular optimization method.

Using the component representation, the second and third equations of (1.43) are handled as follows: we first solve the problem

$$(1 + 2\sigma)\xi_{1,l}^{j+1/2} - \sigma\xi_{2,l}^{j+1/2} = \xi_{1,l}^j,$$

$$-\sigma\xi_{k-1,l}^{j+1/2} + (1 + 2\sigma)\xi_{k,l}^{j+1/2} - \sigma\xi_{k+1}^{j+1/2} = \xi_{k,l}^j, \quad (1.44)$$

$$-\sigma\xi_{N-2,l}^{j+1/2} + (1 + 2\sigma)\xi_{N-1,l}^{j+1/2} = \xi_{N-1,l}^j$$

for a fixed $l = 1, \ldots, N - 1$, and then, for a fixed $k = 1, \ldots, N - 1$:

$$(1 + 2\sigma)\xi_{k,1}^{j+1} - \sigma\xi_{k,2}^{j+1} = \xi_{k,1}^{j+1/2},$$

$$-\sigma\xi_{k,l-1}^{j+1} + (2\sigma + 1)\xi_{k,l}^{j+1} - \sigma\xi_{k,l+1}^{j+1} = \xi_{k,l}^{j+1/2}, \quad (1.45)$$

$$-\rho\xi_{k,N-2}^{j+1} + (1 + 2\sigma)\xi_{k,N-1}^{j+1} = \xi_{k,N-1}^{j+1/2}.$$

To do this, we can use the factorization method.

In conclusion, let us note that the two-dimensional von Neumann problem can be similarly reduced to problem (1.28), which differs from the one we are considering only in the number of components of the solution to be determined: it will not be $(N - 1)^2$ as in the Dirichlet problem, but $(N + 1)^2$.

An important difference between the realization schemes of the difference analog of the von Neumann problem and the Dirichlet problem is that in the von Neumann problem $\alpha(\Lambda) = 0$. Therefore the right-hand side F, as well as the approximate solution ϕ^j must be orthogonal to the vector with identical components at each step. This means that before a new iteration step can be executed we must subtract from each component of the vector ξ^j the constant

$$\frac{1}{(N + 1)^2} \sum_{k,l} \xi^j_{k,l}.$$

For such an orthogonalization procedure, in which vectors with identical components are excluded from the elements of the original Hilbert space thus carrying it into the subspace Φ, and the trial vectors are chosen to lie in Φ, we may choose as the lower spectral bound the lowest nonzero eigenvalue. It is known that the spectrum of the von Neumann problem for the difference analog of the Laplace operator is given by the formula

$$\lambda_{mp} = \frac{4}{h^2}\left(\sin^2 \frac{m\pi}{2(N + 1)} + \sin^2 \frac{p\pi}{2(N + 1)}\right), \qquad m, p = 0, 1, \ldots, N, \quad (1.46)$$

and the spectral bounds are given by

$$\alpha^*(\Lambda) = \frac{8}{h^2}\sin^2 \frac{\pi}{2(N + 1)}, \qquad \beta(\Lambda) = \frac{8}{h^2}\cos^2 \frac{\pi}{2(N + 1)}. \quad (1.47)$$

The parameter σ is given in this case by

$$\sigma = \frac{2}{\sqrt{\alpha^*\beta}}, \quad (1.48)$$

and instead of (1.33) we now have

$$\|\xi^j\| < \alpha^*(\Lambda)\varepsilon \quad (1.49)$$

as the index for terminating the iterative process. Here the differences end.

9.1.4 A Problem of Boundary Conditions

The above approaches to the Poisson equation allow one to make an important general conclusion with regard to forming efficient algorithms for solving boundary-value problems of mathematical physics. It mainly concerns the problem of boundary conditions. In the above, we have shown how to eliminate the boundary conditions imposed on the solution of the problem at hand, and how to modify them in view of the difference analog of the problem. There is a deeper sense in such an approach; if we can remove the boundary conditions from consideration by means of difference methods, we may then proceed without worrying about them any further: they are automatically taken care of by the modified difference equations. This is of importance for stationary problems, and especially for nonstationary problems in which the boundary conditions require careful treatment. In particular, this was the reason why we did not consider in Chapter 7 the splitting-up methods for nonstationary problems (in the differential formulation). This would have required additional work as to the compatibility of the boundary conditions and the decomposed system. From our point of view, it is much simpler first to put the original problem of mathematical physics into correspondence with a system of difference equations (with respect to the spatial variables) and then to eliminate the boundary conditions using the difference analogs of the boundary conditions, the accuracy of which matches that of the difference equations. Having done this, we can next proceed by approximating the equations in time using the splitting-up method or another algorithm. This approach allows us to sidestep the compatibility problem for the boundary conditions as mentioned above, which would run through all stages of forming the numerical algorithm when using splitting-up schemes.

Let us now turn to the following fact, which is closely related to the above problem. Construction of difference schemes for problems of mathematical physics may in some cases benefit from a Fourier series representation of the solution along the eigenelements of the operator from the corresponding difference problem. We have already used this approach many times when analyzing the properties of numerical algorithms. In order to actually apply this method it is necessary, however, that the difference problem be closed with respect to the homogeneous boundary conditions. Nonhomogeneous boundary conditions can be made into homogeneous ones by an auxiliary transformation, which may be conveniently implemented by a method we now will illustrate with a simple example provided by problem (1.1). Thus, consider the system of difference equations (1.2). Let us extend the domain of definition of the solution x_k, $(k = 1, \ldots, N - 1)$, by adding two more net points $x = 0$ $(k = 0)$ and $x = 1$ $(k = N)$, and defining

$$\phi_0 = 0, \qquad \phi_N = 0.$$

Of course, this extension carries only a formal character and has no relation whatsoever to the actual values of the solution at the points $x = 0$ and $x = 1$, which are defined by the nonhomogeneous boundary conditions (1.1). This means that once the problem has been solved, we have to go back and restrict the domain by eliminating the two points. This method permits one to consider the following equivalent problem:

$$\frac{-\phi_{k-1} + 2\phi_k - \phi_{k+1}}{h^2} = g_k \qquad k = 1, 2, \ldots, N - 1,$$

$$\phi_0 = 0, \qquad \phi_N = 0,$$

\hfill (1.50)

where

$$g_k = \begin{cases} \dfrac{a}{h^2} + f_1, & k = 1, \\[2mm] f_k, & k = 2, 3, \ldots, N - 2, \\[2mm] \dfrac{b}{h^2} + f_{N-1}, & k = N - 1. \end{cases} \qquad (1.51)$$

Hence problem (1.1) with nonhomogeneous boundary conditions has been reduced to a homogeneous problem, ready for the Fourier method.

In the case of the nonhomogeneous von Neumann problem (1.10) the domain of the solution is to be extended by adjoining the net points x_{-1} and x_{N+1}.

Eventually we arrive at the problem

$$\frac{-\phi_{k-1} + 2\phi_k - \phi_{k+1}}{h^2} = g_k, \qquad k = 0, 1, 2, \ldots, N,$$

$$\phi_{-1} = \phi_1, \qquad \phi_{N+1} = \phi_{N-1}, \qquad (1.52)$$

where

$$g_k = \begin{cases} f_0 - \dfrac{2a}{h}, & k = 0, \\[2mm] f_k, & k = 1, 2, \ldots, N - 1, \\[2mm] f_N + \dfrac{2b}{h}, & k = N. \end{cases} \qquad (1.53)$$

Having found the solution of (1.52) and (1.53), we may drop the auxiliary components ϕ_{-1} and ϕ_{N+1}.

One can tackle multi-dimensional stationary and nonstationary problems in a similar manner.

If necessary, equations (1.50)–(1.53) can be handled the same way as the nonhomogeneous problems (1.1) and (1.10), i.e., by eliminating the boundary values. Again, we eventually finish up with problems (1.3) and (1.12).

In what follows, we therefore will not consider the nonhomogeneous boundary conditions, since the above-described algorithms permit one to reduce them to homogeneous conditions or to exclude them completely from consideration.

9.2 The Heat Equation

The problem of heat conduction is a typical nonstationary problem of mathematical physics and occupies a prominent place in history. It was the heat equation which stimulated many results of principal importance in numerical mathematics and which led to constructions of first-class algorithms for problems of mathematical physics. These studies go back to the classical work by O'Brien, Hyman and Kaplan (7), where the convergence problem of approximate solutions to the exact ones were taken up. By now, there is a whole series of monographs and original papers devoted solely to the numerical treatment of the heat equation (see, for instance, Saul'ev [3]). Thus, we will restrict our attention to some of the methods which have found the broadest applications.

9.2.1 The One-Dimensional Problem of Heat Conduction

To begin with, consider a simple problem of heat propagation in a homogeneous bar of finite length, heated by internal sources located on the boundary. The problem can be described as follows:

$$\frac{1}{c^2} \cdot \frac{\partial \phi}{\partial t} = \frac{\partial^2 \phi}{\partial x^2} + f(x, t),$$

$$\phi(0, t) = a(t), \qquad \phi(1, t) = b(t), \qquad \phi(x, 0) = \phi^0(x), \tag{2.1}$$

where f, a, b, and ϕ^0 are given sufficiently smooth functions; $c^2 = \text{const}$; the variables x and t run through the domain of definition of the solution $D \times D_t = \{0 \le x \le 1, 0 \le t \le T\}$.

We will solve (2.1) using the finite-difference method. Thus, let us first approximate (2.1) with respect to x with second-order accuracy in $h = \Delta x$. Divide the interval $0 \le x \le 1$ into N subintervals of equal length $h = 1/N$ and denote the points of the grid by x_k $(k = 1, \ldots, N)$. We arrive at the following problem, similar to (1.2):

$$\frac{1}{c^2} \cdot \frac{d\phi_k}{dt} + \frac{-\phi_{k-1} + 2\phi_k - \phi_{k+1}}{h^2} = f_k(t),$$

$$\phi_0 = a(t), \qquad \phi_N = b(t), \qquad \phi_k = \phi_k^0 \quad \text{for} \quad t = 0, \tag{2.2}$$

$$k = 1, 2, \ldots, N - 1,$$

or, using the vector-matrix notation,

$$\frac{1}{c^2} \cdot \frac{d\phi}{dt} + A\phi = g,$$

$$\phi = \phi^0 \quad \text{for} \quad t = 0, \tag{2.3}$$

where A is a positive matrix, and $\phi(t)$ and $g(t)$ are vector functions which have been defined for an arbitrary t in problem (1.4).

Using the ideas from Section 8.1.4, equations (2.2) can be rewritten as

$$\frac{1}{c^2} \cdot \frac{d\phi_k}{dt} + \frac{-\phi_{k-1} + 2\phi_k - \phi_{k+1}}{h^2} = g_k,$$

$$\phi_0 = 0, \qquad \phi_N = 0, \qquad \phi_k = \phi_k^0 \quad \text{for} \quad t = 0, \tag{2.4}$$

where g_k is defined by (1.47).

As we have pointed out in Section 8.1.4, the solution of (2.4) makes sense only at the net points $x_1, \ldots, x_{N-1}$.

Let us now consider the system of ordinary differential equations of (2.4). Integrate each of them with respect to time over the interval $t_j \leq t \leq t_{j+1}$:

$$\frac{\phi_k^{j+1} - \phi_k^j}{\tau} = \frac{\bar{\phi}_{k-1}^j - 2\bar{\phi}_k^j + \bar{\phi}_{k+1}^j}{h^2} + \bar{g}_k^j, \tag{2.5}$$

where

$$\bar{\phi}_k^j = \frac{1}{\Delta t}\int_{t_j}^{t_{j+1}} \phi_k \, dt, \qquad \bar{g}_k^j = \frac{1}{\Delta t}\int_{t_j}^{t_{j+1}} g_k \, dt. \tag{2.6}$$

Here we have adopted the notation

$$\phi_k^j = \phi_k(t_j), \qquad \tau = c^2 \Delta t.$$

The kind of difference equations that we will obtain depends on a particular approximation in (2.6). Consider the following three simple interpolations:

$$\frac{1}{\Delta t}\int_{t_j}^{t_{j+1}} \phi_k \, dt \simeq \begin{cases} \phi_k^j, \\ \phi_k^{j+1}, \\ \frac{1}{2}(\phi_k^j + \phi_k^{j+1}); \end{cases} \qquad \frac{1}{\Delta t}\int_{t_j}^{t_{j+1}} g_k \, dt \simeq \begin{cases} g_k^j, \\ g_k^{j+1}, \\ \frac{1}{2}(g_k^{j+1} + g_k^j). \end{cases}$$

Respectively, they result in the following most widely used difference schemes.

The explicit triangular scheme (.∴.)

$$\frac{\phi_k^{j+1} - \phi_k^j}{\tau} = \frac{\phi_{k-1}^j - 2\phi_k^j + \phi_{k+1}^j}{h^2} + g_k^j. \tag{2.7}$$

The implicit triangular scheme (˙∵˙)

$$\frac{\phi_k^{j+1} - \phi_k^j}{\tau} = \frac{\phi_{k-1}^{j+1} - 2\phi_k^{j+1} + \phi_{k+1}^{j+1}}{h^2} + g_k^{j+1}. \tag{2.8}$$

The Crank–Nicholson scheme (:::)

$$\begin{aligned} \frac{\phi_k^{j+1} - \phi_k^j}{\tau} &= \frac{\phi_{k-1}^{j+1} - 2\phi_k^{j+1} + \phi_{k-1}^{j+1}}{2h^2} \\ &+ \frac{\phi_{k-1}^j - 2\phi_k^j + \phi_{k+1}^j}{2h^2} + \frac{g_k^{j+1} + g_k^j}{2}. \end{aligned} \tag{2.9}$$

The systems of equations (2.7)–(2.9) must be adjoined in addition by the boundary conditions

$$\phi_0^j = 0, \qquad \phi_N^j = 0. \tag{2.10}$$

The triangular scheme (.∴.) can be solved explicitly for the unknown ϕ^{j+1}:

$$\phi^{j+1} = \phi^j + \mu(\phi_{k-1}^j - 2\phi_k^j + \phi_{k+1}^j) + \tau g_k^j,$$

$$\phi_0^j = 0, \qquad \phi_N^j = 0, \qquad \mu = \frac{\tau}{h^2}. \tag{2.11}$$

The implicit triangular scheme ($\because$) has a more complicated realization; we have to solve eventually the following difference equation:

$$-\phi_{k-1}^{j+1} + \left(2 + \frac{1}{\mu}\right)\phi_k^{j+1} - \phi_{k+1}^{j+1} = \frac{1}{\mu}\phi_k^j + h^2 g_k^{j+1},$$

$$\phi_0^{j+1} = 0, \qquad \phi_N^{j+1} = 0.$$
(2.12)

Finally, the Crank–Nicholson scheme ($\vdots$) has the following realization:

$$-\xi_{k-1}^{j+1} + \frac{2(1+\mu)}{\mu}\xi_k^{j+1} - \xi_{k+1}^{j+1} = \frac{2}{\mu}\phi_k^j + h^2 g_k^{j+1/2},$$

$$\xi_0^{j+1} = 0, \qquad \xi_N^{j+1} = 0, \qquad \phi_k^{j+1} = 2\xi_k^{j+1} - \phi_k^j,$$
(2.13)

where

$$g_k^{j+1/2} = \tfrac{1}{2}(g_k^{j+1} + g_k^j).$$

Problems (2.12) and (2.13) can be efficiently solved by the factorization method.

Let us next investigate the stability of the difference schemes of equations (2.7)–(2.9), assuming (2.10). To this end, let us expand the solution into a Fourier series with respect to the complete system of functions $\{\sin n\pi kh\}$, where $h = 1/N$ [the functions clearly satisfy (2.10)]. Let

$$\phi_k^j = \sum_{n=1}^{N-1} \Phi_n^j \sin n\pi kh, \qquad g_k^j = \sum_{n=1}^{N-1} G_n^j \sin n\pi kh,$$
(2.14)

where

$$\Phi_n^j = \frac{1}{q_n}\sum_{k=1}^{N-1} \phi_k^j \sin n\pi kh, \qquad G_n^j = \frac{1}{q_n}\sum_{k=1}^{N-1} g_k^j \sin n\pi kh.$$

$$q_n = \sum_{k=1}^{N-1} \sin^2 n\pi kh.$$

Let us substitute (2.14) into (2.7)–(2.9), multiply the results by $\sin(l n\pi h)$, and then sum them up in l. As a result we obtain the recursive relations for the Fourier coefficients: for the scheme ($\therefore$)

$$\frac{\Phi_n^{j+1} - \Phi_n^j}{\tau} + \lambda_n \Phi_n^j = G_n^j,$$
(2.15)

for the scheme ($\because$)

$$\frac{\Phi_n^{j+1} - \Phi_n^j}{\tau} + \lambda_n \Phi_n^{j+1} = G_n^{j+1},$$
(2.16)

for the scheme ($\vdots$)

$$\frac{\Phi_n^{j+1} - \Phi_n^j}{\tau} + \lambda_n \frac{\Phi_n^{j+1} + \Phi_n^j}{2} = \frac{G_n^{j+1} + G_n^j}{2},$$
(2.17)

where

$$\lambda_n = \frac{4}{h^2} \sin^2 \frac{n\pi h}{2}. \tag{2.18}$$

Solving (2.15)–(2.17), we get correspondingly

$$\Phi_n^{j+1} = (1 - \tau\lambda_n)\Phi_n^j + \tau G_n^j,$$

$$\Phi_n^{j+1} = \frac{1}{1 + \tau\lambda_n} \Phi_n^j + \frac{\tau}{1 + \tau\lambda_n} G_n^{j+1}, \tag{2.19}$$

$$\Phi_n^{j+1} = \frac{1 - \frac{\tau}{2}\lambda_n}{1 + \frac{\tau}{2}\lambda_n} \Phi_n^j + \frac{\tau}{2} \cdot \frac{G_n^{j+1} + G_n^j}{1 + \frac{\tau}{2}\lambda_n}.$$

Note that

$$\frac{4}{h^2} \sin^2 \frac{\pi h}{2} \le \lambda_n \le \frac{4}{h^2} \cos^2 \frac{\pi h}{2} < \frac{4}{h^2}. \tag{2.20}$$

For $(\pi h/2) \ll 1$ we have the asymptotic relation

$$\pi^2 \le \lambda_n \le \frac{4}{h^2}. \tag{2.21}$$

Taking (2.20) into account, we conclude that for

$$\tau \le \frac{h^2}{2 \cos^2 \frac{\pi h}{2}} \tag{2.22}$$

the triangular scheme (.∴.) is stable, since for all n

$$|1 - \tau\lambda_n| < 1. \tag{2.23}$$

Instead of (2.22) one can use the sufficient condition

$$\tau \le \frac{h^2}{2}. \tag{2.24}$$

In the case of the implicit triangular scheme and the Crank–Nicholson scheme we have correspondingly the inequalities

$$\left| \frac{1}{1 + \tau\lambda_n} \right| < 1, \qquad \left| \frac{1 - \frac{\tau}{2}\lambda_n}{1 + \frac{\tau}{2}\lambda_n} \right| < 1,$$

which hold true for any n and $\tau > 0$. This means that, while the scheme (.∴.) requires condition (2.22) for its stability, the schemes (˙∶˙) and (∶∷) are absolutely stable. The Crank–Nicholson scheme has been used most extensively.

In the case of the von Neumann problem for the heat equation, the operator A and the vector function g can be determined exactly as in (1.14). As a result we obtain equations similar to (2.5) and (2.6), which can be solved without any difficulty. At this point let us make the following remark: in the von Neumann problem for the Laplace equation we have "filtered" away from the approximate solution the parasitic component (frequency) corresponding to the eigenvalue $\lambda = 0$. This can no longer be done in the problem of heat conduction; we need to keep this frequency, since it describes the overall increase or decrease in the temperature of the bar due to the external sources.

9.2.2 The Two-Dimensional Problem of Heat Conduction

The problem of heat conduction in two dimensions can be formulated as follows:

$$\frac{1}{c^2}\frac{\partial \phi}{\partial t} - \Delta\phi = f \quad \text{in} \quad D \times D_t$$

$$\phi = g \quad \text{or} \quad \frac{\partial \phi}{\partial n} = g \quad \text{on} \quad \partial D \times D_t \qquad (2.25)$$

$$\phi = s(x, y) \quad \text{in} \quad D \quad \text{for} \quad t = 0,$$

where $D \equiv \{0 \le x \le 1, 0 \le y \le 1\}$.

First of all let us reduce this problem to its finite-difference version in the variables (x, y). The manipulations similar to those used in the two-dimensional Dirichlet problem lead to the following representation:

$$\frac{1}{c^2}\cdot\frac{d\phi}{dt} + (\Lambda_1 + \Lambda_2)\phi = F, \qquad (2.26)$$

$$\phi = s \quad \text{for} \quad t = 0,$$

where the matrices Λ_1, Λ_2 and the vector functions ϕ and F are defined the same way as in problem (1.24), the only difference being that now the components of ϕ and F depend on t. Note that if the heat equation is supplemented with the Dirichlet boundary conditions, we have

$$\Lambda_1 > 0, \qquad \Lambda_2 > 0, \qquad (2.27)$$

whereas in the case of the von Neumann conditions

$$\Lambda_1 \ge 0, \qquad \Lambda_2 \ge 0. \qquad (2.28)$$

Divide the interval $0 \le t \le T$ into subintervals by means of the points t_j. Putting $\tau = c^2 \Delta t$, let us approximate (2.26) using the component splitting-up method. Then $(t_{j-1} \le t \le t_{j+1})$

$$\left(E + \frac{\tau}{2} \Lambda_1 \right) \phi^{j-1/2} = \left(E - \frac{\tau}{2} \Lambda_1 \right) \phi^{j-1},$$

$$\left(E + \frac{\tau}{2} \Lambda_2 \right) (\phi^j - \tau f^j) = \left(E - \frac{\tau}{2} \Lambda_2 \right) \phi^{j-1/2},$$

$$\left(E + \frac{\tau}{2} \Lambda_2 \right) \phi^{j+1/2} = \left(E - \frac{\tau}{2} \Lambda_2 \right) (\phi^j + \tau f^j), \qquad (2.29)$$

$$\left(E + \frac{\tau}{2} \Lambda_1 \right) \phi^{j+1} = \left(E - \frac{\tau}{2} \Lambda_1 \right) \phi^{j+1/2}.$$

Note that the right-hand sides above can be computed in terms of components by the explicit schemes. Therefore, the system of equations in the component form we are dealing with can be represented by means of the three-point schemes similar to (1.39), the role of σ and ξ_k^j being now played by the parameter $\tau/2$ and the right-hand sides in (2.29).

The von Neumann conditions can be handled similarly.

9.3 The Wave Equation

The hyperbolic-type equations play an important role in applications, and the corresponding numerical methods are developed with a sufficient completeness. A distinguishing feature of the hyperbolic equations is the fact that the domains of their solutions are bounded by characteristic cones. Thus, the part of the domain $D \times D_t$ which lies outside such a cone has no influence on the solution. Another feature is that the hyperbolic problems admit the existence of solutions which are not smooth. This latter circumstance is of particular concern to numerical analysts.

Consider the following one-dimensional problem, describing small oscillations of a homogeneous string:

$$\frac{1}{c^2} \cdot \frac{\partial^2 \phi}{\partial t^2} = \frac{\partial^2 \phi}{\partial x^2} + f(x, t),$$

$$\phi(0, t) = a(t), \qquad \phi(1, t) = b(t), \qquad (3.1)$$

$$\phi(x, 0) = p(x), \qquad \frac{\partial \phi}{\partial t}(x, 0) = q(x),$$

where c designates the velocity of propagation of excitations along the string; $a(t)$, $b(t)$, $f(t, x)$, $p(x)$, and $q(x)$ are given functions.

The theory of string oscillations is fairly complete. Here we are concerned with the numerical aspects of the problem. For this purpose let us first transform (3.1) to a system of ordinary differential equations in the time variable, by making use of the difference in the x variable. The method described in Section 9.2 fits well for our present problem. Thus, we obtain the following approximation of (3.1):

$$\frac{1}{c^2} \frac{d^2\phi_k}{dt^2} + \frac{-\phi_{k-1} + 2\phi_k - \phi_{k+1}}{h^2} = g_k(t),$$

$$\phi_0 = 0, \qquad \phi_N = 0, \tag{3.2}$$

$$\phi_k = p_k, \qquad \frac{d\phi_k}{dt} = q_k \quad \text{for} \quad t = 0,$$

where, similarly as in (2.4), the function $g_k(t)$ is defined by (1.4).

As pointed out earlier, the solution $\phi_k(t)$ of (3.2) is only meaningful at the net points $k = 1, 2, \ldots, N - 1$. At the points $k = 0$ and $k = N$ the solution is defined by the boundary conditions from (3.1); the homogeneous conditions in (3.2) are the results of a special extension of the domain of definition of the solution, and do not approximate the function $\phi(x, t)$ at x_0 and x_N. Note that problem (3.2) has an accuracy of second order with respect to $h = \Delta x$.

In order to obtain next the finite-difference approximation of (3.2) with respect to the time variable, let us introduce a system of net points $t = t_i$ (where $t_{i+1} = t_i + \Delta t$), and consider the following two most useful schemes, the explicit "cross" scheme $(\cdot : \cdot)$

$$\frac{\phi_k^{j+1} - 2\phi_k^j + \phi_k^{j-1}}{\tau^2} = \frac{\phi_{k-1}^j - 2\phi_k^j + \phi_{k+1}^j}{h^2} + g_k^j,$$

$$\phi_0^j = 0, \qquad \phi_N^j = 0, \qquad \phi_k^0 = p_k, \tag{3.3}$$

$$\phi_k^1 = p_k + \Delta t q_k + \frac{\tau^2}{2} \left(\frac{p_{k-1} - 2p_k + p_{k+1}}{h^2} + q_k^0 \right),$$

where $\tau = c\Delta t$, and the implicit scheme $(: : :)$, which is analogous to the Crank–Nicholson scheme:

$$\frac{\phi_k^{j+1} - 2\phi_k^j + \phi_k^{j-1}}{\tau^2} = \frac{\phi_{k-1}^{j+1} - 2\phi_k^{j+1} + \phi_{k+1}^{j+1}}{2h^2}$$

$$+ \frac{\phi_{k-1}^{j-1} - 2\phi_k^{j-1} + \phi_{k+1}^{j-1}}{2h^2} + g_k^j,$$

$$\phi_0^{j+1} = 0, \qquad \phi_N^{j+1} = 0, \qquad \phi_k^0 = p_k, \tag{3.4}$$

$$\phi_k^1 = \phi_k^0 + \Delta t q_k + \frac{\tau^2}{\cdot 2} \left(\frac{p_{k-1} - 2p_k + p_{k+1}}{h^2} + q_k^0 \right).$$

It is not difficult to verify by a Taylor series expansion that for sufficiently smooth solutions $\phi(x, t)$ the difference schemes (3.3) and (3.4) approximate the original problem with second-order accuracy with respect to both h and τ.

Problems (3.3) and (3.4) will be solved using the method of Fourier series with respect to the eigenfunctions

$$u_n(k) = \sin n\pi kh, \qquad k, n = 1, 2, \ldots, N - 1,$$

as in (2.14). There result the following recursive relations for the Fourier coefficients: for the explicit "cross" scheme we have

$$\frac{\Phi_n^{j+1} - 2\Phi_n^j + \Phi_n^{j-1}}{\tau^2} + \frac{4}{h^2} \sin^2 \frac{n\pi h}{2} \Phi_n^j = G_n^j,$$

$$\Phi_n^0 = P_n, \ \Phi_n^1 = P_n + \Delta t Q_n + \frac{2\tau^2}{h^2} \sin^2 \frac{n\pi h}{2} P_n + \frac{\tau^2}{2} G_n, \tag{3.5}$$

where Φ_n, G_n, P_n, and Q_n are the Fourier coefficients of the elements ϕ_k, g_k, p_k, and q_k, respectively. For the implicit scheme ($\because$)

$$\frac{\Phi_n^{j+1} - 2\Phi_n^j + \Phi_n^{j-1}}{\tau^2} + \frac{4}{h^2} \sin^2 \frac{n\pi h}{2} \cdot \frac{\Phi_n^{j+1} + \Phi_n^{j-1}}{2} = G_n^j,$$

$$\Phi_n^0 = P_n, \ \Phi_n^1 = P_n + \Delta t Q_n + \frac{2\tau^2}{h^2} \sin^2 \frac{n\pi h}{2} P_n + \frac{\tau^2}{2} G_n^0. \tag{3.6}$$

Our next step is establishing a criterion for countable stability for the problems with smooth input data. For that let us investigate the behavior of the linearly independent solutions of the homogeneous problems (3.5) and (3.6), depending on j. The solutions will be sought in the form

$$\Phi_n^j = A_n \eta_n^j, \tag{3.7}$$

where A_n and η_n are constants (note that η_n^j in the above equation means the jth power of η_n). Putting $G_n^j = 0$ in (3.5) and (3.6) and substituting then (3.7) into these equations, we obtain the following expressions for η_n: in the case of the "cross" scheme ($\cdot \vdots \cdot$)

$$\eta_n^2 - 2(1 - \mu_n^2)\eta_n + 1 = 0, \tag{3.8}$$

and in the case of the implicit scheme ($\because$)

$$\eta_n^2 - \frac{2}{1 + \mu_n^2} \eta_n + 1 = 0, \tag{3.9}$$

where

$$\mu_n^2 = 2 \frac{\tau^2}{h^2} \sin^2 \frac{n\pi h}{2}.$$

Solving the quadratic equations (3.8) and (3.9) we obtain

$$\eta_n = 1 - \mu_n^2 \pm \sqrt{(1 - \mu_n^2)^2 - 1}; \tag{3.10}$$

and

$$\eta_n = \frac{1}{1 + \mu_n^2} \pm \sqrt{\left(\frac{1}{1 + \mu_n^2}\right)^2 - 1}. \tag{3.11}$$

Let us look first at the solution corresponding to the "cross" scheme. It is easy to see that if

$$\mu_n^2 = 2\frac{\tau^2}{h^2} \sin^2 \frac{n\pi h}{2} < 2, \tag{3.12}$$

then

$$|\eta_{n_i}| = 1, \qquad |\eta_{n_i} - \eta_{n_2}| > C\tau, \tag{3.13}$$

and hence the difference scheme is stable. It is necessary that condition (3.12) holds true for all $n = 1, 2, \ldots, N - 1$. The latter is clearly true if τ and h are related as written below:

$$\frac{\tau^2}{h^2} < \frac{1}{\sin^2 \dfrac{\pi h n}{2}} \tag{3.14}$$

or, more simply, if the two parameters satisfy the Courant condition:

$$\frac{\tau}{h} < 1. \tag{3.15}$$

It is not difficult to verify that in the case of the implicit scheme ($\because$) we have

$$|\eta_n| = 1 \tag{3.16}$$

for any $\tau > 0$ and any n. This means that the given scheme is absolutely stable (see Richtmyer and Morton [3]).

In the same fashion one may consider small oscillations of a membrane. In this case we obtain the equation

$$\frac{1}{c^2} \frac{\partial^2 \phi}{\partial t^2} = \frac{\partial^2 \phi}{\partial x^2} + \frac{\partial^2 \phi}{\partial y^2} + f(x, y, t) \quad \text{in} \quad D \times D_t, \tag{3.17}$$

with the boundary condition

$$\phi = g \quad \text{on} \quad \partial D \times D_t, \tag{3.18}$$

and the initial conditions

$$\phi = p, \qquad \frac{\partial \phi}{\partial t} = q \quad \text{in} \quad D \quad \text{for} \quad t = 0.$$

Assuming the input data sufficiently smooth, the solution of this problem in the square D can be found by the finite-difference method, in the same manner as in the case of the string. The method of Fourier series is similar to that we have used earlier for the two-dimensional difference analog of the heat equation. Rather than repeating this route again, we will now turn to a more general problem of the hyperbolic type.

9.4 The Equation of Motion

The equation of motion has become a subject of considerable interest among scientists, mainly because of its significant applications. For example, one of the important building elements of the hydrodynamical process—the propagation of particles along a trajectory—is governed by this equation.

The development of numerical methods for the equation of motion has been triggered by the needs of hydrodynamics and aerodynamics. In the last couple of decades the numerical methods in hydrodynamics have been enriched by a number of interesting ideas due to von Neumann and Richtmyer [7], Dorodnitsyn [3], Godunov [4], Lax [6, 7], Babenko and Rusanov [12], and many others.

Recent stimuli for the development of efficient numerical methods for the equation of motion are coming from the problems of weather forecasting and from ocean dynamics. Thanks to works by Kurihara and Holloway [4], Bryan [4], Marchuk [4], and others, there are now universal, efficient algorithms for solving such problems.

The propagation problem of a substance along the particle trajectories may be considered to belong among the simpler problems of mathematical physics. It can be described by the following equation:

$$\frac{d\phi}{dt} = 0,$$

where

$$\frac{d}{dt} = \frac{\partial}{\partial t} + u\frac{\partial}{\partial x} + v\frac{\partial}{\partial y} + w\frac{\partial}{\partial z}$$

and u, v, and w are the components of the velocity vector $ui + vj + wk$, so that

$$u = \frac{dx}{dt}, \qquad v = \frac{dy}{dt}, \qquad w = \frac{dz}{dt}.$$

The above equation is further restrained by imposing additional conditions, the simplest of which can be written as

$$\phi(x, y, z, 0) = f(x, y, z),$$

assuming the unbounded region.

A similar problem arises in the framework of the general algorithm for numerical solution of equations of hydrodynamics, the theory of radiation and many others. Because of this fact we will discuss thoroughly the possibilities of numerical treatment of such problems.

9.4.1 The Simplest Equations of Motion

When solving problems of hydrodynamics, hydrothermodynamics, weather forecasting, ocean dynamics, and others, we often have to deal with equations describing the transfer of a substance along trajectories. A simple equation of this kind is as follows:

$$\frac{\partial \phi}{\partial t} + u \frac{\partial \phi}{\partial x} = 0 \quad \text{in} \quad D \times D_t,$$

$$\phi = f(x) \quad \text{in} \quad D \quad \text{for} \quad t = 0, \tag{4.1}$$

where u is a given velocity and $f(x)$ represents the initial distribution of ϕ. The region D is taken as the whole real line, and $D_t = \{0 \le t \le T\}$. Assume that $\phi(x, t)$ and $f(x)$ are periodic in x, with the period 2π. If $u = \text{const}$, then the problem (4.1) has an immediate solution

$$\phi(x, t) = f(x - ut), \tag{4.2}$$

assuming f is differentiable. Solution (4.2) describes the propagation of the initial perturbation along the characteristic

$$x - ut = \text{const.}$$

This means that $\phi(x, t) = \text{const}$ on an arbitrary line $x - ut = \text{const}$.

Thus the problem (4.1) defines for $u > 0$ a process of propagation of the perturbation in the direction of the growing x. These well-known facts should be kept in mind when constructing difference analogs of problem (4.1). If the velocity $u = u(x, t)$ is changing, an analytic solution of (4.1) becomes a problem. In this case it is usually necessary to resort to numerical methods based on difference approximations.

Consider a simple difference scheme with $u = \text{const}$ and, to be specific, take $u > 0$. We then have the explicit scheme

$$\frac{\phi_k^{j+1} - \phi_k^j}{\tau} + u \frac{\phi_k^j - \phi_{k-1}^j}{\Delta x} = 0, \tag{4.3}$$

and the implicit scheme

$$\frac{\phi_k^{j+1} - \phi_k^j}{\tau} + u \frac{\phi_k^{j+1} - \phi_{k-1}^{j+1}}{\Delta x} = 0 \tag{4.4}$$

Both schemes are of first-order accuracy in Δx and τ. Indeed, assuming that the solution $\phi(x, t)$ and the initial distribution $f(x)$ are sufficiently smooth, we can expand into the Taylor series in the vicinity of $x = x_k, t = t_j$:

$$\phi(x, t) = (\phi)_k^j + (\phi_t)_k^j(t - t_j) + (\phi_x)_k^j(x - x_k) + \cdots. \tag{4.5}$$

Substituting (4.5) into (4.3) we obtain

$$\frac{\partial \phi}{\partial t} + u \frac{\partial \phi}{\partial x} = \frac{u \Delta x}{2} \cdot \frac{\partial^2 \phi}{\partial x^2} - \frac{\tau}{2} \cdot \frac{\partial^2 \phi}{\partial t^2}. \tag{4.6}$$

In (4.6) we have dropped the higher-order terms. From (4.1) we have

$$\frac{\partial^2 \phi}{\partial t^2} = u^2 \frac{\partial^2 \phi}{\partial x^2}. \tag{4.7}$$

Then expression (4.6) becomes

$$\frac{\partial \phi}{\partial t} + u \frac{\partial \phi}{\partial x} = \frac{u \Delta x - \tau u^2}{2} \cdot \frac{\partial^2 \phi}{\partial x^2} \quad \text{for} \quad x = x_k, t = t_j. \tag{4.8}$$

The above procedure has been suggested by Zhukov.† Some analysis of (4.8) shows, that for

$$\frac{u\tau}{\Delta x} < 1$$

Relation (4.8) can be treated as the heat equation with the domain $x_{k-1} \le x \le x_k, t_j \le t \le t_{j+1}$. Assuming that the terms we have dropped in (4.8) are small, we obtain the equation

$$\frac{\partial \phi}{\partial t} + u \frac{\partial \phi}{\partial x} = \mu \frac{\partial^2 \phi}{\partial x^2},$$

where

$$\mu = \frac{u \Delta x - u^2 \tau}{2}$$

is a so-called *coefficient of artificial* or *"countable" viscosity*. Note, by the way, that if

$$\frac{u\tau}{\Delta x} = 1,$$

then $\mu = 0$, so that all the terms dropped are zero, and the explicit scheme (4.3) has an infinite order of approximation with respect to Δx and τ.

In particular, it is to be noted that if

$$\frac{u\tau}{\Delta x} > 1,$$

we get the equation

$$\frac{\partial \phi}{\partial t} + u \frac{\partial \phi}{\partial x} = -|\mu| \frac{\partial^2 \phi}{\partial x^2}. \tag{4.9}$$

It is easy to see, that for the initial condition

$$\phi = \phi^0(x) \quad \text{for} \quad t = 0 \tag{4.10}$$

† See the monograph by Rozhdestvenskii and Yanenko [2]. This type of analysis was further developed in a number of papers by Yanenko and Shokin [7].

Equation (4.9) results in an ill-posed problem (in the sense of Hadamard). The solution of this problem is unstable with respect to small variations in the initial data. Thus any difference equations for problems like (4.1) must be formed so as to satisfy the well-posedness condition

$$\mu = \frac{u\tau}{\Delta x} \leq 1.$$

Next let us investigate the problem of countable stability of scheme (4.3). To this end consider first the spectral problem

$$(A^h \omega)_k \equiv u \frac{\omega_k - \omega_{k-1}}{\Delta x} = \lambda \omega_k \tag{4.11}$$

on the infinite net interval $D_h = (-\infty < x_k < \infty)$. The solution of (4.11), which is bounded in D_h, is of the form

$$\omega_k = e^{ikp\Delta x}, \tag{4.12}$$

where p is an arbitrary integer. Substituting (4.12) into (4.11), we obtain the following eigenvalue:

$$\lambda_p = \frac{u}{\Delta x} \left(2 \sin^2 \frac{p\Delta x}{2} + i \sin p\Delta x \right). \tag{4.13}$$

Let us write (4.3) in the operator form

$$\frac{\phi^{j+1} - \phi^j}{\tau} + A^h \phi^j = 0. \tag{4.14}$$

The solution of equation (4.14) will be sought in the form

$$\phi^j = \sum_{p=-\infty}^{\infty} \phi_p^j e^{ikp\Delta x}, \tag{4.15}$$

where ϕ_p^j is the Fourier coefficient of ϕ^j. We have

$$\frac{\phi_p^{j+1} - \phi_p^j}{\tau} + \lambda_p \phi_p^j = 0.$$

Hence

$$\phi_p^{j+1} = T_p \phi_p^j,$$

where $T_p = 1 - \tau\lambda_p$ is the *transition factor* of the Fourier coefficients.

Let us find a condition under which the Fourier coefficients do not grow in modulus. For that we may take

$$|1 - \tau\lambda_p| \leq 1. \tag{4.16}$$

This inequality holds if

$$\frac{u\tau}{\Delta x} \leq 1.$$

Indeed,

$$
\begin{aligned}
|1 - \tau\lambda_p|^2 &= \left(1 - \frac{2u\tau}{\Delta x} \sin^2 \frac{p\Delta x}{2}\right)^2 + \left(\frac{u\tau}{\Delta x}\right)^2 \sin^2 p\Delta x \\
&= 1 - 4\frac{u\tau}{\Delta x} \sin^2 \frac{p\Delta x}{2} \left(\frac{u\tau}{\Delta x}\right)^2 \left(4 \sin^4 \frac{p\Delta x}{2} + \sin^2 p\Delta x\right) \\
&= 1 - 4\frac{u\tau}{\Delta x} \sin^2 \frac{p\Delta x}{2} + 4\left(\frac{u\tau}{\Delta x}\right)^2 \sin^2 \frac{p\Delta x}{2} \left(\sin^2 \frac{p\Delta x}{2} + \cos \frac{p\Delta x}{2}\right) \\
&= 1 - 4 \sin^2 \frac{p\Delta x}{2} \left(\frac{u\tau}{\Delta x}\right)\left(1 - \frac{u\tau}{\Delta x}\right) \geq 0.
\end{aligned} \tag{4.17}
$$

The inequality (4.16) is an immediate consequence.

In this manner we arrive at the condition for countable stability of the difference scheme. It is easy to see that in the given case the stability criterion we have established coincides with the condition of well-posedness (4.8).

Let us next discuss the implicit difference scheme (4.4). Using the above method, one can again show that (4.4) is of first-order accuracy in Δx and τ. With the help of the Taylor expansion around $x = x_k$ and $t = t_j$ we get the "asymptotic equation"

$$
\frac{\partial \phi}{\partial t} + u \frac{\partial \phi}{\partial x} = \mu \frac{\partial^2 \phi}{\partial x^2}, \tag{4.18}
$$

where

$$
\mu = \frac{u\Delta x + u^2\tau}{2}.
$$

Already here we can observe the fundamental difference between (4.8) and (4.18). In the latter equation the coefficient of countable viscosity is always positive. Therefore, equation (4.18) with corresponding sufficiently smooth initial data is always well-posed. It is not difficult to show that (4.4) is stable for any ratio of the steps, i.e., it is absolutely stable, since the transition factor of every Fourier coefficient is equal to

$$
T_p = \frac{1}{1 + \tau\lambda_p}.
$$

Hence it follows that

$$
|T_p| \leq 1.
$$

In addition to (4.3) and (4.4) there are yet some other interesting and very convenient difference schemes, for instance:

$$
\frac{\phi_k^{j+1} - \phi_k^j}{\tau} + u \frac{\phi_{k+1}^{j+1} - \phi_{k-1}^{j+1}}{2\Delta x} = 0 \tag{4.19}
$$

or

$$\frac{\phi_k^{j+1} - \phi_k^j}{\tau} + u \frac{\phi_{k+1}^{j+1/2} - \phi_{k-1}^{j+1/2}}{2\Delta x} = 0, \qquad (4.20)$$

where

$$\phi_k^{j+1/2} = \tfrac{1}{2}(\phi_k^{j+1} + \phi_k^j).$$

It is not difficult to show that scheme (4.19) is a first-order approximation in τ, and second order in Δx. The differential equation corresponding to this scheme will have the form of (4.18), where

$$\mu = u^2\tau/2.$$

The stability is determined by the transition operators for the Fourier coefficients. We obtain the spectral problem

$$(A^h\omega)_k \equiv u \frac{\omega_{k+1} - \omega_{k-1}}{2\Delta x} = \lambda\omega_k \qquad (4.21)$$

and we seek its solution in the form of (4.12). There results

$$\lambda_p = i \frac{u}{\Delta x} \sin p\Delta x. \qquad (4.22)$$

From (4.19) it is immediate that the Fourier coefficients must satisfy

$$\frac{\phi_p^{j+1} - \phi_p^j}{\tau} + \lambda_p \phi_p^{j+1} = 0$$

or

$$\phi_0^{j+1} = T_p \phi^j, \qquad (4.23)$$

where

$$T_p = \frac{1}{1 + \tau\lambda_p}.$$

Taking into account (4.22), we get

$$T_p = \frac{1}{1 + i \dfrac{u\tau}{\Delta x} \sin p\Delta x}$$

and consequently

$$|T_p| = \frac{1}{\sqrt{1 + \left(\dfrac{u\tau}{\Delta x}\right)^2 \sin^2 p\Delta x}} \le 1.$$

The absolute countable stability of (4.19) follows from this.

The most interesting scheme in applications is that of Crank–Nicholson (4.20). It is not difficult to verify that this scheme is a second-order approximation in τ and Δx and that it is not dissipative. This means that we have $\mu = 0$ in the differential equation (4.18), and that the terms we have dropped from consideration have the order of $\tau\Delta x$, τ^2, and Δx^2. As far as countable stability is concerned, we have

$$T_p = \frac{1 - i\,\dfrac{u\tau}{2\Delta x}\,\sin p\Delta x}{1 + i\,\dfrac{u\tau}{2\Delta x}\,\sin p\Delta x}.$$

Hence

$$|T_p| = 1,$$

and the scheme is absolutely stable. Let us remark that if, in (4.3), we replace the form

$$u\,\frac{\phi_k^j - \phi_{k-1}^j}{\Delta x}$$

of the difference expression for $u(\partial\phi/\partial x)$ by the form

$$u\,\frac{\phi_{k+1}^j - \phi_k^j}{\Delta x},$$

then the resulting difference scheme (with $u > 0$) becomes unstable, regardless of the relation between the steps. The proof is trivial.

In conclusion, let us consider yet another interesting numerical method for problem (4.1), based on the so called "running-count" scheme. The scheme has been introduced by Landau, Meiman, and Khalatnikov [4], and has the following form:

$$\frac{\phi_k^{j+1} - \phi_k^j}{\tau} + \frac{u}{\Delta x}\left[\left(\frac{\phi_k^{j+1} + \phi_k^j}{2}\right) - \left(\frac{\phi_{k-1}^{j+1} + \phi_{k-1}^j}{2}\right)\right] = 0. \quad (4.24)$$

As can be easily shown, the scheme is a second-order approximation in τ and first order in x. It can be implemented by the following recursive realization:

$$\phi_k^{j+1} = \frac{1 - \dfrac{u\tau}{2\Delta x}}{1 + \dfrac{u\tau}{2\Delta x}}\,\phi_k^j + \frac{\dfrac{u\tau}{2\Delta x}}{1 + \dfrac{u\tau}{2\Delta x}}\,(\phi_{k-1}^{j+1} + \phi_{k-1}^j). \quad (4.25)$$

Following the stability analysis of von Neumann and using the Fourier method it is not difficult to prove that (4.24) is absolutely stable.

In a similar fashion one can construct a "running-count" scheme for the multi-dimensional problem of motion and show its stability under the assumption that the equation has constant coefficients.

In the above we have been assuming throughout that u was a positive constant. If u is negative, we can still replace x by $(-x)$, and obtain equation (4.1). In applications, however, we are particularly interested in the case where $u = u(x, t)$. It can be shown readily by a simple analysis that in this case we may be losing countable stability, even when using implicit dissipative schemes. This is true, in particular, of nonlinear problems. The essence of the matter is as follows: if we expand the solution of the difference scheme and the coefficient $u(x_k, t_j)$ into the Fourier series, the products of the Fourier series will yield both slower and faster harmonic components than those interacting.

As a result of such a process it may happen, that the "energy" due to the round-off errors will be transferred from the low frequencies to the highest frequencies, and the computational process will thus become unstable, despite the fact that the given difference scheme with constant coefficients is countably stable. This kind of instability is usually labeled as *nonlinear*. Sometimes it also appears when dealing with linear problems with variable coefficients. Thus, difference schemes for nonlinear equations, or equations with variable coefficients, which would be stable with respect to arbitrary excitations, are the focal points of current development. In most cases countable instability can be negotiated with the help of dissipative difference schemes, which correspond to a certain choice of the coefficient of countable viscosity μ. Such schemes, however, turn out to be first-order approximations in τ or x or both.

Of particular interest in applications are equations of the form

$$\frac{\partial \phi}{\partial t} + \frac{\partial u \phi}{\partial x} = 0, \tag{4.26}$$

where $u = u(x, t)$.

Below, when discussing the multi-dimensional equations of the type of (4.26), we will show how to obtain absolutely stable, first- (or even second-) order difference schemes for equations of such type (on certain classes of the coefficients).

9.4.2. The Two-Dimensional Equation of Motion with Variable Coefficients

Consider an ensemble of particles moving in the plane (x, y) along given trajectories. In the framework of fluid mechanics the problem can be formulated as follows:

$$\frac{\partial \phi}{\partial t} + u \frac{\partial \phi}{\partial x} + v \frac{\partial \phi}{\partial y} = 0 \quad \text{in} \quad D \times D_t, \tag{4.27}$$

$$\phi(x, y, 0) = g \quad \text{in} \quad D$$

Here $u = u(x, y, t)$ and $v = v(x, y, t)$.

Assume that the components u, v of the velocity vector satisfy at each time instant the continuity condition

$$\frac{\partial u}{\partial x} + \frac{\partial v}{\partial y} = 0. \tag{4.28}$$

Let D be the rectangle $\{0 \leq x \leq a, 0 \leq y \leq b\}$, and suppose that the solution of the problem, as well as the velocity components u and v, are periodic, with identical values on the opposite sides of the rectangle.

Let us rewrite the evolution equation (4.27) in the operator form:

$$\frac{\partial \phi}{\partial t} + A\phi = 0, \tag{4.29}$$

where

$$\phi = g \quad \text{for} \quad t = 0,$$

$$A = u\frac{\partial}{\partial x} + v\frac{\partial}{\partial y}.$$

It is not difficult to show, that A satisfies $(A\phi, \phi) = 0$ in the present case. Indeed, introducing the inner product as usual, we have

$$(A\phi, \phi) = \int_0^a dx \int_0^b dy \left(u\frac{\partial \phi}{\partial x} + v\frac{\partial \phi}{\partial y}\right)\phi. \tag{4.30}$$

Using (4.28), the integrand can be written as

$$\left(u\frac{\partial \phi}{\partial x} + v\frac{\partial \phi}{\partial y}\right)\phi = \frac{\partial u \frac{\phi^2}{2}}{\partial x} + \frac{\partial v \frac{\phi^2}{2}}{\partial y}.$$

Hence

$$(A\phi, \phi) = \int_0^a dx \int_0^b dy \left(\frac{\partial u \frac{\phi^2}{2}}{\partial x} + \frac{\partial v \frac{\phi^2}{2}}{\partial y}\right). \tag{4.31}$$

By the periodicity of the solution on the boundary it now follows that

$$(A\phi, \phi) = 0. \tag{4.32}$$

Let us next attempt to split up the operator A in such a way that each of the elementary operators A_α $(\alpha = 1, 2)$ also satisfy the above condition:

$$(A_\alpha\phi, \phi) = 0. \tag{4.33}$$

The component-by-component splitting-up difference scheme allows us in this case to obtain an absolutely stable difference scheme of second-order accuracy.

The formal decomposition of the operator A into the factors

$$A_1 = u \frac{\partial}{\partial x}, \qquad A_2 = v \frac{\partial}{\partial y} \qquad (4.34)$$

does not satisfy condition (4.33). It is not difficult to verify that

$$(A_1 \phi, \phi) = -\frac{1}{2} \int_0^a dx \int_0^b \phi^2 \frac{\partial u}{\partial x} dy, \qquad (A_2 \phi, \phi) = -\frac{1}{2} \int_0^a dx \int_0^b \phi^2 \frac{\partial v}{\partial y} dy.$$

and therefore the operators A_1 and A_2 can not be taken as elementary operators for the construction of the sequential splitting-up scheme.

Let us take the operators A_1 and A_2 in the following, more complex form:

$$A_1 \phi = u \frac{\partial \phi}{\partial x} + \frac{\phi}{2} \frac{\partial u}{\partial x}, \qquad A_2 \phi = v \frac{\partial \phi}{\partial y} + \frac{\phi}{2} \frac{\partial v}{\partial y}. \qquad (4.35)$$

It is not difficult to see that we now have (4.33) holding for the above operators and that

$$(A_1 + A_2)\phi = u \frac{\partial \phi}{\partial x} + v \frac{\partial \phi}{\partial y} + \frac{\phi}{2} \left(\frac{\partial u}{\partial x} + \frac{\partial v}{\partial y} \right)$$

$$= u \frac{\partial \phi}{\partial x} + v \frac{\partial \phi}{\partial y} = A\phi,$$

i.e., $A_1 + A_2 = A$. Here we have used the fact that the coefficients u and v satisfy equation (4.28).

Hence we have satisfied all the assumptions for the applicability of the splitting-up method and can thus write

$$\frac{\phi^{j-1/2} - \phi^{j-1}}{\tau} + \left(u^j \frac{\partial}{\partial x} + \frac{1}{2} \cdot \frac{\partial u^j}{\partial x} \right) \frac{\phi^{j-1/2} + \phi^{j-1}}{2} = 0,$$

$$\frac{\phi^j - \phi^{j-1/2}}{\tau} + \left(v^j \frac{\partial}{\partial y} + \frac{1}{2} \cdot \frac{\partial v^j}{\partial y} \right) \frac{\phi^j + \phi^{j-1/2}}{2} = 0,$$

$$\frac{\phi^{j+1/2} - \phi^j}{\tau} + \left(v^j \frac{\partial}{\partial y} + \frac{1}{2} \cdot \frac{\partial v^j}{\partial y} \right) \frac{\phi^{j+1/2} + \phi^j}{2} = 0,$$

$$\frac{\phi^{j+1} - \phi^{j+1/2}}{\tau} + \left(u^j \frac{\partial}{\partial x} + \frac{1}{2} \cdot \frac{\partial u^j}{\partial x} \right) \frac{\phi^{j+1} + \phi^{j+1/2}}{2} = 0,$$

$$(4.36)$$

where $t_{j-1} \le t \le t_{j+1}$.

If the functions u, v and the solution ϕ are sufficiently smooth in all their arguments, then (4.36) is a second-order approximation and is absolutely stable in the sense that

$$\|\phi^{j+1}\| = \|\phi^{j-1}\| = \cdots = \|g\|. \qquad (4.37)$$

From this example we may see how the formal splitting into the operators (4.34) may compromise the very idea of splitting; only by introducing additional considerations can we arrive at theoretically justifiable and practically efficient schemes.

With these preliminaries as an introduction, let us now turn to constructing the difference scheme for problem (4.27). To begin with, let us consider feasible ways of approximating the operator A in the space variables x and y. As pointed out in Chapter 2, a convenient approximation of problems of mathematical physics, which conserves the additivity properties and qualitative features of the operator, can be achieved by the coordinate-by-coordinate approximation method.

Suppose that the coefficients u and v are sufficiently smooth, and consider equation (4.27) in the form

$$\frac{\partial \phi}{\partial t} + \frac{\partial u\phi}{\partial x} + \frac{\partial v\phi}{\partial y} = 0 \quad \text{in} \quad D \times D_t,$$

$$\phi = g \quad \text{in} \quad D \quad \text{for} \quad t = 0. \tag{4.38}$$

Consider the operator A defined by the expression

$$A\phi = \frac{\partial u\phi}{\partial x} + \frac{\partial v\phi}{\partial y} - \frac{\phi}{2}\left(\frac{\partial u}{\partial x} + \frac{\partial v}{\partial y}\right). \tag{4.39}$$

Write its difference analog as follows:

$$(A^h\phi)_{k,l} = \frac{u_{k+1,l}\phi_{k+1,l} - u_{k-1,l}\phi_{k-1,l}}{2\Delta x} + \frac{v_{k,l+1}\phi_{k,l+1} - v_{k,l-1}\phi_{k,l-1}}{2\Delta y}$$

$$- \frac{\phi_{k,l}}{2}\left(\frac{u_{k+1,l} - u_{k-1,l}}{2\Delta x} + \frac{v_{k,l+1} - v_{k,l-1}}{2\Delta y}\right). \tag{4.40}$$

Clearly, difference expression (4.40) approximates (4.39) with second-order accuracy in Δx and Δy, provided the function u, v, and ϕ are smooth enough. Equation (4.40) has a serious drawback, though, since the form of the operator A^h disturbs the antisymmetric structure, i.e.,

$$(A^h\phi, \phi) \neq 0. \tag{4.41}$$

This means that our customary approximation becomes unsatisfactory for the purposes of constructing the computational algorithm for (4.27).

We now show that the approximation of (4.39) in the form

$$(A^h\phi)_{k,l} = \frac{u_{k+1/2,l}\phi_{k+1,l} - u_{k-1/2,l}\phi_{k-1,l}}{2\Delta x}$$

$$+ \frac{v_{k,l+1/2}\phi_{k,l+1} - v_{k,l-1/2}\phi_{k,l-1}}{2\Delta y} \tag{4.42}$$

satisfies the fundamental relation

$$(A^h\phi, \phi) = 0, \tag{4.43}$$

and approximates expression (4.38) with second-order accuracy in Δx and Δy. Let us exploit the following approximation of the coefficients:

$$u_{k+1/2,\,l} = u_{k+1,\,l} - \frac{u_{k+1,\,l} - u_{k,\,l}}{2};\qquad v_{k,\,l+1/2} = v_{k,\,l+1} - \frac{v_{k,\,l+1} - v_{k,\,l}}{2};$$

$$u_{k-1/2,\,l} = u_{k-1,\,l} + \frac{u_{k,\,l} - u_{k-1,\,l}}{2};\qquad v_{k,\,l-1/2} = v_{k,\,l-1} + \frac{v_{k,\,l} - v_{k,\,l-1}}{2}.$$

$$(4.44)$$

The substitution of these expressions into (4.42) followed by simple manipulations give

$$(A^h\phi)_{k,\,l} = \frac{u_{k+1,\,l}\phi_{k+1,\,l} - u_{k-1,\,l}\phi_{k-1,\,l}}{2\Delta x} + \frac{v_{k,\,l+1}\phi_{k,\,l+1} - v_{k,\,l-1}\phi_{k,\,l-1}}{2\Delta y}$$

$$- \frac{\phi_{k,\,l}}{2}\left(\frac{u_{k+1,\,l} - u_{k-1,\,l}}{2\Delta x} + \frac{v_{k,\,l+1} - v_{k,\,l-1}}{2\Delta y}\right)$$

$$- (\Delta x^2 R_{k,\,l} + \Delta y^2 Q_{k,\,l}),$$

$$(4.45)$$

where the quantities $R_{k,\,l}$ and $Q_{k,\,l}$ approach the limits

$$R_{k,\,l} \to \frac{1}{4}\cdot\frac{\partial}{\partial x}\left(\frac{\partial u}{\partial x}\cdot\frac{\partial\phi}{\partial x}\right),\qquad Q_{k,\,l} \to \frac{1}{4}\cdot\frac{\partial}{\partial y}\left(\frac{\partial v}{\partial y}\cdot\frac{\partial\phi}{\partial y}\right)$$

as $\Delta x \to 0$ and $\Delta y \to 0$.

Assume now, that the coefficients $u_{k,\,l}$, $v_{k,\,l}$ satisfy the following relation:

$$\frac{u_{k+1,\,l} - u_{k-1,\,l}}{2\Delta x} + \frac{v_{k,\,l+1} - v_{k,\,l-1}}{2\Delta y} = O(h^2). \qquad (4.46)$$

If the coefficients u, v and the solution ϕ have bounded third derivatives, with respect to x and y, it can be seen that (4.45) [with (4.46)] differs from (4.40) by the same second-order term as does (4.40) from (4.39). Hence (4.42) approximates (4.39) with second-order accuracy relative to Δx and Δy.

We will show that the operator A^h we have constructed satisfies condition (4.43), and, moreover, that each of the operators A_1^h, A_2^h defined by

$$A_1^h\phi = \frac{u_{k+1/2,\,l}\phi_{k+1,\,l} - u_{k-1/2,\,l}\phi_{k-1,\,l}}{2\Delta x},$$

$$A_2^h\phi = \frac{v_{k,\,l+1/2}\phi_{k,\,l+1} - v_{k,\,l-1/2}\phi_{k,\,l-1}}{2\Delta y},$$

$$(4.47)$$

also satisfy the corresponding equation

$$(A_\alpha^h\phi,\ \phi) = 0. \qquad (4.48)$$

To this end, we introduce the inner product

$$(a, b) = \sum_k \sum_l a_{k,l} b_{k,l} \Delta x \Delta y,$$

for vector quantities a, b; with its help we may write

$$(A_1^h \phi, \phi) = \tfrac{1}{2} \sum_k \sum_l \Delta y \left(\sum_k u_{k+1/2,l} \phi_{k+1,l} - \sum_k u_{k-1/2,l} \phi_{k-1,l} \right) \phi_{k,l}, \tag{4.49}$$

$$(A_2^h \phi, \phi) = \tfrac{1}{2} \sum_l \sum_k \Delta x \left(\sum_l v_{k,l+1/2} \phi_{k,l+1} - \sum_l v_{k,l-1/2} \phi_{k,l-1} \right) \phi_{k,l}.$$

After a little arrangement we obtain the equations (4.48). Equation (4.43) follows immediately from (4.48). We have thus constructed the necessary spatial approximations. Our next task is the time reduction of the system of ordinary differential equations

$$\frac{d\phi^h}{dt} + A^h \phi^h = 0 \quad \text{in} \quad D_h \times D_t,$$

$$\phi^h = g^h \quad \text{in} \quad D_h \quad \text{for} \quad t = 0, \tag{4.50}$$

where

$$A^h = A_1^h + A_2^h,$$

ϕ^h is a vector function with the components $\phi_{k,l}$ and A_α^h satisfies condition (4.48). This means that the problem (4.50) can be solved using the splitting-up method. Dropping the unimportant index h, we obtain the following system on the interval $t_{j-1} \leq t \leq t_{j+1}$:

$$\frac{\phi^{j-1/2} - \phi^{j-1}}{\tau} + A_1^j \frac{\phi^{j-1/2} + \phi^{j-1}}{2} = 0,$$

$$\frac{\phi^j - \phi^{j-1/2}}{\tau} + A_2^i \frac{\phi^j + \phi^{j-1/2}}{2} = 0,$$

$$\frac{\phi^{j+1/2} - \phi^j}{\tau} + A_2^j \frac{\phi^{j+1/2} + \phi^j}{2} = 0, \tag{4.51}$$

$$\frac{\phi^{j+1} - \phi^{j+1/2}}{\tau} + A_1^j \frac{\phi^{j+1} + \phi^{j+1/2}}{2} = 0.$$

The problem (4.27) has been reduced to a system of simple one-dimensional equations, which can be solved by the factorization method for three-point difference equations. The results of numerical experiments on the schemes described in this section will be found in a paper by Marchuk, Rivin, and Yudin [4].

9.4.3 The Multi-Dimensional Equation of Motion

Consider now the multi-dimensional equation, describing the hydrody-namical motion of a substance along a trajectory,

$$\frac{\partial \Phi}{\partial t} + \sum_{\alpha=1}^{n} v_\alpha \frac{\partial \Phi}{\partial x_\alpha} = 0 \quad \text{in} \quad D \times D_t,$$

$$\Phi(x, 0) = f(x) \quad \text{in} \quad D. \tag{4.52}$$

Suppose that the coefficients in (4.52) satisfy the continuity conditions

$$\sum_{\alpha=1}^{n} \frac{\partial v_\alpha}{\partial x_\alpha} = 0 \quad \text{in} \quad D \times D_t. \tag{4.53}$$

Problems (4.52) and (4.53) can be rewritten in the divergence form

$$\frac{\partial \Phi}{\partial t} + \sum_{\alpha=1}^{n} \frac{\partial v_\alpha \Phi}{\partial x_\alpha} = 0 \quad \text{in} \quad D \times D_t,$$

$$\Phi(x, 0) = f(x) \quad \text{in} \quad D. \tag{4.54}$$

The form (4.54) is of primary importance when using splitting-up methods.

To begin with, let us consider the difference equation obtained for (4.54) by using the Crank–Nicholson scheme:

$$\frac{\Phi^{j+1} - \Phi^j}{\tau} + \sum_{\alpha=1}^{n} \frac{\partial u_\alpha^j \Phi^{j+1/2}}{\partial x_\alpha} = 0, \qquad t_j \le t \le t_{j+1}, \tag{4.55}$$

where

$$\Phi^{j+1/2} = \tfrac{1}{2}(\Phi^{j+1} + \Phi^j). \tag{4.56}$$

Also, in (4.55) we have used certain approximations of the coefficients v_α. It is natural to choose either first- or second-order accuracy with respect to τ. For instance, in order to obtain the first order we may take

$$u_\alpha^j = v_\alpha(x, t_j),$$

while the second order is obtained by the choice

$$u_\alpha^j = \frac{v_\alpha(x, t_{j+1}) + v_\alpha(x, t_j)}{2}.$$

Denote

$$A^j \Phi = \sum_{\alpha=1}^{n} A_\alpha^j \Phi, \qquad A_\alpha^j \Phi = \frac{\partial u_\alpha^j \Phi}{\partial x_\alpha}.$$

Then (4.55) can be written in the form

$$\frac{\Phi^{j+1} - \Phi^j}{\tau} + A^j \Phi^{j+1/2} = 0, \qquad \Phi^{j+1/2} = \tfrac{1}{2}(\Phi^{j+1} + \Phi^j)$$

or,

$$\left(E + \frac{\tau}{2} A^j\right)\Phi^{j+1} = \left(E - \frac{\tau}{2} A^j\right)\Phi^j. \tag{4.57}$$

Suppose (for the sake of simplicity) that the solution of the problem is periodic relative to the n-dimensional rectangle D. Define the inner product

$$(a, b) = \int_D ab \, dD.$$

Then it can be easily verified that

$$(A^j\Phi, \Phi) = 0. \tag{4.58}$$

Solve for Φ^{j+1} in equation (4.57):

$$\Phi^{j+1} = \left(E + \frac{\tau}{2} A^j\right)^{-1}\left(E - \frac{\tau}{2} A^j\right)\Phi^j.$$

With the help of the Kellogg lemma and (4.58) we find

$$\|\Phi^{j+1}\| = \|\Phi^j\|. \tag{4.59}$$

Consider next the approximation of (4.55) in the spatial coordinates. To this end, project the region D on D_h and write (we drop from the symbols u, Φ the indices which do not change)

$$\frac{\Phi^{j+1} - \Phi^j}{\tau} + \Lambda^j\Phi^{j+1/2} = 0, \tag{4.60}$$

where

$$\Lambda^j = \sum_{\alpha=1}^{n} \Lambda_\alpha^j, \tag{4.61}$$

$$\Lambda_\alpha^j\Phi = \frac{1}{2\Delta x_\alpha}(u_{\alpha, k_\alpha + 1/2}^j \Phi_{k_\alpha + 1} - u_{\alpha, k_\alpha - 1/2}^j \Phi_{k_\alpha - 1}), \tag{4.62}$$

and the index k_α corresponds to the net points of the variable x_α. Introduce the inner product

$$(a, b) = \sum_{k_1 \cdots k_n} a_{k_1 k_2 \cdots k_n} b_{k_1 k_2 \cdots k_n} \Delta x_1 \Delta x_2 \cdots \Delta x_n \tag{4.63}$$

It is not difficult to verify that

$$(\Lambda^j\Phi, \Phi) = 0. \tag{4.64}$$

Consequently, using (4.64), we have

$$\|\Phi^{j+1}\| = \|\Phi^j\| = \cdots = \|f\|,$$

where Φ^j solves (4.60). Thus, exploiting the approximation of A^j by Λ^j from (4.60)–(4.62), we have again obtained an absolutely stable scheme.

Let us now investigate the approximation of A^j by Λ^j. In this connection consider again the elementary operators A^j_α and Λ^j_α. Below, the coefficients in $\Lambda^j_\alpha \Phi$ will be taken as follows:

$$u^j_{\alpha, k_\alpha + 1/2} = u^j_{\alpha, k_\alpha + 1} - \tfrac{1}{2}(u^j_{\alpha, k_\alpha + 1} - u^j_{\alpha, k_\alpha}),$$
$$u^i_{\alpha, k_\alpha - 1/2} = u^j_{\alpha, k_\alpha - 1} + \tfrac{1}{2}(u^j_{\alpha, k_\alpha} - u^j_{\alpha, k_\alpha - 1}). \tag{4.65}$$

Then we have

$$\Lambda^j_\alpha \Phi = \frac{u^j_{\alpha, k_\alpha + 1}\Phi_{\alpha, k_\alpha + 1} - u^j_{\alpha, k_\alpha - 1}\Phi_{k_\alpha - 1}}{2\Delta x_\alpha} - \frac{\Phi_{k_\alpha}}{2}\frac{u^j_{\alpha, k_\alpha + 1} - u^j_{\alpha, k_\alpha - 1}}{2\Delta x_\alpha} - \frac{\Delta x^2_\alpha}{4}R^j_\alpha, \tag{4.66}$$

where

$$R^j_\alpha = \frac{1}{\Delta x^3_\alpha}[(u^j_{\alpha, k_\alpha + 1} - u^j_{\alpha, k_\alpha})(\Phi_{k_\alpha + 1} - \Phi_{k_\alpha}) - (u^j_{\alpha, k_\alpha} - u^j_{\alpha, k_\alpha - 1})(\Phi_{k_\alpha} - \Phi_{k_\alpha - 1})]. \tag{4.67}$$

Recalling the assumption on the smoothness of the solution and the coefficients $u(x, t)$, we have that

$$R^j_\alpha \to \frac{\partial}{\partial x_\alpha}\left(\frac{\partial u^j}{\partial x_\alpha} \cdot \frac{\partial \Phi}{\partial x_\alpha}\right)$$

as $\Delta x_\alpha \to 0$. It follows from (4.66) that, generally speaking, the operator Λ^j_α does not approximate the operator A^j_α.

Consider the full operator Λ^j and investigate the expression $\Lambda^j \Phi$. We have

$$\Lambda^j \Phi = \sum_{\alpha = 1}^n \frac{u^j_{\alpha, k_\alpha + 1}\Phi_{k_\alpha + 1} - u^j_{\alpha, k_\alpha - 1}\Phi_{k_\alpha - 1}}{2\Delta x_\alpha}$$
$$- \sum_{\alpha = 1}^n \frac{\Phi_{k_\alpha}}{2} \cdot \frac{u^j_{\alpha, k_\alpha + 1} - u^j_{\alpha, k_\alpha - 1}}{2\Delta x_\alpha} - \frac{1}{4}\sum_{\alpha = 1}^n \Delta x^2_\alpha R^j_\alpha. \tag{4.68}$$

Assume that this difference version of the continuity equation is such that

$$\sum_{\alpha = 1}^n \frac{u_{\alpha, k_\alpha + 1} - u_{\alpha, k_\alpha - 1}}{2\Delta x_\alpha} = O(h^2) \tag{4.69}$$

where

$$h = \max_\alpha \{\Delta x_\alpha\}.$$

In this case the second sum in (4.68) becomes zero, and we obtain

$$\Lambda^j \Phi = \sum_{\alpha = 1}^n \frac{u^j_{\alpha, k_\alpha + 1}\Phi_{k_\alpha + 1} - u^j_{\alpha, k_\alpha - 1}\Phi_{k_\alpha - 1}}{2\Delta x_\alpha} + O(h^2). \tag{4.70}$$

Equation (4.70) thus implies that the full operator Λ^j approximates A with an accuracy up to the second order in all spatial variables.

Let us now turn to the splitting of (4.54). For this consider the following two-cycle scheme:

$$\left(E + \frac{\tau}{2}\Lambda_\alpha^j\right)\Phi^{j-[(n-\alpha)/n]} = \left(E - \frac{\tau}{2}\Lambda_\alpha^j\right)\Phi^{j-[(n-\alpha+1)/n]},$$

$$\alpha = 1, 2, \ldots, n, \qquad (4.71)$$

$$\left(E + \frac{\tau}{2}\Lambda_\alpha^j\right)\Phi^{j+[(n-\alpha+1)/n]} = \left(E - \frac{\tau}{2}\Lambda_\alpha^j\right)\Phi^{j+[(n-\alpha)/n]},$$

$$\alpha = n, n - 1, \ldots, 1.$$

Here it is necessary that the coefficients in the operator Λ_α^j be chosen according to the equations $u^j = v(x, t_j)$. This will guarantee the second-order approximation of the system on every interval

$$t_{j-1} \le t \le t_{j+1}. \qquad (4.72)$$

The splitting-up method of (4.71) is in addition absolutely stable. Indeed, since the approximation of the operators Λ_α^j preserves the condition

$$(\Delta_\alpha^j \Phi, \Phi) = 0, \qquad (4.73)$$

it is not difficult to see that the Kellogg lemma implies

$$\|T_\alpha^j\| = 1, \qquad \alpha = 1, 2, \ldots, n$$

and consequently

$$\|\Phi^{j-[(n-\alpha)/n]}\| = \|\Phi^{j-[(n-\alpha+1)/n]}\|, \qquad \alpha = 1, 2, \ldots, n,$$

$$\|\Phi^{j+[(n-\alpha+1)/n]}\| = \|\Phi^{j+[(n-\alpha)/n]}\|, \qquad \alpha = n, n - 1, \ldots, 1.$$

Eliminating the intermediate values of $\|\Phi\|$, we obtain

$$\|\Phi^{j+1}\| = \|\Phi^{j-1}\|. \qquad (4.74)$$

The above discussed method of solving problems of hydrodynamical motion can be easily generalized to the case of quasi-linear equations of hydrodynamics. The only thing to be done is to supplement in addition a good scheme for extrapolating the coefficients u_α at time t_j from their preceding history.

In our development we have chosen a second-order approximation relative to the spatial coordinates. The extension of the algorithm to higher-order approximations is also possible. To see this, let

$$\Lambda_\alpha \Phi = \sum_{m=1}^{p} \beta_m \frac{\frac{1}{2}(u_{k+m} + u_k)\Phi_{k+m} - \frac{1}{2}(u_k + u_{k-m})\Phi_{k-m}}{2(m\Delta x_\alpha)} \qquad (4.75)$$

where β_m satisfy the following system of equations:

$$\sum_{m=1}^{p} \beta_m = 1; \qquad \sum_{m=1}^{p} m^2 \beta_m = 0, \ldots; \qquad \sum_{m=1}^{p} m^{2p-2} \beta_m = 0. \qquad (4.76)$$

Then, if $x_\alpha = x_{k_\alpha}$, we have

$$\sum_{\alpha=1}^{n} \Lambda_\alpha \Phi = \left(\sum_{\alpha=1}^{n} \frac{\partial u_\alpha \Phi}{\partial x_\alpha} \right)_{x_\alpha = x_{k_\alpha}} + O(h^{2p}).$$

At the same time we assume that the coefficients satisfy the following relation:

$$\sum_{\alpha=1}^{n} \sum_{m=1}^{p} \beta_m \frac{u_{\alpha,k_\alpha+m} - u_{\alpha,k_\alpha-m}}{2m\Delta x_\alpha} = O(h^{2p}). \qquad (4.77)$$

Using further the splitting up algorithm of (4.71), the solution (4.54) can be obtained with an accuracy up to $O(h^{2p} + \tau^2)$.

The given algorithm can be very easily generalized to the case where the hydrodynamical fluid is compressible. In this case, the fundamental equations are as follows:

$$\frac{\partial \rho \Phi}{\partial t} + \sum_{\alpha=1}^{n} \frac{\partial \rho v_\alpha \Phi}{\partial x_\alpha} = 0,$$

$$\frac{\partial \rho}{\partial t} + \sum_{\alpha=1}^{n} \frac{\partial \rho v_\alpha}{\partial x_\alpha} = 0. \qquad (4.78)$$

Noting that ρ is an essentially positive function, we may bring the system (4.68) into the form

$$\frac{\partial \sqrt{\rho} \Phi}{\partial t} + \sum_{\alpha=1}^{n} \left(\tfrac{1}{2} u_\alpha \frac{\partial \sqrt{\rho} \Phi}{\partial x_\alpha} + \frac{1}{2} \frac{\partial \sqrt{\rho} u_\alpha \Phi}{\partial x_\alpha} \right) = 0,$$

$$\frac{\partial \sqrt{\rho}}{\partial t} + \sum_{\alpha=1}^{n} \left(\tfrac{1}{2} u_\alpha \frac{\partial \sqrt{\rho}}{\partial x_\alpha} + \frac{1}{2} \frac{\partial \sqrt{\rho} u_\alpha}{\partial x_\alpha} \right) = 0. \qquad (4.79)$$

(Sinyaev [4], Marchuk, Dymnikov, and others [4], Lenenko [4]). Writing $\sqrt{\rho} \Phi = \psi$, we approximate (4.79) by the expression

$$\frac{\psi^{j+1} - \psi^j}{\tau} + \sum_{\alpha=1}^{n} \frac{(u_\alpha)_{k_\alpha+1/2} \psi_{k_\alpha+1}^{j+1/2} - (u_\alpha)_{k_\alpha-1/2} \psi_{k_\alpha-1}^{j+1/2}}{2\Delta x_\alpha} = 0,$$

$$\frac{\sqrt{\rho^{j+1}} - \sqrt{\rho^j}}{\tau} + \sum_{\alpha=1}^{n} \frac{(u_\alpha)_{k_\alpha+1/2} \sqrt{\rho_{k_\alpha+1}^{j+1/2}} - (u_\alpha)_{k_\alpha-1/2} \sqrt{\rho_{k_\alpha-1}^{j+1/2}}}{2\Delta x_\alpha} = 0, \qquad (4.80)$$

where

$$(u_\alpha)_{k_\alpha+1/2} = \frac{(u_\alpha)_{k_\alpha+1} + (u_\alpha)_{k_\alpha}}{2},$$

$$\psi^{j+1/2} = \frac{\psi^{j+1/2} + \psi^j}{2}.$$

It is easily seen that (4.80) approximates (4.79) with second-order accuracy in τ and h and, moreover, the relations

$$\|\psi^{j+1}\| = \|\psi^j\|, \qquad \|(\sqrt{\rho^{j+1}})\| = \|(\sqrt{\rho^j})\|$$

are satisfied. To solve (4.80) we apply a method like that of (4.71).

9.5 The Neutron Transport Equation

In this section we apply the splitting-up method to an urgent problem of mathematical physics—the theory of radiative transfer. Since our aim is basically methodological we confine our attention to some elementary problems which are, nevertheless, of practical interest. Although we have chosen to use the splitting-up method as our fundamental mathematical machinery, the underlying ideas for constructing difference analogs of the transport problem may be applied in other approaches.

9.5.1 The Nonstationary Equation

We consider an elementary problem in transport theory with a plane-parallel geometry; taking account of the initial conditions the problem statement is as follows:

$$\frac{1}{c}\frac{\partial \phi}{\partial t} + \mu\frac{\partial \phi}{\partial z} + \sigma\phi = \frac{\sigma_s}{2}\int_{-1}^{1}\phi\,d\mu + f, \tag{5.1}$$

$$\phi = 0 \quad \text{for} \quad z = 0, \mu > 0, \tag{5.2}$$

$$\phi = 0 \quad \text{for} \quad z = H, \mu < 0,$$

$$\phi = \phi^0 \quad \text{for} \quad t = 0, \tag{5.3}$$

where $\phi(z, \mu, t)$ is the particle density at the point z, supposed moving with velocity c (we shall take $c \equiv 1$ in our later work) at an angle v to the Oz axis, at the epoch t; $\mu = \cos v$; $f(z, \mu, t)$ represents given sources of radiation; and the functions $\sigma(z)$, $\sigma_s(z)$ are piecewise-continuous and satisfy the inequalities

$$0 < \sigma_0 \le \sigma \le \sigma_1 < \infty, \qquad 0 \le \sigma_s \le \sigma_s' < \infty,$$

$$0 < \sigma_{c0} \le \sigma_c = \sigma - \sigma_s.$$

We now use some transformations introduced by Kuznetsov [17]; similar transformations for the one-dimensional transport equation were introduced by Vladimirov [17]. We denote the solution of (5.1) for $\mu > 0$ by ϕ^+ and for $\mu < 0$ by ϕ^-. Then the transport equation may be written as a system of two equations:

$$\frac{\partial \phi^+}{\partial t} + \mu \frac{\partial \phi^+}{\partial z} + \sigma \phi^+ = \frac{\sigma_s}{2} \int_0^1 (\phi^+ + \phi^-) \, d\mu' + f^+,$$

$$\frac{\partial \phi^-}{\partial t} - \mu \frac{\partial \phi^-}{\partial z} + \sigma \phi^- = \frac{\sigma_s}{2} \int_0^1 (\phi^+ + \phi^-) \, d\mu' + f^-. \tag{5.4}$$

The boundary conditions for the functions ϕ^+ and ϕ^- are as follows:

$$\phi^+(z, \mu) = 0 \quad \text{for} \quad z = 0,$$

$$\phi^-(z, \mu) = 0 \quad \text{for} \quad z = H. \tag{5.5}$$

Now we first add these two equations, and then subtract one from the other, obtaining two equations as a result:

$$\frac{\partial u}{\partial t} + \mu \frac{\partial v}{\partial z} + \sigma u = \sigma_s \int_0^1 u \, d\mu' + g, \tag{5.6}$$

$$\frac{\partial v}{\partial t} + \mu \frac{\partial u}{\partial z} + \sigma v = r,$$

where

$$u = \tfrac{1}{2}(\phi^+ + \phi^-), \qquad v = \tfrac{1}{2}(\phi^+ - \phi^-),$$

$$g = \tfrac{1}{2}(f^+ + f^-), \qquad r = \tfrac{1}{2}(f^+ - f^-).$$

It is easily seen that the boundary conditions (5.5) go over into the equations

$$u + v = 0 \quad \text{for} \quad z = 0,$$

$$u - v = 0 \quad \text{for} \quad z = H \tag{5.7}$$

and the initial data are

$$u = u^0, \qquad v = v^0 \qquad \text{for} \quad t = 0. \tag{5.8}$$

The problems (5.6)–(5.8) may be put into operator form. Consider the vector functions w, w^0, and F, and the operator A:

$$w = \left\| \begin{matrix} u \\ v \end{matrix} \right\|, \qquad \omega^0 = \left\| \begin{matrix} u^0 \\ v^0 \end{matrix} \right\|, \qquad F = \left\| \begin{matrix} g \\ r \end{matrix} \right\|,$$

$$A = \left\| \begin{matrix} \sigma - \sigma_s \int_0^1 d\mu' & \mu \dfrac{\partial}{\partial z} \\ \mu \dfrac{\partial}{\partial z} & \sigma \end{matrix} \right\|. \tag{5.9}$$

Consider also the Hilbert space of functions $L_2(D)$ on $D = [0, H] \times [0, 1]$, with the scalar product

$$(a, b) = \sum_{i=1}^{2} \int_0^1 d\mu \int_0^H a^i b^i \, dx \qquad (5.10)$$

where a^i, b^i are the components of the vector functions a and b.

Further, single out the subspace Φ of vector functions that satisfy the condition

$$(Aw, w) < +\infty. \qquad (5.11)$$

Note that the scalar product is a function of the time; we shall not make special mention of this fact in what follows. We require that the components of the vector function w be continuous in D and have absolutely continuous first derivatives $\partial w / \partial z$. Note that the smoothness of u and v in D is an immediate consequence of the requirement that ϕ^+ and ϕ^- shall be smooth in D. Finally, we select from the space Φ the set vector functions satisfying the conditions (5.7) and having absolutely continuous first derivatives with respect to time. This subspace we denote by Φ^0. Clearly, it is the domain of definition of the operator

$$L \equiv \frac{\partial}{\partial t} + A.$$

Now we arrive at the following problem

$$\frac{\partial w}{\partial t} + Aw = F \quad \text{on} \quad D \times [0, T],$$

$$w = w^0 \quad \text{for} \quad t = 0 \quad \text{on} \quad D, \qquad (5.12)$$

with

$$F(t) \in L_2(D \times [0, T]), \qquad w^0 \in \Phi, w(t) \in \Phi^0.$$

It is not hard to verify that the inequality

$$(Aw, w) > 0 \qquad (5.13)$$

holds on the functions belonging to Φ^0, which are also in the domain of definition of the operator A.

It is known (Vladimirov [17]) that A is positive definite, i.e.,

$$(Aw, w) \geq \gamma(w, w), \qquad (5.14)$$

where γ is a positive constant related to the characteristic geometrical dimensionality of the region.

We now pass to the difference approximation of (5.6)–(5.8) with respect to the spatial variable z. To this end, we introduce two systems of net points: the fundamental system $\{z_k\}_{k=0}^N$, $z_0 = 0$, $z_N = H$, and the auxiliary system $\{z_{k+1/2}\}_{k=0}^{N-1}$. The points in these two systems are serially interspersed, i.e., $z_{k-1/2} < z_k < z_{k+1/2}$.

We integrate the first of the equations (5.6) with respect to z, with the limits $(z_0, z_{1/2})$, $(z_{k-1/2}, z_{k+1/2})$ $(k = 1, \ldots, N - 1)$, $(z_{N-1/2}, z_N)$, and the second equation with the limits (z_{k-1}, z_k) $(k = 1, \ldots, N)$. Then those equations take the form

$$\frac{\partial}{\partial t} \int_{z_0}^{z_{1/2}} u\, dz + \mu \int_{z_0}^{z_{1/2}} \frac{\partial v}{\partial z}\, dz + \int_{z_0}^{z_{1/2}} \sigma u\, dz = \int_{z_0}^{z_{1/2}} dz \int_0^1 \sigma_s u\, d\mu' + \int_{z_0}^{z_{1/2}} g\, dz,$$

$$\frac{\partial}{\partial t} \int_{z_0}^{z_1} v\, dz + \mu \int_{z_0}^{z_1} \frac{\partial u}{\partial z}\, dz + \int_{z_0}^{z_1} \sigma v\, dz = \int_{z_0}^{z_1} r\, dz,$$

. . . .

$$\frac{\partial}{\partial t} \int_{z_{k-1/2}}^{z_{k+1\,2}} u\, dz + \mu \int_{z_{k-1/2}}^{z_{k+1/2}} \frac{\partial v}{\partial z}\, dz + \int_{z_{k-1/2}}^{z_{k+1/2}} \sigma u\, dz$$

$$= \int_{z_{k-1/\pi}}^{z_{k+1\,2}} dz \int_0^1 \sigma_s u\, d\mu' + \int_{z_{k-1/2}}^{z_{k+1/2}} g\, dz, \tag{5.15}$$

$$\frac{\partial}{\partial t} \int_{z_k}^{z_{k+1}} v\, dz + \mu \int_{z_k}^{z_{k+1}} \frac{\partial u}{\partial z}\, dz + \int_{z_k}^{z_{k+1}} \sigma v\, dz = \int_{z_k}^{z_{k+1}} r\, dz,$$

. . . .

$$\frac{\partial}{\partial t} \int_{z_{N-1}}^{z_N} v\, dz + \mu \int_{z_{N-1}}^{z_N} \frac{\partial u}{\partial z}\, dz + \int_{z_{N-1}}^{z_N} \sigma v\, dz = \int_{z_{N-1}}^{z_N} r\, dz,$$

$$\frac{\partial}{\partial t} \int_{z_{N-1/2}}^{z_N} u\, dz + \mu \int_{z_{N-1/2}}^{z_N} \frac{\partial v}{\partial z}\, dz + \int_{z_{N-1/2}}^{z_N} \sigma u\, dz$$

$$= \int_{z_{N-1/2}}^{z_N} dz \int_0^1 \sigma_s u\, d\mu' + \int_{z_{N-1/2}}^{z_N} g\, dz.$$

Now introduce the following notation:

$$\Delta z_0 = z_{1/2} - z_0, \qquad \Delta z_k = z_{k+1/2} - z_{k-1/2}, \qquad k = 1, \ldots, N - 1,$$

$$\Delta z_N = z_N - z_{N-1/2}, \qquad \Delta z_{k-1/2} = z_k - z_{k-1}, \qquad k = 1, \ldots, N,$$

$$h = \max(\Delta z_0, \Delta z_N, \Delta z_k, \Delta z_{k+1/2}), \tag{5.16}$$

$$\sigma_k = \frac{1}{\Delta z_k} \int_{z_{k-1/2}}^{z_{k+1/2}} \sigma \, dz,$$

$$\sigma_{sk} = \frac{1}{\Delta z_k} \int_{z_{k-1/2}}^{z_{k+1/2}} \sigma_s dz, \quad \sigma_{k+1/2} = \frac{1}{\Delta z_{k+1/2}} \int_{z_k}^{z_{k+1}} \sigma \, dz, \qquad (5.17)$$

$$g_k = \frac{1}{\Delta z_k} \int_{z_{k-1/2}}^{z_{k+1/2}} g \, dz, \quad r_{k+1/2} = \frac{1}{\Delta z_{k+1/2}} \int_{z_k}^{z_{k+1}} r \, dz.$$

Then if: (1) u and v are continuous for almost all values of z and μ in D; and (2) the functions σ, σ_s, g, and r are piecewise-continuous with discontinuities of the first kind (if any) at the points z_k, we may use the methods presented in Section 2.3 (and take account of the boundary conditions) to arrive at the following difference approximation to the equations (5.6):

$$\frac{\partial u_0}{\partial t} + \mu \frac{v_{1/2} - u_0}{\Delta z_0} + \sigma_0 u_0 = \sigma_{s0} \int_0^1 u_0 \, d\mu' + g_0,$$

$$\frac{\partial v_{1/2}}{\partial t} + \mu \frac{u_1 - u_0}{\Delta z_{1/2}} + \sigma_{1/2} v_{1/2} = r_{1/2},$$

$$\cdots\cdots\cdots\cdots\cdots\cdots\cdots\cdots\cdots\cdots\cdots\cdots\cdots\cdots\cdots\cdots$$

$$\frac{\partial u_k}{\partial t} + \mu \frac{v_{k+1/2} - v_{k-1/2}}{\Delta z_k} + \sigma_k u_k = \sigma_{sk} \int_0^1 u_k \, d\mu' + g_k,$$

$$(5.18)$$

$$\frac{\partial v_{k+1/2}}{\partial t} + \mu \frac{u_{k+1} - u_k}{\Delta z_{k+1/2}} + \sigma_{k+1/2} v_{k+1/2} = r_{k+1/2},$$

$$\cdots\cdots\cdots\cdots\cdots\cdots\cdots\cdots\cdots\cdots\cdots\cdots\cdots\cdots\cdots\cdots$$

$$\frac{\partial v_{N-1/2}}{\partial t} + \mu \frac{u_N - u_{N-1}}{\Delta z_{N-1/2}} + \sigma_{N-1/2} v_{N-1/2} = r_{N-1/2},$$

$$\frac{\partial u_N}{\partial t} + \mu \frac{u_N - v_{N-1/2}}{\Delta z_N} + \sigma_N u_N = \sigma_{sN} \int_0^1 u_N \, d\mu' + g_N.$$

Let $M_h(0, 2N)$ be the Hilbert space of the vector functions $a = (a_0, a_{1/2}, \ldots, a_{N-1/2}, a_N)$ with the scalar product and norm

$$(a, b) = \sum_{i=0}^{2N} \int_0^1 \Delta z_{i/2} a_{i/2} b_{i/2} \, d\mu,$$

$$\|a\| = (a, a)^{1/2}, \qquad a, b \in M_h(0, 2N).$$

Introduce the vector functions

$$\phi = (u_0, v_{1/2}, u_1, \ldots, u_{N-1}, v_{N-1/2}, u_N),$$

$$F = (g_0, r_{1/2}, g_1, \ldots, g_{N-1}, r_{N-1/2}, g_N),$$

$$\phi^{(0)} = (u_0^{(0)}, v_{1/2}^{(0)}, u_1^{(0)}, \ldots, u_{N-1}^{(0)}, v_{N-1}^{(0)}, u_N^{(0)})$$

and the matrix operator $A = L - S$, where L and S have the form

$$
L = \left\|
\begin{array}{ccccccc}
\dfrac{\mu}{\Delta z_0} + \sigma_0 & \dfrac{\mu}{\Delta z_0} & 0 & \cdots & 0 & 0 & 0 \\[2mm]
\dfrac{-\mu}{\Delta z_{1/2}} & \sigma_{1/2} & \dfrac{\mu}{\Delta z_{1/2}} & \cdots & 0 & 0 & 0 \\[2mm]
\multicolumn{7}{c}{\cdots\cdots\cdots\cdots\cdots\cdots\cdots\cdots\cdots\cdots\cdots\cdots} \\[2mm]
0 & 0 & 0 & \cdots & & \sigma_{N-1/2} & \dfrac{\mu}{\Delta z_{N-1/2}} \\[2mm]
0 & 0 & 0 & \cdots & 0 & \dfrac{-\mu}{\Delta z_N} & \dfrac{\mu}{\Delta z_N} + \sigma_N
\end{array}
\right\|,
$$

$$
S = \operatorname{diag}\left(\sigma_{s,\,i/2} \int_0^1 \gamma_{i/2}\, d\mu\right), \qquad i = 0, \ldots, 2N,
$$

$$
\gamma_{i/2} = \begin{cases} 1, & \text{if } i/2 \text{ is an integer or zero,} \\ 0, & \text{if } i/2 \text{ is a fraction.} \end{cases}
$$

Then we can write (5.18) in operator form:

$$
\frac{d\phi}{dt} + A\phi = F, \qquad t \in [0, T], \tag{5.19}
$$

$$
\phi = \phi^{(0)} \quad \text{for} \quad t = 0.
$$

We now examine certain properties of the operator problem (5.19). First we note that S is self-adjoint in $M_h(0, 2N)$ and is positive because of our assumptions about the initial data. We now show that A and L are positive definite on M_h. For, suppose, $w \in M_h$; then

$$
\begin{aligned}
(Lw, w) &= \int_0^1 \left(\mu\, \frac{w_{1/2} + w_0}{\Delta z_0} + \sigma_0 w_0\right) \Delta z_0 w_0\, d\mu \\[2mm]
&+ \int_0^1 \left(\mu\, \frac{w_1 - w_0}{\Delta z_{1/2}} + \sigma_{1/2} w_{1/2}\right) \Delta z_{1/2} w_{1/2}\, d\mu \\[2mm]
&+ \sum_{i=1}^{N-1} \int_0^1 \left\{\Delta z_i \left(\mu\, \frac{w_{i+1/2} - w_{i-1/2}}{\Delta z_i} + \sigma_i w_i\right) w_i \right. \\[2mm]
&\quad + \left. \Delta z_{i+1/2}\left(\mu\, \frac{w_{i+1} - w_i}{\Delta z_{i+1/2}} + \sigma_{i+1/2} w_{i+1/2}\right) w_{i+1/2}\right\} d\mu \\[2mm]
&+ \int_0^1 \Delta z_N \left(\mu\, \frac{w_N - w_{N-1/2}}{\Delta z_N} + \sigma_N w_N\right) w_N\, d\mu \\[2mm]
&= \sum_{i=0}^{2N} \int_0^1 \Delta z_{i/2}\, \sigma_{i/2} w_{i/2}\, d\mu + \int_0^1 \mu(w_0^2 + w_N^2)\, d\mu,
\end{aligned}
\tag{5.20}
$$

$$
(Lw, w) \geq \gamma \|w\|^2, \qquad \gamma = \text{const} > 0,
$$

and

$$(Aw, w) = \int_0^1 \mu(w_0^2 + w_N^2) \, d\mu$$

$$+ \sum_{i=0}^{2N} \Delta z_{i/2} \int_0^1 \left(\sigma_{i/2} w_{i/2}^2 - \sigma_{si/2} w_{i/2} \int_0^1 \gamma_{i/2} w_{i/2} \, d\mu \right) d\mu \geq \sigma_{co} \|w\|^2.$$

This proves our assertion.

We now find two *a priori* estimates for the solution of (5.19). We multiply (5.19) scalarly by the function ϕ; then we integrate the resulting equation over the interval $(0, t)$ and obtain

$$\tfrac{1}{2}\|\phi\|^2(t) + \int_0^t (A\phi, \phi) \, dt = \int_0^t (F, \phi) \, dt^1 + \tfrac{1}{2}\|\phi^{(0)}\|^2$$

$$\leq \left(\int_0^t \|F\|^2 \, dt^1 \right)^{1/2} \left(\int_0^t \|\phi\|^2 \, dt^1 \right)^{1/2} + \tfrac{1}{2}\|\phi^{(0)}\|^2$$

$$\leq C \left(\int_0^t \|F\|^2 \, dt^1 \right)^{1/2} \left(\tfrac{1}{2}\|\phi\|^2(t) + \int_0^t (A\phi, \phi) \, dt^1 \right)^{1/2}$$

$$+ \tfrac{1}{2}\|\phi^{(0)}\|^2, \tag{5.21}$$

$$C = \text{const} > 0.$$

Using the inequality

$$|a \cdot b| \leq \frac{1}{4\varepsilon} a^2 + \varepsilon b^2, \qquad \varepsilon > 0,$$

we get from the inequalities (5.21) an *a priori* estimate for the solution ϕ,

$$\|\phi\|^2(t) + \int_0^t (A\phi, \phi) \, dt^1 \leq C \left(\int_0^T \|F\|^2 \, dt^1 + \|\phi^{(0)}\|^2 \right), \tag{5.22}$$

where $C > 0$ is a constant not depending on t or ϕ.

We use (5.22) to prove that (5.19) is uniquely soluble. The smoothness of ϕ with respect to t will be higher by one order than that of the corresponding vector function F. We shall not dwell on this question, however, but will assume *a priori* that (5.19) has a unique solution ϕ having the smoothness just described.

Let us now represent the operator A in the form $A = B + D$, where

$$D = \left\|
\begin{array}{cccccc}
\dfrac{\mu}{\Delta z_0} + \sigma_0 & 0 & 0 & \cdots & 0 & 0 \\
0 & \sigma_{1/2} & 0 & \cdots & 0 & 0 \\
\cdots\cdots\cdots\cdots\cdots & & & \cdots\cdots\cdots\cdots\cdots & & \\
0 & 0 & 0 & \cdots & \sigma_{N-1/2} & 0 \\
0 & 0 & 0 & \cdots & 0 & \dfrac{\mu}{\Delta z_N} + \sigma_N
\end{array}
\right\|.$$

Clearly D is a positive definite operator on $M_h(0, 2N)$. Therefore, the equation in (5.19) is equivalent to the following:

$$\frac{d}{dt}(D^{-1}v) + D^{-1/2}AD^{-1/2}v = D^{-1/2}F, \tag{5.23}$$

where $v = D^{1/2}\phi$. We multiply (5.23) scalarly by v and integrate the result with respect to t over the interval $(0, t_1)$. Then we find the relation

$$\tfrac{1}{2}\|\phi\|^2(t_1) + \int_0^{t_1}(D^{-1/2}AD^{-1/2}v, v)\,dt = \int_0^{t_1}(D^{-1/2}F, v)\,dt + \tfrac{1}{2}\|\phi^0\|^2,$$

from which we derive the inequality

$$\tfrac{1}{2}\|\phi\|^2(t_1) + \int_0^{t_1}(D^{-1/2}\tilde{D}D^{-1/2}v, v)\,dt$$

$$\leq \tfrac{1}{2}\|\phi\|^2(t_1) + \int_0^{t_1}(D^{-1/2}AD^{-1/2}v, v)\,dt$$

$$\leq \left(\int_0^{t_1}\|D^{-1/2}F\|^2\,dt\right)^{1/2}\left(\int_0^{t_1}\|v\|^2\,dt\right)^{1/2} + \tfrac{1}{2}\|\phi^{(0)}\|^2, \tag{5.24}$$

after making a few simple estimates. Here

$$\tilde{D} = \begin{Vmatrix} \dfrac{\mu}{\Delta z_0} + \sigma_{c,0} & 0 & \cdots & 0 & 0 \\ 0 & \sigma_{c,1/2} & \cdots & 0 & 0 \\ \cdots & \cdots & \cdots & \cdots & \cdots \\ 0 & 0 & \cdots & \sigma_{c,N-1/2} & 0 \\ 0 & 0 & \cdots & 0 & \dfrac{\mu}{\Delta z_N} + \sigma_{c,N} \end{Vmatrix}.$$

Since

$$\int_0^{t_1}(D^{-1/2}\tilde{D}D^{-1/2}v, v)\,dt \geq \int_0^{t_1}dt\int_0^{t_1}\left(\Delta z_0\frac{\mu + \Delta z_0\sigma_{c0}}{\mu + \Delta z_0\sigma_0}v_0\right.$$

$$\left. + \Delta z_{1/2}\frac{\sigma_{c,1/2}}{\sigma_{1/2}}v_{1/2} + \cdots + \Delta z_N\frac{\mu + \Delta z_N\sigma_{c,N}}{\mu + \Delta z_N\sigma_N}v_N^2\right)$$

$$\times\,d\mu \leq C_1\int_0^{t_1}\|v\|^2\,dt,$$

where the constant $C_1 > 0$ does not depend on v or $\Delta z_{1/2}$, we obtain from (5.24) a second a priori estimate of ϕ:

$$\tfrac{1}{2}\|\phi(t_1)\|^2 + C_1 \int_0^{t_1} \|v\|^2 \, dt$$

$$\leq \left(\int_0^{t_1} \|D^{-1/2}F\|^2 \, dt\right)^{1/2} \left(\int_0^{t_1} \|v\|^2 \, dt\right)^{1/2} + \tfrac{1}{2}\|\phi^{(0)}\|^2, \qquad (5.25)$$

$$\|\phi(t_1)\|^2 + \int_0^{t_1} \|D^{1/2}\phi\|^2 \, dt \leq C\left(\int_0^{t_1} \|D^{-1/2}F\|^2 \, dt + \|\phi^{(0)}\|^2\right),$$

where the constant $C > 0$ is independent of ϕ and $z_{i/2}$.

Now we use (5.25) to estimate the error with which the solution of (5.19) approximates the vector function

$$\phi_T = (u(z_0, \mu, t), v(z_{1/2}, \mu, t), \ldots, v(z_{N-1/2}, \mu, t), u(z_N, \mu, t))$$

constructed from the values of the exact solution of (5.6). We assume that the solution and the initial data of (5.6)–(5.8) are sufficiently smooth with respect to their variables, that the mesh sizes of the net are uniform (i.e., $h = H/N$, $z_i = ih$, $z_{i-1/2} = (i - \tfrac{1}{2})h$) and that h is small enough. The error estimates obtained for this comparatively narrow class of functions allow us to expect that the approximation method we are using will be effective in practical problems.

Note that the approximation errors ε_i $(i = 0, N)$, in the first and last equations of (5.18) are of first order in h and the remaining $\varepsilon_{i/2}$ are of order $O(h^2)$. Accordingly, replacing ϕ in (5.25) by $\phi_T - \phi$ and F by

$$\varepsilon = (\varepsilon_0, \varepsilon_{1/2}, \ldots, \varepsilon_{N-1/2}, \varepsilon_N),$$

we obtain the desired estimate

$$\|\phi_T - \phi\|^2(t_1) + \int_0^{t_1} \|D^{1/2}(\phi_T - \phi)\|^2 \, dt$$

$$\leq C \int_0^{t_1} \|D^{-1/2}\varepsilon\|^2 \, dt$$

$$\leq C \int_0^T dt \int_0^1 \left[\frac{\varepsilon_0^2 h^2}{2(2\mu + h\sigma_0)} + \frac{\varepsilon_{1/2}^2 h}{\sigma_{1/2}} + \cdots + \frac{\varepsilon_{N-1/2}^2 h}{\sigma_{N-1/2}} + \frac{\varepsilon_N^2 h^2}{2(2\mu + h\sigma_N)}\right] d\mu$$

$$= O(h^4) + O(h^4) \int_0^1 \frac{d\mu}{(2\mu + h\sigma_0)} + O(h^4) \int_0^1 \frac{d\mu}{2\mu + h\sigma_N}$$

$$= O\left(h^4 \ln \frac{1}{h}\right), \qquad (5.26)$$

$$\max_t \|\phi_T - \phi\|(t) + \left(\int_0^T \|D^{1/2}(\phi_T - \phi)\|^2 \, dt\right)^{1/2} = O\left(h^2 \ln^{1/2} \frac{1}{h}\right).$$

The above analysis, and the a priori estimates, are due to Agoshkov.

The inequality (5.25) can be used to obtain an estimate of the error in the solution of (5.19) under weaker constraints on the smoothness of the solution and initial data of the problem.

Let us turn to the formulation of the splitting-up method for the solution of (5.19). We introduce two operators A_1 and A_2, acting on $M_h(0, 2N)$:

$$
A_1 = \left\|
\begin{array}{cccccc}
\dfrac{\mu}{\Delta z_0} & \dfrac{\mu}{\Delta z_0} & 0 & 0 & 0 & 0 \\[2ex]
-\dfrac{\mu}{\Delta z_{1/2}} & 0 & \dfrac{\mu}{\Delta z_{1/2}} & 0 & 0 & 0 \\[2ex]
\multicolumn{6}{c}{\cdots\cdots\cdots\cdots\cdots\cdots\cdots\cdots\cdots\cdots\cdots\cdots\cdots} \\[1ex]
0 & 0 & 0 \cdots & -\dfrac{\mu}{\Delta z_{N-1/2}} & 0 & \dfrac{\mu}{\Delta z_{N-1/2}} \\[2ex]
0 & 0 & 0 \cdots & 0 & -\dfrac{\mu}{\Delta z_N} & \dfrac{\mu}{\Delta z_N}
\end{array}
\right\|.
$$

$$
A_2 = \operatorname{diag}\left(\sigma_{i/2} - \sigma_{s,\,i/2} \int_0^1 d\mu'\gamma_{i/2} \right),
$$

and then

$$
A = A_1 + A_2.
$$

It is easy to show that A_1 is positive definite and, using the relation

$$
\int_0^1 \left(\sigma_{i/2} a_{i/2}^2 - \sigma_{s,\,i/2} a_{i/2} \int_0^1 \gamma_{i/2} a_{i/2}\, d\mu^1 \right) d\mu \geq \sigma_{c,\,0} \int_0^1 a_{i/2}^2\, d\mu,
$$

to show also that A_2 is positive definite on $M_h(0, 2N)$, i.e.,

$$
(A_1 a, a) \geq 0, \qquad (A_2 a, a) \geq \gamma \|a\|^2, \qquad \gamma = \text{const} > 0.
$$

The definiteness conditions allow us to formulate a solution algorithm on the basis of the two-cycle component-by-component splitting-up method (cf. Section 5.3). We consider the following problem.

On the interval $t_{j-1} \leq t \leq t_j$:

$$
\left(E + \frac{\tau}{2} A_1^h \right) \phi^{j-2/3} = \left(E - \frac{\tau}{2} A_1^h \right) \phi^{j-1},
$$

$$
\left(E + \frac{\tau}{2} A_2^h \right) \phi^{j-1/3} = \left(E - \frac{\tau}{2} A_2^h \right) \phi^{j-2/3},
$$

(5.27)

on the interval $t_{j-1} \leq t \leq t_{j+1}$:

$$
\phi^{j+1/3} = \phi^{j-1/3} + 2\tau F^l,
$$

(5.28)

and on the interval $t_j \le t \le t_{j+1}$:

$$\left(E + \frac{\tau}{2} A_2^h\right) \phi^{j+2/3} = \left(E - \frac{\tau}{2} A_2^h\right) \phi^{j+1/3},$$

$$\left(E + \frac{\tau}{2} A_1^h\right) \phi^{j+1} = \left(E - \frac{\tau}{2} A_1^h\right) \phi^{j+2/3}, \tag{5.29}$$

where F^j is a vector with the components

$$F_i^j = \frac{1}{(t_{j+1} - t_{j-1})} \int_{t_{j-1}}^{t_{j+1}} F_i \, dt, \qquad t_j = j\tau, \tau = T/y.$$

It follows from the properties of A_1, A_2 that the scheme (5.27)–(5.29), given the necessary smoothness of the solution, approximates (5.19) with accuracy of order $O(\tau^2)$ and is stable:

$$\max_j \|\phi^{(j)}\| \le C\left(\|\phi^{(0)}\| + \max_j \|F^{(j)}\|\right), \tag{5.30}$$

where $C > 0$ is a constant. From the approximation and from the inequality (5.30) we also have that

$$\max_j \|\phi(t_j) - \phi^{(j)}\| \le O(\tau^2). \tag{5.31}$$

If the estimate (5.26) holds, the solution of the problem (5.27)–(5.29) converges in the metric of the space $M_h(0, 2N)$ to the exact solution of the original problem as $h, \tau \to 0$, and

$$\max \|\phi_T(t_j) - \phi^{(j)}\| \le O\left\{h^2 \left|\ln^{1/2} \frac{1}{h}\right| + \tau^2\right\}, \tag{5.32}$$

where

$$\phi_T(t_j) = (u(z_0, t_j, \mu), v(z_{1/2}, \mu, t_j), \ldots, v(z_{N-1/2}, \mu, t_j), u(z_N, \mu, t_j)).$$

Let us pause to consider the solution of (5.27)–(5.29). Since the matrix operator

$$\left(E + \frac{\tau}{2} A_1\right)$$

is tri-diagonal, the solution of the first equation in (5.27) and the second in (5.29) gives us no trouble for arbitrary fixed μ. The matrix operator

$$\left(E + \frac{\tau}{2} A_2\right)$$

is diagonal, so that the solution of the second equation in (5.27) reduces to a computation according to the formulas

$$\phi_i^{j-1/3} = \frac{1}{1 + \dfrac{\tau\sigma_i}{2}} \left[\left(1 - \frac{\tau\sigma_i}{2}\right)\phi_i^{j-2/3} + \frac{\tau\sigma_{s,i}}{1 + \dfrac{\tau\sigma_{c,i}}{2}} \int_0^1 \phi_i^{j-2/3}\, d\mu \right],$$

$$i = 0, 1, \ldots, N, \quad (5.33)$$

$$\phi_{i-1/2}^{j-1/3} = \frac{1 - \dfrac{\tau\sigma_i}{2}}{1 + \dfrac{\tau\sigma_i}{2}} \phi_{i-1/2}^{j-2/3}, \qquad i = 1, \ldots, N.$$

Similar formulas arise for the solution of the first problem in (5.29), and we see that we have defined a numerical algorithm for the solution of the system (5.27)–(5.29).

Now we complete the difference approximation with respect to μ. We divide the interval $0 \le \mu \le 1$ into subintervals $\Delta\mu_l$ by the net points μ_l to yield the best approximation of the intervals in (5.33), on the given class of solutions. Let

$$\int_0^1 \psi(\mu)\, d\mu \simeq \sum_{l=1}^m s_l\psi_l, \qquad \psi_l = \psi(\mu_l),$$

where the s_l are weights in the selected quadrature formulas. Replacing the integrals in (5.27)–(5.29) by the quadratures and considering the system for $\mu = \mu_l$ $(l = 1, \ldots, m)$, we arrive at a system of linear algebraic equations approximating the original problem:

$$\left(E + \frac{\tau}{2}A_{1,l}\right)\phi_l^{j-2/3} = \left(E - \frac{\tau}{2}A_{1,l}\right)\phi_l^{j-1},$$

$$\left(E + \frac{\tau}{2}A_{2,l}\right)\phi_l^{j-1/3} = \left(E - \frac{\tau}{2}A_{2,l}\right)\phi_l^{j-2/3},$$

$$\phi_l^{j+1/3} = \phi_l^{j-1/3} + 2\tau F_l^j, \qquad (5.34)$$

$$\left(E + \frac{\tau}{2}A_{2,l}\right)\phi_l^{j+2/3} = \left(E - \frac{\tau}{2}A_{2,l}\right)\phi_l^{j+1/3},$$

$$\left(E + \frac{\tau}{2}A_{1,l}\right)\phi_l^{j+1} = \left(E - \frac{\tau}{2}A_{1,l}\right)\phi_l^{j+2/3},$$

$$\phi_l^0 = \phi_l^{(0)},$$

where the $\phi_l^{j-1}, \ldots, \phi_l^{j+1}$ are vectors of dimension $(2N + 1)$ $(l = 1, \ldots, m)$,

$$\phi_l^{(0)} = \phi^{(0)}(\mu_l), \qquad A_{1,l} = A_1(\mu_l),$$

and the operator $A_{2,l}$ is defined by the formulas

$$(A_{2,l}\phi_l^j)_{i/2} = \sigma_{i/2}\phi_{l,i/2}^j - \sigma_{s,i/2}\sum_{k=1}^m s_k\phi_{k,i/2}^j, \qquad i = 0, \ldots, 2N.$$

As we have already noted, the first and last equations in (5.34) are easily solved, for instance by the sweep method; the second and fourth are solved by use of the formulas

$$\phi_{1,i}^{j-1/3} = \frac{1}{1 + \dfrac{\tau\sigma_i}{2}} \left[\left(1 - \frac{\tau\sigma_i}{2}\right) \phi_{i,i}^{j-2/3} + \frac{\tau\sigma_{si}}{1 + \dfrac{\tau\sigma_{ci}}{2}} \sum_{k=1}^{m} s_k \phi_{k,i}^{j-2/3} \right],$$

$$i = 0, \ldots, N, \quad (5.35)$$

$$\phi_{1,i-1/2}^{j-1/3} = \frac{1 - \dfrac{\tau\sigma_i}{2}}{1 + \dfrac{\tau\sigma_i}{2}} \phi_{1,i-1/2}^{j-2/3}, \qquad i = 1, \ldots, N,$$

for the second equation and by the same formulas for the fourth after replacing j by $j + 1$.

This completes the definition of a numerical algorithm for solving the nonstationary transport equation. As a result, we arrive at an absolutely stable scheme, with accuracy of the order

$$O(h^2(\ln 1/h)^{1/2} + \tau^2)$$

on smooth solutions. The accuracy with respect to μ depends on the choice of quadrature formula. (Note that in the scheme that we have contemplated, the value $\mu_1 = 0$ is admissible as a net point.)

Up to now we have not restricted the choice of parameters. However, we note that if we apply the factorization method to (5.34), its stability will be guaranteed by the choice $\tau < \min_i (\Delta z_{i/2})$. Therefore in practical problems this condition imposes a restriction on the choice of the time-mesh-size.

Multi-dimensional transport problems may be treated in a similar fashion.

9.5.2 The Transport Equation in Self-Adjoint Form

Consider the stationary problem corresponding to (5.6), (5.7) under the assumption that the source function f is even with respect to the angular variable μ. Then $r \equiv 0$ and we have the system

$$\begin{cases} \mu \dfrac{\partial v}{\partial z} + \sigma u = \sigma_s \displaystyle\int_0^1 u \, d\mu' + g, \\[2mm] \mu \dfrac{\partial u}{\partial z} + \sigma v = 0, \\[2mm] u + v = 0 \quad \text{for} \quad z = 0, \, u - v = 0 \quad \text{for} \quad z = H. \end{cases}$$

The second equation yields an expression for v, which we substitute in the first equation and in the boundary conditions. Then we have the transport equations in self-adjoint form:

$$\begin{cases} -\mu \dfrac{\partial}{\partial z} \dfrac{\mu}{\sigma} \dfrac{\partial u}{\partial z} + \sigma u = \sigma_s \displaystyle\int_0^1 u \, d\mu' + g, \\[2mm] u - \dfrac{\mu}{\sigma} \dfrac{\partial u}{\partial z}\bigg|_{z=0} = u + \dfrac{\mu}{\sigma} \dfrac{\partial u}{\partial z}\bigg|_{z=H} = 0 \end{cases} \tag{5.36}$$

We may apply the Ritz method for solution, since the operator in this system is symmetric and positive definite. If we choose a basis of the form

$$Q_k(z) = \frac{1}{\sqrt{h}} \begin{cases} 1 - \displaystyle\int_z^{z_k} \sigma(\xi) \, d\xi \bigg/ \int_{z_{k-1}}^{z_k} \sigma(\xi) \, d\xi, & z \in (z_{k-1}, z_k), \\[3mm] 1 - \displaystyle\int_{z_k}^{z} \sigma(\xi) \, d\xi \bigg/ \int_{z_k}^{z_{k+1}} \sigma(\xi) \, d\xi, & z \in (z_k, z_{k+1}), \\[3mm] 0, & z \notin (z_{k-1}, z_{k+1}), \end{cases} \tag{5.37}$$

$$k = 0, 1, \ldots, N,$$

the Ritz method coincides with the method of integral identities discussed in Chapter 2, as applied to the diffusion equation. We wrote it there in variational form under the condition that the approximate solution was constructed with the functions (5.37).

Agoshkov, however, in a study of the arguments and approaches applicable in the method of integral identities, has found a rather easy way to estimate the rate of convergence of the approximate solution in a broad class of functions that includes the exact solutions to problems important in practice. As an illustration we consider a second algorithm.

Multiply (5.36) by $Q_i(z)$ and integrate with respect to z over the interval $[0, H]$, observing the boundary conditions. The result is a system of integral identities:

$$\frac{\mu u(z_0, \mu)}{\sqrt{h}} + \mu^2 \left[\frac{u(z_0, u) - u(z_1, \mu)}{\sqrt{h} \int_{z_0}^{z_1} \sigma(z) \, dz} \right] + (\sigma u, Q_0) = (Su, Q_0) + (g, Q_0),$$

$$\mu^2 \left[\frac{u(z_i, \mu) - u(z_{i+1}, \mu)}{\sqrt{h} \int_{z_i}^{z_{i+1}} \sigma(z) \, dz} + \frac{u(z_i, \mu) - u(z_{i-1}, \mu)}{\sqrt{h} \int_{z_{i-1}}^{z_i} \sigma(z) \, dz} \right] + (\sigma u, Q_i)$$

$$= (Su, Q_i) + (g, Q_i), \qquad i = 1, \ldots, N-1,$$

$$\frac{\mu u(z_N, \mu)}{\sqrt{h}} + \mu^2 \left[\frac{u(z_N, \mu) - u(z_{N-1}, \mu)}{\sqrt{h} \int_{z_{N-1}}^{z_N} \sigma(z) \, dz} \right] + (\sigma u, Q_N) = (Su, Q_N) + (g, Q_N),$$

where

$$Su = \sigma_S \int_0^1 d\mu' u(z, \mu'), \qquad (u, v) = \int_0^H dz \, u(z, \mu) v(z, \mu),$$

and the net $\{z_i\}$ may be nonuniform.

Introduce an interpolant of the exact solution in the form

$$u_1(z, \mu) = \sum_{i=0}^{N} u(z_i, \mu)\sqrt{h}\, Q_i(z).$$

Then the system of integral identities may be written as

$$\left(\frac{\mu}{\sigma}\frac{\partial u_I}{\partial z}, \frac{\mu\partial Q_i}{\partial z}\right) + \mu(u_I Q_i|_{z=0} + u_I Q_i|_{z=H})$$

$$+ (\sigma u, Q_i) = (Su, Q_i) + (g, Q_i), \qquad i = 0, \dots, N. \qquad (5.38)$$

We seek the approximate solution in the form

$$u_h(z, \mu) = \sum_{i=0}^{N} a_i(\mu) Q_i(z),$$

where the unknown functions $a_i(\mu)$ are defined by the system of integral equations

$$\left(\frac{\mu}{\sigma}\frac{\partial u^h}{\partial z}, \frac{\mu\partial Q_i}{\partial z}\right) + \mu(u^h Q_i|_{z=0} + u^h Q_i|_{z=H})$$

$$+ (\sigma u^h, Q_i) = (Su^h, Q_i) + (g, Q_i), \qquad i = 0, \dots, N. \qquad (5.39)$$

To obtain an *a priori* estimate of the approximate solution we multiply each of the equations (5.39) by the corresponding function $a_i(\mu)$, add the products for $i = 0, \dots, N$, and integrate with respect to μ over the interval $\mu \in [0, 1]$. The result is the equation

$$\int_0^1 d\mu\left[\left(\frac{\mu}{\sigma}\frac{\partial u^h}{\partial z}, \mu\frac{\partial u^h}{\partial z}\right) + \mu(u^h u^h|_{z=0} + u^h u^h|_{z=H})\right.$$

$$\left. + (\sigma u^h, u^h) - (Su^h, h^h)\right] = \int_0^1 d\mu(g, u^h),$$

from which, since

$$\int_0^1 d\mu[(\sigma u^h, u^h) - (Su^h, u^h)] \geq \sigma_{co}\int_0^1 d\mu(u^h, u^h),$$

we derive the inequalities

$$\int_0^1 d\mu\left[\left(\frac{\mu}{\sigma}\frac{\partial u^h}{\partial z}, \mu\frac{\partial u^h}{\partial z}\right) + \mu(u^{h^2}|_{z=0} + u^{h^2}|_{z=H}) + \sigma_{co}(u^h, u^h)\right]$$

$$\leq \int_0^1 d\mu(g, u^h) \leq \left(\int_0^1 d\mu(g, g)\right)^{1/2} \cdot \left(\int_0^1 d\mu(u^h, u^h)\right)^{1/2}.$$

Accordingly,

$$\int_0^1 d\mu\left[\left(\frac{\mu}{\sigma}\frac{\partial u^h}{\partial z}, \mu\frac{\partial u^h}{\partial z}\right) + \mu(u^h|_{z=0} + u^h|_{z=H}) + \sigma_{co}(u^h, u^h)\right]$$

$$\leq C \cdot \int_0^1 d\mu(g, g). \qquad (5.40)$$

This *a priori* estimate guarantees the unique solvability of the system (5.39) and also the stability of the algorithm.

We now estimate the approximation error. To this end, we consider an identity that follows from (5.38) and (5.39):

$$\int_0^1 d\mu \left[\left(\frac{\mu}{\sigma} \frac{\partial(u_I - u^h)}{\partial z}, \mu \frac{\partial(u_I - u^h)}{\partial z} \right) + \mu(u_I - u^h)^2|_{z=0} + \mu(u_I - u^h)^2|_{z=H} \right.$$

$$\left. + (\sigma(u - u^h), u - u^h) - (S(u - u^h), u - u^h) \right]$$

$$= \int_0^1 \mu[(\sigma(u - u^h), u - u_I) - (S(u - u^h), u - u_I)].$$

Since

$$\int_0^1 d\mu[(\sigma(u - u^h), u - u^h) - (S(u - u^h), u - u^h)]$$

$$\geq \sigma_{c0} \int_0^1 d\mu(u - u^h, u - u^h),$$

$$\int_0^1 d\mu[(\sigma(u - u^h), u - u_I) - (S(u - u^h), u - u_I)]$$

$$\leq \sigma_1 2 \left[\int_0^1 d\mu((u - u^h), u - u^h) \right]^{1/2} \times \left[\int_0^1 d\mu((u - u_I), u - u_I) \right]^{1/2},$$

our identity yields the following inequality:

$$\int_0^1 d\mu \left[\left(\frac{\mu}{\sigma} \frac{\partial(u_I - u^h)}{\partial z}, \frac{\mu\partial(u_I - u^h)}{\partial z} \right) + \mu(u_I - u^h)^2|_{z=0} + \mu(u_I - u^h)^2|_{z=H} \right.$$

$$\left. + \sigma_{c0}(u - u^h, u - u^h) \right] \leq C \int_0^1 d\mu(u - u_I, u - u_I).$$

Now we make use of a result from approximation theory. (Agoshkov [17], On the smoothness of solutions of the transport equation and approximate methods of constructing solutions.) Namely, if we assume that

$$\sigma(z) \in L_\infty(0, H), \qquad \sigma_s \in L_\infty(0, H), \qquad g \in L_\infty((0, H) \times (0, 1)),$$

we obtain the following estimate:

$$\int_0^1 d\mu(u - u_I, u - u_I) \leq C \cdot h^{1/2} \|g\|_{L_\infty}$$

This yields the desired estimate of the approximation error

$$\int_0^1 d\mu \left[\left(\frac{\mu}{\sigma} \frac{\partial(u_I - u^h)}{\partial z}, \frac{\mu\partial(u_I - u^h)}{\partial z} \right) + \mu(u_I - u^h)^2 |_{z=0} \right.$$

$$\left. + \mu(u_I - u^h)^2 |_{z=H} + \sigma_{c0}(u - u^h, u - u^h) \right] \leq Ch\|g\|_{L^\infty}^2$$

which is, therefore, obtained with no unrealistic constraint on the smoothness of the exact solution $u(z, \mu)$.

We note in conclusion that the left-hand side of the system (5.39) contains the symmetric tri-diagonal matrix $A(\mu)$ which, because of the given assumptions concerning the coefficients in the problem, will be diagonally dominant for all $\mu \in [0, 1]$. Hence, and from the condition $\sigma(z) - \sigma_s(z) \geq \sigma_{c0} > 0$, it follows that the eigenvalues of the matrix $A(\mu)$ for arbitrary fixed $\mu \in [0, 1]$, and the eigenvalues of the system (5.39), will be nonzero positive constants. This fact guarantees the stability of the algorithm we have under consideration, and also of the sweep method, if we apply it to the inversion of the matrix A for a fixed value of μ in the solution of (5.39) by quadratures and an iterative method.

CHAPTER 10
A Review of the Methods of Numerical Mathematics

The development of large-scale computers has formed a basis for algorithmic constructions and extensive mathematical experiments in many areas of science and technology, thereby attracting a new generation of scientists to problems of numerical mathematics. The valuable experience accumulated by solving applied problems has later been used for constructing effective methods and algorithms for numerical mathematics.

In this chapter, we will briefly review the fundamental directions in numerical mathematics as of the present time, and indicate the trends of their development.

10.1 The Theory of Approximation, Stability, and Convergence of Difference Schemes

Wide applications of the finite-difference method in differential equations of mathematical physics have made it necessary to study in detail those properties of difference equations which have direct bearing on the performance of difference schemes, particularly the stability and convergence properties.

The development of the theory of stability and convergence started in response to the discovery that the difference scheme for a well-posed differential problem can become unstable (or ill-posed). An unstable scheme is sensitive to the round-off errors arising in the numerical process and can lead to a solution which differs considerably from the one corresponding to the differential problem.

456

This particular feature of difference equations triggered intensified theoretical investigations regarding the relation between convergence on one side and stability on the other.

In the mid-1950s, Lax [6, 7], Richtmyer [3, 6, 7], Ryabenkii and Filippov [6], and Meiman [4] formulated—almost simultaneously and from different viewpoints—the following fundamental result which became known as the equivalence theorem: consider a well-posed linear homogeneous differential problem and an approximating difference scheme; in order for the solution of the difference problem to converge to the solution of the original differential problem it is necessary and sufficient that the difference scheme be stable. The final formulation of this theorem and its proof for an abstract evolution equation in a Banach space were given by Ryabenkii and Filippov [6]. Richtmyer [3] generalized the equivalence theorem to the case of nonhomogeneous linear differential equations. The equivalence theorem is formulated in terms of one single norm. Convergence with respect to other norms can be established on the bases of the imbedding theorems of Sobolev [1]. With the initial data having stronger smoothness assumptions, the requirement for stability of the scheme can be weakened, as was originally pointed out by Ryabenkii and Filippov [6]. This idea has been later reflected in the equivalence theorem due to Strang [7] in connection with the concept of weak stability.

Regarding the choice of efficient stability criteria one must primarily refer to von Neumann and Richtmyer [7]. They formulated the so-called local criterion of stability. This criterion, however, holds true only for equations with constant coefficients describing self-adjoint problems. This circumstance prompted intensive search for the limits of applicability of the local criterion.

Lax and Nirenberg [6, 7] have developed a stability theory for hyperbolic difference schemes in terms of the so-called symbol of the difference scheme. In the case of explicit difference equations the symbol coincides with the usual transition matrix obtained by the Fourier method; here the local stability criterion holds true if the coefficients have bounded second derivatives with respect to x.

Strang [7] has formulated a convergence theorem for systems of quasilinear hyperbolic equations under the condition of local stability of the difference equations corresponding to the first variation of the differential system, and also assuming a sufficient smoothness of the solution.

The development of difference schemes with variable coefficients is associated with the concept of dissipativity. Here we note first of all the work by Kreiss [6]. His theorems relate the order of dissipativity of the difference equations approximating systems of hyperbolic equations to the order of their accuracy. It is assumed that the matrix coefficients of the difference equations are Hermitian and Lipschitz continuous in x.

An interesting approach to the problem of stability has been given by Yanenko and Shokin [7]: instead of the difference equation we consider some accompanying differential equation, the so-called first-differential approximation, which, if ill-posed, implies the instability of the difference scheme.

A very important class of difference schemes consists of those with positive coefficients; it was considered by Friedrichs [2]. He introduced a general concept of positive schemes and established an L^2-stability criterion.

Godunov and Ryabenkii [6, 7] have introduced the notion of a spectrum of a family of difference operators. This concept allows them to form necessary conditions for stability of difference schemes which nicely expose the roots of instability. A new concept, the kernel of the spectrum of the family of difference operators, was introduced.

The above authors gave estimates of the norms of powers of the operators from the family in terms of the radii of the spectral kernels. These estimates turn out to be uniform on the family and can be conveniently used for stability investigations.

All the stability criteria we have considered so far can be classified as spectral criteria, since they are based on the spectral properties of difference operators. These criteria can be used to establish L^2-convergence. It is preferable, however, to have the convergence in norm of the C-space. The related results can be found in Serdyukova [6], Thomee [6], Samarskii [7], and others.

Among the nonspectral approaches to the stability problem regarding the difference analogs of parabolic or hyperbolic equations we point out a fairly general theory due to Samarskii [3, 6, 7], based on energy inequalities and *a priori* estimates. This theory contains necessary and sufficient conditions of stability for a wide class of two-layer and three-layer schemes. The conditions have the form of inequalities relating the coefficients operators of the difference schemes. They are very constructive, and in addition to the stability analysis they permit one to devise new stable schemes.

The energy method has been highly developed in recent years. The idea of the method is to choose a norm, in which the vector solution grows from step to step no faster than $1 + O(\Delta t)$, which means stability in that norm. Later on, the proof is reduced to showing the equivalence of this norm to L^2-norm.

The energy method goes back to Courant, Friedrichs, and Lewy [7]. It was successfully developed by Ladyzhenskaya [7], Lees [7], Lax [7], Kreiss [6], Samarskii [3], Konovalov [15], and others.

Kreiss [6] has thoroughly studied the stability problem for hyperbolic problems. He has established sufficient conditions for the stability of difference analogs under very general assumptions concerning the initial data.

The development of the theory of approximation and convergence in the general framework of functional analysis has been provided by Kantorovich and Akilov [1], who considered wide classes of operator equations with a special emphasis on the problem of numerical solution of integral equations.

Of importance for the theory of convergence is the closure theory due to Sobolev. It is widely used for the purposes of theoretical justification of approximation methods in mathematical physics.

10.2 Numerical Methods for Problems of Mathematical Physics

The concepts of approximation, stability, and convergence have provided the necessary basis for wide research of efficient difference schemes for problems of mathematical physics. The algorithms of the finite-difference methods combine, as a rule, constructions of the difference analogs and methods for their solutions. Therefore, the advances in the constructive theory of finite-difference methods depend on a mutually coordinated development of the two fields.

In trying to summarize the vast growth in the development of finite difference methods in the recent years, we will provisionally distinguish the following important trends.

Constructions of Difference Schemes

A particular trend is related to the development of methods for constructing conservative difference schemes based on the laws of conservation, characteristic of the majority of physical processes. The starting point for devising conservative difference schemes is to write the balance equations for a single cell of the net region and to apply subsequently the quadrature and interpolation formulas. Performing then the necessary transformations and summing over the net region, we obtain equations satisfying the integral laws of conservation.

Such approaches have been considered by Ladyzhenskaya [7], who constructed difference operators with discontinuous coefficients which have an identical form for any internal point of the region. In order to get a solid theoretical base for the algorithms, the concept of generalized solutions is used and proves that the solution of the difference problem is represented by a certain functional converging to the functional of the differential problem as $h \to 0$.

A class of conservative difference schemes in hydrodynamics, based on explicit difference approximations, have been considered by Godunov [4] and Lax and Wendroff [6]. Of considerable importance for hydrodynamical problems has been the method of integral relations suggested by Dorodnitsyn [3] and developed by Belotserkovskii, Chushkin [4], and others, which uses a partial difference approximation of the equations in a divergence form, based on the method of straight lines. These methods played a fundamental role in forming the general approach to difference schemes for quasi-linear equations. Interesting general approaches to problems of hydrodynamics have been also suggested by Babenko and Rusanova [12], Fromm [4], Crowley [4], Kuropatenko [4].

Much attention has been paid recently to the construction of highly accurate solutions for the problems of mathematical physics. There are two main directions of study. The first works toward an increase in the accuracy with which the difference equations are approximated, and has been pursued by Samarskii [3], Yanenko and Balliullin [3], Mitchell [3], and others. The second direction is connected with the use of rather inaccurate difference equations on a sequence of nets with varying mesh size. This has been called the Richardson extrapolation method, and has been studied by Volkov [4], Fox [4], Shaidurov [4], and others.

The above methods as applied to elliptic and parabolic equations with discontinuous coefficients have been studied by Tikhonov and Samarskii [4] and others, by means of the integro-interpolation method.

Variational Methods

In recent years there has been a marked interest in variational methods for problems of mathematical physics. The variational methods of Ritz, Galërkin, Treftz, and others have long occupied an important place in numerical mathematics. These methods are especially effective for computing functionals of the solutions. It has turned out that functionals can be obtained highly accurately even for comparatively poor approximations. The most complete justification of these methods has been given by Mikhlin [1], who has established the necessary and sufficient conditions of stability of the variational methods for problems with energy norms. The active development of variational methods has also displayed some of their deficiencies related to the difficulties in constructing the test functions which would reflect special features of the solution of the problem, and which would give satisfactory approximations of the solution when taken in small numbers.

A new twist in methods of constructing difference equations of mathematical physics has been brought about by combining variational methods with a special design of test functions, taken identically zero outside some rather small regions belonging to the domain of definition of the solution: The first results of Courant [5], Oganesyan and Rukhovets [5], Lions and Teman [5], Aubin [5], Birkhoff, Schultz and Varga [5], Bramble and Saulev [4] proposed the method of fictive domains, studied later by Mignot fundamentally enlarged the applicability of variational methods to problems with self-adjoint and nonself-adjoint operators, and increased interest in schemes of high-order accuracy based on finite element methods, spline approximations, etc. The development of these methods is due primarily to Babuška [5], Strang and Fix [5], Zlámal [5], Douglas and Dupont [5], Rivkind [4], and others.

Multi-Dimensional Stationary Problems

An intensive development of methods for solving linear algebraic equations with Jacobi and tri-diagonal block matrices resulted in a number of excellent numerical algorithms for stationary processes based on the factorization of the corresponding difference operator. We mention in particular various methods of matrix factorization studied by Keldysh, Gel'fand and Loku-tsievskii [12], Babenko and Rusanov [12], Chentsov and Godunov [12], Abramov [12], and others.

Vishik, Sobolev, and Lyusternik have studied the solution of boundary problems in regions with a complicated geometry. On the basis of their work Saulev [4] proposed the method of fictive domains, studied later by Mignot [2], Kopchenov [4], Rukhovets [4], Konovalov [4], and others. Lebedev [4] and Rivkind [4] have studied the difference equations that approximate the boundary value problem arising in the method of fictive domains.

There has been a recent intensive development of noniterative methods for solving difference equations corresponding to differential problems admitting a separation of variables. The fast Fourier transform and the method of cyclic reduction yield a significant decrease in the computational effort. Work in this direction is represented by papers of Cooley and Tukey [13], Buzbee, Golub, and Nilson [12], Hockney [13], and others.

At the same time that the exact factorization methods were being developed, there has been a development of approximate factorization, in which the factorization of the operator is combined with the successive approximation method. A need for such algorithms emerged as early as the problems of mathematical physics were being reduced to large algebraic systems. The first results of Buleev [15] and Baker and Oliphant [15] triggered the development of new methods for the multi-dimensional problems based on the fast-converging processes.

The early 1960s were marked by a major contribution in computational mathematics associated with the names of Douglas, Peaceman, and Rachford, who suggested the so-called alternating-direction method. The success of the method was ensured by the use of a simple reduction of a multi-dimensional problem to a sequence of one-dimensional problems with Jacobi matrices easily manageable by a computer. Essentially the method can be viewed as an iteration method in which the computations are optimized by a special choice of the contraction operator consisting of a product of simpler operators and a number of free relaxation parameters. The succession of inversions of the simple operators is implemented by the linear factorization method. Such iteration schemes are very economical and effective. An increase in the amount of numerical operations is insignificant in comparison with the explicit method of Richardson. The alternating-direction method has considerably influenced constructions of algorithms in various areas of applied mathematics and the investigations of nonlocal and block-iterative processes. The theoretical investigations devoted to this method can be found in Douglas [15], Birkhoff, Varga, and Young [15], Wachspress [15], Kellogg [15], Vorobyev [15], and others.

New methods are being developed by considering homogeneous and nonhomogeneous approximations. In the case of nonhomogeneous approximations, any of the auxiliary problems may not possibly approximate the original problem; but as a whole, and using special norms, such approximations hold good. These methods have become known as the splitting-up methods. They have been studied by the Soviet mathematicians Yanenko [3, 15], Dyakonov [3, 15], Samarskii [3, 15], Saulyev [3], Marchuk [15], and others.

A large volume of research regarding the splitting-up methods has been devoted to choosing the parameters to be optimized by means of spectral or variational techniques, cf. Samarskii [3, 15], Dyakonov [15], Il'in [3, 15], and others.

Various aspects of the alternating direction method and the splitting-up method have been considered in papers by Andreev [15], Windlund [15], Fairweather and Mitchell [15], and others.

Multi-Dimensional Nonstationary Problems

Experience with the one-dimensional problems has formed a basis for devising algorithms for more complex problems of mathematical physics. An important stage in the development of the methods for nonstationary two-dimensional problems is the alternating-direction method based on the homogeneous approximations. Originally the method was applied to multi-dimensional parabolic equations, and it since has been widely used in many problems of mathematical physics.

Further advances in multi-dimensional nonstationary problems are related to the splitting-up techniques, based as a rule on the nonhomogeneous difference approximations of the original problems. Essentially, the splitting-up method consists in a reduction of a complicated operator to simpler operators. Thus the problem of solving a given equation is replaced by a sequence of problems with a simpler structure. Only at the end must the difference schemes satisfy at the same time the conditions of approximation and stability. This allows flexibility in constructing the schemes for virtually all fundamental problems of mathematical physics. In the case of explicit schemes the splitting-up method was suggested by Godunov and Bagrinovskii

[15]. In the implicit case splitting-up schemes have been given by Yanenko [15], Dyakonov [15], Samarskii [15], and others. These methods have found wide applications in diverse problems and have stimulated a more general approach to problems of mathematical physics on the basis of the so-called weak approximation, introduced by Yanenko [3, 15] and Samarskii [3, 15]. It has turned out that the splitting-up method can be viewed as the weak approximation of the original equation by a simpler one. Convergence criteria for the weak approximation method are given by the Yanenko–Demidov theorem [15] and can also be found in Lebedev [15] and Dyakonov [3, 15]. The method has found natural applications in problems of hydrodynamics, meteorology, oceanology, radiative transfer theory, etc. (see Marchuk [3, 17] and Yanenko [3]).

Another original scheme, widely applied in hydrodynamics, meteorology, and oceanology is the predictor–corrector method of Lax and Wendroff, in which the corrector is represented by an explicit difference scheme. This scheme is conditionally stable, has a very simple realization, and is second-order accurate with respect to all the variables. The detailed exposition is given in the book by Richtmyer and Morton [3].

Various variants of the predictor–corrector method, as based on implicit difference approximations, have been introduced by Bryan [4], Douglas [15], Sofronov [12], Marchuk and Yanenko [15], and others. It turns out that all these schemes are in a sense equivalent, differing only by their realizations. In the last of the sources just quoted the predictor is taken as an implicit splitting-up scheme with first-order accuracy and with the factorized operator. In problems of hydrodynamics the predictor is usually taken as an implicit dominated scheme.

Of particular interest is the method of decomposition and decentralization formulated by Lions and Temam [5]. It is similar to the splitting-up methods and to weak approximations.

The Particles-in-the-Cell Method

Recently there has been an intensive development of a new method for solving multi-dimensional problems of mathematical physics heralded by Harlow [19]. This method has become known as the method of large particles. It is widely used for computing the multi-dimensional hydrodynamical flows of highly compressible liquids with large relative displacements and colliding separation surfaces. The essence of the method is as follows: using a weak approximation at every small time interval, the equations of hydrodynamics are reduced to two simpler systems. The first system describes the mutual adaptation of the hydrodynamical fields with no account of the advection terms and is integrated by usual methods on a fixed Euler net. The second system describes the transport of the fluid in a

Lagrangian coordinate system. In order to solve the second system we use a phenomenological simplification of the fluid model, replacing it by a system of particles in every cell of the Euler system so that the overall balance of the mass, impulse, and energy of the particles in the cell coincides with the corresponding characteristics of the fluid. As soon as the trajectory (computed individually) of a particle, "bearing" certain mass, crosses the borders of a cell, the mass, impulse, and energy corresponding to this particle are subtracted from the abandoned cell and added to the new cell in which the particle is now located. The Harlow scheme is based on explicit solution methods at the first and second stages. It is globally conditionally stable and is especially fruitful at the first stages of the implicit schemes. In this case the stability criterion of the overall system coincides with the well-known Courant condition. Absolutely stable schemes of the particle method have not been obtained so far. The next few years should bring progress in this regard.

Recently Dyachenko [19], Belotserkovskii and Davydov [19], and Yanenko, Anuchina, Petrenko, and Shokina [19] have given various modifications of this method, which considerably reduce the density and the damping fluctuations characteristic of the method with improved stability. The above authors have also considered various realization schemes.

It is hoped that absolutely stable methods in the absence of the fluctuations will permit one to extend the applications of this technique to low-compressible liquids and to multi-dimensional problems in general.

The Monte Carlo Method

This method, suggested by von Neumann and Ulam, has been in development for more than two decades. The initial optimism for this method has eventually yielded to an equally unjustified pessimism. The heart of the matter is that it became clear at the early stages of the development that the efficiency of the Monte Carlo method would depend on computer speed. Millions of operations per second are required to implement the statistical tests for reducing the mean-squares error of the result.

Nevertheless, in spite of the difficulties in implementing this method on an average-size computer, or possibly thanks to them, the theory of the method has been significantly improved and its efficiency increased. The most important improvements have been achieved by bringing in the conditional probabilities of the processes and the statistical weights, defined by means of the information about the solutions of the adjoint equations related to the key functionals of the problem. Such methods reduce the dispersion error by an order of one or even two, and they consequently cut down the required solution time by an order of one or two as compared to methods of direct statistical modeling.

The third-generation computers have now brought in the necessary basis for the reactivation of this method. The Monte Carlo method has now a solid position in the theory of radiative transfer, public-service systems, quadrature and interpolation processes, integral equations, and algebraic equations. Lately the method has been tried on the nonlinear Boltzmann equation, for problems of linear programming, etc.

In connection with the Monte Carlo method important contributions were made by Vladimirov [18], Bakhvalov [18], Sobol [18], Chentsov [18], Fano, Spencer and Berger [18], Ermakova and Zolotukhina [18], Mikhailova [18], Buslenko and Golenko et al. [18], and others. Simple and universal as it stands, the Monte Carlo method will become, undoubtedly, an important tool of computational mathematics.

10.3 Conditionally Well-Posed Problems

An important role in the numerical treatment of problems of mathematical physics is played by the concept of well-posed problems, a concept first introduced by Hadamard at the beginning of the century. While there are a variety of classical problems of mathematical physics which are well posed in the sense of Hadamard, a deeper understanding of various problems in natural sciences and technology requires solving the so-called conditionally well-posed problems. Tikhonov [16] has formulated natural requirements on the ill-conditioned (in the sense of Hadamard) problems. Essentially, these conditions require that the solution of the problem be *a priori* assumed to exist in a given compact. In order to establish well-posedness, it is necessary to prove uniqueness.

Well-posed problems have been extensively studied by Lavrentiev [16]. Various aspects in the theory of the conditionally well-posed problems of mathematical physics have been considered by John [16], Mergelyan [16], Douglas [16], Krein [16], and others.

Tikhonov [16] has introduced the concept of regularization; that is, in place of the unbounded operator which gives the exact formula for the solution of the ill-posed problem, we consider a sequence (the regularization family) of continuous operators such that, on every element belonging to the existence domain, the corresponding sequence converges to the solution.

An interesting approach to ill-conditioned problems is provided by the theory of probability. Such investigations in their most complete form can be found in Lavrentiev and Vasyl'ev [16]. In the framework of probability one defines the concept of stability and constructs the optimal (in a sense) algorithms for various classes of problems, under some assumptions on the probabilistic properties of the input data errors and of the required solutions.

Lattes and Lions [16] have formulated a numerical method for inverse evolution problems using quasi-inversion. In their method, the evolution equation is supplemented by an additional regularization operator with a small parameter. This operator coincides with the product of the original operator and its adjoint. The small parameter is chosen by considering certain optimal estimates. The method is very simple as far as its realization is concerned.

In a joint paper with Atanbayev [16], the present author has developed a method for solving conditionally well-posed problems of the evolution type by means of the residual method on the whole space-time domain of definition of the solution. In this method one regularizes the problem by choosing the optimal number of steps of the iterative process on the basis of an *a priori* error estimate for the input data.

Morozov [14, 16] has made a very full study of the theory of improperly-posed problems and the regularization method.

As evidenced by the tendencies in solving conditionally well-posed problems, the techniques used are closely related to those of optimization of numerical processes.

10.4 Numerical Methods in Linear Algebra

There is an ever-increasing interest in solving large systems of linear algebraic equations with both sparse and dense matrices, in solving ill-conditioned systems, and in spectral problems for matrices of arbitrary structure. At the same time considerable attention has been paid to the possibility of exploiting the *a priori* as well as the *a posteriori* information about the problem in the course of the solution process. Computer technology has greatly influenced the process of reviewing the old linear algebraic methods and has stimulated interest in new computer-oriented algorithms.

Direct Methods of Linear Algebra

By direct methods of linear algebra we usually mean methods which solve a given problem in a finite number of arithmetic operations. Direct methods play a prominent role in solving systems of linear equations, inversions of matrices, and computing determinants. Direct methods can be used to decompose the original matrix into a product of two matrices each of which is easily invertible.

A classical example is the Gauss elimination method. Other distinguished examples fall under the heading of conjugate gradients methods: the method of conjugate gradients of Hestenes and Stiefel [11] and the residual method of Lanczos [3]. We can also trace back to the methods based on orthogonalization.

Lately there has been a considerable development of direct methods by Faddeev, Faddeeva and Kublanovskaya [8], **Bauer** [8], Householder [3],

Wilkinson [8], Henrici [4], Forsythe and Moler [8], Golub [9], Voevodin [8], and others.

Ill-conditioned systems still remain a problem. They are closely connected with the conditionally well-posed problems of mathematical physics. The difficulty here is the strong sensitivity of the solution to the errors in the matrix and in the right side of the system. Although there is already a number of significant results, we are still at the start of an extensive scientific development aimed at a general theory.

Iterative Methods

Iterative methods represent an important tool for solving problems of linear algebra. Their development has led to a number of good, computer-oriented algorithms. This progress was spurred first of all by the necessity to deal with problems of mathematical physics, economics, and control theory, which reduce eventually to high-order systems with matrices of special kinds. In such cases the direct methods are more or less powerless, although the progress in hardware increases their chances.

At present we can distinguish certain directions in the construction of iterative methods. We restrict our attention to only two of them. The first relies on the spectral characteristics of the operators involved. Methods of this type can be described as follows: one constructs an iterative process with the transition matrix depending on a set of parameters. These parameters can be taken either identical for all steps and chosen by minimizing the spectral radius of the transition matrix; or they are chosen at each step in such a way that the error vector approaches zero as fast as possible, uniformly in the initial approximations. In both cases we use the *a priori* information about the spectra of the corresponding matrices. The choice of such parameters is an integral part of the optimization of the numerical algorithm. As a rule, the main problem is in finding the corresponding spectral bounds.

Advances in spectral optimization of numerical algorithms have stimulated the formulations of a number of problems as exemplified below. It is to be kept in mind that the spectral optimization methods are especially convenient when dealing with a whole family of problems having the same operator and different input data. Increasingly greater interest has been recently paid to the Lanczos transformation [3] of matrices of arbitrary structure; it takes the original problem into one with a symmetric matrix, the spectrum of which consists of two intervals symmetrically placed about the origin. Such a symmetrization (along with some other techniques) can be used to accelerate convergence by exploiting the polynomials with the smallest deviation from zero on the above intervals. Another problem arises in connection with optimization of processes involving matrices with the spectra consisting of many intervals. Here we need first to develop methods for determining these intervals.

The second direction makes use of variationl principles. Methods of this class are used to implement a sequential minimization of certain functionals (quadratic as a rule), the minimum of which is achieved at the required solution. The foundation of the variational approach have been laid by Kantorovich [11], Lanczos [3], Hestenes and Stiefel [11], Krasnoselskii and Krein [11], and others. Among the recent contributions we note Petryshin [9, 10], Forsythe [11], Danial [11], Marchuk and Kuznetsov [3, 11], Godunov and Prokopov [11], Lebedev [9], and others.

The advantages of the variational methods of the steepest descent type and the iterative processes of minimal residuals stem from the fact that the relaxation parameters can be chosen by using the *a posteriori* information generated at each step. The rate of convergence of these methods does not drop below the one corresponding to Chebyshev methods. It is also very important that such methods converge, disregarding whether the matrix is symmetric or not (assuming it is positive). A number of efficient residual-type methods have been proposed recently.

An important circumstance which is still delaying development of nonstationary variational methods is the necessity to store a larger amount of transient information, much larger than that for the Chebyshev optimization methods.

Recently there is a trend to combine the spectral and variational techniques. Lebedev has formulated the conditions under which the iterative process has a nonimprovable estimate of the number of arithmetic operations required. There is also a probabilistic technique of choosing the optimal number of iterations. A number of interesting results in this area have been obtained by Vorob'ev [9]. The overrelaxation method of Young and Frankel [10] did not loose its great appeal, despite its becoming a classic. This method has been generalized by Wasov and Forsythe [3], Varga [3], Isaacson and Keller [3], and others. A systematic review of the iteration methods can be found in the book by Marchuk and Lebedev [17].

Much study has been devoted to iterative methods for solving linear systems with singular matrices. Marchuk and Kuznetsov [8, 11] proposed a general approach to the study of the convergence of stationary and nonstationary iterative processes in the case of compatible equation systems. This approach yielded not only a wider applicability of known iterative methods, but also a potential for the development of a class of new methods, termed the matrix analogs of the method of fictive domains (cf. Kuznetsov and Matsokin [4]). Iterative methods for the solution of incompatible systems have been proposed by Kuznetsov [8] and others.

Let us next turn to the iterative processes for the full eigenvalue problem with general matrices. We consider only the power methods since this trend has been recently enriched by some significant result due to Wilkinson [8], Bauer [8], Collatz [3], Voevodin [8], Frencis [8], Kublanovskaya [8], Eberlein [8], and many others.

The power methods are based on sequential transformations of the original matrix to one with easily obtainable eigenvalues, such as diagonal or triangular matrices, or a block-triangular matrix (the diagonal blocks being at most second order). For this purpose one can use either unitary similarity transformations (Jacobi method, QR-algorithm) or similarity transformations with triangular matrices (LR-algorithm).

Until recently the existing algorithms for the eigenvalue problem, such as the Jacobi algorithm (see Faddeev and Faddeeva [8] and Rutishauser [10]) required symmetric matrices. The discovery of the QR-algorithms (see Kublanovskaya [8] and Frencis [8]) and the generalized method of rotations (see Voevodin [8]) permit one to consider problems with arbitrary matrices. At present the emphasis is put on the development of various modifications of the QR-algorithm.

Advances in the eigenvalue problem are also due in part to the research related to nuclear reactor design. From this comes the interest in iterative methods for solving partial eigenvalue problems with nonnegative matrices. The foundations of these methods were built by Perron and Frobenius, and considerably developed by Varga [3], Traub [8] and others.

Round-Off Error Analysis

Until recently computational methods were judged by the number of arithmetic operations and the memory size required for their realization. Now we have to consider one more factor, namely the accuracy. In other words, the analysis of round-off errors has become an integral part of the algorithm.

The development in this area has been initiated by von Neumann. A systematic study regarding round-off errors has been done by Wilkinson [8]. The basis of his mathematical apparatus is the method of equivalent perturbations, which is used for obtaining the norm-estimates of all linear algebraic transformations. He has also obtained the norm-estimates of the equivalent perturbations for many methods.

The development of the equivalent perturbations method has been paralleled by that of statistical round-off error analysis. The origins of the latter are traced to Bakhvalov [8], Voevodin [8], Kim [8], and others. Statistical methods will play unquestionably an important role in the analysis of round-off errors in numerical schemes.

Complexes of Standard Programs

Achievements in numerical linear algebra have prompted the development of high-performance standard programs for solving systems of linear equations and finding eigenvalues. The journal *Numerische Mathematik*, for instance, has already come up with a large number of various procedures which are widely used for solving general problems of linear algebra, as well as special problems of mathematical physics, economics, and procedures related to matrices of special kinds.

This problem will surely attract the attention of researchers; the goal should be to develop a universal numerical system for solving problems of linear algebra. There are at least two trends already in progress in the pursuit of this goal. One is a thorough elaboration of complexes of algorithms and programs for the general problems; the other consists in developing universal methods adaptable to the special features of the given families of problems. Both trends are extremely interesting since they pave the way to a universal, problem-oriented software for the computers of fourth and later generations.

10.5 Optimization Problems in Numerical Methods

An important goal of computational mathematics is to find the fastest and most economical algorithms, in other words, optimization of numerical algorithms. In dealing with optimization problems under some given constraints, one usually has to rely on general mathematical theorems and estimate the lowest possible cost of solving a specific problem from a given class. Solution of a single isolated optimization problem usually does not give any practical answer. Nevertheless, if one knows how to find the conditionally extremal solution, i.e., the best technique of solving the local problem given the numerical possibilities and means, then we actually have the general problem in our hands. This concept of numerical optimization has been formulated by Bakhvalov [20], Nikol'skii [20], Babuška and Sobolev [20], and it sufficiently reflects the essence of the present problem.

Yet in many cases the optimal algorithm cannot be constructed, although it may be possible to construct a nearly optimal one. This situation is typical, for instance, of asymptotically optimal algorithms. We note that at the present time the theory of asymptotic estimates is in fact an effective tool for optimizing algorithms for various classes of problems.

The most developed subject from the computational point of view is unfortunately the theory of quadrature formulas due to Sobolev [1, 20] and Babuška [20]. Here the problem of estimating the quadrature formulas has been reduced to the minimization of a linear functional of errors. The above authors have obtained the quadrature formulas estimates on classes of periodic functions and infinitely differentiable functions. The methods are based on the asymptotic approximation estimates. The theory of quadrature formulas is also the subject of the contributions by Bakhvalov [20] and Sobol; they consider the optimal convergence estimates for the quadrature processes, the integration methods of the Monte Carlo type, and also the problem of designing the best techniques of numerical integration.

A different, number theoretical approach to quadrature formulas has been suggested by Vinogradov [2] and further pursued by Korobov [20], and Frolov [20]. The formulas obtained are exact on finite trigonometric polynomials. The loss estimates are derived on the class of periodic functions.

Kolmogorov [20] has introduced a number of general set-theoretical notions which can be used to estimate the bounds for the amount of numerical operations needed for solving a given problem. These estimates are of special importance for finding algorithms in the case where the upper and lower asymptotics diverge. He gave an estimate for problems involving linear differential operators with compact inverses. His estimate permits one to find algorithms which are asymptotically close to the optimal algorithms (in the sense of the number of arithmetic operations).

Bakhvalov [20] has studied a complex of algorithms for problems of mathematical physics by means of finite difference methods. In particular he has given lower estimates on the amount of operations needed for the solution of the Dirichlet problem for the Laplace equation. By a special choice of the net one can get also an upper estimate. The above kinds of estimates have been also obtained for the Peaceman–Rachford method [15] and the Douglas–Rachford method [15]. The method of Young and Frankel [10] and some other have also been considered.

An interesting trend in optimization has been developed by Lebedev [9, 15], by considering various difference methods. For fundamental minimization functionals he takes the so-called value of the algorithm and the entropy. The method has been used in transport theory.

In optimization problems we often have to abstract from many factors such as rounding-off of numbers in the realization process, various details in the implementation of arithmetic operations in the registers of a particular computer, etc. On the other hand, these factors sometimes determine the efficiency of the algorithm at hand. But this takes us to optimization problems regarding the computational process itself.

The theory of computational processes and their optimization have been dealt with in many contributions by Babuška [20], Dahlquist [20], Henrici [3], and others. Babuška, Vitásek, and Práger [3] have introduced the concept of α_k-sequences of numerical processes, reflecting thereby the fact that the accuracy of the computations should increase as a power of the increasing length of the computational process. In connection with the theory of α_k-sequences the concepts of local and global stability of numerical processes have been introduced. These concepts allow one to analyze large classes of actual algorithms of numerical mathematics.

Moor [20], Nickel [20], and others have introduced recently a new trend in estimating the accuracy of an actual algorithm; it has become known as interval arithmetic. The main purpose of interval arithmetic is to obtain *a posteriori* round-off-error estimates which can be analyzed by a double procedure on a single computer.

10.6 Optimization Methods

Problems with a constrained extremum have been known since antiquity. The developmental history of this field of learning may be (very conditionally) apportioned among various branches of mathematics in a series of stages. At the outset only separate problems were studied, principally of geometric origin. Some were highly significant and, suitably generalized, are still of interest today. The solution of the isoperimetric problem was known even to the ancient Greeks (Xenodorus). The general method of solving analytical problems with constrained maxima and minima is due to Lagrange. The technique of Lagrange multipliers yields a standard approach to the solution of many problems without subjecting them to a highly complicated individual study. However, only problems on smooth manifolds may be solved in this way. The boundaries of manifolds, imposed by inequalities, must be studied separately.

The next stage may be linked to the name of Minkovsky. We point, in particular, to his theorem on the possibility of obtaining all the consequences of a system of inequalities by combining them with nonnegative coefficients, and on the finite generability of the solution of a system of inequalities. These theorems essentially lay the foundation for the whole theory of linear programming. The actual founding of linear programming began much later, with the work of Kantorovich and Dantzig. The active development of methods of optimization began at that time. Studies on linear, convex, general nonlinear, and dynamic programming appeared in several countries. The majority were devoted to methods for numerical solution of these problems, and to their various applications. The work continues, primarily in the direction of finding correct and economical methods for the development of algorithms, and implementing them on electronic computers as packages of applied programs suitable for the solution of broad classes of practical problems.

The contemporary theory of optimal control is founded on questions that apparently were first seriously propounded in connection with the control of rocket flight. From the point of view of general theory, however, the results obtained were of a partial nature. Most of the problems lead to a nonclassical variational formulation—namely they are problems with constraints.

The first very important results were obtained in the theory of systems demanding minimal control time. The basic theory of processes optimized for linear systems with respect to high speed action was laid down in 1952–55, mainly in the work of Feldbaum.

In 1956 Pontrjagin formulated a principle leading to a solution of the general problem of finding the optimal control process with respect to speed of action. This, called the maximum principle, was first introduced for various types of systems and, in particular, was proved by Gamkrelidze for linear systems. Boltyanskii completely proved the Pontrjagin maximum principle in its role as a necessary condition for optimality with respect to speed of action. The principle was then extended to the general case—the minimization of a functional represented by the integral of a function of the variables of a system.

Any discussion of questions touching on the theory of optimal processes must include mention of the many papers by Bellman, The method of dynamic programing, as developed by him, provides a new machinery for solving variational problems and is closely linked to the Pontrjagin maximum principle.

As we noted in Chapter 8, the theory of optimal control includes problems with delays, with discrete time, with parameters, with isoperimetric constraints, and with problems of optimal control for partial differential equations. Papers based on other concepts and on the latest developments in optimal control theory are listed in Section 22 of the bibliography.

Many problems of mathematical physics admit a natural variational formulation, in which the task reduces to a search for the extremum of a functional. The variational approach allows us to remove the requirements that the solution be smooth when smoothness is not called for by the physical nature of the phenomenon under study; moreover, it allows us to construct an automatically stable difference scheme (cf. Chapter 2).

In practice, we often encounter problems of extremals on sets much more restricted than the traditional sets; the corresponding functionals may fail to be as smooth as needed for the classical variational computations. In this case we may apply the so-called variational inequalities, which yield solutions of highly complicated problems of mechanics and physics, heretofore insoluble.

Variational inequalities arise in various branches of the mechanics of continuous media, in free-boundary problems, in many problems of optimal control, and in others. Theoretical questions arising from this approach have been developed primarily by French mathematicians. A systematic exposition of the numerical methods for studying variational inequalities arising in various applications may be found in Glowinski, Lions, and Tremolières, *Analyse Numerique des Inequations Variationnelles*, Vols. 1 and 2, Dunod, Paris, 1976.

10.7 Some Trends in Numerical Mathematics

Advances in computer technology in recent years have had a significant effect on many areas of numerical sciences which show merging tendencies. The relationship between the hardware, methods of numerical and applied mathematics, the theory of automatic programming, and the theory of languages has become so close that choosing a strategy for solving a specific problem is presently of paramount significance. Although the optimization of the individual components of the numerical process is the core of the theory, as before, the attention becomes ever increasingly focused on the optimization of the process as a whole.

The optimization of numerical processes is presently, without any doubt, one of the central problems of computational sciences; it stimulates explorations of new numerical algorithms and methods of their realizations.

The next trend is related to the shift from solving particular problems to solving classes of problems and the standardization of algorithms. There is a need for systematization and ordering of the large flow of the computer-processed information. The valuable experience accumulated by solving problems of sciences and technology makes it possible in many cases to aim at universal methods suitable for handling wide classes of problems of the same type. A rational strategy in this direction depends on what class of problems we deal with. For diverse and rarely repeated problems we need to construct universal numerical algorithms which would adapt to the optimal regime by using the *a posteriori* information. In the case of frequently repeated problems, a reasonable strategy is a careful elaboration of special algorithms. These two approaches complement one another and form a basis for an economical exploitation of resources in creating an effective software system. The first steps in the development of the theory of universal optimal algorithms, self-adjusting to the optimal (in some sense) regime, have already been made, as well as the projections regarding the further development.

Several factors have contributed to the development of a new methodology—the creation of packages of applied programs aimed at the goals of science and technology. These factors include: the standardization of programs and algorithms; great progress in the use of computers for solving engineering, economic, and control problems; the growth in the power of computing techniques; and the raising of the level of systems integrity of computers. There can be no doubt that these factors will continue to develop actively, and will influence the methods of numerical.mathematics.

The computer systems of new generations, featuring high speeds and large memories, are becoming an effective storage place of the valuable and immediately accessible information. In addition, the multi-access systems permit one to materialize new dialog forms of the man-computer interaction. Therefore, the standardization of software in general and the numerical algorithms in particular is becoming an urgent problem.

The development in software confronted the numerical mathematics with a number of new problems, such as the problem of constructing a net for a complicated region, which would in some sense uniformly cover the domain of definition of the solution of the given problem. While this problem is nearly solved in the case of two-dimensional regions, it is still in a beginning stage in three and more dimensions. The problem of information inputs and outputs on a computer has also created a number of problems, namely with regard to the graphical representations of the information. This triggered, for instance, the development of new interpolation methods on various classes of functions.

Computer achievements in the area of analytic mappings have facilitated practical possibilities for solving problems of mathematical physics using well-developed methods of the theory of continuous argument. These methods will penetrate more and more extensively into software as the necessary hardware for implementing the analytic operations grows. Advances in the computer realizations of analytic transformations will bring new possibilities in the development of effective methods for problem solving.

Finally, let us note that the rate of growth in the development of numerical mathematics is determined by the level of research in the fundamental areas of mathematics. The importance of basic research and the rate with which it develops have increased significantly in the era of technological progress. Only harmonious research in all branches of mathematics will create necessary and favorable conditions for the spontaneous development of mathematics and its applications.

References

1 Functional Analysis and Numerical Mathematics

Anselone, P. H. M.: *Collectively Compact Operator Approximation Theory and Applications to Integral Equations*. Englewood Cliffs, N.J.: Prentice-Hall, 1967.

Balakrishnan, A. V.: *Applied Functional Analysis*. New York: Springer-Verlag, 1976.

Collatz, L.: *Functional Analysis and Numerical Mathematics*. New York: Academic Press, 1966.

Kantorovich, L. V.: Functional analysis and applied mathematics. *Usp. Mat. Nauk*, **3**, 6 (1948) [Russian].

Kantorovich, L. V. and Akilov, G. P.: *Functional Analysis in Normed Spaces*. New York: Macmillan, 1964.

Keldysh, M. V. and Lidskii, V. B.: On the spectral theory of non-self-adjoint operators. In: *Trudy IV Vsesoyuz. Mat. S'ezda, Vol. I*. Moscow, Izd. AN SSSR, 1963 [Russian].

Kolmogorov, A. N. and Fomin, S. V.: *Introductory Real Analysis*. Englewood Cliffs, N.J.: Prentice-Hall, 1970.

Kolmogorov, A. N. and Fomin, S. V.: *Elements of Function Theory and Functional Analysis*. Moscow: Nauka, 1976 [Russian].

Krasnosel'skii, M. A., Vainikko, G. M., Zabreiko, P. P., Rutickii, Ya. B., and Stetsenko, V. Ya.: *Approximate Solutions of Operator Equations*. Moscow: Nauka, 1969 [Russian].

Krein, S. G.: *Linear Differential Equations in Banach Space*. Providence, R. I.: American Mathematical Society, 1971.

Lavrentiev, M. A. and Shabat, B. V.: *Methods of Functions of a Complex Variable*. Moscow: Fizmatgiz, 1965 [Russian].

Lions, J. L.: *Equations Differentielles Opérationelles*. Berlin–Göttingen–Heidelberg: Springer-Verlag, 1961.

Lyusternik, L. A. and Sobolev, V. I.: *Elements of Functional Analysis*. New York: Ungar, 1964.

Mikhlin, S. G.: *The Minimum of a Quadratic Functional*. Moscow: Gostekhizdat, 1952.

Mikhlin, S. G.: *Variational Methods in Mathematical Physics*. New York: Macmillan, 1964.

Mikhlin, S. G. and Smolickii, Kh. L.: *Approximate Methods for Solving Differential and Integral Equations*. Moscow: Nauka, 1965.

Natanson, I. P.: *Constructive Function Theory*. New York: Ungar, 1964.

Natanson, I. P.: *Theory of Functions of a Real Variable*. Moscow: Nauka, 1974.

Nikol'skii, S. M.: *Approximation of Functions of Many Variables and Imbedding Theorems*. Moscow: Nauka, 1969 [Russian].

Sobolev, S. L.: *Applications of Functional Analysis in Mathematical Physics*. Providence, R. I.: American Mathematical Society, 1963.

Sobolev, S. L.: *Lectures on Two-Dimensional Numerical Integration Formulas*. Novosibirsk: Izd. Novosib. Univ., 1964, 1965 [Russian].

Sobolev, S. L.: *Introduction to the Theory of Cubature Formulae*. Moscow: Nauka, 1974.

Varga, R.: *Functional Analysis and Approximation Theory in Numerical Analysis*. Philadelphia: The Society for Industrial and Applied Mathematics, 1971.

Weinstein, A. and Stenger, W.: *Methods of Intermediate Problems for Eigenvalues: Theory and Ramifications*. London–New York: Academic Press, 1972.

2 Partial Differential Equations and Mathematical Physics

Bitsadze, A. V.: *Boundary Value Problems for Elliptic Equations of the Second Order*. Moscow: Nauka, 1966 [Russian].

Bossavit, A.: *Regularisation d'Equations Variationnelles et Applications*. Centre National de la Recherche Scientifique, Institut Blaise Pascal, 1970.

Courant, R.: *Partial Differential Equations*.

Courant, R. and Hilbert, D.: *Methods of Mathematical Physics*. New York: Interscience, 1953.

Fichtenholtz, G. M.: *Differential and Integral Computation, Vol. 3*. Moscow: Fizmatgiz, 1963 [Russian].

Friedrichs, K.: Nonlinear hyperbolic differential equations for functions of two independent variables. *Amer. J. Math.*, **70** (1948).

Godunov, S. K.: *Equations of Mathematical Physics*. Moscow: Nauka, 1971 [Russian].

Kantorovich, L. V. and Krylov, V. I.: *Approximate Methods of Higher Analysis*. Moscow–Leningrad: Fizmatgiz, 1962 [Russian].

Kondrat'ev, V. A.: Boundary problems for elliptic equations in domains having conic or angular points. *Trudy Moskovskogo matematiceskogo obscestva*, **16** (1967).

Ladyzhenskaya, O. A.: *Mixed Problems for Hyperbolic Equations*. Moscow: Gostekhizdat, 1953 [Russian].

Lavrentiev, M. M.: *Variational Methods for Boundary Value Problems for Equations of Elliptic Type*. Moscow: Izd. AN SSSR, 1952 [Russian].

Lavrentiev, M. M.: *On Some Ill-Posed Problems of Mathematical Physics*. Novosibirsk: Izd. Sib. Otd, AN SSSR, 1962 [Russian].

Lions, J. L.: *Quelques Méthodes de Résolution des Problèmes aux Limites non Linéaires*. Paris: Dunod, 1969.

Lions, J. L. and Magenes, E.: *Nonhomogenous Boundary Value Problems and Applications*. Berlin–New York: Springer-Verlag, 1972, 1973.

Mignot, A., Methodes d'approximation des solutions de problemes aux limites. *Rend. del sem. Mat. della Univ. di Padova*, **11** (1968).

Miller, J. H. (Ed.): Topics in numerical analysis. In: *Proceedings of the Royal Irish Academy Conference on Numerical Analysis, 1972*. London–New York: Academic Press, 1973.

Miller, J. H. (Ed.): Topics in numerical analysis. In: *Proceedings of the Royal Irish Academy Conference on Numerical Analysis, 1974*. New York–London: Academic Press, 1975.

Petrovskii, I. G.: *Lectures on Partial Differential Equations*. New York: Interscience, 1954.

Rozhdestvenskii, B. L. and Yanenko, N. N.: *Systems of Quasi-Linear Equations*. Moscow: Nauka, 1968 [Russian].

Smirnov, V. I.: *Lectures in Higher Mathematics*. Vols. *1–5*. Moscow: Gostekhizdat, 1948 [Russian].

Sobolev, S. L.: *Equations of Mathematical Physics*. Moscow: Nauka, 1966 [Russian].

Tikhonov, A. N. and Samarskii, A. A.: *Equations of Mathematical Physics*. Moscow: Nauka, 1966 [Russian].

Vekua, I. N.: *New Methods for Elliptic Equations*. Moscow: Gostekhizdat, 1948 [Russian].

Visik, M. I. and Lyusternik, L. A.: Asymptotic behavior of solutions of differential equations with large or fast-growing coefficients and boundary conditions. *Usp. Mat. Nauk*, **15**, 4 (1964) [Russian].

Vladimirov, V. S.: *Equations of Mathematical Physics*. Moscow: Nauka, 1967 [Russian].

Vladimirov, V. S.: *Generalized Functions in Mathematical Physics*. Moscow: Nauka, 1976.

3 Numerical Methods (Monographs and Textbooks)

Aubin, J. P.: *Approximate Solution of Elliptic Boundary Problems*.

Babuška, I., Vitásek, E., and Práger, M.: *Numerical Processes for Solving Differential Equations*. New York: Interscience, 1966.

Bakhvalov, N. S.: *Numerical Methods, Vol. 1*. Moscow: Nauka, 1973 [Russian].

Balakrishnan, A. V. and Neustadt, L. W.: *Computing Methods in Optimization Problems*. New York: Academic Press, 1964.

Bellman, R., Kalaba, R., and Lockett, J.: *Numerical Inversion of the Laplace Transform*. New York: American Elsevier, 1966.

Berezin, I. S. and Zhidkov, N. P.: *Numerical Methods, Vols. 1–2*. Moscow: Fizmatgiz, 1962, 1966 [Russian].

Blum, E. K.: *Numerical Analysis and Computation: Theory and Practice*. London: Addison-Wesley, 1972.

Bramble, J. H. (Ed.): Numerical solution of partial differential equations. In: *Proceedings of a Symposium Held at the University of Maryland*. London–New York: Academic Press, 1966.

Cea, J.: *Optimization Théorie et Algorithmes*. Paris: Dunod, 1971.

Collatz, L.: *The Numerical Treatment of Differential Equations*. Berlin: Springer-Verlag, 1960.

Dahlquist, G. and Bjorck, A.: *Numerical Methods*. Englewood Cliffs, N.J.: Prentice-Hall, 1974.

Dorodnitsyn, A. A.: On a numerical method for certain nonlinear problems of aerodynamics. In: *Trudy III Vsesoyuz. Mat. S'ezda, Vol. 2*. Moscow: Izd. AN SSSR, 1956 [Russian].

Dorodnitsyn, A. A.: *Lectures on Numerical Methods for Equations of Viscous Liquids*. Moscow: Izd. VC AN SSSR, 1969 [Russian].

Drobishevich, V. I., Dymnikov, V. P., and Rivin, G. S.: *Problems in Numerical Mathematics*. Moscow: Nauka, 1980.

Dyakonov, E. G.: *Iterative Methods for Solving Difference Analogs of Boundary Value Problems for the Equations of Elliptic Type*. Kiev: Izd. Isnt. Kibernet. AN USSR, 1970 [Russian].

Dyakonov, E. G.: *Difference Methods for the Solution of Boundary Problems, Vol. 1 (Stationary Problems)*. Moscow: Izd. Mosc. Univ., 1971. *Vol. 2 (Nonstationary Problems)*, Moscow: Izd. Mosc. Univ. 1972 [Russian].

Fox, L. and Mayers, D. F.: *Computing Methods for Scientists and Engineers*. Oxford, 1968.

Gel'fond, A. O.: *The Method of Finite Differences*. Moscow: Nauka, 1967 [Russian].

Godunov, S. K.: *Difference Equations for Equations of Gas Dynamics*. Novosibirsk: Izd. Novosib. Univ., 1962 [Russian].

Godunov, S. K. and Ryaben'kii, V. S.: *Introduction to the Theory of Difference Schemes*. Moscow: Fizmatgiz, 1962 [Russian].

Godunov, S. K. and Ryaben'kii, V. S.: *Difference Schemes*. Moscow: Nauka, 1973 [Russian].

Godunov, S. K., Zabrodin, A. V., Ivanov, M. Ya. et al.: *Numerical Solution of Multi-Dimensional Problems in Gas Dynamics*. Moscow: Nauka, 1976 [Russian].

Henrici, P.: *Error Propagation for Difference Methods*. New York: Wiley, 1963.

Householder, A. S.: *Principles of Numerical Analysis*. New York: McGraw-Hill, 1953.

Il'in, V. P.: *Difference methods for the solution of elliptic equations*. Novosibirsk: Izd. Novosib. Univ., 1970 [Russian].

Il'in, V. P.: *Numerical Methods for Problems of Electro-Optics*. Novosibirsk: Nauka, 1974 [Russian].

Isaacson, E. and Keller, H. B.: *Analysis of Numerical Methods*. New York: Wiley, 1966.

Kantorovich, L. V. and Krylov, V. I.: *Approximate Methods in Numerical Analysis*. Moscow–Leningrad: Fizmatgiz, 1962 [Russian].

Keller, H. B.: *Numerical Methods for Two-Point Boundary-Value Problems*. New York: Blaisdell, 1968.

Konovalov, A. N.: *Numerical Solution of Problems of the Theory of Elasticity*. Novosibirsk: Izd. Novosib. Univ., 1968 [Russian].

Lanczos, C.: *Applied Analysis*. Englewood Cliffs, N.J.: Prentice-Hall, 1956.

Lions, J. L. and Marchuk, G. I.: *Sur les Methodes Numeriques en Sciences Physiques et Economiques*. Paris: Dunod, 1974.

Marchuk, G. I.: *Design of Nuclear Reactors*. Moscow: Atomizdat, 1961 [Russian].

Marchuk, G. I.: *Numerical Methods in Weather Prediction*. Leningrad: Gidrometizdat, 1967 [Russian].

Marchuk, G. I.: Methods and problems of computational mathematics. In: *Proceedings of the International Congress of Mathematicians*. Nice, 1970.

Marchuk, G. I.: *Methods of Numerical Analysis*. New York–Heidelberg–Berlin: Springer-Verlag, 1975.

Miller, J. and Strang, G.: Matrix theorems for partial differential and difference equations. *Math. Scand.*, **18**, 2 (1966).

Mitchell, A. R.: *Computational Methods in Partial Differential Equations*. London: Wiley, 1970.

Mysovskikh, I. P.: *Lectures on Numerical Methods*. Moscow: Fizmatgiz, 1962 [Russian].

Mysovskikh, I. P.: Also in *Numerical Problems in Applied Mathematics* (edited by M. M. Lavrentiev. Novosibirsk: Nauka, 1975 [Russian].

Nikol'skii, S. M.: *Quadratic Formulae*. Moscow: Nauka, 1974 [Russian].

Polozhii, G. N.: *Numerical Solution of Two-Dimensional and Three-Dimensional Problems of Mathematical Physics and Functions with the Discrete Argument*. Kiev: Izd. Kiev. Univ., 1962 [Russian].

Pontryagin, L. S., Boltianskii, V. G., Gamkrelidze, R. V., and Mischenko, E. F.: *Mathematical Theory of Optimal Processes*. New York: Interscience, 1962.

Richtmyer, R. D.: *Difference Methods for Initial Value Problems*. New York: Interscience, 1957.

Richtmyer, R. D. and Morton, K. W.: *Difference Methods for Initial Value Problems*. New York: Wiley, 1967.

Samarskii, A. A.: *Introduction to the Theory of Difference Schemes.* Moscow: Nauka, 1971 [Russian].

Samarskii, A. A. and Andreev, V. B.: *Difference Methods for Elliptic Equations.* Moscow: Nauka, 1976 [Russian].

Samarskii, A. A. and Gulin, A. V.: *Stability of Difference Schemes.* Moscow: Nauka, 1975 [Russian].

Samarskii, A. A. and Popov, Yu. P.: *Difference Schemes in Gas Dynamics.* Moscow: Nauka, 1975 [Russian].

Saul'ev, V. K.: *Integration of the Equations of Parabolic Type by the Method of Nets.* Moscow: Fizmatgiz, 1960 [Russian].

Strang, G. and Fix, J.: *Theory of the Finite Elements Method.*

Traub, J. F.: *Iterative Methods for the Solution of Equations.* Englewood Cliffs, N.J.: Prentice-Hall, 1964.

Valiullin, A. N.: High precision schemes for problems of mathematical physics. In: *Lectures for Students of Novosibirsk University.* Novosibirsk: Izd Novosib. Univ., 1973 [Russian].

Varga, R. S.: *Matrix Iterative Analysis.* New York, 1963.

Vorob'ev, Yu. V.: *Method of Moments in Applied Mathematics.* Moscow: Fizmatgiz, 1958 [Russian].

Wasov, W. R. and Forsythe, G. E.: *Finite Difference Methods for Partial Differential Equations.* New York: Wiley, 1960.

Yanenko, N. N.: *The Method of Fractional Steps for Solving Multi-Dimensional Problems of Mathematical Physics.* Novosibirsk: Nauka, 1967 [Russian].

Yanenko, N. N.: *Introduction to Difference Schemes of Mathematical Physics.* Novosibirsk: Izd. Novosib. Univ., 1968 [Russian].

4 The Method of Nets

Belotserkovskii, O. M. and Chushkin, P. I.: The numerical method of integral relations *Zh. Vych. Matem. Matem. Fiz.*, **2**, 5 (1962) [Russian].

Bryan, K.: A Scheme for numerical integration of the equations of motion on an irregular grid free of nonlinear instability. *Monthly Weather Review*, **94**, 1 (1966).

Chudov, L. A. and Kudryavtsev, V. P.: On round-off errors of difference methods for elliptic equations and systems with initial conditions. In: *Numerical Methods in Gas Dynamics.* Moscow: Izd. Mosk. Univ., 1963 [Russian].

Crowley, W.: Second-order numerical advection. *J. Comp. Phys.*, **1**, 4 (1967).

Demyanovich, Yu. K.: Method of nets for certain problems of mathematical physics. *Doklady AN SSSR*, **159**, 2 (1964) [Russian].

Fichera, G.: Further development in the approximation theory of eigenvalues. In: *Numerical Solutions of Partial Differential Equations II, SYNSPADE 1970.* New York–London: Academic Press, 1971.

Fox, L., Henrici, P., and Moler, C.: Approximations and bounds for eigenvalues of elliptic operators. *SIAM J. Numer. Anal.*, **4**, 1 (1967).

Fromm, J. E.: Numerical methods for computing nonlinear time-dependent buoyancy of air in rooms. *J. Res. Dev. IBM*, **15**, 5 (1971).

Godunov, S. K. and Prokopov, G. P.: On the computation of conformal representations and the construction of difference nets. *Zh. Vych. Matem. Matem. Fiz.*, **7**, 5 (1967) [Russian].

Godunov, S. K. and Semendyaev, K. A.: Difference methods for the solution of problems in gas dynamics. *Zh. Vych. Matem. Matem. Fiz.*, **2**, 7 (1962) [Russian].

Godunov, S. K. and Zaborodin, A. V.: On difference schemes of second-order accuracy for multi-dimensional problems. *Zh. Vych. Matem. Matem. Fiz.*, **2**, 4 (1962) [Russian].

Joyce, D. C.: Survey of extrapolation processes in numerical analysis: *SIAM Review*, **13**, 4 (1971)

Keller, H.: A new difference scheme for parabolic problems. In: *Numerical Solutions of Partial Differential Equations, II, SYNSPADE 1970*. New York–London: Academic Press, 1971.

Kellogg, R.: Singularities in interface problems. In: *Numerical Solutions of Partial Differential Equations II, SYNSPADE 1970*. New York–London: Academic Press, 1971.

Konovalov, A. N.: The method of fictive domains in problems involving torsion. In: *Numerical Methods in the Mechanics of Continuous Media*. Novosibirsk: Izd. VC Sib. Otd. AN SSSR, 1973 [Russian].

Kopchenov, V. D.: Approximate solutions of the Dirichlet problem by the method of fictive domains. *Differential Equations*, **4**, 1 (1968) [Russian].

Kurihara, Y. and Holloway, I.: Numerical integration of a nine-level global primitive equations model formulated by the box method. *Monthly Weather Review*, **95**, 8 (1967).

Kuropatenko, V. F.: A method of constructing difference schemes for numerical solution of the equations of gas dynamics. *Izv. Vuzov. Matematika*, **3**(28) (1962) [Russian].

Kuznetsov, Yu. A. and Matsokin, A. M.: Solution of the Helmholtz equation by the method of fictive domains. In: *Numerical Methods of Linear Algebra*. Novosibirsk: Izd. VC Sib. Otd. AN SSSR, 1972 [Russian].

Kuznetsov, Yu. A. and Matsokin, A. M.: *A Matrix Analogue of the Method of Fictive Domains, with Applications*. Novosibirsk: Izd. VC Sib. Otd. AN SSSR, 1977 [Russian].

Kuznetsov, Yu. A. and Shaydurov, V. V.: On uniform convergence of difference schemes. In: *Numerical Methods of Linear Algebra*. Novosibirsk: Izd. VC Sib. Otd. AN SSSR, 1972 [Russian].

Landau, L. D., Meiman, N. N., and Khalatnikov, I. M.: Numerical methods for solving partial differential equations based on the method of nets. *Trudy III Vsesoyuz. Mat. S'ezda, Vol. II*. Moscow: Izd. AN SSSR, 1956 [Russian].

Lebedev, V. I.: The method of nets for the equations of Sobolev type. *Doklady AN SSSR*, **114**, 6 (1957) [Russian].

Lebedev, V. I.: On the method of nets for a system of partial differential equations. *Izv. AN SSSR, Ser. Mat.*, **22**, 5 (1958) [Russian].

Lebedev, V. I.: On the Dirichlet and Neumann problems on triangular and hexagonal nets. *Doklady AN SSSR*, **138**, 1 (1961) [Russian].

Lyusternik, L. A.: On difference approximations of the Laplace operator. *Usp. Mat. Nauk*, **IX**, 2 (1954) [Russian].

Marchuk, G. I., Dymnikov, V. P., Galin, V. Ya., et al.: *A Hydrodynamic Model of the General Circulation of the Atmosphere and the Ocean*. Novosibirsk: 1975 [Russian].

Marchuk, G. I., Rivin, G. S., and Yudin, M. I.: Numerical experiments with balancing schemes. *Izv. AN SSSR, Ser. FAO*, **11** (1973) [Russian].

Marchuk, G. I. and Shaidurov, V. V.: On the numerical solution of an evolution problem with a bounded operator. *Doklady AN SSSR*, **216**, 1 (1974) [Russian].

Marchuk, G. I. and Shaidurov, V. V.: Increasing the accuracy of projective-difference schemes. In: *Lecture Notes in Computer Science, Vol. 11*. Berlin–Heidelberg–New York: Springer-Verlag, 1974.

Matsokin, A. M.: On the development of the method of fictive domains. In: *Numerical Methods of Linear Algebra*. Novosibirsk: Izd. VC Sib. Otd. AN SSSR, 1972 [Russian].

Matsokin, A. M.: Automation of the triangulation of a region with a smooth boundary in the solution of an elliptic equation. Preprint No. 15 for the seminar: *Numerical Methods in Applied Mathematics*, VC Sib. Otd. AN SSSR, 1975 [Russian].

Matsokin, A. M.: On the construction and solution of variational-difference equations. Autoref. Dand. Dissert., Novosibirsk: Izd. Sib. Otd. VC AN SSSR, 1975 [Russian].

Matsokin, A. M.: The variational-difference method for solving elliptic equations in a three-dimensional region. In: *Variational-Difference Methods in Mathematical Physics.* Novosibirsk: Izd. VC Sib. Otd. AN SSSR, 1976 [Russian].

Penenko, V. V.: Numerical aspects in problems of the mathematical modelling of the dynamics of atmospheric processes. Autoref. Doct. Dissert., Novosibirsk, 1976 [Russian].

Raviart, P. A.: Sur l'approximation de certaines équations d'évolution linéaires et nonlinéaires. *J. Mathem. Pures Appliq.*, **46**, 1 (1967).

Richardson, L. F.: The approximate arithmetical solution by finite differences of physical problems involving differential equations, with an application to the stress in a masonry dam. *Philos. Trans. Roy. Soc., London, Ser. A*, **210** (1910).

Rivkind, V. Ya.: An approximate method for the Dirichlet problem, and on rate estimates for convergence of solutions of difference equations to solutions of elliptic equations with discontinuous coefficients. *Vestnik Leningradskogo Univ., Ser. Mat.*, **3** (1964) [Russian].

Rivkind, V. Ya.: On rate convergence estimates for homogeneous difference schemes corresponding to elliptic and parabolic equations with discontinuous coefficients. In: *Problems of Mathematical Analysis.* Leningrad: Izd. Leningr. Univ., 1966 [Russian].

Rukhovets, L. A.: A remark on the method of fictive domains. *Differential Equations*, **3**, 4 (1967) [Russian].

Samarskii, A. A.: On monotone difference schemes for elliptic and parabolic equations in the case of non-self-adjoint elliptic operator. *Zh. Vych. Matem. Matem. Fiz.*, **5**, 3 (1965) [Russian].

Samarskii, A. A.: On accuracy of the method of nets for the Dirichlet problem in an arbitrary region. *Apl. Mat.* **10**, 3 (1965) [Russian].

Samarskii, A. A.: Some problems in the theory of difference schemes. *Zh. Vych. Matem. Matem. Fiz.*, **6**, 4 (1966) [Russian].

Saul'ev, V. K.: On a method for automating the solution of boundary problems on fast computers. *Doklady AN SSSR*, **142**, 3 (1962) [Russian].

Saul'ev, V. K.: On the solution of some boundary problems on fast computers by the method of fictive domains. *Sibirsk. Mat. Journ.*, **4**, 4 (1963) [Russian].

Shaidurov, V. V.: On a method for increasing the accuracy of difference equations. In: *Numerical Methods in the Mechanics of Continuous Media.* Novosibirsk: Izd. VC Sib. Otd. AN SSSR, 1972 [Russian].

Tikhonov, A. N. and Samarskii, A. A.: On difference schemes for equations with discontinuous coefficients. *Doklady AN SSSR*, **108**, 3 (1956) [Russian].

Tikhonov, A. N. and Samarskii, A. A.: On homogeneous difference schemes. *Zh. Vych. Matem. Matem. Fiz.*, **1**, 1 (1961) [Russian].

Tikhonov, A. N. and Samarskii, A. A.: Homogeneous difference schemes on non-uniform nets. *Zh. Vych. Matem. Matem. Fiz.*, **2**, 6 (1962) [Russian].

Urvanets, A. L. and Shaidurov, V. V.: Increasing the accuracy of an approximate solution of the quasi-linear Poisson equation. In: *Variational-Difference Methods in Mathematical Physics.* Novosibirsk: Izd. VC Sib. Otd. AN SSSR, 1976 [Russian].

Valitskii, Yu. N.: On the convergence of difference approximations of eigenvalues and eigenvectors of a two-dimensional elliptic operator. *Doklady AN SSSR*, **198**, 2 (1971) [Russian].

Valiullin, A. N. and Yanenko, N. N.: Economical difference schemes of high precision for poly-harmonic equations. *Izv. Sib. Otd. AN SSSR, Ser. Tekn. Nauk*, **13**, 3 (1967) [Russian].

Volkov, E. A.: Investigation of a way to increase the precision of the method of nets for solving the Poisson equation. In: *Numerical mathematics, No.. 1.* Moscow: Izd. VC AN SSSR, 1957 [Russian].

Volkov, E. A.: Solving the Dirichlet problem by the method of increased accuracy, Parts I and II. *Differential Equations*, **1**, 7 and 8 (1965) [Russian].

Volkov, E. A.: The method of nonuniform nets for finite and infinite regions with conic points: *Differential Equations*, **10**, 2 (1966) [Russian].

Volkov, E. A.: The method of nets for the Laplace equation on finite and infinite regions with piecewise continuous boundaries. Dissertation, Moscow, 1967.

Wachpress, E. L.: The numerical solution of boundary value problems. In: *Mathematical Methods for Digital Computers*, New York: 1960.

Yanenko, N. N., Suchkov, V. A., and Pogodin, Yu. Ya.: On a difference solution of the heat equation in the curvilinear coordinates. *Doklady AN SSSR*, **128**, 5 (1959) [Russian].

5 Variational-Difference Methods

Agoshkov, V. I.: On the variational form of an integral identity due to G. I. Marchuk. Preprint VC Sib. Otd. AN SSSR, Novosibirsk, 1977.

Agoshkov, V. I.: A generalized formulation of the method of integral identities. Preprint VC Sib. Otd. AN SSSR, Novosibirsk, 1979 [Russian].

Aubin, J. P.: Approximation des espaces des distributions et des opérateurs différentiels. *Bull. Soc. Math. France, Memoire*, **12** (1967).

Aubin, J. P.: Behavior of the error of the approximate solutions of boundary value problems for linear elliptic equations by Galërkin's and finite difference methods. *Ann. Scuola Norm. Super., Pisa*, **21**, 4 (1967).

Aubin, J. P.: Best approximations of linear operators in Hilbert space. *SIAM J. Numer. Anal.*, **5**, 3 (1968).

Aubin, J. P. and Burchard, H. G.: Some aspects of the method of hypercircle applied to elliptic variational problems. In: *Numerical Solutions of Partial Differential Equations II, SYNSPADE 1970*. New York–London: Academic Press, 1971.

Babuška, I: The finite element method for elliptic differential equations. In: *Numerical Solutions of Partial Differential Equations II, SYNSPADE 1970*. New York–London: Academic Press, 1971.

Babuška, I.: The rate of convergence for finite element method. *SIAM J. Numer. Anal.*, **8**, 2 (1971).

Birkhoff, G., Schultz, M. H., and Varga, R. S.: Hermite interpolation in one and two variables with applications to partial differential equations. *Numer. Math.*, **11**, 3 (1968).

Bramble, J.: A second-order finite difference analog of the first biharmonic boundary value problem. *Numer. Math.*, **9**, 3 (1966).

Bramble, J. and Hubbard, B.: On the formulation of finite difference analogues of the Dirichlet problem for Poisson's equation. *Numer. Math.*, **4**, 4 (1962).

Bramble, J. and Schatz, A.: On the numerical solution of elliptic value problems by least-squares approximations of the data. In: *Numerical Solutions of Partial Differential Equations II, SYNSPADE 1970*. New York–London: Academic Press, 1971.

Cea, J.: Approximation opérationelle des problèmes aux limites. *Ann. Inst. Fourier, Grenoble*, **14**, 2 (1964).

Cea, J. and Glovinski R.: Sur des methodes d'optimisation par relaxation. *Revue Francaise d'Automatique, Informatique, et Recherche Operationnelle, 1973, R-3, S-32*.

Cea, J. and Glovinski, R.: Methodes numeriques pour l'écoulement laminaire d'un fluide rigide viscoplastique incompressible. *Inf. J. Comp. Math., Ser. B*, **3** (1974).

Courant, R.: Variational methods for the solutions of problems of equilibrium and variations. *Bull. Amer. Math. Soc.*, **49** (1943).

Douglas, J. and Dupont, T.: Alternating-direction Galërkin methods on rectangles. In: *Numerical Solutions of Partial Differential Equations II, SYNSPADE 1970*. New York–London: Academic Press, 1971.

Duvaur, G. and Lions, J. L.: *Les Inéquations en Mécanique et en Physique*. Paris: Dunod, 1972. English translation: *Grundlehren der Mathematik, 219*. Berlin–Heidelberg–New York: Springer-Verlag, 1976.

Federova, O. A.: A variational-difference scheme for the homogeneous diffusion equation. *Matem. zametki*, **17**, 6 (1975) [Russian].

Glovinski, R.: Introduction to the approximation of elliptic variational inequalities. *Report 76006, Laboratoire d'Analyse Numerique de l'Universite de Paris*, **6** (1976).

Glovinski, R., Lions, J. L., and Tremolières, R.: *Analyse Numerique des Inéquations Variationnelles, Vol. 1, No. 2*. Paris: Dunod, 1976.

Glovinski, R. and Marroco, A.: Sur l'approximation, par elements finis d'ordre un, et la resolution par penalisation—dualite, d'une classe de problemes de Dirichlet nonlineares. *Revue Francaise d'Automatique, Informatique, et Recherche Operationelle, 1975, R-2*.

Hubbard, B.: Remarks on the convergence in the discrete Dirichlet problem. In: *Numerical Solutions of Partial Differential Equations* (edited by J. H. Bramble). New York–London: Academic Press, 1965.

Keldysh, M. V.: On the Galërkin method for boundary value problems. *Izv. AN SSSR, Ser. Matem.*, **6** (1942) [Russian].

Lebedev, V. I.: Difference analogs for orthogonal expansions of the fundamental differential operators and certain boundary value problems of mathematical physics. *Zh. Vych. Matem. Matem. Fiz.*, **4**, 3 and 4 (1964) [Russian].

Lions, J. L. and Teman, R.: Une méthode d'éclament des opérateurs et des constraintes en calcul des variations. *C. R. Acad. Sci., Paris*, **263** (1966).

Oganesyan, L. A.: Numerical plate calculations. In: *Solution of engineering Problems on Electronic Computers*. Leningrad: 1963 [Russian].

Oganesyan, L. A.: A variational-difference scheme on a regular net for the Dirichlet problem. *Zh. Vych. Matem. Matem. Fiz.*, **11**, 6 (1971) [Russian].

Oganesyan, L. A., Rivkind, V. Ya., and Rukhovets, L. A.: Variational-difference methods for solving elliptic equations, Parts I and II. In: *Differential Equations and Their Applications, Vols. 5 and 8* Vilnius, 1974 [Russian].

Oganesyan, L. A. and Rukhovets, L. A.: On variational difference schemes for second-order linear elliptic equations in a two-dimensional region with the piecewise continuous boundary. *Zh. Vych. Matem. Matem. Fiz.*, **8**, 1 (1968) [Russian].

Oganesyan, L. A. and Rukhovets, L. A.: A study of the rate of convergence of variational-difference schemes for the second-order elliptic equations in a two-dimensional region with smooth boundary. *Zh. Vych. Matem. Matem. Fiz.*, **9**, 5 (1969) [Russian].

Rukhovets, L. A.: A study of the rate of convergence of variational-difference schemes for elliptic equations of second order. Dissertation, Leningrad, 1970 [Russian].

Shaidurov, V. V.: Richardson's extrapolation for the projective-difference Sturm–Liouville problem. In: *Variational-Difference Methods in Mathematical Physics*. Novosibirsk: Izd. VC Sib. Otd. AN SSSR, 1974 [Russian].

Smelov, V. V.: *Approximation of Piecewise-Smooth Functions by Trigonometric Polynomials and the Use of the Latter in Variational Methods*. Novosibirsk: Izd. VC Sib. Otd. AN SSSR, 1975 [Russian].

Strang, G.: The finite element method and approximation theory. In: *Numerical Solutions of Partial Differential Equations II, SYNSPADE 1970*. New York–London: Academic Press, 1971.

Strang, G. and Fix, G.: A Fourier analysis of the finite element variational method. Preprint (1970).

Zienkiewicz, O.: *The Finite Element Method in Structural and Continuum Mechanics*. London: McGraw-Hill, 1967.

Zlámal, M.: On the finite element method. *Numer. Math.*, **12**, 5 (1968).
Zlámal, M.: On some finite element procedures for solving second-order boundary value problems. *Numer. Math.*, **14**, 1 (1969).

6 The Theory of Stability of Difference Schemes

Federov, M. V.: On *c*-stability of the Cauchy problem for difference equations and partial differential equations. *Zh. Vych. Matem. Matem. Fiz.*, **7**, 3 (1967) [Russian].
Filippov, A. F.: On stability of difference equations. *Doklady AN SSSR*, **100**, 6 (1955) [Russian].
Il'in, A. M.: Stability of difference schemes for the Cauchy problem for systems of partial differential equations. *Doklady AN SSSR*, **164**, 3 (1965) [Russian].
Keller, H. B. and Thomee, V.: Unconditionally stable difference methods for mixed problems for quasi-linear hyperbolic systems in two dimensions. *Comm. Pure Appl. Math.*, **15**, 1 (1962).
Kreiss, H. O.: Uber die Stabilitätesdefinition für Differenzengleichungen die partielle Differentialgleichungen approximieren. *Nordisk Tikskr. Informations Behandlung*, **2**, 2 (1963).
Kreiss, H. O.: On difference approximations of the dissipative type for hyperbolic differential equations. *Comm. Pure Appl. Math.*, **17**, 3 (1964).
Kreiss, H. O.: Initial boundary value problems for partial differential and difference equations in one space dimension. In: *Numerical Solutions of Partial Differential Equations II, SYNSPADE 1970*. New York–London: Academic Press, 1971.
Lax, P. D.: On stability of finite difference approximations for hyperbolic equations with variable coefficients. *Matematika* (Selected translations), **6**, 3 (1962) [Russian].
Lax, P. D. and Nirenberg, L.: On stability of difference schemes: Exact form of Garding's inequality. *Matematika* (Selected Translations), **11**, 6 (1967) [Russian].
Lax, P. D. and Wendroff, B.: On the stability of difference schemes with variable coefficients. *Comm. Pure Appl. Math.*, **15**, 4 (1962).
Richtmyer, R.: On the nonlinear instability of difference schemes. In: *Some Problems of Numerical and Applied Mathematics*. Novosibirsk: Nauka, 1966 [Russian].
Ryabenki, V. S. and Filippov, A. F.: *On Stability of Difference Equations*. Moscow: Gostekhizdat, 1956 [Russian].
Samarskii, A. A.: Necessary and sufficient conditions of stability of two-layer difference schemes. *Doklady AN SSSR*, **181**, 4 (1968) [Russian].
Serdyukova, S. I.: *C*-stability of explicit difference schemes with constant real l_2-stable coefficients. *Zh. Vych. Matem. Matem. Fiz.*, **3**, 2 (1963) [Russian].
Strang, G.: Difference methods for mixed boundary value problems. *Duke Math. J.*, **27**, 2 (1960).
Thomee, V.: Generally unconditionally stable difference operators. *SIAM J. Numer. Anal.*, **4**, 1 (1967).

7 Stability and Convergence

Andreev, B. B.: On the convergence of difference schemes approximating the second and the third boundary value problems of elliptic equations. *Zh. Vych. Matem. Matem. Fiz.*, **8**, 6 (1968) [Russian].
Courant, R., Friedrichs, K., and Lewy, H.: Über die partiellen Differenzengleichungen der mathematischen Physik. *Math. Ann.*, **100**, 32 (1928).
Du Fort, E. C. and Frankel, S. P.: Stability conditions in the numerical treatment of parabolic differential equations. *Math. Tables Other Aids Comput.*, **7**, 43 (1953).
Godunov, S. K. and Ryabenkii, V. S.: Canonical forms of systems of linear ordinary difference equations with constant coefficients. *Zh. Vych. Matem. Matem. Fiz.*, **3**, 2 (1963) [Russian].

Godunov, S. K. and Ryabenkii, V. S.: Spectral stability criteria of boundary value problems for non-self-adjoint difference equations. *Usp. Matem. Nauk*, **XVIII**, 3 (111) (1963) [Russian].

John, F.: On the integration of parabolic equations by difference methods. I. Linear and quasi-linear equations for the infinite interval. *Comm. Pure Appl. Math.*, **5**, 2 (1952).

Ladyzhenskaya, O. A.: The method of finite differences in the theory of partial differential equations. *Usp. Matem. Nauk*, **XII**, 5 (1957) [Russian].

Lax, P. D. and Richtmyer, R. D.: Survey of the stability of linear finite difference equations. *Comm. Pure. Appl. Math.*, **9**, 2 (1956).

Lax, P. D. and Wendroff, B.: System of Conservation laws. *Comm. Pure Appl. Math.*, **13**, 2 (1960).

Lees, M.: *A priori* estimate for the solution of difference approximations to parabolic partial differential equations. *Duke Math. J.*, **27**, 3 (1960).

Lees, M.: Energy inequalities for the solution of differential equations. *Trans. Amer. Math. Soc.*, **94**, 1 (1960).

Lions, J. P.: Equations différentielles opérationnelles dans les espaces de Hilbert. Centro Int. Mat. Estivo, Varenna (1963). (Equazioni Differenziali Astratte, Cremonese, Roma (1963).

von Neumann, J. and Richtmyer, R. D.: A method for the numerical calculation of hydrodynamic shocks. *J. Appl. Physics*, **21**, 3 (1950).

O'Brien, G. G., Hyman, M. A., and Kaplan, S.: A study of the numerical solution of partial differential equations. *J. Math. Phys.*, **29**, 4 (1951).

Phillips, N. A.: *The Atmosphere and the Sea in Motion. Scientific Contributions to the Rossby Memorial Volume*. The Rockefeller Institute, 1959.

Ryabenkii, V. S.: Structure of spectra of families of non-self-adjoint difference operators. In: *Materials of Joint Soviet–American Symposium on Partial Differential Equations*. Novosibirsk, 1963 [Russian].

Ryabenkii, V. S.: The spectrum of a family of difference operators on functions on the net graph. *Zh. Vych. Matem. Matem. Fiz.*, **7**, 6 (1967) [Russian].

Samarskii, A. A.: Some problems in the general theory of difference schemes. In: *Partial Differential Equations* (Papers of the symposium honoring the sixtieth birthday of academician S. L. Sobolev). Moscow: Nauka, 1970 [Russian].

Sobolev, S. L.: Some remarks on numerical solutions of integral equations. *Izv. AN SSSR, Ser. Matem.*, **20**, 4 (1956) [Russian].

Strang, G.: Accurate partial difference methods. I. Linear Cauchy problem. *Arch. Rational Mech. Anal.*, **12**, 5 (1963).

Strang, G.: Implicit difference methods for initial boundary value problems. *J. Math. Anal.*, **16**, 1 (1966).

Thomee, V.: On the rate of convergence of difference schemes for hyperbolic equations. In: *Numerical Solutions of Partial Differential Equations II, SYNSPADE 1970*. New York–London: Academic Press, 1971.

Wendroff, B.: Well-posed and stable difference operators. *SIAM J. Numer. Anal.* **5**, 1 (1968).

Widlund, O. B.: Stability of parabolic difference schemes in the maximum norm. *Numer. Math*, **8**, 2 (1968).

Yanenko, N. N. and Boyarintsev, Yu. E.: On the convergence of difference schemes for the heat equation with variable coefficients. *Doklady AN SSSR*, **139**, 6 (1961) [Russian].

Yanenko, N. N. and Shokin, Yu. I.: On the relation between well-posed first differential approximations and the stability of difference schemes for hyperbolic systems. *Matematicheskie zametki*, **4**, 5 (1968) [Russian].

Yanenko, N. N. and Shokin, Yu. I.: On well-posed first differential approximations of difference schemes. *Doklady AN SSSR*, **182**, 4 (1968) [Russian].

8 Numerical Methods of Linear Algebra

Abramov, A. A.: Ideas of perturbation theory in some algorithms of linear algebra. In: *Numerical Methods of Linear Algebra, Vol. 1.* Moscow: VC AN SSSR, 1968 [Russian].

Bakhvalov, N. S.: On the question of the hypothesis of the independence of determination errors in numerical integration. *Zh. Vych. Matem. Matem. Fiz.*, **4**, 3 (1964) [Russian].

Bakhvalov, N. S.: Foundations of Numerical Mathematics. Moscow: Izd. Mosc. Univ., 1970 [Russian].

Bauer, F. L. and Fike, C. T.: Norms and exclusion theorems. *Numer. Math.*, **2**, 3 (1960).

Bellman, R.: *Introduction to Matrix Analysis.* New York: McGraw-Hill 1970.

Dorodnitsyn, A. A.: On the problem of computing eigenvalues and eigenvectors of matrices. *Doklady AN SSSR*, **126**, 6 (1959) [Russian].

D'yakonov, E. G.: On the solution of some elliptic difference equations. *J. Inst. Math. Applics.*, **7** (1971).

Eberlein, P.: A Jacobi-like method for the automatic computation of eigenvalues and eigenvectors of an arbitrary matrix. *J. Soc. Ind. Appl. Math.*, **10** (1962).

Fadheev, D. K.: On some sequential polynomials useful in the construction of iterative methods for the solution of systems of linear algebraic equations. *Vestnik LGU. Ser. matem.*, **7**, 2 (1958) [Russian].

Faddeev, D. K. and Faddeeva, V. N.: *Computational Methods of Linear Algebra.* San Francisco: H. W. Freeman, 1963.

Faddeev, D. K. and Faddeeva, V. N.: Numerical methods of linear algebra. In: *Notes of Science Seminars at LOMI, Vol. 54.* Leningrad: Nauka, 1975 [Russian].

Faddeev, D. K., Faddeeva, V. N., and Kublanovskaya, V. N.: Linear algebraic systems with rectangle matrices. In: *Numerical Methods of Linear Algebra.* Moscow: Nauka, 1968 [Russian].

Flanders, G. A. and Shortley, G.: Numerical determination of fundamental modes. *J. Appl. Phys.*, **21**, (1950).

Forsythe, G. E. and Moler, C. B.: *Computer Solution of Linear Algebraic Systems.* Englewood Cliffs, N.J.: Prentice-Hall, 1967.

Frencis, J.: The QR-transformation, Parts I and II. *Computer J.*, **4** (1961, 1962).

Gantmakher, F. R.: *The Theory of Matrices.* New York: Chelsea, 1959.

Ikramov, Kh. D.: *Matrix Norms and Jacobi-Like Methods.* Moscow: Izv. Mosc. Univ., 1969 [Russian].

Il'in, V. P.: On some estimates in the method of conjugate gradients. *Zh. Vych. Matem. Matem. Fiz.*, **16**, 4 (1976) [Russian].

Kellogg, R. and Noderer, L.: Sealed iterations and linear equations. *SIAM J.*, **8**, 4 (1960).

Kim, G.: *On the Distribution of the Round-Off Errors for Iteration Methods for Linear Algebraic Systems.* Moscow: Izd. Mosc. Univ., 1969 [Russian].

Kublanovskaya, V. N.: Orthogonal transformations in algebraic problems. Dissertation, Leningrad, 1972 [Russian].

Kuznetsov, Yu. A.: Iterative methods for solving incompatible systems of linear equations. In: *Some Problems of Numerical and Applied Mathematics.* Novosibirsk: Nauka, 1975 [Russian].

Kuznetsov, Yu. A.: Iterative methods for solution of noncompatible systems of linear equations. In: *Lecture Notes in Economics and Mathematical Systems, 134.* Berlin–Heidelberg–New York: Springer-Verlag, 1976.

Kuznetsov, Yu. A. and Marchuk, G. I.: Iterative methods for the solution of systems of linear equations with singular matrices. *Acta Universitatis Carolinse—Mathematica et Physica.* Prague: 1974 [Russian].

Kuznetsov, Yu. A. and Marchuk, G. I.: Stationary iterative methods for the solution of systems of linear equations with singular matrices. In: *Gatlinburg VI, Symposium on Numerical Algebra, Conference Manuscripts.* Munich: 1974.

Marchuk, G. I. and Kuznetsov, Yu. A.: Iterative methods and quadratic functionals. In: *Methods of Numerical Mathematics.* Novosibirsk: Nauka, 1975 [Russian].

Marek, I.: Iterations for linear bounded operators and Kellogg's process. Dissertation, Prague, 1962 [Czech].

Marek, I.: On iteration of linear bounded operators and the convergence of Kellogg's iteration process. *Czech Math. J.,* **12** (1962).

Nemchinov, S. V. and Libov, S. L.: A direct method of increasing accuracy of solutions of boundary value problems for the Helmholtz equation on a rectangular net region. *Zh. Vych. Matem. Matem. Fiz.,* **4,** 4 (1964) [Russian].

Samarskii, A. A. and Nikolaev, E. S.: *Methods for Solving Network Equations.* Moscow: Nauka, 1978 [Russian].

Stiefel, E.: Kernel polynomials in linear algebra and their numerical applications. *NBS, Appl. Math., Ser. 49,* 1 (1958).

Strang, G.: *Linear Algebra and Its Applications.* New York: Academic Press, 1976.

Traub, J.: *Iterative Methods for the Solution of Equations.* Englewood Cliffs, N.J.: Prentice-Hall, 1964.

Voevodin, V. V.: *Numerical Methods of Algebra. Theory and Algorithms.* Moscow: Nauka, 1966 [Russian].

Voevodin, V. V.: Round-off errors and stability in direct methods of linear algebra. Moscow: Izd. Mosc. Univ., 1969 [Russian].

Wilkinson, J.: *The Algebraic Eigenvalue Problem.* Oxford: Clarendon Press, 1965.

Wilkinson, J. H. and Reinsch, C.: *Linear Algebra.* Berlin–Heidelberg–New York: Springer-Verlag, 1971.

Young, D. M.: *Iterative Solution of Large Linear Systems.* London: Academic Press, 1971.

9 Optimization of Iterative Processes by Spectral Methods

Abramov, A. A.: On an acceleration method for iterative processes. *Doklady AN SSSR,* **74,** 6 (1950) [Russian].

Bakhvalov, N. S.: On the convergence of a relaxation method under the natural constraints on the elliptic operator. *Zh. Vych. Matem. Matem. Fiz.,* **6,** 5 (1966) [Russian].

Collatz, L.: Fahlerabschatzung für das Iterationsverfahren zur Auflösung linearer Gleichungssysteme. *Z. Angew. Math. Mech.,* 22 (1942).

Dyakonov, E. G.: Constructions of iteration methods using spectrum-equivalent operators. *Zh. Vych. Matem. Matem. Fiz.,* **6,** 1, 4 (1966) [Russian].

Fedorenko, R. P.: Solution of difference elliptic equations by the relaxation method. *Zh. Vych. Matem. Matem. Fiz.,* **1,** 5 (1961) [Russian].

Fedorenko, R. P.: On the rate of convergence of an iterative process. *Zh. Vych. Matem. Matem. Fiz.,* **4,** 3 (1964) [Russian].

Gavurin, M. K.: An application of the best approximation polynomials for improving convergence of iterative processes. *Usp. Matem. Nauk,* **V,** 3 (1950) [Russian].

Gavurin, M. K.: Nonlinear functional equations and continuous analogs of iterative methods. *Izv. Vuzov. Matematika,* **5,** 6 (1958) [Russian].

Golub, G. H. and Varga, R. S.: Chebyshev semi-iterative methods, successive overrelaxation iterative methods and second-order Richardson iterative methods. Parts I and II. *Numer. Math.,* **3,** 2 (1961).

Ivanov, V. K.: On the convergence of iterative processes for linear algebraic systems. *Izv. AN SSSR. Ser. Matem.,* 4 (1939) [Russian].

Juncosa, M. L. and Milliken, T. M.: On the increase of convergence rates of relaxation procedures for elliptic partial difference equations. *J. Assos. Comp. Math.*, **7**, 1 (1960).

Lanczos, C.: An iteration method for the solution for the eigenvalue problem for linear differential and integral operators. *J. Res. Nat. Bur. Stand.*, **45**, 1 (1950).

Lanczos, C.: Chebyshev polynomials in the solution of large-scale linear systems. *Proc. Assoc. Comput. Math.*, Toronto, September, 1952.

Lebedev, V. I.: On iteration methods of solving operator equations with the spectrum formed by several segments, *Zh. Vych. Matem. Matem. Fiz.*, **9**, 6 (1969) [Russian].

Lebedev, V. I.: On a construction of the *P*-operation in the *KP*-method. *Zh. Vych. Matem. Matem. Fiz.*, **9**, 4 (1969) [Russian].

Levedev, V. I.: Iterative methods with Chebyshev parameters for determining the largest eigenvalue and the corresponding eigenfunction. *Zh. Vych. Matem. Matem. Fiz.*, **17**, 1 (1977) [Russian].

Lebedev, V. I. and Finogenov, S. A.: On the order of choosing the iteration parameters in the Chebyshev cyclic iteration method. *Zh. Vych. Matem. Matem. Fiz.*, **11**, 2 (1971) [Russian].

Lebedev, V. I. and Finogenov, S. A.: Solution of the problem of ordering the parameters in Chebyshev iterative methods. *Zh. Vych. Matem. Matem. Fiz.*, **13**, 1 (1973) [Russian].

Lebedev, V. I. and Finogenov, S. A.: On the use of ordered Chebyshev parameters in iterative methods. *Zh. Vych. Matem. Matem. Fiz.*, **16**, 4 (1976) [Russian].

Marchuk, G. I. and Sarbasov, K. E.: On a method of solving stationary problems. *Doklady AN SSSR*, **182**, 1 (1968) [Russian].

Ostrowski, A. M.: On the linear iteration procedures for symmetric matrices. *Univ. Roma, Inst. Naz. Alta Math. Rend. Mat. e Appl.*, **14**, 1 and 2 (1954).

Petryshin, W.: On a general iterative method for the approximate solution of linear operator equations. *Math. Comput.*, **17**, 1 (1963).

Reich, I.: On the convergence of the classical iterative method of solving linear simultaneous equations. *Ann. Math. Statist.*, **20**, 3 (1949).

Vorob'ev, Yu. B.: A random iterative process. *Zh. Vych. Matem. Matem. Fiz.*, **4**, 6 (1964); and **5**, 5 (1965) [Russian].

Zolotarev, E. I.: Application of elliptic functions to problems of functions least and most different from zero. In: *Notes of the Russian Academy of Sciences, 1877* [Russian].

10 The Over-Relaxation Method

Broyden, C. G.: Some generalizations of the theory of successive over-relaxation. *Numer. Math.*, **6**, 4 (1964).

Broyden, C. G.: On convergence criteria for the method of successive over-relaxation. *Math. Comut.*, **18**, 85 (1964).

Evans, D. J.: Note on the linear over-relaxation factor for small mesh size. *Comput. J.*, **5**, 1 (1962).

Evans, D. J. and Forington, C. V.: An iterative process for optimizing symmetric successive over-relaxation. *Comput. J.*, **6**, 3 (1963).

Faddeev, D. K.: On the over-relaxation method in systems of linear equations. *Izv. Vuzov. Matematika*, **5** (1958) [Russian].

Garabedian, P.: Estimation of the relaxation factor for small mesh size. *Math. Tables Aid Comp.*, **10**, 56 (1956).

Gastinel, N.: Sur le meilleur des paramètres de sur-relaxation (Procédé de Peaceman-Rachford). *Chiffres*, **5**, 2 (1962).

Golub, G. H.: The use of Chebyshev matrix polynomials in the iterative solution of linear equations compared with the method of successive over-relaxation. Doct. Thesis, Univ. of Illinois, **133** (1959).

Hageman, L. A. and Kellogg, R. B.: Estimating optimum over-relaxation parameters. *Math. Comput.*, **22**, 101 (1968).

Klimova, E. G. and Rivin, G. S.: On methods for solving the Buleev–Marchuk equation. *Izv. AN SSSR, ser. FAO*, **15**, 4 (1979) [Russian].

Linn, M. S.: On the round-off error in the method of successive over-relaxation. *Math. Comput.*, **18**, 85 (1964).

Ostrowski, A. M.: On over- and under-relaxation in the theory of the cyclic single step iteration, *Math Tables Aid Comp.*, **7**, 43 (1953).

Petryshin, W.: On generalized over-relaxation method for the approximate solution of operator equations in Hilbert space. *SIAM J.*, **10**, 4 (1962).

Petryshin, W. V.: On the extrapolated Jacobi or simultaneous displacement method in the solution of matrix and operator equations. *Math. Comput.*, **19**, 89 (1965).

Rutishauser, H.: The Jacobi method for real symmetric matrices. *Numer. Math.*, **9**, 1 (1966).

Sheldon, J.: On the numerical solution of elliptic difference equations. *Math. Tables Aids Comput.* **9** (1955).

Varga, R. S.: *P*-cyclic matrices: The generalization of the Young–Francel successive over-relaxation scheme. *Pacif. J. Math.*, **9** (1959).

Varga, R. S.: Orderings of the successive over-relaxation scheme. *Pacif. J. Math.*, **9** (1959).

Young, D. M.: Iterative methods for solving partial difference equations of elliptic type. *Trans. Amer. Math. Soc.*, **74** (1954).

Young, D. M.: A bound for the optimum relaxation factor for the successive over-relaxation method. *Numer. Math.*, **16**, 5 (1971).

Young, D. M.: Convergence properties of the symmetric and unsymmetric successive over-relaxation methods and related methods. *Math. Comput.*, **24**, 112 (1971).

11 The Gradient Methods

Birman, M. Sh.: Some estimates for the steepest descent method. *Usp. Matem. Nauk*, **V**, 3 (1950) [Russian].

Danial, J. W.: The conjugate gradient method for linear and nonlinear operator equations. *SIAM J. Numer. Anal.*, **4**, 1 (1967).

Danial, J. W.: Convergence of the conjugate gradient method with computationally convenient modifications. *Numer. Math.*, **10**, 2 (1967).

Forsythe, G. E.: On the asymptotic directions of the *s*-dimensional optimum gradient method. *Numer. Math.*, **11**, 1 (1968).

Forsythe, A. I. and Forsythe, G. E.: Punched card experiments with accelerated gradient methods for linear equations. Contributions to the solution of linear equations and the determination of eigenvalues. *NBS Appl. Math., Ser. 39* (1954).

Forsythe, G. E. and Motzkin, T. S.: Asymptotic properties of the optimum gradient method. *Bull Amer. Math. Soc.*, **57**, 2 (1951).

Forsythe, G. E. and Motzkin, T. S.: Acceleration of the optimum gradient method. *Bull. Amer. Math. Soc.*, **57**, 4 (1951).

Fridman, V. M.: New methods for solving linear operator equations. *Doklady AN SSSR*, **128**, 3 (1959) [Russian].

Godunov, S. K. and Prokopov, G. P.: Variational approach to large systems of linear equations arising in strongly elliptic problems. Preprint of the Inst. Appl. Math. AN SSSR, Moscow, 1968 [Russian].

Godunov, S. K. and Prokopov, G. P.: On the difference Laplace equation. *Zh. Vych. Matem. Matem. Fiz.*, **9**, 2 (1969) [Russian].

Gorbenko, N. I. and Il'in, V. P.: On alternating directions gradient methods. In: *Problems of Numerical and Applied Mathematics*. Novosibirsk: Nauka, 1975 [Russian].

Hestenes, M. R. and Stiefel, E.: Method of conjugate gradients for solving linear systems. *J. Res. NBS*, **49** (1952).

Kantorovich, L. V.: On the steepest descent method. *Doklady AN SSSR*, **56**, 3 (1947) [Russian].

Krasnosel'skii, M. A. and Krein, S. G.: An iterative process with minimal residuals. *Matem. Sb.*, **31** (1952) [Russian].

Kuznetsov, Yu. A.: On the theory of iterative processes. *Doklady AN SSSR*, **184**, 2 (1969) [Russian].

Kuznetsov, Yu. A.: On the symmetrization of iterative processes. In: *Numerical Methods of Linear Algebra*. Novosibirsk: Izd. VC Sib. Otd. AN SSSR, 1969 [Russian].

Kuznetsov, Yu. A.: Some problems in the theory and applications of iterative methods. Dissertation, Novosibirsk, 1969 [Russian].

Lanczos, C.: Solution of the system of linear equations by minimized iterations. *J. Res. NBS*, **49**, 1 (1952).

Marchuk, G. I. and Kuznetsov, Yu. A.: On optimal iterative processes. *Doklady AN SSSR*, **181**, 6 (1968) [Russian].

Marchuk, G. I. and Kuznetsov, Yu. A.: Some problems in the theory of multi-step processes. In: *Numerical Methods of Linear Algebra*. Novosibirsk: Izd. VC Sib. Otd. AN SSSR, 1969 [Russian].

Marchuk, G. I. and Kuznetsov, Yu. A.: On solution of systems of linear equations by iterative methods. In: *Problems of Accuracy and Efficiency of Numerical Algorithms*, *V. 1*. Kiev: Izd. Inst. kibernetiki AN USSR, 1969 [Russian].

Samokish, B. A.: A study of the rate of convergence of the steepest descent method. *Usp. Matem. Nauk*, **XII**, 1 (1957) [Russian].

12 The Factorization Method

Abramov, A. A. and Andreev, V. B.: An application of the factorization method for finding periodic solutions of differential and difference equations. *Zh. Vych. Matem. Matem. Fiz.*, **3**, 2 (1963) [Russian].

Ains, E.: *Ordinary Differential Equations*. Moscow: ONTI, 1939.

Bakhvalov, N. S.: On the accumulation of computational errors in the numerical solution of differential equations. In: *VI, Moscow University*, **1** (1962) [Russian].

Buleev, N. I.: A numerical method for solving two- and three-dimensional diffusion equations. *Matem. Sb.*, **51**, 2 (1960) [Russian].

Buleev, N. I.: The method of incomplete factorization for the equation of two- and three-dimensional equations of diffusion type. *Zh. Vych. Matem. Matem. Fiz.*, **10**, 4 (1970) [Russian].

Buzbee, V., Golub, G., and Nilson, E.: On direct methods for solving Poisson's equations. *SIAM J. Numer. Anal.*, **7**, 4 (1970).

Degtyarev, L. M. and Favorskii, A. P.: Flow variant of the factorization method. *Zh. Vych. Matem. Matem. Fiz.*, **8**, 3 (1968) [Russian].

Degtyarev, L. M. and Favorovskii, A. P.: A flow variant of the factorization method for difference problems with strongly varying coefficients. *Zh. Vych. Matem. Matem. Fiz.*, **9**, 1 (1969) [Russian].

Fage, M. K.: On the factorization method. *Doklady AN SSSR*, **191**, 2 (1970) [Russian].

Gel'fand, I. M. and Lokutsievskii, O. B.: The factorization method for difference equations. In: *Introduction to the Theory of Difference Schemes* (S. K. Godunov and V. S. Ryabenkii, Eds.), Moscow: Fizmatgiz, 1962 [Russian].

Godunov, S. K.: The orthogonal factorization method for systems of difference equations. *Zh. Vych. Matem. Matem. Fiz.*, **2**, 6 (1962) [Russian].

Ogneva, V. V.: The factorization method for solving difference equations. *Zh. Vych. Matem. Matem. Fiz.*, **7**, 4 (1967) [Russian].

Oliphant, T. A.: An implicit numerical method for solving two-dimensional time-dependent diffusion problems. *Quart. Appl. Math.*, **XIX**, 3 (1961).

Rusanov, V. V.: Stability of the matrix factorization method. In: *Numerical Mathematics, No. 6*. Moscow: 1960 [Russian].

Sofronov, I. D.: On the factorization method for solving boundary value problems for difference equations. *Zh. Vych. Matem. Matem. Fiz.*, **4**, 2 (1964) [Russian].

Sofronov, I. D.: A diagonal matrix factorization scheme for the heat equation. *Zh. Vych. Matem. Matem. Fiz.*, **5**, 2 (1965) [Russian].

Vladimirov, V. S.: Approximate solution of a boundary problem for a second-order differential equation. *Prikl. matem. i mekh.*, **19**, 3 (1955) [Russian].

13 The Fast Fourier Transform

Bingham, C., Godfrye, M. D., and Tukey, J.: Modern techniques of power spectrum estimation. *IEEE Trans., Audio and Electroacoustics*, AU-15 (1967).

Cooley, J. W., Lewis, P. A., and Welch, P. D.: The fast fourier transform algorithms and applications. IBM Research Paper RC-1743, Feb. (1967).

Cooley, J. W. and Tukey, J. W.: An algorithm for machine calculation of complex Fourier series. *Math. Comp.*, **19**, 90 (1965).

Gold, B. and Rader, C. M.: *Digital Processing of Signals*. New York: McGraw-Hill, 1969.

Helms, R. D.: Fast Fourier transform method for computing difference equations and simulating filters. *IEEE Trans.*, AU-15 (1967).

Hockney, R. W.: A fast direct solution of Poisson's equation using Fourier analysis. *J. Assoc. Comp. Mach.*, **12**, 1 (1965).

Kaneko, T. and Liu, B.: Accumulation of roundoff error in fast Fourier transforms. *J. Assoc. Comput. Mach.*, **17** (1970).

Klauder, J. R., Price, A. C., Darlington, S., and Albersheim, W. J.: The theory and design of chirp radars. *Bell Syst. Tech. J.*, **39** (1960).

Kuznetsov, Yu. A. and Matsokin, A. M.: A solution of the Helmholtz equation by the method of fictive regions. In: *Numerical Methods of Linear Algebra*. Novosibirsk: Izd. VC Sib. Otd. AN SSSR, 1972 [Russian].

Nemchinov, S. V.: An application of the method of nets to boundary value problems for partial differential equations with periodic boundary conditions. In: *Dynamic Meteorology*. Tashkent: Nauka, 1965 [Russian].

Segeth, K.: *Roundoff Errors in the Fast Computation of Discrete Convolutions*. Prague: Math. Ustav, CSAV, 1979.

Tukey, J. W.: An introduction to the calculations of numerical spectrum analysis. In: *Spectral Analysis in Time Series* (edited by E. Harris). New York: Wiley, 1967.

14 Interpolation Using Spline Functions

Alberg, J. H., Nilson, E. N., and Walsh, J. L.: Extremal orthogonal lines. *J. Math. Anal. Appl.*, **12**, 1 (1965).

Alberg, J. H., Nilson, E. N., and Walsh, J. L.: *The Theory of Splines and Their Applications*. New York: Academic Press, 1967.

Anan'in, A. Z., Smelov, V. V., and Vasilenko, V. A.: An effective process for transforming a variational smoothing problem to a linear algebraic system. Preprint VC Sib. Otd. AN SSSR, Vol. 28. Novosibirsk, 1976 [Russian].

Anselon, P. M. and Laurent, P. J.: A general method for construction of interpolating or smoothing spline functions. *Numer. Math.*, **12**, 1 (1968).

Atteia, M.: Généralisation de la définition et des propriétés des "spline fonctions". *C. R. Acad. Sci., Paris*, **260** (1965).

Belonosov, A. S. and Tsetsokho, B. A.: A computational algorithm and procedures for smoothing functions defined approximately at the nodes of an irregular net in the plane. In: *Ill-Posed Problems of Mathematical Physics and Problems of Interpretation of Geophysical Experiments (Mathematical Problems of Geophysics)*. Novosibirsk: Izd. VC Sib. Otd. AN SSSR, 1976 [Russian].

Birkhoff, G. and Garabedian, P.: Smooth surface interpolation. *J. Math. Phys.*, **39**, 3 (1960).

de Boor, C.: Bicubic spline interpolation. *J. Math. Phys.*, **41**, 2 (1962).

Holladey, J. C.: Smoothest curve approximation. *Math. Tables Aid Comp.*, **11**, 60 (1957).

Lebedev, V. I.: On an interpolation method in n-dimensional space for arbitrary nodes and some quadratic formulae. Preprint VC Sib. Otd. AN SSSR, Vol. 10. Novosibirsk, 1975 [Russian].

Mikhalevich, Yu. I. and Omel'chenko, O. K.: *Procedures of Piecewise Polynomial Interpolation of Functions of One and Two Arguments*. Novosibirsk: Izd. VC Sib. Otd. AN SSSR, 1970 [Russian].

Morozov, V. A.: On the choice of parameter for the solution of functional equations by the method of regularization. *Doklady AN SSSR*, **175**, 6 (1967) [Russian].

Morozov, V. A.: Theory of splines and the problem of stable computation of the values of an unbounded operator. *Zh. Vych. Matem. Matem. Fiz.*, **11**, 3 (1971) [Russian].

Privovarova, N. B. and Pukhnacheva, T. P.: Smoothing of experimental data by local splines. Preprint VC Sib. Otd. AN SSSR, Vol. 9, Novosibirsk, 1975 [Russian].

Reinsch, C. H.: Smoothing by spline functions. *Numer. Math.*, **10**, 4 (1967).

Ryabenkii, V. S.: *Local Formulae for the Smooth Completion and Smooth Interpolation of Functions Given Their Values at the Nodes of a Nonuniform Rectangular Net.* Moscow: IPM AN SSSR, 1974 [Russian].

Schoenberg, I. J.: Contributions to the problem of approximation of equidistant data by analytic functions. *Quart. Appl. Math.*, **4** (1946).

Schumaker, L. L.: *Approximation by Splines: Theory and Application of Spline Functions*. New York–London: Academic Press, 1969.

Tsetsokho, Va. A., Belonosov, A. S., and Belonosova, A. V.: On a method for smooth approximation of a function of several variables. Preprint VC Sib. Otd. AN SSSR, Vol. 8. Novosibirsk, 1974 [Russian].

Vasilenko, V. A.: Convergence of splines in a Hilbert space. In: *Numerical Methods of Mechanics of Continuous Media*. Novosibirsk: Izd. VC Sib. Otd. AN SSSR, **3**, 3 (1972) [Russian].

Vasilenko, V. A.: Convergence of operator-interpolating splines. In: *Variational-Difference Methods in Mathematical Physics*. Novosibirsk: Izd. VC Sib. Otd. AN SSSR, 1973 [Russian].

Vasilenko, V. A.: Smoothing of splines on subspaces and compactness theorems. In: *Numerical Methods in the Mechanics of Continuous Media*. Novosibirsk: Izd. VC Sib. Otd. AN SSSR, **5**, 5 (1974) [Russian].

Vasilenko, V. A.: Finite-dimensional approximation in the method of least squares. In: *Numerical Methods of Mathematical Physics, Vol. 2*. Novosibirsk: Izd. VC Sib. Otd. AN SSSR, 1975 [Russian].

Vasilenko, V. A. and Perelomov, E. M.: Spline interpolation in rectangular regions with chaotically distributed nodes. In: *Machine Graphics and Their Application*. Novosibirsk: Izd. VC Sib. Otd. AN SSSR, 1973 [Russian].

Walsh, J. L., Alberg, J. H., and Nilson, E. N.: Best approximation properties of the spline fit. *J. Math. Mech.*, **11**, 2 (1962).

Yanenko, N. N. and Kvasov, B. I.: Iterative methods of constructive polycubic spline functions. *Numer. Methods Fluid Mech.* (*Informal bull.*), **1**, 3 (1970) [Russian].
Zav'alov, Yu. S.: Interpolation by cubic multi-responses. In: *Computational Systems, Vol. 38.* Novosibirsk: Izd. Inst. Math. Sib. Otd. AN SSSR, 1970 [Russian].
Zav'alov, Yu. S.: Interpolation by bi-cubic multiresponses. In: *Computational Systems, Vol. 38.* Novosibirsk: Izd. Inst. Math. Sib. Otd. AN SSSR, 1970 [Russian].
Zav'alov, Yu. S.: An extremal property of the cubic multi-responses and the smoothing problem. In: *Computational Systems, 42,* Novosibirsk: Izd. Inst. Math. Sib. Otd. AN SSSR, 1970 [Russian].
Zav'alov, Yu. S.: Interpolation by multicubic splines. In: *Computational Systems, Vol. 65.* Novosibirsk: 1975.

15 Splitting-Up Methods

Andreev, V. B.: On the splitting-up difference schemes for general p-dimensional parabolic equations of second order with mixed derivatives. *Zh. Vych. Matem. Matem. Fiz.*, **7**, 2 (1967) [Russian].
Bagrinovskii, K. A. and Godunov, S. K.: Difference methods for multi-dimensional problems. *Doklady AN SSSR*, **115**, 3 (1957) [Russian].
Baker, G. A.: An implicit numerical method for the n-dimensional heat equation. *Quart. Appl. Math.*, **17**, 4 (1960).
Baker, G. A. and Oliphant, T. A.: An implicit numerical method for solving the two-dimensional heat equation. *Quart. Appl. Math.*, **17**, 4 (1960).
Bensoussan, A.: Pure decentralization for inter-related payoffs. Symposium on Optimization, Los Angeles, 1971.
Bensoussan, A., Lions, J. L., and Teman, R.: Sur les méthodes de décomposition de décentralisation et de coordination et application. Cahiers IRIA n° a, Tome 2, Paris, 1972.
Birkhoff, G. and Varga, R. S.: Implicit alternating direction methods. *Trans. Amer. Math. Soc.*, **92**, 1 (1959).
Birkhoff, G., Varga, R., and Young, D.: Alternating direction implicit methods In: *Advances in Computing*, 3. New York–London: Academic Press, 1962.
Buleev, N. I.: A numerical method for the two- and three-dimensional diffusion equations. *Matem. Sb.*, **51**, 2 (1960) [Russian].
Douglas, J. and Gunn, J. I.: Two high-order correct difference analogues for the equation of multi-dimensional heat flow. *Math. Comput.*, **17**, 81 (1963).
Douglas, J. and Gunn, J. E.: A general formulation of alternating direction methods. Part I. Parabolic and hyperbolic problems. *Numer. Math.*, **6**, 5 (1964).
Douglas, J. and Jones, B. F.: On predictor-corrector methods for nonlinear parabolic differential equations. *J. Soc. Ind. Appl. Math.*, **11**, 1 (1963).
Douglas, J., Kellogg, R. B., and Varga, R. S.: Alternating direction methods for n space variables. *Math. Comput.*, **17**, 83 (1963).
Douglas, J. and Pearcy, C. M.: On convergence of alternating directions procedures in the presence of singular operators. *Numer. Math.*, **5**, 2 (1963).
Douglas, J. and Rachford, H.: On the numerical solution of the heat conduction problems in two and three space variables. *Trans. Amer. Math. Soc.*, **82**, 2 (1956).
Dupont, T.: A factorization procedure for the solution of elliptic difference equations. *SIAM J. Numer. Anal.*, **5**, 4 (1968).
D'akonov, E. G.: The alternating direction method for solving systems of equations in finite differences. *Doklady AN SSSR*, **138**, 2 (1961) [Russian].
D'akonov, E. G.: On certain difference schemes for boundary value problems. *Zh. Vych. Matem. Matem. Fiz.*, **2**, 1 (1962) [Russian].
D'akonov, E. G.: Difference schemes with split-up operators for multi-dimensional stationary problems. *Zh. Vych. Matem. Matem. Fiz.*, **2**, 4 (1962) [Russian].

D'akonov, E. G.: Solution of some multi-dimensional problems of mathematical physics using nets. Dissertation, Moscow, 1962 [Russian].

D'akonov, E. G. and Lebedev, V. I.: The splitting-up method for the third boundary value problem. In: *Numerical Methods and Programming, Vols. V and IV*. Moscow: Izd. Mosc. Univ., 1967 [Russian].

Fairweather, G. and Mitchell, A. R.: Some computational results of an improved ADI method for the Dirichlet problem. *Comput. J.*, **9**, 3 (1966).

Fryazinov, I. V.: On difference schemes for Poisson's equation in polar, cylindrical, and spherical coordinates. *Zh. Vych. Matem. Matem. Fiz.*, **11**, 5 (1971) [Russian].

Gunn, G. E.: The solution of elliptic difference equations by semi-explicit iterative techniques. *SIAM J. Numer. Anal.*, **2**, 1 (1965).

Hubbard, B. I.: Alternating direction schemes for the heat equation in a general domain. *SIAM J. Numer. Anal.*, **2**, 3 (1966).

Il'in, V. P.: On splitting up of difference equations of the parabolic and elliptic types. *Sib. Math. J.*, **VI**, 1 (1965) [Russian].

Il'in, V. P.: On explicit alternating direction schemes. *Izv. Sib. Otd. AN SSSR, Ser. tekhn. nauk*, **13**, 3 (1967) [Russian].

Kellogg, R. B.: Another alternating direction implicit method. *J. Soc. Ind. Appl. Math.*, **11**, 4 (1963).

Kellogg, R. B.: An alternating direction method for operator equations. *J. Soc. Ind. Appl. Math.*, **12**, 4 (1964).

Kellogg, R. B. and Spanier, J.: On optimal alternating direction parameters for singular matrices. *Math. Comput.*, **19**, 91 (1965).

Konovalov, A. N.: The method of fractional steps for solving the Cauchy problem for the multi-dimensional wave equation. *Doklady AN SSSR*, **147**, 1 (1962) [Russian].

Konovalov, A. N.: The application of the splitting-up method in dynamical problems of the theory of elasticity. *Zh. Vych. Matem. Matem. Fiz.*, **4**, 4 (1964) [Russian].

Konovalov, A. N.: *Filtering Problems of Multi-phase Incompressible Liquids*. Novosibirsk: Izd. Novosib. Univ., 1972 [Russian].

Kuznetsov, B. G.: Numerical methods for solving some problems of viscous liquids. *Fluid Dynamics Trans.*, **4**, 1969.

Lees, M.: Alternating direction methods for hyperbolic differential equations. *J. Soc. Ind. Appl. Math.*, **10**, 4 (1960).

Lees, M.: Alternating direction and semi-explicit difference methods for parabolic partial differential equations. *Numer. Math.*, **3**, 5 (1961).

Lions, P. L. and Mercier, B.: Splitting algorithms for the sum of two nonlinear operators. *Rapport Interne No. 29, Janvier 1978*. Centre de Mathematiques Appliques, Paris, France.

Marchuk, G. I.: On the theory of the splitting-up method. In: *Numerical Solutions of Partial Differential Equations II, SYNSPADE 1970*. New York–London: Academic Press, 1971.

Marchuk, G. I. and Sultangazin, U. M.: The splitting-up method for the transport equation. *Zh. Vych. Matem. Matem. Fiz.*, **5**, 5 (1965) [Russian].

Marchuk, G. I. and Yanenko, N. N.: The application of the splitting-up method (fractional steps) to problems of mathematical physics. In: *Some Problems of Numerical and Applied Mathematics*. Novosibirsk: Nauka, 1966 [Russian].

Peaceman, D. W. and Rachford, H. H.: The numerical solution of parabolic and elliptic differential equations. *SIAM J.*, **3**, 1 (1955).

Samarskii, A. A.: On an economical difference method for multi-dimensional parabolic equations in arbitrary regions. *Zh. Vych. Matem. Matem. Fiz.*, **2**, 5 (1962) [Russian].

Samarskii, A. A.: On convergence of the fractional steps methods for the heat equation. *Zh. Vych. Matem. Matem. Fiz.*, **2**, 6 (1962) [Russian].

Samarskii, A. A.: Locally one-dimensional difference schemes on nonuniform nets. *Zh. Vych. Matem. Matem. Fiz.*, **3**, 3 (1963) [Russian].

Samarskii, A. A.: On an economical algorithm for numerical solutions of differential and algebraic equations. *Zh. Vych. Matem. Matem. Fiz.*, **4**, 3 (1964) [Russian].

Samarskii, A. A.: Economical difference schemes for hyperbolic systems with mixed derivatives and their applications to the equations of the theory of elasticity. *Zh. Vych. Matem. Matem. Fiz.*, **5**, 1 (1965) [Russian].

Samarskii, A. A.: Additive schemes. In: *Papers of the International Symposium of Mathematicians in Moscow, 1966* [Russian].

Teman, R.: Sur la stabilité et la convergence de la méthode des pas fractionnaires. *Annali. di Mat. Pura Appl.*, **IV**, 79, 1968.

Teman, R.: Quelques méthodes de décomposition en analyse numérique. *Acte du Congres Intern. Math.*, V. 3, 1970.

Varga, R.: Some results in approximation theory with applications to numerical analysis. In: *Numerical Solutions of Partial Differential Equations II, SYNSPADE 1970*. New York–London: Academic Press, 1971.

Vorob'yev, Yu. V.: A random iterative process in the alternating direction method. *Zh. Vych. Matem. Matem. Fiz.*, **8**, 3 (1968) [Russian].

Wachspress, E. L.: Optimum alternating direction implicit iteration parameters for a model problem. *SIAM J.*, **10**, 2 (1962).

Wachspress, E. L.: Extended application of alternating direction implicit iteration model problem theory. *SIAM J.*, **11**, 3 (1963).

Wachspress, E. L.: *Iterative Solution of Elliptic Systems and Applications to the Neutron Diffusion Equations of Reactor Physics*. Englewood Cliffs, N.J.: Prentice-Hall, 1966.

Wachspress, E. L. and Habetler, G. J.: An alternating direction implicit iteration technique. *SIAM J.*, **8**, 2 (1960).

Widlund, O.: On the rate of an alternating direction implicit method in a noncommutative case. *Math. Comp.*, **20**, 96 (1966).

Widlund, O.: On the effects of scaling of the Peaceman–Rachford method. *Math. Comput.*, **25**, 113 (1971).

Yanenko, N. N.: On a difference method for the multi-dimensional heat equation. *Doklady AN SSSR*, **125**, 6 (1959) [Russian].

Yanenko, N. N.: On economic implicit schemes (the method of fractional steps). *Doklady AN SSSR*, **134**, 5 (1960) [Russian].

Yanenko, N. N.: On implicit difference methods for the multi-dimensional heat equation. *Izv. Vuzov. Matematika*, **4**, 23 (1961) [Russian].

Yanenko, N. N.: On convergence of the splitting-up method for the heat equations with variable coefficients. *Zh. Vych. Matem. Matem. Fiz.*, **2**, 5 (1962) [Russian].

Yanenko, N. N.: On the weak approximation of differential equations. *Sib. Math. J.*, **V**, 6 (1964) [Russian].

Yanenko, N. N. and Demidov, G. V.: The method of weak approximation as a constructive method for the Cauchy problem. In: *Some Problems of Numerical and Applied Mathematics*. Novosibirsk: Nauka, 1966 [Russian].

16 Conditionally Well-Posed Problems and Some Inverse Problems of Mathematical Physics

Anikonov, Yu. E.: *Methods for the Study of Multi-Dimensional Inverse Problems of Differential Equations*. Novosibirsk: Nauka, 1978 [Russian].

Berezanskii, Yu. M.: On the uniqueness of the determination of Schrodinger's equation by its spectrum. *Doklady AN SSSR*, **93**, 4 (1953) [Russian].

Bucheim, A. L.: On a class of operators of a Volterra equation of the first kind. *Funkts. analiz*, **6**, 1 (1972) [Russian].

Bucheim, A. L.: Operator equations of a Volterra equation in the scales of Banach spaces. *Doklady AN SSSR*, **242**, 2 (1978) [Russian].

Douglas, J.: On the relation between stability and convergence in the numerical solution of linear parabolic and hyperbolic differential equations. *J. Soc. Ind. Appl. Math.*, **4**, 1 (1956).

Faddeeva, V. N.: Motion in systems with ill-conditioned matrices. *Zh. Vych. Matem. Matem. Fiz.*, **5**, 5 (1965) [Russian].

Fedotov, A. M.: Two approaches to the study of conditionally well-posed problems with random errors in the initial data. Preprint 80 VC Sib. Otd. AN SSSR, Novosibirsk, 1978 [Russian].

Frank, L. S. and Chudov, L. A.: Difference methods for ill-posed Cauchy problems. In: *Numerical Methods in Gas Dynamics.* Moscow: Izd. Mosc. Univ., 1965 [Russian].

Fuks, K.: Perturbation theory in neutron multiplication problems. *Prob. Phys. Soc.*, **62**, 791 (1949).

Goncharskii, A. V., Cherepashuk, A. M., and Yagoda, A. G.: *Numerical Methods for Solving Inverse Equations of Astrophysics.* Moscow: Nauka, 1978 [Russian].

Ivanov, V. K.: On ill-posed problems. *Matem. Sb.*, **61**, 2 (1963) [Russian].

Ivanov, V. K., Vasin, V. V., and Tanana, V. P.: *Theory of Linear Ill-Posed Problems and its Applications*, Moscow: Nauka, 1978 [Russian].

John, F.: Differential equations with approximate and improper data. Lectures. New York University, 1955.

Kadomtsev, B. B.: On the effect function in the theory of radiative transfer. *Doklady AN SSSR*, **113**, 3 (1957) [Russian].

Krein, S. G.: On classes of certain well-posed problems. *Doklady AN SSSR*, **114**, 6 (1957) [Russian].

Krein, S. G. and Prozorovskaya, O. I.: On approximate methods for ill-posed problems. *Zh. Vych. Matem. Matem. Fiz.*, **3**, 1 (1963) [Russian].

Landis, E. M.: On some properties of the solutions of elliptic equations. *Doklady AN SSSR*, **107**, 4 (1956) [Russian].

Lavrentiev, M. M.: On the Cauchy problem for the Laplace equation. *Doklady AN SSSR*, **102**, 2 (1956) [Russian].

Lavrentiev, M. M.: Formulation of certain ill-posed problems of mathematical physics. In: *Some Problems of Numerical and Applied Mathematics*, Novosibirsk: Nauka, 1966 [Russian].

Lavrentiev, M. M.: Numerical solution of conditionally properly posed problems. In: *Numerical Solutions Of Partial Differential Equations II, SYNSPADE 1970.* New York–London: Academic Press, 1971.

Lavrentiev, M. M.: *Conditionally Ill-Posed Problems for Differential Equations.* Novosibirsk: Izd. Novosib. Univ., 1973.

Lavrentiev, M. M., Romanov, V. G., and Shishatskii, S. P.: *Ill-posed Problems of Mathematical Physics and Analysis.* Moscow: Nauka, 1980 [Russian].

Lavrentiev, M. M., Romanov, V. G., and Vasil'ev, V. G.: *Multi-Dimensional Inverse Problems for Differential Equations.* Novosibirsk: Nauka, 1969 [Russian].

Lavrentiev, M. M. and Vasil'ev, V. G.: On some ill-posed problems of mathematical physics. *Sib. Math. J.*, **VII**, 3 (1966) [Russian].

Lions, J. L. and Lattes, R.: *The Method of Quasi-Reversibility: Application to Partial Differential Equations.* New York: American Elsevier, 1969.

Magnitskii, N. A.: On a method for regularizing a Volterra equation of the first kind. *Zh. Vych. Matem. Matem. Fiz.*, **15**, 5 (1975) [Russian].

Marchenko, V. A.: *Sturm–Liouville Operators and Their Applications.* Kiev: Naukova dumka, 1978.

Marchuk, G. I.: Equations for the value of information from the meteorological satellites and formulations of inverse problems. *Kosmicheskie Issledovania*, **11**, 3 (1964) [Russian].

Marchuk, G. I.: Formulation of the theory of perturbations for complicated models. *Appl. Math. Optim.*, **2**, 1 (1975).

Marchuk, G. I.: Optimally accurate methods of solving renewal problems. Preprint 10 VC Sib. Otd. AN SSSR, Novosibirsk, 1976 [Russian].

Marchuk, G. I. and Atanbayev, S. A.: Some problems in global regularization. *Doklady AN SSSR*, **190**, 3 (1970) [Russian].

Marchuk, G. I. and Drobyshev, Yu. P.: Some problems in the linear theory of measurements. *Avtometriya*, 1967, No. 3 [Russian].

Marchuk, G. I. and Orlov, V. V.:On the conjugate functions theory. In: *Neitronnaya Fizika*. Moscow: Gosatomizdat, 1961 [Russian].

Marchuk, G. I. and Vasil'ev, V. G.: On an approximate solution of operator equations of the first order. *Doklady AN SSSR*, **199**, 4 (1970) [Russian].

Mergel'yan, S. N.: The harmonic approximation and the approximate solution of the Cauchy problem for the Laplace equation. *Usp. Matem. Nauk*, **XI**, 5 (1956) [Russian].

Morozov, V. A.: Methods of solving unstable problems (Lecture notes). Rotaprint Mosc. Univ., 1967 [Russian].

Mukhametov, R. G.: The regeneration problem in an anisotropic Riemannian metric in an n-dimensional region. Preprint 136 VC Sib. Otd. AN SSSR, Novosibirsk, 1978 [Russian].

Prilepko, A. I. Inverse problems in potential theory. *Matem. zametki*, **14**, 5 (1973) [Russian].

Romanov, V. G.: *Inverse Problems for Hyperbolic Equations*. Novosibirsk: Nauka, 1972 [Russian].

Romanov, V. G.: *Inverse Problems for Differential Equations (Inverse Kinematic Problems in Seismology)*. Novosibirsk: Izd. Novosib. Univ., 1978 [Russian].

Sergeev, V. O.: Regularization of the Volterra equation of the first kind. *Doklady AN SSSR*, **197**, 3 (1971) [Russian].

Shishatskii, S. P.: On a method for approximate solution of ill-posed Cauchy problems for the evolution equation. In: *Mathematical Problems of Geophysics, Vol. 3*, Novosibirsk: Izd. VC Sib. Otd. AN SSSR, [Russian].

Tikhonov, A. N.: On stability of inverse problems. *Doklady AN SSSR*, **39**, 5 (1943) [Russian].

Tikhonov, A. N.: On the solution of ill-posed problems using the method of regularization. *Doklady AN SSSR*, **151**, 3 (1963) [Russian].

Tikhonov, A. N.: On regularization of ill-posed problems. *Doklady AN SSSR*, **153**, 1 (1963) [Russian].

Tikhonov, A. N. and Arsenin, V. Ya.: Methods for solving ill-posed problems. *Doklady AN SSSR*, **153**, 1 (1963) [Russian].

Usachev, L. N.: An equation of the cost of the kinetic reactor neutrons and the perturbation theory. In: *Reactor Design and the Theory of Reactors*. Moscow: Izd. AN SSSR, 1955 [Russian].

17 Numerical Methods in the Transport Theory

Agoshkov, V. I.: On the smoothness of solutions of the transport equation and approximate methods for constructing solutions, Parts I and II. In: *Differential and Integro-Differential Equations*. Novosibirsk: Izd. VC Sib. Otd. AN SSSR, 1977 [Russian].

Agoshkov, V. I.: Some singularities of the solution of the transport equation, and the calculation of these in the course of constructing basis functions. Preprint 58 VC Sib. Otd. AN SSSR, Novosibirsk, 1977 [Russian].

Agoshkov, V. I.: Solution of the transport equation in the X-Y geometry by the method of integral identities. Preprint 159 VC Sib. Otd. AN SSSR, Novosibirsk, 1979 [Russian].

Bardos, P. G.: Equations du premier ordre à coefficients réels. *Ann. Scient. Ec. Norm. Sup.*, *4ᵉ serie*, 3 (1970).

Bogolyubov, N. N.: *Problems of the Dynamical Theory in Statistical Physics*. Moscow: Gostekhtheoryzdat, 1946 [Russian].

Germogenova, T. A.: On the convergence of certain approximate methods for the transport equation. *Doklady AN SSSR*, **181**, 3 (1968) [Russian].

Germogenova, T. A.: Generalized solutions of boundary value problems for the transport equation. *Zh. Vych. Matem. Matem. Fiz.*, **9**, 3 (1969) [Russian].

Godunov, S. K.: The energy integral and accuracy estimates of approximate eigenvalues. *Zh. Vych. Matem. Matem. Fiz.*, **11**, 5 (1971) [Russian].

Godunov, S. K. and Sultangazin, U. M.: On the dissipativity of the boundary conditions of V. S. Vladimirov for a symmetric system of the method of spherical harmonics. *Zh. Vych. Matem. Matem. Fiz.*, **11**, 3 (1971) [Russian].

Gol'din, V. Ya.: The quasi-diffusion method for the kinetic equation. *Zh. Vych. Matem. Matem. Fiz.*, **4**, 6 (1964) [Russian].

Jorgens, K.: An asymptotic expansion in the theory of neutron transport. *Comm. Pure Appl. Math.*, **11**, 2 (1958).

Karlson, B. and Bell, Dzh.: Solving the transport equation by the S_n method. In: *Physics of Nuclear Reactors* (Transl. from English), Moscow, 1963.

Kuznetsov, E. S. and Marchuk, G. I.: Numerical methods in the theory of radiative transfer. In: *Trudy IV Vsesoyuz. Mat. S'ezda, Vol. II*. Moscow: Izd. AN SSSR, 1964 [Russian].

Lebedev, V. I.: On solving kinetic problems in the theory of transfer. Doct. Dissertation, Novosibirsk, 1967 [Russian].

Lebedev, V. I.: On the *KP*-method and the difference schemes for the kinetic equations. In: *Numerical Methods in the Transport Theory*. Moscow: Atomizdat 1969 [Russian].

Marchuk, G. I.: *Numerical Methods in the Design of Nuclear Reactors*. Moscow: Atomizdat, 1958 [Russian].

Marchuk, G. I. and Kochergin, V. P.: An effective method for two-dimensional diffusion equation using the cells of rectangular and hexagonal forms. *Atomic Energy*, **18**, 6 (1965) [Russian].

Marchuk, G. I. and Lebedev, V. I.: *Numerical Methods in Neutron Transport Theory*. Moscow: Atomizdat, 1971 [Russian].

Marchuk, G. I. and Sultangazin, U. M.: On convergence of the splitting-up method for equations of radiative transfer. *Doklady AN SSSR*, **161**, 1 (1965) [Russian].

Marchuk, G. I. and Sultangazin, U. M.: On solving the kinetic equation of radiative transfer by the splitting-up method. *Doklady AN SSSR*, **163**, 4 (1965) [Russian].

Marchuk, G. I. and Yanenko, N. N.: Solution of the multi-dimensional kinetic equation by the splitting-up method. *Doklady AN SSSR*, **157**, 6 (1964) [Russian].

Marek, I.: On a problem of mathematical physics. *Appl. Math.*, **11**, 89 (1966).

Nikolaishvili, Sh. S.: An approximate solution of the transport equation by the method of moments. *Atomnaya energia*, **9**, 2 (1961) [Russian].

Nikolaishvili, Sh. S.: On the solution of the one-velocity transport equation using the Ivon–Martens approximations. *Atomnaya energia*, **20**, 4 (1966) [Russian].

Shikhov, S. B.: Some problems of the mathematical theory of the critical state of a reactor. *Zh. Vych. Matem. Matem. Fiz.*, **7**, 1 (1967) [Russian].

Smelov, V. V.: *Lectures on the Theory of Neutron Transfer*. Moscow: Atomizdat, 1972 [Russian].

Sultangazin, U. M.: Differential Properties of Solutions of the Mixed Cauchy Problem for the Nonstationary Kinetic Equation. Preprint VC Sib. Otd. AN SSSR, Novosibirsk, 1971 [Russian].

Sultangazin, U. M.: Concerning the weak approximation method for the equations of spherical harmonics. Preprint VC Sib. Otd. AN SSSR, Novosibirsk, 1971 [Russian].

Sultangazin, U. M.: Weak convergence of the spherical harmonics method. Preprint VC Sib. Otd. AN SSSR, Novosibirsk, 1971 [Russian].

Vladimirov, V. S.: The numerical solution of the kinetic equation for a sphere. In: *Numerical Mathematics, No. 3.* Moscow: Izd. AN SSSR, 1958 [Russian].

Vladimirov, V. S.: On some variational methods of approximate solution of the transport equation. In: *Numerical Mathematics, No. 7.* Moscow: Izd. AN SSSR, 1961 [Russian].

Vladimirov, V. S.: Mathematical problems in the one-velocity theory of particle transport. *Trudy Mat. Inst. AN SSSR*, **61**(1961).

18 The Monte Carlo Method

Bakhvalov, N. S.: Optimality of estimates of the rate of convergence of quadratic processes and of Monte Carlo methods for integration on classes of functions. In: *Numerical Methods for Solving Differential and Integral Equations and Quadrature Formulae.* Moscow: Nauka, 1964 [Russian].

Buslenko, N. P., Golenko, D. I., Sobol, I. M., Sragovich, V. G., and Shreyder, Yu. A.: *The Monte Carlo Method.* Moscow: Fizmatgiz, 1962 [Russian].

Ermakov, S. M.: *The Monte Carlo Method and Related Problems.* Moscow: Nauka, 1971 [Russian].

Ermakov, S. M. and Mikhailov, G. A.: *Statistical Modelling.* Moscow: Nauka, 1976 [Russian].

Ermakov, S. M. and Zolotukhin, V. G.: The polynomial approximations and the Monte Carlo method. *Teoria Veroyatnostei Prim.*, **5**, 4 (1960) [Russian].

Fano, U., Spencer, L., and Berger, M.: *Gamma Radiative Transfer.* Moscow: Gosatomizdat, 1963 [Russian transl. from English].

Gel'fand, I. M., Frolov, A. S., and Chentsov, N. N.: Computation of the continual intervals by the Monte Carlo method. *Izv. Vuzov. Matematika*, **5** (1958) [Russian].

Kertis, D.: The Monte Carlo methods for the iterations of linear operators. *Usp. Matem. Nauk*, **XII**, 5 (1957) [Russian].

Marchuk, G. I., Mikhailov, G. A., Nazaraliev, M. A., et al.: *The Monte Carlo Method in Atmospheric Optics.* Novosibirsk: Nauka, 1976 [Russian].

Metropolis, M. and Ulam, S.: The Monte Carlo method. *J. Amer. Stat. Assoc.*, **44**, 247 (1949).

Mikhailov, G. A.: *Some Problems in the Theory of the Monte Carlo Method.* Novosibirsk: Nauka, 1974 [Russian].

Sobol', I. M.: *Numerical Monte Carlo Methods.* Moscow: Nauka, 1973 [Russian].

Span'e, J. and Gelbard, Z.: *The Monte Carlo Method and Problems of Neutron Transport.* Moscow: Atomizdat, 1972 [Russian].

Vladimirov, V. S.: Application of the Monte Carlo method to find the smallest eigenvalue and the corresponding eigenfunction of a linear integral operator. *Teor. veryatn. i ee primen.*, **1** (1956) [Russian].

Vladimirov, V. S. and Sobol, I. M.: Computation of the smallest eigenvalue of the Peierls equation by the Monte Carlo method. In: *Numerical Mathematics, No. 3.* Moscow: Izd. AN SSSR, 1958 [Russian].

19 The Method of Large Particles

Belotserkovskii, O. M. and Davydov, Yu. M.: The nonstationary method of "large particles" in the gas-dynamical design. *Zh. Vych. Matem. Matem. Fiz.*, **11**, 1 (1971) [Russian].

Dyachenko, V. F.: On a new numerical method for nonstationary problems of gas dynamics with two spatial variables. *Zh. Vych. Matem. Matem. Fiz.*, **5**, 4 (1965) [Russian].

Harlow, F.: The "particle in the cell" method for the problems of hydrodynamics. In: *Numerical Methods in Hydrodynamics*. Moscow: Mir, 1967 [Russian].

Vedeshkina, K. A., Levina, Z. F., Lomnev, S. P., Prudkovskii, G. P., Rastopchina, T. V., Ruben, G. V., and Yurchenko, V. V., *Solving Problems by the Method of "Large Particle."* Moscow: Izd. VC AN SSSR, 1970 [Russian].

Yanenko, N. N., Anuchina, N. N., Petrenko, V. E., and Shokin, Yu. I.: On the computation methods in gas dynamics with large deformations. *Numerical Methods in Continuum Mechanics*, **1**, 1 (1970) [Russian].

20 Optimization of Algorithms

Babuška, I. and Sobolev, S. L.: Optimization of numerical methods. *Appl. Mat.*, **10**, 2 (1965) [Russian].

Bakhvalov, N. S.: An estimate of the amount of computation required for the approximate solution of problems. In: Appendix IV to Godunov and Ryaben'kii, *Introduction to the Theory of Difference Schemes*. Moscow: Fizmatgiz, 1962 [Russian].

Bakhvalov, N. S.: On optimal methods of problem solving. *Appl. Mat.*, **13**, 1 (1968) [Russian].

Chernousko, F. L., Banichuk, N. V., and Petrov, V. M.: The numerical solution of variational and boundary value problems by the local variations method. *Zh. Vych. Matem. Matem. Fiz.*, **6**, 6 (1966) [Russian].

Dahlquist, G.: Convergence and stability in the numerical integration of ordinary differential equations. *Math. Scand.*, **4**, 1 (1956).

Frolov, K. K.: On the connection between quadrature formulae and subnets of nets of integer vectors. *Doklady AN SSSR*, **232**, 1 (1977) [Russian].

Kolmogorov, A. N.: Discrete automata and finite algorithms. In: *Trudy IV Vsesoyuz. Mat. S'ezda, Vol. I.* Moscow: Izd. AN SSSR, 1963 [Russian].

Korobov, N. M.: Computation of multiple integrals by the method of optimal coefficients. *Vest. Mosc. Univ., Ser. Mat.*, **4**, 1959 [Russian].

Moiseev, N. N.: Numerical methods using variations in the state space and some control problems of large systems. In: *Papers of the International Congress of Mathematicians, Moscow, 1966* [Russian].

Moiseev, N. N. and Krasovskii, N. N.: A theory of optimal controllable systems. *Izv. AN SSSR, Tekh. Kibernetika*, **5** (1967) [Russian].

Moor, R.: *Interval Analysis*. Englewood Cliffs, N.J.: Prentice-Hall, 1966.

Nickel, K.: Über die Notwendigkeit einer Fehlerschranken—Arithmetik für Rechnenautomaten. *Numer. Math.*, **9**, 1 (1966).

Nickel, K.: Bericht über neue Karlsruher Ergebnisse bei der Fehlererfassung von numerischen Prozessen. *Appl. Mat.*, **13**, 2 (1968).

Vinogradov, I. M.: On an estimate of trigonometric sums. *Izv. AN SSSR, Ser. Matem.*, **29**, 3 (1965) [Russian].

21 Numerical Methods of Conditional Optimization

Abadie, J. and Carpenter, J.: Generalization of the reduced gradient method to the case of nonlinear constraints. In: *Optimization*. London: Academic Press, 1969.

Arrow, K. J.: Hurvitz, L., and Uzawa, H.: *Studies in Linear and Nonlinear Programing*. Stanford: Stanford University Press, 1958.

Boot, J. C. G.: *Quadratic Programming*, Amsterdam: North Holland, 1964.

Bulavskii, V. A., Zvyagina, P. A., and Yakovleva, M. A.: *Numerical Methods of Linear Programming*. Moscow: Nauka, 1977 [Russian].

Carrol, C. W.: The created response-surface technique for optimizing nonlinear restrained systems. *Oper. Res.*, **9**, 2 (1961).

Cea, J.: *Optimization—Theory and Algorithms*. Moscow: Mir, 1973 [Russian].

Conn, R.: Constrained optimization using a differentiable penalty function. *SIAM J. Numer. Anal.*, **10**, 4 (1973).

Courant, R.: Variational methods for the solution of problems of equilibrium and vibrations. *Bull. Amer. Math. Soc.* **49**, 1 (1943).

Danilin, Yu. M.: Minimization of nonlinear functionals in problems with constraints. *Kibernetika*, **3** (1970) [Russian].

Dantzig, G. *Linear Programming, Applications and Generalization*. Moscow: Progress, 1966.

Dem'yanov, V. F.: Minimization of functions on bounded convex sets. *Kibernetika*, **6** (1965) [Russian].

Dem'yanov, V. F.: *Approximate Methods for Solving Extremal Problems*. Leningrad: Izd. Lening. Univ., 1968 [Russian].

Duvaur, G. and Lions, J. L.: *Inequalities in Mechanics and Physics*. Paris: Dunod, 1972. English translation: *Grundlehren der mathematik, 219*. Berlin–Heidelberg–New York: Springer-Verlag, 1976.

Eremin, I. I.: On the penalty method in convex programming. *Kibernetika*, **4** (1967) [Russian].

Fletcher, R.: A general quadratic programming algorithm. *J. Inst. Maths. Applics.* **7** (1971).

Frank, M. and Wolfe, R.: An algorithm for quadratic programming. *Naval Res. Logist. Quart.*, **3**, 1, 2 (1956).

Ganzhela, I. F.: An algorithm for constrained descent. *Zh. Vych. Matem. Matem. Fiz.*, **10**, 1 (1970) [Russian].

Gass, S.: *Linear Programming: Methods and Applications*. New York: McGraw-Hill, 1969.

Gilbert, E. G.: An iterative procedure for computing a quadratic form on a convex set. *SIAM J. Control*, **4**, 1 (1966).

Glovinski, R.: Introduction to the approximation of elliptic variational inequalities. *Report 76006, Laboratoire d'Analyse Numerique de l'Universite de Paris*, **6** (1976).

Glovinski, R., Lions, L., and Tremolières, R.: *Analyse Numerique des Inéquations Variationnelles, Vols. 1 and 2*. Paris: Dunod, 1976.

Goldstein, A. A.: Convex programming in Hilbert space. *Bull. Amer. Math. Soc.*, **70**, 5 (1964).

Hadley, D.: *Nonlinear and Dynamic Programming*. Moscow: Mir, 1967 [Russian].

Kantorovich, L. V.: *Mathematical Methods in the Organization and Planning of Production*. Leningrad: Izd. Lening. Univ., 1939 [Russian].

Kaplan, A. A.: On the problem of implementing the method of permissible directions. *Tr. Inst. Math. Sib. Otd. AN SSSR*, **5**, 22 (1972) [Russian].

Karmanov, V. G.: *Lectures on Mathematical Programming*. Moscow: Izd. Mosc. Univ., 1971 [Russian].

Karmanov, V. G.: *Mathematical Programming*. Moscow: Nauka, 1975 [Russian].

Kuhn, H. W. and Tucker, A. W.: Nonlinear programming. In: *Proceedings of the Second Berkeley Symposium on Mathematical Statistics and Probability*. Berkeley: University of California Press, 1951.

Levetin, E. S. and Polyak, B. T.: Methods for minimization in the presence of constraints. *Zh. Vych. Matem. Matem. Fiz.*, **6**, 5 (1966) [Russian].

Lions, J. L. and Stampacchia, G.: Variational inequalities. *Comm. Pure Appl Math.*, **XX** (1967).

Lions, J. L. and Teman, R.: Une method d'eclatement des operateurs et des contraintes en calcul des variations. *C. R. Acad. Sci., Paris*, **263** (1966).

Lootsma, F. A.: Constrained optimization via penalty functions. *Philips Res. Repts.*, **23**, 5 (1968).

Moiseev, N. N., Ivanilov, Yu. P., and Stolyarova, E. M.: *Methods of Optimization*. Moscow: Nauka, 1978 [Russian].

Morozov, V. A.: Linear and nonlinear ill-posed problems. *Mat. analiz*, **II** (1973) [Russian].

Polyak, B. T.: The conjugate gradient method in extremal problems. *Zh. Vych. Matem. Matem. Fiz.* **9**, 4 (1969) [Russian].

Powell, M. J. D.: A method for nonlinear constraints in minimization problems. In: *Optimization*. London: Academic Press, 1969.

Pshenichnyi, B. N.: An algorithm for the general mathematical programming problem. *Kibernetika*, **5** (1970) [Russian].

Pshenichnyi, B. N. and Danilin, Yu. M.: *Numerical Methods in Extremal Problems*. Moscow: Nauka, 1975 [Russian].

Roseh, J. B.: The gradient projection method for nonlinear programming. I. Linear constraints, II. Nonlinear constraints. *SIAM J.* **8**, 1 (1960); **9**, 4 (1961).

Rosenbloom, P.: The method of steepest descent. *Proc. Symp. Appl. Math.*, **6** (1956).

Rubenshtein, G. Sh.: *Finite-Dimensional Optimization Models—Lectures*. Novosibirsk: Izd. Novosib. Univ., 1970 [Russian].

Sargent, R. W. and Murtagh, B. A. H.: Projection methods for nonlinear programming. *Math. Prog.*, **4** (1973).

Tikhonov, A. N.: On ill-posed problems of optimal planning. *Zh. Vych. Matem. Matem. Fiz.*, **6**, 1 (1966) [Russian].

Vulf, F.: New methods in nonlinear programming. In: *Applications of Mathematics in Economic Research, Vol. 3*. Moscow: Mysl, 1968 [Russian].

Yudin, D. B. and Goldstein, E. G.: *Linear Programming, Theory and Finite Methods*. Moscow: Fizmatgiz, 1963 [Russian].

Zangwill, W. I.: *Nonlinear Programming: a Unified Approach*. Englewood Cliffs: N.J.: Prentice-Hall, 1969.

Zoutendijk, G.: *Methods of Feasible Directions*. Amsterdam: Elsevier, 1960.

22 Theory of Optimal Control (Dynamic Programming and the Maximum Principle)

Balakrishnan, A.: *Introduction to the Theory of Optimization in Hilbert Space*. Moscow: 1974 [Russian].

Bellman, R.: *Dynamic Programming*. Princeton: Princeton University Press, 1957.

Bellman, R. and Dreyfus, S. E.: *Applied Dynamic Programming*. Princeton: Princeton University Press, 1962.

Boltyanskii, V. G.: *Mathematical Methods of Optimal Control*. Moscow: Nauka, 1969 [Russian].

Boltyanskii, V. G.: *Optimal Control of Discrete Systems*. Moscow: Nauka, 1973 [Russian].

Butkovskii, A. G.: *Methods of Control by Systems with Distributed Parameters*. Moscow: Nauka, 1975 [Russian].

Chernous'ko, F. L. and Banichuk, V. P.: *Variational Problems of Mechanics and Control*. Moscow: Nauka, 1973 [Russian].

Eneev, T. M.: On the application of the gradient method in problems of optimal control. *Kosmicheskie issledovaniya*, **4**, 5 (1966) [Russian].

Fedorenko, R. P.: *Approximate Solution of the Problem of Optimal Control.* Moscow: Nauka, 1978 [Russian].

Fel'dbaum, A. A.: *Fundamental Theory of Optimal Automatic Systems.* Moscow: Nauka, 1966 [Russian].

Gabasov, R. and Kirilova, F. M.: On the problem of extending the Pontrjagin maximum principle to discrete systems. *Avtomatika I telemekhanika,* **27,** II (1966) [Russian].

Krasovskii, N. N.: *Theory of the Control of Motion.* Moscow: Nauka, 1968 [Russian].

Letov, A. M.: *Dynamics of Flight and Control.* Moscow: Nauka, 1969 [Russian].

Li, E. B. and Markus, L.: *Fundamentals of the Theory of Optimal Control.* Moscow: Nauka, 1972 [Russian].

Lions, J.-L.: *Optimal Control by Systems Described by Partial Differential Equations.* Moscow: Mir, 1972 [Russian].

Lur'e, K. A.: *Optimal Control in the Problems of Mathematical Physics.* Moscow: Nauka, 1977 [Russian].

Moiseev, N. N.: *Elements of the Theory of Optimal Systems.* Moscow: Nauka, 1975 [Russian].

Pontrjagin, L. S., Boltyanskii, V. G., Gamkrelidze, R. V., and Mischenko, E. F.: *The Mathematical Theory of Optimal Processes.* Moscow: Nauka, 1976 [Russian].

Young, L.: *Lectures on Variational Computation and the Theory of Optimal Control.* Moscow: Mir, 1974 [Russian].

Index of Notation

Index

507